Fungal Biology

Series Editors

Vijai Kumar Gupta
ERA Chair VALORTECH
Estonian University of Life Sciences
Tartu, Estonia

Maria G. Tuohy
School of Natural Sciences
National University of Ireland Galway
Galway, Ireland

About the Series

Fungal biology has an integral role to play in the development of the biotechnology and biomedical sectors. It has become a subject of increasing importance as new fungi and their associated biomolecules are identified. The interaction between fungi and their environment is central to many natural processes that occur in the biosphere. The hosts and habitats of these eukaryotic microorganisms are very diverse; fungi are present in every ecosystem on Earth. The fungal kingdom is equally diverse, consisting of seven different known phyla. Yet detailed knowledge is limited to relatively few species. The relationship between fungi and humans has been characterized by the juxtaposed viewpoints of fungi as infectious agents of much dread and their exploitation as highly versatile systems for a range of economically important biotechnological applications. Understanding the biology of different fungi in diverse ecosystems as well as their interactions with living and non-living is essential to underpin effective and innovative technological developments. This series will provide a detailed compendium of methods and information used to investigate different aspects of mycology, including fungal biology and biochemistry, genetics, phylogenetics, genomics, proteomics, molecular enzymology, and biotechnological applications in a manner that reflects the many recent developments of relevance to researchers and scientists investigating the Kingdom Fungi. Rapid screening techniques based on screening specific regions in the DNA of fungi have been used in species comparison and identification, and are now being extended across fungal phyla. The majorities of fungi are multicellular eukaryotic systems and therefore may be excellent model systems by which to answer fundamental biological questions. A greater understanding of the cell biology of these versatile eukaryotes will underpin efforts to engineer certain fungal species to provide novel cell factories for production of proteins for pharmaceutical applications. Renewed interest in all aspects of the biology and biotechnology of fungi may also enable the development of "one pot" microbial cell factories to meet consumer energy needs in the 21st century. To realize this potential and to truly understand the diversity and biology of these eukaryotes, continued development of scientific tools and techniques is essential. As a professional reference, this series will be very helpful to all people who work with fungi and should be useful both to academic institutions and research teams, as well as to teachers, and graduate and postgraduate students with its information on the continuous developments in fungal biology with the publication of each volume.

More information about this series at http://www.springer.com/series/11224

Ahmed M. Abdel-Azeem
Editor

Recent Developments on Genus *Chaetomium*

Springer

Editor
Ahmed M. Abdel-Azeem
Botany Department
Faculty of Science
Suez Canal University
Ismalilia, Egypt

ISSN 2198-7777 ISSN 2198-7785 (electronic)
Fungal Biology
ISBN 978-3-030-31611-2 ISBN 978-3-030-31612-9 (eBook)
https://doi.org/10.1007/978-3-030-31612-9

This Springer imprint is published by the registered company Springer Nature Switzerland AG
The registered company address is: Gewerbestrasse 11, 6330 Cham, Switzerland

*This volume is dedicated to my late parents, **Mohamed Ahmed Abdel-Azeem** and **Aida Elsayed Elayoty**, both of whom always had confidence in me and offered me encouragement and support at every stage in my life*
*Also I dedicated this work to my brother **Mohamed Mohamed Ahmed Abdel-Azeem** and my sister **Fatma Soliman El-Hanafy** for their unlimited support all the time*

Ismailia, Egypt *Ahmed Mohamed Ahmed Abdel-Azeem*

Foreword

Bioprospection of functional microbial diversity from unique and undisturbed areas for the search of new microorganisms having potential to produce bioactive secondary metabolites is an emerging area of research. *Chaetomium* species are well known to produce antagonistic potential against several plant pathogens, and the genus has been reported to have more than 350 species. The important mechanisms for their antagonistic potential are the production of lytic enzyme and secondary metabolites. The genus is widely reported to have diverse biological activities like helping in biodegradation of municipal waste, and as efficient antioxidant agents and industrial enzyme productions. It is also reported to be a contaminant for human infections like type I allergic reactions and general infections. For example, *Chaetomium atrobrunneum* results in infections in immunocompromised peoples.

Recent Developments on Genus Chaetomium covers most of the important topics starting from the diversity, the identification, and the role of genus *Chaetomium* in natural product biosynthesis. Each chapter provides the future goals in the subject which will be helpful for the researchers and academicians to take the work forward. This volume contains 16 book chapters, and each chapter has colored illustrations which will be easy to follow by the readers.

I am confident and strongly believe that the readers will find the contents in this volume as a comprehensive coverage of various important aspects of exploration and exploitation of *Chaetomium* and their role in sustainable development.

Bhim Pratap Singh
Department of Biotechnology
Aizawl, Mizoram University
Mizoram, India

Preface

Chaetomium Kunze (*Chaetomiaceae*, *Sordariales*) was established by Gustav Kunze with *Chaetomium globosum* Kunze as its generic type. The species within this genus are cosmopolitan and usually found in soil and air, and grow on plant debris. This ascomycetous genus is characterized by ostiolate ascomata usually covered with hairs or setae and clavate, fusiform or cylindrical, fasciculate, evanescent asci, and brown to gray-brown, single-celled ascospores with one or two germ pores.

Chaetomium has been connected to several anamorphic genera, such as *Acremonium*, *Botryotrichum*, *Chrysosporium*, *Histoplasma*, *Humicola*, *Phialophora*, *Scopulariopsis*, and *Scytalidium*. All these genera were described later than *Chaetomium* except for *Acremonium*, and the generic names *Botryotrichum*, *Histoplasma*, *Phialophora*, *Scopulariopsis*, and *Scytalidium* were used in much less frequencies than *Chaetomium*.

Phylogenetic studies by various investigators provided evidence that the genus *Chaetomium* was not monophyletic. Also, phylogenetic assessment and taxonomic revision remained to be done for these anamorphic genera, as most of them appeared to be polyphyletic.

A number of species produce mycotoxins, including sterigmatocystin, chaetoglobosins, chaetochromin, and mollicellins. Medical cases involving *Chaetomium* species are quite rare, and all are categorized in ACDP Hazard Group 1.

The taxonomy of *Chaetomium* has been studied by several authors. Since the establishment of the genus, more than 400 species have been described, many of which were synonymized/excluded and only 273 *Chaetomium* species were accepted according to the Index Fungorum Partnership (IFP 2019). Due to the diversity of species and of inhabiting environments, *Chaetomium* spp. might conceive diverse biosynthetic gene clusters, which transform into various secondary metabolites (the fungi languages) to adapt to different ecological environments. Until now, more than 200 compounds with a wide range of bioactive effects have been isolated from *Chaetomium* spp., but compared with its richness of species, more bioactive secondary metabolites might be found in this member of fungi. Therefore, together with its ubiquitous nature, these species have great significant impact on ecosystems, agriculture, food production, biotechnology, and human and animal health.

The objective of this volume on the recent developments on genus *Chaetomium* is to keep the readers informed about the recent developments in research of the genus *Chaetomium* and the challenges for the researchers to look in the upcoming years. It is very important to shed the light on one of the largest genera of Ascomycetes and to attract the reader's attention toward the recent developments on genus *Chaetomium*.

This book comprises 16 chapters on genus *Chaetomium* and is divided into two parts: the first deals with biology and biotechnology, which include the diversity of genus *Chaetomium* in different ecological habitats, while the second deals with *Chaetomium*'s metabolites which show the diversity of the genus, both positive and negative. Besides the taxonomic information, some ecological aspects, like distribution and substrate/host preferences, have also been dealt with in some chapters, where appropriate. The book chapters cover the wide applications of *Chaetomium* taxa recovered from different ecosystems, the methods of identification, and characterization of their products.

Thus this book presents as much comprehensive information concerning genus *Chaetomium* as possible. I express my sincere gratitude to all the contributors for their valuable contributions and support throughout. I extend my sincere thanks to the research team working in Systematic Mycology Laboratory, Department of Botany, Faculty of Science, Suez Canal University, for their hard work.

(December, 2019) Ahmed M. Abdel-Azeem

Acknowledgments

I am grateful to all of those academicians and scientists with whom I have had the pleasure to work during this project. They happily agreed to share their work on one of the most important genera in Ascomycetes, genus *Chaetomium*, in this volume.

Nobody has been more important to me in the pursuit of this project than the members of my family. I would like to thank my wife, Eman Hussein, and my sons, Mohamed and Abdelwahid, who provide unending inspiration.

I am equally thankful to Springer Publishing for their full cooperation during the production of the volume. In particular, I am thankful to the series editors, Dr. Vijai Kumar Gupta and Prof. Maria G. Tuohy, for accepting our proposal and providing their full support and encouragements. I am also thankful to the production team of Springer Nature for all their efforts for publishing the volume on time. I admit that it is quite possible to have some mistakes in the text inadvertently, and I take responsibility for the mistakes, and please feel free to inform me the same.

I would like to express my sincere thanks to Prof. Bhim Partap Singh, Molecular Microbiology and Systematics Laboratory, Department of Biotechnology, Mizoram University, Aizawl, Mizoram, India, for his kind support during the preparation of this work.

I am also thankful to my team of young active and energetic researchers in Systematic Mycology Laboratory, Department of Botany, Faculty of Science, Suez Canal University, Egypt. All of them have supported me during all stages and progress of this project.

Ismailia, Egypt Ahmed M. Abdel-Azeem

Contents

About the Editor

Ahmed M. Abdel-Azeem is currently working as academic staff member for the Botany Department, Faculty of Science, Suez Canal University, and as a mycologist with particular interest in the ecology, taxonomy, biology, and conservation of fungi, especially on the members of the phylum Ascomycota. His research includes isolation, identification, and taxonomic assessments of these fungi with particular emphasis on those which produce bioactive materials from different ecological habitats. Most recently, his interests are on the effects of climate change on fungi. This in turn has led him to become involved in fungal conservation. He is a member of the IUCN Species Survival Commission Specialist Group for Cup Fungi, Truffles, and their Allies and also the founder of the Arab Society for Fungal Conservation. In 2014 and 2016, he proposed a good candidate for the celebration of Egypt's National Fungus Day on 20 February. He, with the help of international societies, agencies, and mycologists, decreed the Egypt's National Fungus Day in Bibliotheca Alexandrina on 20 February 2016 for the first time. He has received various grants, fellowships, and national and international projects, e.g., EOL Fellowship in 2011, Mohamed bin Zayed Species Conservation Fund in 2014 and 2018, and National Geographic Society Fund in 2016. In 2018, he was elected to be a member in the Executive Committee of International Mycological Association (IMA) for the next 4 years. He was hired for his experience in taxonomy, ecology, biology, and conservation of fungi to study the fungi in ancient air of unveiled Cheops Solar Boat Project and fungi that degraded ancient wood in Abydos Middle Cemetery Project. He studied the biodiversity of macrofungi in Romania, the UK, the USA, Finland, Sweden, Italy, Greece, Puerto Rico, Malta and Poland. He is the editor in chief of *Microbial Biosystems Journal* (*MBJ*) and a reviewer of more than seven international journals. He has published more than 62 research paper journals, 7 book chapters in the books published by international publishers, and 5 books.

Part I
Chaetomium: Biology to Biotechnology

Chapter 1
Taxonomy and Biodiversity of the Genus *Chaetomium* in Different Habitats

Ahmed M. Abdel-Azeem

1.1 Introduction

Chaetomium was established by Gustav Kunze based on *C. globosum* as its type species (Fig. 1.1). This ascomycetous genus is characterized by superficial ascomata usually covered with hairs or setae (Hawksworth and Wells 1973) (Fig. 1.2); membranaceous peridium, consisting of several pseudoparenchymatous layers; asci that are clavate or fusiform (with biseriately arranged ascospores) or sometimes cylindrical (with uniseriately arranged ascospores), thin-walled, evanescent, and without apical structures; scarce paraphyses that disappear before ascocarps mature (von Arx et al. 1986); and ascospores that are brown or gray-brown (never opaque or black), one celled, with one or sometimes two germ pores, and exuding as a dark, black, sticky mass (Hess et al. 1967; Millner et al. 1977).

After scattered initial studies on the genus, Ames (1963) and Seth (1970) delimited *Chaetomium* species mainly on the basis of the shapes of ascospores and ascomatal hairs. Hawksworth and Wells (1973) emphasized the type of ascomatal setae ornamentation. Their study was the first in which species diagnoses were based entirely on the examination of type or authentic material. Sörgel (1960) and Dreyfuss (1975) provided partial revisions in which peridium structure and shapes of asci and ascospores were taken as primary criteria to delimit species. Von Arx et al. (1986) provided a more comprehensive revision, which largely followed the concepts of Sörgel and Dreyfuss and the unpublished thesis by Carter (1982) but tended to underestimate differences in hair morphology.

Members of the genus *Chaetomium* are cosmopolitan and prevalent components of different ecosystems in a wide range of environmental and climatic zones (von Arx et al. 1986) because they can colonize a wide variety of substrates. *Chaetomium* species are well known as coprophilous, seed, and soil fungi (Abdel-Azeem 2003;

A. M. Abdel-Azeem (✉)
Botany Department, Faculty of Science, Suez Canal University, Ismailia, Egypt
e-mail: ahmed_abdelazeem@science.suez.edu.eg

A. M. Abdel-Azeem (ed.), *Recent Developments on Genus Chaetomium*,
Fungal Biology, https://doi.org/10.1007/978-3-030-31612-9_1

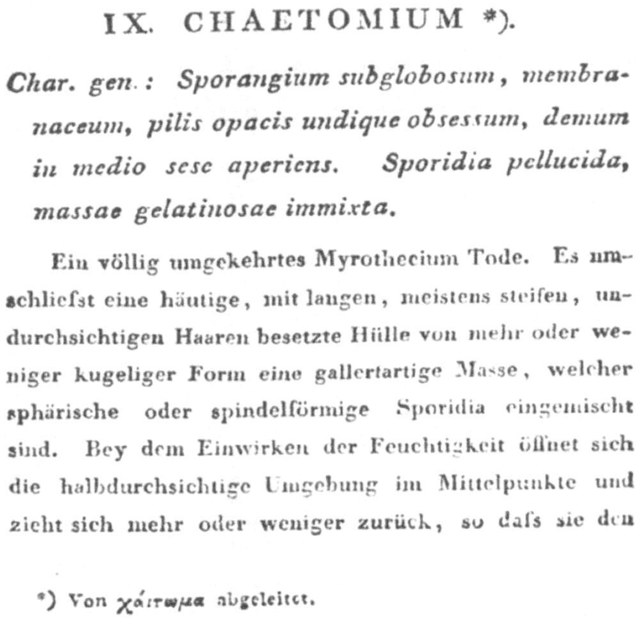

IX. CHAETOMIUM *).

Char. gen.: Sporangium subglobosum, membranaceum, pilis opacis undique obsessum, demum in medio sese aperiens. Sporidia pellucida, massae gelatinosae immixta.

Ein völlig umgekehrtes Myrothecium Tode. Es umschliefst eine häutige, mit langen, meistens steifen, undurchsichtigen Haaren besetzte Hülle von mehr oder weniger kugeliger Form eine gallertartige Masse, welcher sphärische oder spindelförmige Sporidia eingemischt sind. Bey dem Einwirken der Feuchtigkeit öffnet sich die halbdurchsichtige Umgebung im Mittelpunkte und zieht sich mehr oder weniger zurück, so dafs sie den

*) Von χάιτωμα abgeleitet.

Fig. 1.1 A part of Kunze original description of genus *Chaetomium* in 1817

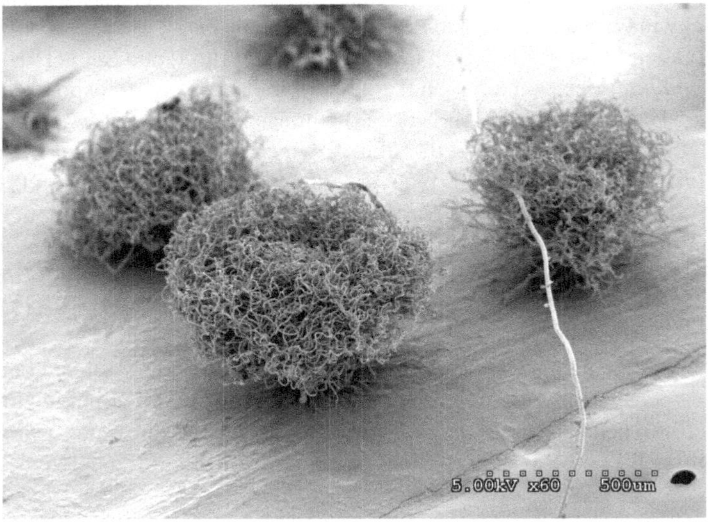

Fig. 1.2 SEM of *Chaetomium* ascomata showing peridial hairs ©Abdel-Azeem 2019

Somrithipol 2004; Somrithipol et al. 2004), colonize living plant tissues (endophytic) (Abdel-Azeem and Salem 2012; Salem and Abdel-Azeem 2014; Abdel-Azeem et al. 2016a, 2018a) and archeological wood (Abdel-Azeem et al. 2019), and are also found in organic compost (Abdel-Azeem 2003).

Chaetomium taxa degrade cellulose and other organic materials and act as antagonist against plant fungal pathogens (Soytong et al. 2001). *C. globosum* is reported by several researchers to be a strong cellulose decomposer (Umikalsom et al. 1997, 1998), which exhibited a very effective antagonistic activity against various soil microorganisms (Aggarwal et al. 2004; Dhingra et al. 2003; Soytong et al. 2001). Recently *Chaetomium* and its species have drawn much attention to be used to manage several economically important diseases. *C. globosum* has been identified as a potential antagonist of *Cochliobolus sativus* (Biswas et al. 2012). Ascospore suspensions and culture extracts of *C. globosum* reduced infection of apple seedlings by *Venturia inaequalis* (Cullen and Andrews 1984). Some isolates of *C. globosum* produce antibiotics that can suppress damping-off of sugar beet caused by *Pythium ultimum* (Walther and Gindrat 1988; Di Pietro et al. 1992). *C. cupreum* and *C. globosum* reduce leaf spot disease of corn caused by *Curvularia lunata*, rice blast disease caused by *Magnaporthe oryzae*, sheath blight disease of rice caused by *Rhizoctonia solani*, and tomato wilt disease caused by *Fusarium oxysporum* f.sp. *lycopersici* (Soytong 1992a, b).

In view of the aforementioned information, the idea of this chapter was evolved with a main objective of surveying *Chaetomium* taxa in various habitats and reporting on their frequency of occurrence and substrate preference, if present, hopefully to draw a preliminary picture about the taxonomy and biodiversity of this genus worldwide.

1.2 Taxonomic History

Since the discovery of the genus *Chaetomium* (Kunze and Schmidt 1817), several attempts have been made to study the detailed morphology for species identification. Some important and significant keys affected significantly taxa identification of *Chaetomium* are mentioned chronologically in this chapter. Furthermore, the molecular advancement for the identification of *Chaetomium* species is also provided along with the history of development of DNA barcoding for economically important fungi.

1.2.1 Cultural Conditions of Chaetomium

Tschudy (1937a) studied the morphology of some species, viz., *C. globosum*, *C. cochliodes*, *C. funicolum*, and *C. elatum*, using cultural characters to determine the species identity. He observed no production of perithecia in *C. globosum* and

C. elatum when grown on media containing peptone. He also found that these species had very markedly different appearances on a medium with dextrose and peptone and malt-dextrose-starch medium (MDS). On MDS medium luxuriant growth was observed along with the formation of abundant perithecia within a short time. Whereas in the case of the dextrose-peptone medium, however, there was sufficient mycelial growth, but no perithecia were formed. Further, he studied the effects of carbohydrates to assess the growth and development of *Chaetomium* fungus using cellulose, starch, dextrine, maltose, glucose, inulin, levulose, sucrose, galactose, arabinose, xylose, and the pentosan containing gum and gum arabic. All the species produced perithecia on cellulose agar and pure agar, but on starch media *C. indicum*, *C. funicolum*, *C. cochliodes*, and *C. elatum* failed to produce perithecia. However, dextrin, a decomposition product of starch, was the only carbohydrate tested, on which all the species were able to produce perithecia. These marked differences in carbohydrate metabolism and the morphogenic changes on the different media led to experimentation with other nutrients. Out of the 75 different media tested, few produced marked differences and others inhibited the growth entirely. Finally, he concluded that the MDS medium (dextrose, 1.0 g.; starch, 1.0 g.; malt syrup, 2.0 g.; agar, 2.0 g.; water, 100.0 ml) is the best for the growth and development of many species of *Chaetomium*.

Further, Tschudy, in 1937b extended his experimentation to study the appearance together with the abnormalities in development under various conditions of cultivation. He found the abnormalities of *Chaetomium* species on different media. The important observation was that the perithecia were produced submerged in the agar and lack terminal setae. The hypothesis was that this abnormality was due to lack of O2 or superabundance of CO2, but it was found faulty. Inhibition of development of terminal setae and perithecia by peptone and by alcohol was among the facts discovered. The effect of pH was found to have little effect on perithecial development.

Plomley (1959) studied the colony characters of *Chaetomium* species. Parameters such as the growth of the hypha, size of the colony, and density of the hyphae within the colony were considered for his study. His observations were increase in length of individual hyphae at constant rate and exponential growth through branching in a mass of hyphae. He also studied the germination of the spore until a constant rate of marginal expansion is set up. He observed the lag phase of growth followed by acceleration phase and then an exponential phase, and finally a steady state is set up in the margin so that it expands at constant rate. From the overall observations, he concluded that the colony characters can be described on the basis of growth of the hyphae.

Sharma and Pandey (2010) studied the mycelial growth rate, colony character, and sporulation pattern of 10 fungal isolates including one species of *Chaetomium*, grown on three different culture media, viz., Potato Dextrose Agar (PDA), Czapek-Dox + Yeast Extract Agar (CYA), and Lignocellulose Agar (LCA) after 7 days of incubation at 25 ± 1 °C. The colony diameter, culture characteristics (texture, surface and reverse coloration, zonation), and sporulation of selected test fungi were greatly influenced by the type of growth medium used. LCA exhibited comparatively higher mycelial growth in six test fungi, whereas all the ten isolates revealed heavy sporulation on this culture medium. *C. funicola* and *Fusarium oxysporum* showed highest growth on CYA medium.

They further suggested that these results would be useful for fungal taxonomic studies. Millner in 1977 tried to establish taxonomic relationship by correlating the radial growth responses of this fungus at varied temperature ranges. According to him the maximum growth rates of thermophiles were found significantly greater than those of the non-thermophiles. Further he used the temperature data to support other salient features of the genus in establishing the synonymy of *C. verrucichaeta* with *C. abuense*, *C. virginicum* with *C. thermophile* var. *coprophile*, and *C. erraticum* with *C. gracile*.

Prokhorov and Linnik (2011) studied the growth rates and development of *C. globosum*, *C. funicola*, *C. elatum*, and *C. spirale* under different temperatures (17–20 °C, 25 °C, 27 °C, 30 °C, and 33–35 °C) and carbon source media (glucose, saccharose, mannite, lactose, amylum, and cellulose). The optimal growth temperature was 25–27 °C for all the studied species, while the temperature range of 33–35 °C inhibited the colonies' growth.

Growth dynamics and colony shape and morphology and development of mycelium and ascocarps varied greatly in different *Chaetomium* species with regard to the carbon source media.

Germination studies of ascospore of *Chaetomium* were performed by Buston et al. in 1966, and they found 30–35% of the ascospore germination of *C. globosum* on a complete synthetic medium. Later in 1975, Chapman and Fergus (1975) also studied the germination of the ascospores of the *C. globosum* on different media. The germination percentage of ascospores on cornmeal, Czapek-Dox, lima bean, malt extract, oak wilt, oatmeal, and potato dextrose were 86, 2, 96, 86, 43, 4, and 99, respectively. The cardinal temperatures found for germination were minimum 4–10°C, optimum 24–38°C, and the highest percent germination was found at 38°C. Fergus and Delwiche (1975) reported poor germination (8% or less) of ascospores of *C. rectopilium* on a synthetic agar medium containing all of the major nutrients and also on media containing a single nutrient. Further, they observed better germination (89%) on malt extract agar than the potassium acetate. The cardinal temperatures for ascospore germination (determined on potassium acetate agar) were found to be as follows: minimum, 12–16°C; optimum, 46°C; and maximum, 48–50°C. A high percentage of ascospore germination was observed on malt extract agar at pH 4.9 to 7.0, but none were germinated at pH 3.5. They also found that the thermal death point for *Chaetomium* fungus was 60–62°C. They ascertained that the ascospores were not thermoduric, as they were surviving only up to 8 hours when exposed to 55 °C and surviving only 20 min when exposed to 60°C.

1.2.2 *Morphological Characterization of* Chaetomium *Species*

The genus *Chaetomium* was established by Kunze in 1817, and Corda (1837) improved the original description by the discovery of asci in the centrum, but he could not understand the true nature of the ascus. He observed that it was in some way responsible for supporting the spores. Van Teighem (1875) described a coiled ascogonium subtended by a ramifying hyphal branch which formed the perithecial wall. In the following year, Van Teighem (1876) studied *C. murorum*, *C. indicum*, and several other new species described. Zopf contributed a well-illustrated monograph describing 10 species of *Chaetomium* in 1881, and Chivers recognized 28 distinct species in 1915. The form of the perithecial hairs was used as the major separation character by the earlier workers. Skolko and Groves (1948) suggested that the shape and size of the perithecium and spores, in addition to the conspicuous characteristics of the perithecial hairs, provide a more practical basis for species classification. By 1953, Skolko and Groves (1953) recognized 53 species, and Ames in 1961 described 85 distinct species of *Chaetomium.*

Whiteside (1957) studied the development of fruiting body from the perithecial initials of genus *Chaetomium.* Two patterns of ascogonial development were observed, which suggests that the genus may not be the single phylogenetic unit that is indicated by the use of one generic name. Most of the species studied were found to have perithecial initials similar to those of *C. globosum*, but *C. brasiliense* possesses a second distinct type that apparently has not been described previously for the genus. The first stage of perithecial initiation in *C. globosum* involves a specialized bent hyphal branch, the ascogonium, having the same diameter as the hypha from which it originates. The ascogonium may be sessile, or it may occur at the end of one or a few cells. The tip elongates and coils itself, the apex twining tightly above or in between other coils in an irregular course. While the ascogonium is developing, several additional specialized hyphal branches originate from adjacent mycelial cells and form the proximal cells of stalked ascogonia. These hyphal branches ramify around the coiled ascogonial structure, forming the perithecial walls and subsequently develop into perithecial hairs. Whiteside concluded that there is no relationship between pattern of ascogonial development and the origin of terminal perithecial hairs. Later Cooke worked on perithecial development of *C. erraticum* (Cooke 1969a) and *C. funicolum* (Cooke 1969b) in 1969 and of *C. trilaterale* in 1970.

Wicklow (1979) related the presence of hair ornamentation with respect to their successful survival on dung or plant material in comparison with other fungi. He observed that larvae of the sciarid fly, *Lycoriella mali*, avoided ascocarps of *C. bostrycodes* on rabbit feces but actively consumed the sporocarps of other coprophilous fungi. The author suggested that the ornamentation on perithecia in *Chaetomium* and related genera might have evolved as a mechanical deterrent against predation by arthropod detritivores, thus enabling them to successfully sporulate on the surface of nutrient-rich, relatively short-lived substrates (i.e., dung, plant detritus).

Ellis (1981) used electron microscopy for the first time for the observation of *Chaetomium* to establish taxonomic relationship among the species. He examined terminal hair ornamentation and ascocarp morphology using scanning electron microscopy in five thermophilic *Chaetomium* species, viz., *C. thermophile*, *C. gracile*, *C. erraticum*, *C. rectopilium*, and *C. cellulolyticum*. However, he found that the terminal hair ornamentation has little use in the separation of species. Later in the year 1988, similar kind of studies was attempted by Figueras and Guarro (1988) on *C. malaysiense* to study the ascocarp ontogeny and provided a detailed illustration of ascocarp development. In 2005, Ahammed et al. studied the morphological variability of *C. globosum*. They observed that under SEM, the shape of perithecia could not be ascertained, as they were densely covered with stiff hairs having depositions. Berkson (1966) studied the cytomorphology of the ascogenous hyphae in four species of *Chaetomium* and illustrated only two types of ascus development. In *C. aureum* and *C. murorum*, the ascus arose from an ascogenous cell, and no definite croziers were observed. On the other hand, in *C. dolichotrichum* and *C. caprinum*, croziers in the ascogenous hyphae were regularly observed, and the ascus developed from the penultimate cell of the crozier.

Ascospore was considered as one of the important characters for *Chaetomium* species identity. Several authors used this character for species discrimination. Seth (1972) provided a key based on the shape of the ascospores. The shape of the asci and the structure of the ascospores have been introduced as characters to delimit apparently more homogeneous taxa by Sörgel (1960, 1961) Aue and Muller (1967), and Dreyfuss (1975).

Millner et al. (1977) have paid more attention to the germ pore of the ascospores, which may be small or distinct, flat or protuberant, and apical, subapical, or slightly lateral and occasionally surrounded by a thickened and/or darkened wall for studying large number of taxa. Dreyfuss (1975) observed ascospores with two germ pores, one at each end, in some species. von Arx et al. (1986) reevaluated the genus *Chaetomium* based on the size, shape, symmetry, and number of germ pores of ascospores. Ahammed et al. (2005) observed ascospores of *C. globosum* which were lemon-shaped, smooth surfaced with apical papillae under SEM. Doveri (2013) reevaluated the genus on the basis of phenotypic criteria too.

Useful Criteria Used in Identification of *Chaetomium* (Moustafa and Abdel-Azeem 2005)

- Ability for growth at 40 °C or more.
- Peridial hairs (see Fig. 1.3).
- Ascus morphology (cylindrical or clavate).
- Ascospore shape (see Fig. 1.4).
- Ascospore arrangement (uniseriate, irregular).
- Germ pores (see Fig. 1.5).
- In addition to some secondary characters such as:

  (a) Presence or absence of anamorph.
  (b) Colony color and exudates.

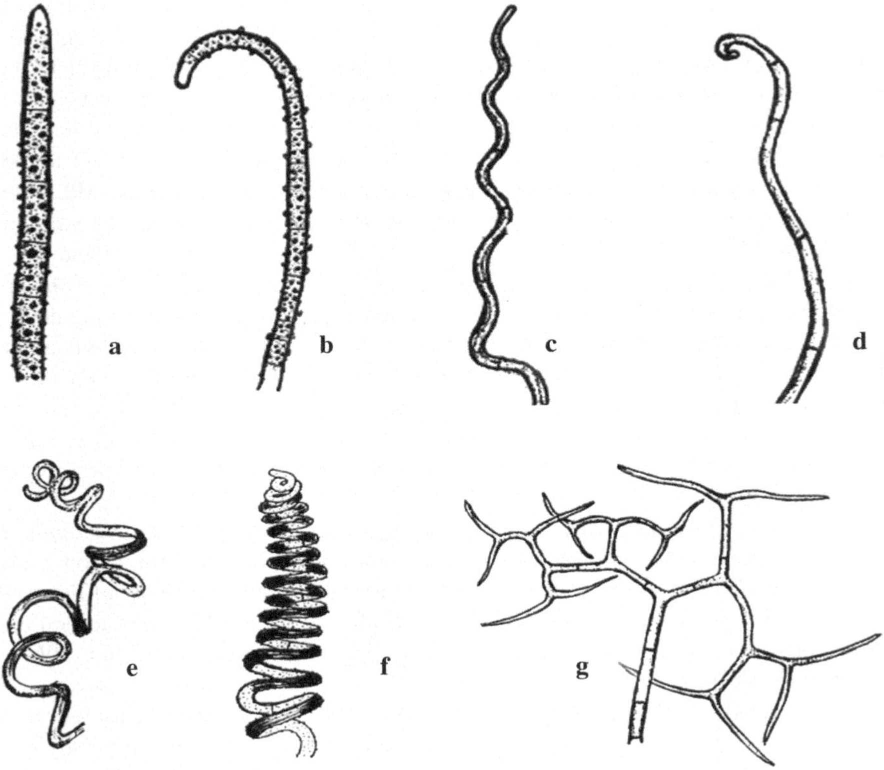

Fig. 1.3 Ascomatal hairs of *Chaetomium*. (**a**) straight; (**b**) arcuate; (**c**) undulate; (**d**) undulate with coiled tips; (**e**) sinuous to loosely coiled; (**f**) compactly coiled; and (**g**), dichotomously branched

1.3 Methodology of Studying *Chaetomium* Biodiversity

1.3.1 Phenotypic Studies

Microscopic features of *Chaetomium* teleomorph are important part of the species concept. However, many debatable taxonomic schemes in several sections of the genus have resulted due to the occurrence of much morphological variation. Reevaluation of *Chaetomium* based on phenotypic characters such as the size, shape, symmetry, and number of germ pores of ascospores has been considered by von Arx et al. (1986) and Doveri (2013) as shown in Figs. 1.3, 1.4, and 1.5. The preliminary identification of species can be performed with the aid of taxonomic keys and descriptions available in the literature (Ames 1969; von Arx et al. 1986; Gené and Guarro 1996; Decock and Hennebert 1997; Udagawa et al. 1997; Rodríguez et al. 2002; Wang and Zheng 2005a, b; Asgari and Zare 2011, Doveri 2013, 2016). Furthermore, all these phenotypic features have to be determined by

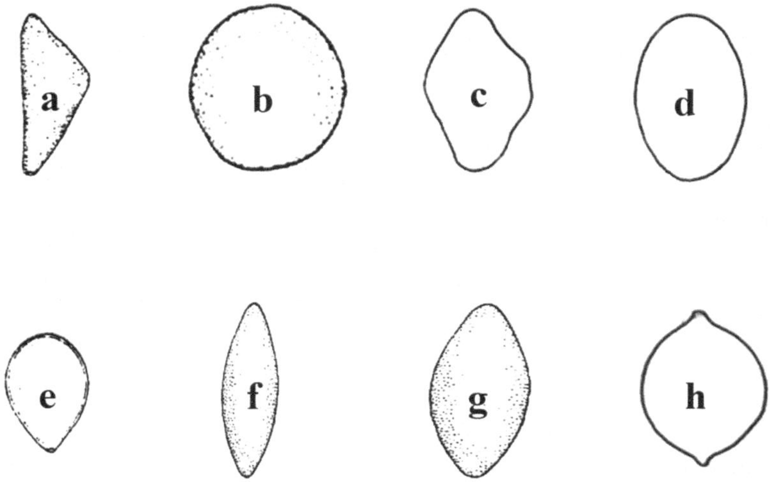

Fig. 1.4 Ascospore shape. (**a**) triangular; (**b**) globose; (**c**) irregular with lateral bulge; (**d**) elliptical; (**e**) almond-shaped; (**f**) narrow fusiform; (**g**) broad fusiform; and (**h**) limoniform

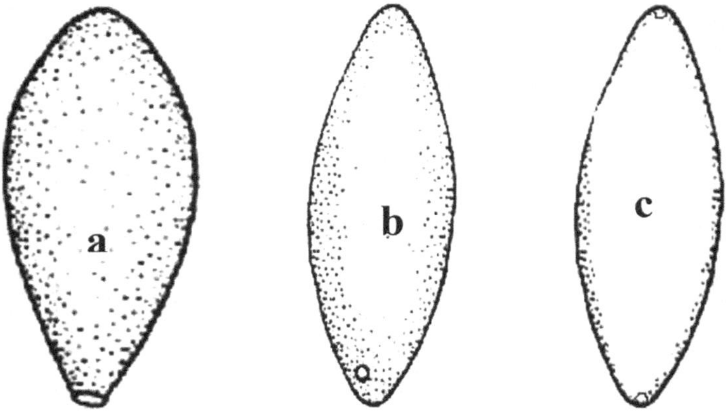

Fig. 1.5 Germ pores. (**a**) single (apical); (**b**) single (subapical); and (**c**) two apical germ pores

trained mycologists under standardized laboratory conditions to obtain an accurate identification (Abdel-Azeem 2003; Blanchette et al. 2017; Abdel-Azeem et al. 2018b). However, without professional expertise this may often lead to incorrect description; therefore, the use of biochemical and molecular methods is recommended.

According to Index Fungorum Partnership (IFP 2019), there are 273 *Chaetomium* species accepted (see species list in Appendix of Chap. 4). The work developed by von Arx et al. (1986) is considered as the basis for the classical taxonomic studies of the genus, which is followed in important articles and monographs on the genus (e.g., Doveri 2004). However, recent studies with molecular approaches have revealed a high number of undescribed species on this genus, especially when they are considered more potential substrates for the occurrence of genus (Wang et al. 2014, 2016a, b; Zhang et al. 2016) (please check Table 4.2 – Chap. 4).

Updated World Key to *Chaetomium* by Abdel-Azeem After von Arx et al. (1986) and Doveri (2008, 2013)

(In bold type new species, new combinations, or taxa accepted in *Chaetomium*)

(1) Ascospores triangular in frontal view, brown at maturity .... **118.**

(1∗) Ascospores not triangular in frontal view, brown, or gray, or bluish grey ....0.**2.**

(2) Ascospores cylindrical, with attenuated ends, 9–11 × 2.5–3 μm. Colonies often with red or orange exudates. Peridium of a *textura intricata*. Ascomatal hairs seta-like, dichotomously branched ....*C. fusum.*

(2∗) Ascospores not cylindrical, broader ....0.**3.**

(3) Asci 2- to 6-spored. Ascospores ellipsoidal, biumbonate, bilaterally flattened, 11–15 × 7–9 × 5–6 μm. Peridium of a *textura angularis*. Hairs flexuous or undulate, branched at the apex. Colonies with yellow exudates ....*C. tetrasporum.*

(3∗) Asci 8-spored ....0.**4.**

(4) Ascospores oblate or hemispherical, 6–7.5 × 4–5 μm, lacking a distinct germ pore. Peridium of a *textura angularis*. Ascomatal hairs spirally coiled .... *C. oblatum.*

(4∗) Ascospores not oblate or larger ....0.**5.**

(5) Asci clavate, obovate or fusiform. Ascospores biseriate or conglobate ....0.**6.**

(5∗) Asci narrowly cylindrical, ascospores uniseriate ....0.**108.**

(6) Ascospores spherical or subspherical, sometimes with two germ pores ....0.**7.**

(6∗) Ascospores different in shape ....0.**13.**

(7) Ascospores with one (occasionally two) germ pore and usually a regular outline ....0.**8.**

(7∗) Ascospores with two (or more) germ pores and often an irregular outline ... .0.**12.**

(8) Ascospores 5.2–6.3 × 5.5–7.8 μm. Terminal hairs irregularly or dichotomously branched ....***C. mesopotamicum*** Abdullah and Zora 1993.

(8∗) Ascospores larger ....0.**9.**

(9) Ascospores globose, 10–13 μm in diameter. Peridium of a *textura angularis*. Terminal hairs flexuous to undulate .... *C. globosporum.*

(9∗) Ascospores nearly spherical to broadly ellipsoidal or ovoidal ....0.**10.**

(10) Ascomata subglobose. Peridium of a *textura epidermoidea*. Terminal hairs flexuous to undulate, 4–6 μm diam. at the base. Ascospores 9–14 × 7–11 × 9–9.5 μm, with a paler longitudinal band and a subapical to lateral germ pore ....***C. tarraconense*** Stchigel et al. in Rodríguez et al. 2002.

(10∗) Ascomata obovoidal. Peridium with a different *textura*. Ascospores without a longitudinal band ....0.**11.**

(11) Peridium of a *textura angularis* or *intricata*. Terminal hairs flexuous, undulate or hypha-like, 4–7.5 in diameter near the base. Ascospores 10–12.3 × 9–11.3 × 7–9 μm, with an apical or slightly subapical germ pore .... ***C. interruptum*** Asgari and Zare 2011.

(11∗) Peridium of a *textura intricata*. Terminal hairs of two types: short, rigid, and straight or long, sinuous, and branched, both types 3–5 μm in diameter near the base. Ascospores 8–11 × 8–9.5 μm, with a subapical or sometimes lateral germ pore ....***C. umbratile*** Udagawa et al. 1997

(12) Ascospores 12–16 × 10–14 × 8–12 μm, subspherical to ovoidal, often bilaterally flattened, with two germ pores. Ascomatal hairs flexuous, or undulate, or hyphoid. Peridium of a *textura angularis*. Colonies occasionally with pale green exudates. Homothallic .... *C. megalocarpum.*

(12∗) Ascospores 15–22 × 11–16 μm, broadly ellipsoidal with rounded ends to ovoidal, pyriform, or subglobose, with two or more germ pores. Hairs hyphoid. Peridium of a *textura epidermoidea*. Colonies often with yellow exudates. Heterothallic ....*C. nozdrenkoae.*

(13) (From 6) Ascospores with one germ pore (occasionally with a paler spot at the end opposite to the germ pore) .....0.**14.**

(13∗) Ascospores at least partly with two germ pores or lacking distinct germ pores ....0.**94.**

(14) Germ pore apical ....0.**15.**

(14∗) Germ pore subapical or lateral ....0.**75.**

(15) Ascospores rhomboid with umbonate ends, bilaterally flattened, 9–11 × 6.5–8 × 4.5–6 μm. Peridium pale, of a *textura epidermoidea*. Ascomatal hairs flexuous .... *C. umbonatum.*

(15∗) Not with the above characters ....0.**16.**

(16) Ascospores limoniform or irregular, usually biapiculate, bilaterally flattened, 9–14 × 7–12 μm. Peridium with an outer layer of *textura intricata* ....0.**17.**

(16∗) Not with the above characters ....0.**24.**

(17) Ascomata with a very pale, yellowish peridium. Terminal hairs sparse, delicate, wavy, pale cream. Ascospores pink, red in mass, 7–10 × 7–7.5 .... *C. cruentum.*

(17∗) Peridium pigmented. Terminal hairs dense. Ascospores brown at maturity ....0.**18.**

(18) Ascospores rather irregular, often with a lateral bulge ....0.**19.**

(18∗) Ascospores regular, limoniform, without a lateral bulge ....0.**20.**

(19) Colonies with yellow-green exudates. Ascomata olivaceous in reflected light. Ascomatal hairs flexuous, undulate, or coiled. Ascospores with a lateral bulge, 9–11 × 7–9 × 6–7 μm .... *C. madrasense.*

(19∗) Colonies without exudates. Ascomata yellow in reflected light. Hairs hyphoid. Ascospores irregular, 8–12 × 5–7.
μm ....*C. citrinum.*

(20) Homothallic. Terminal hairs flexuous, undulate, or coiled. Phialoconidia absent ....0.**21.**

(20∗) Homothallic or heterothallic. Hairs branched or flexuous when unbranched. Phialoconidia present or absent ......0.**22.**

(21) Colonies usually with yellow-green or red exudates. Ascospores 9–12 × 8–10 × 6–8 μm ......,,,,,,,,.. *C. globosum* (and its relatives).

(21∗) Colonies without exudates. Ascospores 12–13.5 × 8–10 × 6–7.5 μm ....***C. undulatulum*** Asgari and Zare 2011.

(22) Heterothallic. Colonies lacking pigmented exudates. Terminal hairs unbranched, flexuous. Ascospores 11–15 × 8–11 × 7–8.5 μm. Phialoconidia present .... *C. subaffine* (and *C. angustispirale*).

(22∗) Terminal hairs dichotomously branched ....0.**23.**

(23) Homothallic. Phialoconidia present. Colonies without exudates. Hairs usually branched at right angles. Ascospores 9–11.5 × 6.5–8.5 × 5.5–6.5 μm .... ***C. rectangulare*** Asgari and Zare 2011.

∗23∗) Homothallic or eterothallic. Phialoconidia present, or absent in homothallic isolates. Colonies often with green or yellow exudates. Hairs irregularly branched. Ascospores 11–14 × 8–11 × 7–9 μm .... *C. elatum.*

(24) (From 16) Ascospores 6–13 × 5–10 μm, limoniform, bilaterally flattened, umbonate or apiculate at the ends, brown (exceptionally pale gray) at maturity. Ascomata often ampulliform or pyriform and with a conical or cylindrical neck;

peridium of a *textura angularis* (rarely and partly covered with a *textura intricata*) or occasionally *epidermoidea* ....0.**25.**

(24∗) Not with the above characters ....0.**35.**

(25) Ascomata with a long neck formed of joined setae ....0.**26.**

(25∗) Ascomata without such a neck ....0.**28.**

(26) Neck less than 250 μm long. Ascospores 8–11 × 7.5–9 × 5–6 μm with an apical germ pore .... *C. malayense.*

(26∗) Neck 300–2300 μm long ....0.**27.**

(27) Ascospores limoniform, 9–12 × 8–10 × 7–8 μm, with an apical or slightly subapical germ pore .... *C. longicolleum.*

(27∗) Ascospores limoniform to quadrangular in frontal view, 6–6.5 × 5–6 × 3.5–4 μm, with an apical germ pore .... ***C. cuyabenoense*** Decock and Hennebert 1997.

(28) Ascomata with apical setae few in number. Ascospores 8–10 × 7–8.5 × 5–6 μm, biapiculate .... *C. seminudum.*

(28∗) Ascomata with dense apical setae or hairs ....0.**29.**

(29) Ascomata subglobose to ovoid, with a very large (up to 50 μm diam.) ostiole. Terminal hairs arcuate or recurved at the apex, yellow to pale brown, unbranched. Peridium of a *textura angularis* to *epidermoidea.* Ascospores 10–13 × 7–10 × 6.5–7.5 μm, biapiculate .... ***C. macrostiolatum*** Stchigel et al. in Rodríguez et al. 2002

(29∗) Ascospores smaller. Terminal hairs darker ....0.**30.**

(30) Ascomata spherical or obovate ....0.**31.**

(30∗) Ascomata ovate to ampulliform ....0.**32.**

(31) Ascomata spherical. Peridium of a *textura angularis*. Ascomatal hairs flexuous or undulate, often branched. Ascospores 7.5–9 × 6–7 × 4–5 μm, brown, with an apical germ pore .... *C. sphaerale.*

(31∗) Ascomata obovate. Peridium of a *textura angularis*, partly covered with an outer layer of *textura intricata*. Terminal hairs unbranched, constricted at the septa and articulated, loosely coiled at their tips. Ascospores 7–8.5 × 6–7.2 × 5–6 μm, pale gray, with an apical or slightly subapical germ pore ....***C. floriforme*** Gené and Guarro 1996.

(32) Ascomata ovate or ampulliform. Hairs spirally coiled. Ascospores 6–7 × 4.5–6 × 3.5–4.5 μm .... *C. subspirale.*

(32∗) Ascomata ampulliform. Hairs not spirally coiled. Ascospores larger on average ....0.**33.**

(33) Terminal hairs seta-like, straight, not branched. Ascospores 6.5–9.5 × 5.5–6.5 × 3.5–4.5 µm, strongly umbonate. Asci both clavate and cylindrical in the same ascoma .... ***C. ampulliellum*** X.W. Wang in Wang and Zheng 2005a, b.

(33∗) Terminal hairs often branched or circinate. Ascospores slightly umbonate. Asci always clavate ....0.**34.**

(34) Hairs seta-like or flexuous, often branched or circinate at their apex. Ascospores 6.5–8.5 × 6–7.5 × 5–6 µm ....

....

.... *C. homopilatum.*

(34∗) Hairs seta-like, apically branched. Ascospores 8–9.5 × 4.5–6 µm ....

....*C. ampullare.*

(35) (From 24) Ascospores 5.5–10 × 5–8 µm, biapiculate or angular, bilaterally flattened, dextrinoid when young, pale gray-olivaceous at maturity. Ascomata often with a darkened disc around the ostiole...0.**36.**

(35∗) Ascospores not dextrinoid when young, usually brown at maturity. Ascomata lacking a darkened disc ....0.**42.**

(36) Ascomata small, spherical or oblate, pale. Ascomatal hairs partly short and verrucose, partly long, flexuous or undulate, and smooth. Peridium of a *textura angularis*..0.**37.**

(36∗) Ascomata ovate, ampulliform or cylindrical, rarely spherical. Hairs usually spirally coiled at their apex and.

verrucose, often with some branches. Peridium of a *textura angularis* or *cephalothecoidea* ....0.**38.**

(37) Ascospores 5.5–6.5 × 5–6 × 3.5–5 µm, evenly pigmented ....

.... *C. causiiforme.*

(37∗) Ascospores 7–9 × 6–7 × 5–6 µm, darker near the ends ....

.... *C. anguipilium.*

(38) Ascospores quadrangular in frontal view, 6.5–7.5 × 6–7 × 4–5 µm. Peridium of a *textura cephalothecoidea* ....

....*C. quadrangulatum.*

(38∗) Ascospores not quadrangular in frontal view ....

....0.**39.**

(39) Ascomata ampulliform. Upper peridial portion a *textura prismatica* of dark, elongated cells in vertical rows. Ascospores 5.5–7.5 × 5–6.5 × 4–5.5 µm ....

.... *C. robustum.*

(39∗) Upper peridial portion not formed of vertically elongated, dark cells ....

....0.**40.**

(40) Ascospores 5.5–7.5 × 5–6.5 × 4–5.5 µm ....

.... *C. bostrychodes.*

(40∗) Ascospores longer ....

....0.**41.**

(41) Ascomata ovate or ampulliform. *Textura cephalothecoidea*. Ascospores 7–9.5 × 6–7 × 5–6 μm .... *C. convolutum.*

(41∗) Ascomatal spherical. *Textura angularis*. Ascospores 9–10.5 × 7–8 × 6–7 μm ....*C. hexagonosporum.*

(42) (From 35) Ascospores ovate, 5–8 × 4–6 μm, brown when mature, thick-walled. Peridium of a *textura angularis* often with an outer layer of *textura intricata*. Ascomatal hairs seta-like and/or dichotomously branched ....0.**43.**

(42∗) Not with the above characters ....0.**46.**

(43) Ascomatal hairs mainly seta-like, thick and stiff, with some lateral branches when old. Ascospores 6–8 × 4–5 μm, bilaterally not flattened .... *C. spinosum.*

(43∗) Ascospores bilaterally flattened ....0.**44.**

(44) Ascomatal hairs seta-like, unbranched. Ascomata occasionally not ostiolate. Ascospores 5.5–6.5 × 3.5–4.5 × 3–3.5 μm... *C. variostiolatum.*

(44∗) Ascomatal hairs at least partly dichotomously branched. Ascomata ostiolate ....0.**45.**

(45) Ascomatal hairs partly seta-like, partly branched. Ascospores 6–7.5 × 4–5.5 × 3–4 μm .... *C. funicola.*

(45∗) Nearly all hairs branched. Ascospores 5.5–7 × 4–5 × 3.5–4.5 μm .... *C. indicum* (and its relatives).

(46) Peridium orange or pink. Ascomatal hairs much coiled. Ascospores ovate or pyriform, bilaterally flattened, brown, 12–16 × 9–11 × 7–10 μm .... *C. semen-citrulli.*

(46∗) Not with the above characters ....0.**47.**

(47) Ascomata nearly smooth, with a short cylindrical or conical neck. Peridium pale, reddish-brown. Ascospores irregular in shape, 12–13 × 7–8 μm, greenish-brown, darkened around the germ pore .... *C. hamadae.*

(47∗) Not with the above characters ....0.**48.**

(48) Ascospores 11–23 × (5) 6–18 μm, ellipsoidal or fusiform ....0.**49.**

(48∗) Ascospores smaller on average ....0.**63.**

(49) Ascospores usually more than 20 μm long ....0.**50.**

(49∗) Ascospores less than 20 μm long ....0.**51.**

(50) Ascomata brown. Peridium of a *textura angularis*. Ascomatal hairs brown, undulate, rarely branched, strongly constricted at their septa and articulate. Ascospores broadly fusiform, 20–22 × 16–18 μm, with a protuberant, subapical (sometimes apical) germ pore, 2–2.5 µm wide .... ***C. sinaiense*** Mo ustafa and Ezz-Eldin 1989.

(50∗) Ascomata pale ochraceous. Peridium of a *textura epidermoidea* or indistinctly *angularis*. Hairs pale and short or absent. Ascospores fusiform, 18–23 × 9–12 µm, with an apical and smaller germ pore .... *C. deceptivum.*

(51) Ascospores 17–20 × 11–13 µm, often ovoidal and occasionally with umbonate ends, laterally not flattened. Ascomata yellow or ochraceous. Peridium pale, with an outer layer of *textura intricata*. Ascomatal hairs hyphoid, sparse, flexuous ... . *C. vitellinum.*

(51∗) Ascospores smaller .... 0.**52.**

(52) Colonies with yellow exudates. Ascomata non-ostiolate when young. Peridium translucent, a *textura angularis* of pale yellow or brown cells. Ascomata hairs hyphoid, often absent. Ascospores ellipsoidal, often irregular, 12–17 × 7–10 µm ....*C. irregulare.*

(52∗) Not with the above characters .... 0.**53.**

(53) Peridium of a *textura angularis*. Ascomatal hairs seta-like, dark brown, unbranched. Ascospores fusiform, often inequilateral, 14–16 × 5–6 µm, dark brown when mature, paler at the ends ....*C. dreyfussii.*

(53∗) Not with the above characters .... 0.**54.**

(54) Ascomata up to 140 µm in diameter .... 0.**55.**

(54∗) Ascomata larger .... 0.**57.**

(55) Terminal hairs tight spirally coiled all over the length. Ascospores fusiform, usually equilateral, with a truncated end around the germ pore, 11.5–13.5 × 5.7–7 µm ....***C. truncatulum*** Asgari and Zare 2011.

(55∗) Terminal hairs not with such features. Ascospores larger on average, often inequilateral, without a truncate end 0.**56.**

(56) Ascomatal hairs flexuous, undulate or spirally coiled, yellow green or olivaceous in reflected light. Ascospores ellipsoidal-fusiform, 13–16 × 7–8 µm ....*C. subspirilliferum.*

(56∗) Ascomatal hairs hyphoid, straight or flexuous, dark olivaceous gray in reflected light. Ascospores fusiform, 13–15× 6.5–8 µm ....*C. hispanicum.*

(57) Anamorph *Histoplasma*-like. Ascomatal hairs straight to coiled. Ascospores ellipsoidal to narrowly ellipsoidal, 11.2–16 × 7.2–9.6 µm .... ***C. histoplasmoides*** Carris & Glawe 1987.

(57∗) Anamorph different or lacking ....0.**58.**

(58) Anamorph *Botryotrichum*. Ascomatal hairs flexuous or apically circinate. Ascospores ellipsoidal with attenuated ends, 13–17 × 8–10 μm .... *C. piluliferum.*

(58∗) Botryose aleurioconidia absent or rare ....0.**59.**

(59) Ascomatal hairs spirally coiled, partly at least ....0.**60.**

(59∗) Ascomatal hairs not spirally coiled ....0.**61.**

(60) Terminal hairs tight spirally coiled all over the length. Ascospores fusiform, usually equilateral, with a truncated end around the germ pore, 11.5–13.5 × 5.7–7 μm ....***C. truncatulum*** Asgari and Zare 2011.

(60∗) Terminal hairs rather delicate, flexuous to spirally coiled. Ascospores ellipsoidal-fusiform, often inequilateral, without a truncated end, 12–17 × 6–8 μm .... *C. succineum.*

(61) Peridium pale brown, a *textura intricata*. Hairs hyphoid, irregularly branched, pale brown. Ascospores fusiform, 14–16.5 × 5–7 μm .... ***C. nepalense*** (Udagawa and Sugiy 1982) Arx.

(61∗) Peridial wall darker, with a different *textura*. Ascospores ellipsoidal with attenuated ends. Hairs brown ....0.**62.**

(62) *Textura intricata* to *epidermoidea*. Ascomatal hairs flexuous, undulate or apically circinate. Ascospores 13–17 × 7–9 μm ....*C. murorum* (and its relatives).

(62∗) *Textura cephalothecoidea*. Hairs straight or slightly curved, apically open circinate. Ascospores 15–17.5 × 8.5–10.5 μm .... *C. ancistrocladum.*

(63) (From 48) Ascomatal hairs undulate or coiled; ascospores ellipsoidal or ovate, occasionally irregular or bilaterally flattened.. ....0.**64.**

(63∗) Not with the above characters ....0.**67.**

(64) Ascomata dark in reflected light. Peridium of a *textura intricata* or *epidermoidea*. Ascomatal hairs undulate or spirally coiled. Ascospores ovoidal, 5–8 × 4–5 μm, bilaterally not flattened .... *C. nigricolor.*

(64∗) Not with the above characters.. ....0.**65.**

(65) Ascospores ovoidal, bilaterally flattened, 8–11 × 6–9 μm. Peridium of a *textura angularis*. Hairs undulate or much spirally coiled ....*C. uniporum.*

(65∗) Ascospores **not** bilaterally flattened ....0.**66.**

(66) Ascospores irregular in size and shape, 8–12 × 5–6.5 μm. Peridium often of a *textura cephalothecoidea*. Ascomatal hairs horizontally expanding, 4–6 μm in diameter .... *C. repens.*

(66∗) Ascospores regular, ellipsoidal or ovate, 8.5–11 × 5–6.5 μm. Peridium of a *textura angularis*. Ascomatal hairs, 2.5–4 μm broad, with long coils ....,............ .... *C. spiralotrichum.*

(67) Terminal hairs straight. Ascospores fusiform, ellipsoidal or reniform ....0.**68.**

(67∗) Terminal hairs flexuous, arcuate or hyphoid, occasionally with a circinate tip; ascospores often fusiform ....0.**70.**

(68) Ascospores fusiform to broadly fusiform, 9.5–12 × 5–7 μm, with an apical or subapical germ pore. Colonies with abundant olive-green exudates .... ***C. olivicolor*** K. Rodr. et al. in Rodríguez et al. 2002.

(68∗) Ascospores different in shape. Colonies not described in the protologues or lacking exudates ....0.**69.**

(69) Lateral hairs arcuate. Ascospores ellipsoidal in frontal view, reniform or lunate in side view, 5–6.5 × 2.5–3 × 2.5–3 μm, with an apical germ pore. Colonies without exudates .... ***C. novae-caledonicum*** Udagawa et al. 1994.

(69∗) Lateral hairs not arcuate. Ascospores reniform to slightly curved or sometimes with one flattened side, 8–13 × 4.5–6 μm, with one apical germ pore (rarely two). Exudates not described.. ***C. subcurvisporum*** Abdullah and Al-Bader 1989.

(70) Ascospores reniform or lunate in frontal view. Hairs with copper or red encrustations ....0.**71.**

(70∗) Not with the above characters ....0.**72.**

(71) Colonies with a pink exudate. Terminal hairs circinate, with 1–3 convolutions at the apex. Ascospores lunate, 8.5–10 × 5–7 μm .... ***C. lunasporium*** Udaiyan and V.S. Hosag. 1991.

(71∗) Colonies with a red exudate. Terminal hairs circinate or coiled. Ascospores reniform or lunate, 7–10 × 4.5–6 μm.. .... *C. cupreum.*

(72) Ascomatal hairs more than 5 μm diam., closely septate ....0.**73.**

(72∗) Hairs 2–5 μm in diameter, not closely septate.. ....0.**74.**

(73) Ascomatal hairs 5–7 μm in diameter, arcuate, circinate, or coiled at the apex. Ascospores ellipsoidal, 8.5–10.5 × 5–6 μm.. .... *C. lentum.*

(73∗) Ascomatal hairs 6–9 μm in diameter, arcuate, apically incurved, or circinate. Ascospores ellipsoidal, often asymmetrical, 9–11 × 7–9 μm, occasionally with a single germ pore, usually with two .... *C. turgidopilosum.*

(74) Ascomatal hairs flexuous, thick-walled, smooth; ascospores 8–9.5 × 5.5–6.5 μm, with a distinct germ pore .... *C. cymbiforme.*

(74*) Not with the above characters ....0.**75.**

(75) Ascomatal hairs flexuous, apically undulate or curved, thin, verrucose. Peridium of a *textura angularis*. Ascospores. 9–11 × 5–5.7 μm, with a small germ pore .... *C. barilochense.*

(75*) Not with the above characters ....0.**76.**

(76) (From 14 and 75) Peridium of a *textura angularis*. Ascospores with a nearly apical or slightly subapical germ pore...0.**77.**

(76*) Peridium with an outer layer of *textura intricata*. Ascospores with a lateral or distinctly subapical germ pore ....0.**85.**

(77) Hairs seta-like, straight, dark in reflected light. Ascospores fusiform to elongated pyriform, 9–11 × 4–6 μm .... *C. atrobrunneum.*

(77*) Hairs not seta-like, not black in reflected light ....0.**78.**

(78) Colonies with an olive-green exudate. Terminal hairs straight, unbranched. Ascospores fusiform to broadly fusiform, 9.5–12 × 5–7 μm .... ***C. olivicolor*** K. Rodr. et al. in Rodríguez et al. 2002.

(78*) Colonies lacking exudates or with differently colored exudates. Terminal hairs different ....0.**79.**

(79) Colonies black. Ascomatal hairs undulate, rarely branched, constricted at septa and articulate. Ascospores broadly fusiform, 20–22 × 16–18 μm, with a protuberant, subapical (sometimes apical) germ pore, 2–2.5 μm wide.......... .... ***C. sinaiense*** Moustafa and Ezz-Eldin 1989.

(79*) Hairs not articulate, different in shape. Ascospores and germ pore smaller..0.**80.**

(80) Colonies dark. Ascomatal hairs pale, orange or ochraceous, arcuate or flexuous. Ascospores ellipsoidal or ovate, 7–9 × 5–7 μm ....*C. lucknowense.*

(80*) Colonies pale. Hairs darker, brown to black. Ascospores longer on average ....0.**81.**

(81) Colonies with exudates ....0.**82.**

(81*) Colonies lacking pigmented exudates ....0.**83.**

(82) Exudates yellow-green. Ascomatal hairs arcuate, partly undulate. Ascospores ellipsoidal or broadly fusiform, 11–15 × 6–8.5 μm, occasionally with two germ pores .... *C. gracile.*

(82∗) Exudates orange. Hairs arcuate or undulate, circinate. Ascospores ellipsoidal, often asymmetrical, 11–13 × 6–7 μm .... *C. chiversii.*

(83) Terminal hairs spirally coiled. Ascospores 10–11 × 5.5–6.5 μm, inequilaterally fusiform to navicular ....***C. iranianum*** Asgari and Zare 2011.

(83∗) Terminal hairs arcuate or flexuous. Ascospores ellipsoidal-fusiform with (sub)acute ends, equilateral or slightly inequilateral at most ....0.**84.**

(84) Ascospores 10–12.5 × 6–7 μm. Hairs 4–6 μm in diameter near the base .... *C. muelleri.*

(84∗) Ascospores 7–10 × 4.5–6 μm. Hairs 2.5–4 (−5) μm near the base .... *C. carinthiacum.*

(85) (From 76) Ascomatal hairs of two types: seta-like and dark or pale, thinner, branched. Ascospores ellipsoidal-fusiform, 8–12 × 5–7 μm... *C. cuniculorum.*

(85∗) Hairs unbranched, undulate or coiled ....0.**86.**

(86) Ascospores up to 11 × 6.5 μm ....0.**87.**

(86∗) Ascospores larger on average ....0.**88.**

(87) Ascospores 10–11 × 5.5–6.5 μm, inequilaterally fusiform to navicular. Terminal hairs spirally coiled ....***C. iranianum*** Asgari and Zare 2011.

(87∗) Ascospores ovoidal or broadly fusiform, 8–11 × 5–6.5 μm. Ascomatal hairs of two types: long, flexuous, apically circinate or coiled; short, tapering, often arranged in brushes ....*C. gelasinosporum.*

(88) Ascospores 12–15 × 7.5–9.5 μm, ellipsoidal. Terminal hairs coiled. Aleurioconidia often present ......*C. gangligerum.*

(88∗) Ascospores narrower on average. Aleurioconidia absent ....0.**89.**

(89) Colonies with orange exudates. Ascomata black in reflected light, with much coiled hairs. Ascospores fusiform or narrowly ovate, 10–12.5 × 6.5–8 μm .... *C. jabalpurense.*

(89∗) Orange exudates absent. Ascomata not black in reflected light ....0.**90.**

(90) Colonies lacking exudates. Hairs undulate or spirally coi led...**91.**

(90∗) Colonies with yellow or yellow-green exudates ....0.**92.**

(91) Ascospores fusiform or ovate, 12.5–14 × 6.5–7.5 μm .... *C. perlucidum.*

(91∗) Ascospores fusiform or elongate-ellipsoidal, 1 symmetrical or sometimes asymmetrical, 12.3 (6.5–16.2) × 7.6 (6.5–9.7) μm .... ***C. jatrophae*** Sharma et al. 2013.

(92) Ascospores fusiform or narrowly ovate, 10–13 × 6–7 μm. Ascomatal hairs flexuous, undulate or coiled .... *C. raii.*

(92∗) Ascospores larger ....0.**93.**

(93) Ascomatal hairs spirally coiled, brown. Ascospores fusiform, 14–17 × 7–8 μm .... *C. fusisporum.*

(93∗) Hairs flexuous or undulate, pale, thin. Ascospores elongate ellipsoidal or fusiform**,** 14–19 × 6–8 μm *C. jodhpurense.*

(94) (From 13) Ascospores fusiform, 22–27 × 8–9 μm, without distinct germ pores. Ascomatal hairs straight or flexuous. Aleurioconidia present .... *C. piluliferoides.*

(94∗) Ascospores at least partly with two germ pores ....0.**95.**

(95) Ascospores with two subapical germ pores, ellipsoidal-fusiform, 17–21 × 10–12 μm. Peridium of a *textura intricata* or *epidermoidea*. Hairs straight or arcuate .... *C. retardatum.*

(95∗) Ascospores with apical germ pores ....0.**96.**

(96) Ascomata and ascomatal hairs pinkish-purple, the latter hyphoid, pale brown, unbranched. Ascospores broadly fusiform to musiform, 15.5–21 × 9–11.5 μm, with a wall sometimes darkened around the pores........................ .... ***C. purpurascens*** (Udagawa and Y. Sugiy. 1982) Arx.

(96∗) Ascomata and ascomatal hairs not pinkish-purple ....0.**97.**

(97) Ascospores fusiform or navicular, 18–23 × 8–13 μm. Hairs flexuous and tapering. Colonies often with red exudates .... *C. megasporum.*

(97∗) Ascospores smaller ....0.**98.**

(98) Ascospores reniform or irregular in shape ....0.**99.**

(98∗) Ascospores different and regular in shape ....0.**100.**

(99) Ascospores reniform to slightly curved or sometimes with one flattened side, 8–13 × 4.5–6 μm, with one apical germ pore (rarely two). Terminal hairs straight, unbranched .... ***C. subcurvisporum*** Abdullah and Al-Bader 1989.

(99∗) Ascospores ellipsoidal or reniform, often irregular in shape, 10–13 × 7–9 μm, with two germ pores. Hairs flexuous ....*C. variosporum.*

(100) Ascospores fusiform, 15–18 × 5–8.5 μm ....0.**101.**

(100∗) Ascospores shorter ....0.**103.**

(101) Hairs spirally coiled. Ascospores symmetrical, 15–18 × 6.5–8 μm, darkened around the germ pores *C. mareoticum.*

(101∗) Hairs not spirally coiled ....0.**102.**

(102) Terminal hairs flexuous or undulate. Ascospores often asymmetrical, 15–18 × 7.5–8.5 μm .... ***C. biporatum*** Cano and Guarro 1987.

(102∗) Terminal hairs straight, flexuous, or undulate. Ascospores usually symmetrical, 13–19 × 5–7.5 μm .... *C. fusiforme.*

(103) Ascomatal hairs 6–9 μm in diameter, arcuate, apically incurved, or circinate. Ascospores ellipsoidal, often asymmetrical, 9–11 × 7–9 μm, occasionally with a single germ pore ....*C. turgidopilosum.*

(103∗) Ascomatal hairs 2–5 μm in diameter ....0.**104.**

(104) Colonies usually with red exudates. Terminal hairs arcuate, apically circinate, or coiled ....0.**105.**

(104∗) Colonies with orange or yellow-green exudates. Hairs also of other ypes ....0.**106.**

(105) Hairs olivaceous green in reflected light. Ascospores inequilaterally fusiform or navicular, 8–12 × 4–7 μm .... *C. aureum.*

(105∗) Hairs red or orange-red in reflected light. Ascospores fusiform or slightly inequilaterally fusiform, 11.1–13.5 × 4.3–5 μm .... ***C. siamense*** Pornsuriya and Soytong in Ponsuriya et al. 2011.

(106) Colonies with orange exudates. Hairs arcuate, apically circinate. Ascospores broadly fusiform or navicular, often asymmetrical, 11–14 × 7–9 μm .... *C. flavigenum.*

(106∗) Colonies with yellow-green exudates ....0.**107.**

(107) Ascomata spherical. Hairs arcuate, partly undulate, with a swollen basal cell. Ascospores ellipsoidal or broadly fusiform, 11–15 × 6–8.5 μm, usually with a single germ pore, occasionally with two .... *C. gracile.*

(107∗) Ascomata ampulliform. Hairs straight, seta-like. Ascospores ellipsoidal-fusiform, often asymmetrical, 9–14 × 5.5–8 μm, occasionally with a single germ pore .... *C. virescens.*

(108) (From 5) Thermophilic. Ascomatal hairs flexuous, dichotomously or irregularly branched. Ascospores globose in frontal view, broadly ovoidal in side view, 7–9.5 × 6–8.5 × 4–6 μm, with an apical germ pore ....*C. thermophilum.*

(108∗) Mesophilic, occasionally thermotolerant or psychrotolerant ....0.**109.**

(109) Ascospores broadly ovoidal or irregular, often inequilateral, 13–18 × 10–14 × 9–12 μm, with an apical germ pore. Ascomatal hairs undulate .... *C. uniapiculatum.*

(109∗) Ascospores smaller ....0.**110.**

(110) Ascospores lacrymiform, bilaterally flattened, 9–11 × 7–8 × 6–7 μm, with a subapical or nearly apical germ pore. Terminal hairs flexuous or undulate, often branched .... *C. senegalense.*

(110∗) Ascospores different in shape ....0.**111.**

(111) Ascospores (sub)globose to broadly ovoidal, sometimes with a subapical or lateral germ pore. Hairs spirally coiled ....0.**112.**

(111∗) Ascospores different in shape, with an apical germ pore. Ascomatal hairs hyphoid, often branched ....0.**117.**

(112) Ascospores often globose or ovoidal, bilaterally flattened, irregular in shape, 10–13 × 9–10 μm, with a subapical or lateral germ pore. Terminal hairs unbranched or very rarely branched, with up to 20 turns .... ***C. myricicola*** Horie and Udagawa 1990.

(112∗) Ascospores smaller, usually regular in shape, with an apical germ pore ... .0.**113.**

(113) Colonies restricted, with yellow exudates. Ascospores broadly ovoidal, often biapiculate, bilaterally flattened, 7–11 × 6–9 × 4–6.5 μm .... *C. crispatum.*

(113∗) Colonies expanding, lacking yellow exudates. Ascospores slightly smaller ....0.**114.**

(114) Hairs 5–8 μm in diameter, coiled and with coiled branches. Ascospores broadly ovoidal, 7–8.5 × 6–7 × 5–6 μm .... *C. medusarum.*

(114∗) Ascomatal hairs 2.5–5 μm broad, unbranched ....0.**115.**

(115) Ascomata with a dark apical area. Ascospores subglobose-ellipsoidal, rarely ovoidal, bilaterally flattened, 5.5–9 × 5–7.5 × 4.5–5.5 μm .... ***C. acropullum*** X.W. Wang in Wang and Zheng 2005a, b.

(115∗) Ascomata without a darker apical area. Ascospores broadly ovoidal ....0.**116.**

(116) Ascospores 8–9.5 × 7–8 × 6–7 μm. Ascomatal hairs with numerous coils ... . *C. mollicellum.*

(116∗) Ascospores 7–8.5 × 6–7 × 5–6 μm. Hairs flexuous and undulate or with three to five coils ....*C. brasiliense.*

(117) Colonies sometimes with yellow-brown or orange-brown exudates. Peridium dark brown, a *textura angularis* covered with an outer layer of *textura intricata*. Hairs about 1 μm in diameter. Asci cylindrical to clavate. Ascospores limoniform, bilaterally flattened, biapiculate, 8.5–10 × 6.5–7.5 × 5–6 μm.

Scytalidium-like resting ascospores sometimes produced .... ***C. luteum*** (J.N. Rai and J.P. Tewari 1964) P.F. Cannon.

(117∗) Colonies often with orange exudates. Peridium sulfur-yellow in reflected light, of a *textura intricata* tending to *epidermoidea*. Hairs up to 3 μm in diameter. Asci cylindrical. Ascospores broadly fusiform to rhomboid, sometimes slightly inequilateral, 11–13 × 6–7.5 μm. Anamorph *Sporothrix*-like ....***C. strumarium*** (J.N. Rai et al. 1964) P.F. Cannon.

(118) (From 1) Peridium of a *textura angularis*, almost glabrous or with some tapering hairs. Ascospores 10–12 × 7–8.5 × 5–6 μm, not dextrinoid. Conidia absent .... *C. microascoides.*

(118∗) Peridium cephalothecoid with dark seta-like hairs. Ascospores 9–12 × 5–7 × 4–5 μm, dextrinoid when young. Conidia present. .... *C. trigonosporum.*

1.3.2 Numerical Taxonomy:

Genus *Chaetomium* has been the target of the numerical taxonomy for the first time by Seth in 1983. In his study, he concentrated on the studies carried out by previous authors, namely, Chivers (1912, 1915), Ames (1961), Udagawa (1960a, b), and Seth (1972), all of whom have given importance to the characters of the terminal hairs as the most important taxonomic characters simply because they are the most obvious structure in the fungus showing variations. He added to the terminal hairs the shapes and size of ascospores because he considered them as the second most variable structure, which one can easily observe under a microscope.

In his study, he examined several hundred Fungarium specimens and cultures of various species in detail and their characters noted as clearly as possible. In order to make his study, a wide selection of characters of the genus, numbering up to 66, were examined (Tables 1.1 and 1.2). So as to randomize the selection deliberately because of the larger number of characters, the nearer one can approach a random selection of the characters of the genome. Then these characters were coded mainly using the simplest forms of coding (+ and −), but certain quantitative characters were coded as character states from 1 to 9. The data thus obtained were recorded in a table, in a matrix form. He processed this information at Atlas Computer Centre, Birmingham (UK). An abridged version of the cluster analysis method of Legendre and Rogers (1972) was used to calculate the similarity indices and to produce a dendrogram using the weighted pair-group method outlined in Sokal and Sneath (1963).

Seth's (1983) results showed that the printout received from the computer center was interesting in that it produced a dendrogram (Fig. 1.6) and phenon lines at 70%, 80%, 90%, and 100%, which showed groupings, based on similarities, hitherto never produced by any of the work done by taxonomists using classical methods. For example, Seth (1972) produced the following groupings:

1. Terminal hairs dichotomously branched.
2. Terminal hairs with or without side branches.
3. Terminal hairs flexed to long undulate, at times branched.
4. Terminal hairs arcuate.
5. Terminal hairs sinuous to loosely coiled.
6. Terminal hair spirally coiled.
7. Terminal hairs in reversed loops.
8. Perithecia elongate.

These groupings have some *Chaetomium* sp. which appear in more than one group, but by using a computer, two distinct groups A and B were produced, based on the following characters:

Group A – Terminal hairs of two kinds, rarely of three kinds. Ascospores olive brown to dark brown sometimes filled with refractive globules. Ovoid to lemon-shaped, may or may not be apiculated at ends.

Table 1.1 Characters of the genus *Chaetomium* used by Seth (1983)

Perithecia	**Terminal hairs**	**Lateral hairs**	50. Monostichous.
1. On dung.	19. Black.	33. Black.	51. Biseriate.
2. On rotten plant materials.	20. Olive green.	34. Olive brown.	52. Deliquescent.
3. On cellulose.	21. Dark below faint above.	35. Dark below faint above.	**Ascospores**
4. Black.	22. Septate.	36. Septate.	53. Hyaline.
5. Olive green.	23. Roughened.	37. Roughened.	54. Olive brown.
6. Yellow green.	24. One kind.	38. One kind.	55. Apiculate at one end.
7. Olive brown.	25. Two kinds.	39. Two kinds.	56. Apiculate at both ends.
8. Pink.	26. Three kinds.	40. Three kinds.	57. Lemon-shaped.
9. Ostiolate.	27. Branched.	41. Branched.	58. Globose.
10. Globose.	28. Swollen at the base.	42. Swollen at the base.	59. Sub-globose.
11. Sub-globose.	29. Spiraled.	43. Spiraled.	60. Fusiform.
12. Vase shaped.	30. Arcuate.	44. Arcuate.	61. Angular.
13. Superfacial.	31. Straight.	45. Straight.	62. Almond.
14. Partially immersed.	32. Thickness.	46. Thickness.	63. Ovoid.
15. Rhizoids.		**Asci**	64. Length.
16. Psuedoparenchymatous.		47. Unitunicate.	65. Breadth.
17. Length.		48. Cylindrical.	66. Cirrhi.
18. Breadth.		49. Clavate.	

Table 1.2 Scoring classes (characters and character state) used by Seth (1983)

Characters	1 Present or absent	2 1–5 µm	3 6–10 µm	4 11–15 µm	5 16–20 µm	6 50–250 µm	7 251–450 µm	8 451–650 µm	9 651–850 µm	10 851–1050 µm
Perithecia										
1. On dung.										
2. On rotten plant materials.										
3. On cellulose.										
4. Black.										

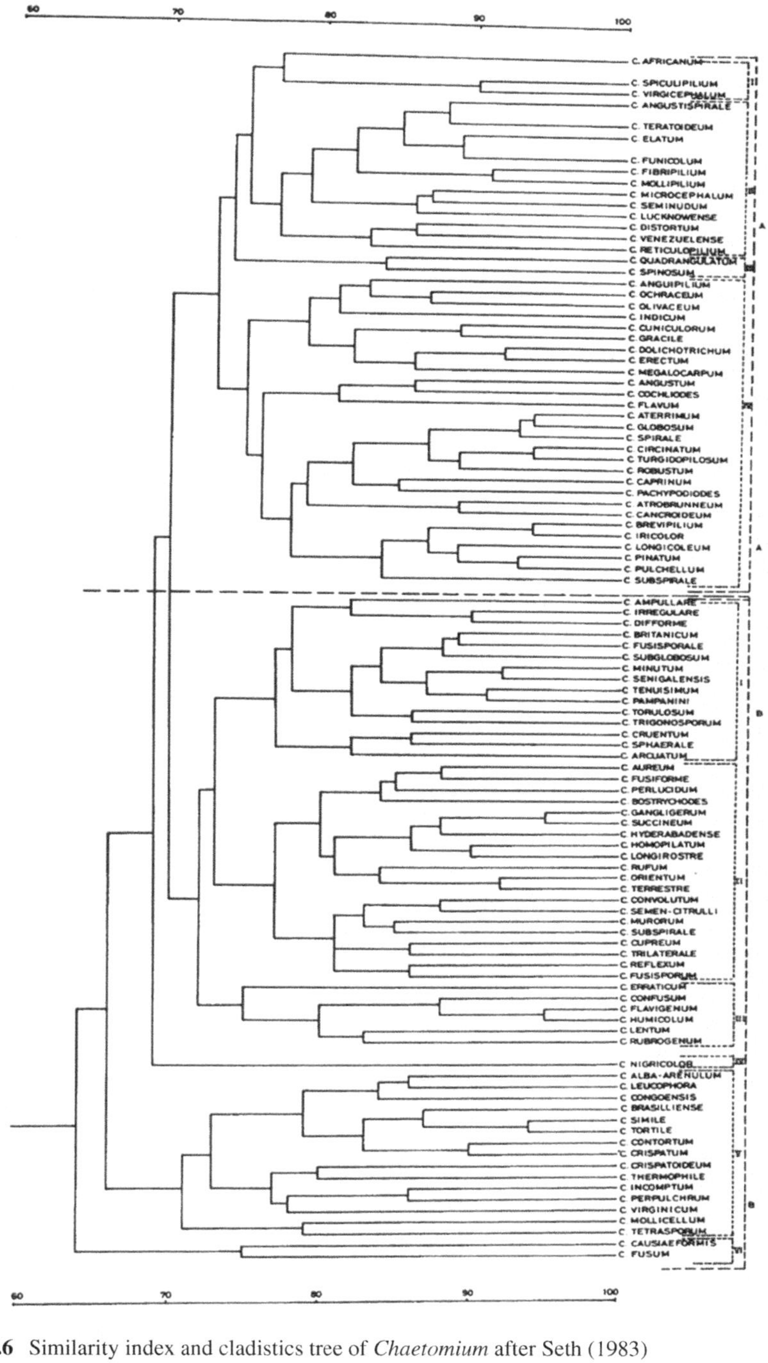

Fig. 1.6 Similarity index and cladistics tree of *Chaetomium* after Seth (1983)

Group B – Terminal hairs mostly of one kind. Ascospores olive brown to dark brown, sometimes filled with refractive globules, ovoid to lemon-shaped, may or may not be apiculate or umbonate at the ends.

Group A has a further four subgroups based on the following characters:

Subgroup 1 – Ascospores light brown to olive brown, broadly ovoid to elliptical with sub-apiculate ends.
Subgroup 2 – Ascospores light brown to olive brown, broadly ovoid to lemon-shaped. When ovoid, apiculate or subapiculate at ends.
Subgroup 3 – Ascospores hyaline to greenish olive filled with refractive globules, ovoid.
Subgroup 4 – Terminal hairs can be of two and rarely of three kinds. Ascospores olive green to olive brown mostly lemon-shaped, sometimes ovate with apiculate ends.

Group B consists of six subgroups. It is based on the following characters:

Subgroup 1 – Ascospores ellipsoid to globose, apiculate at one or both ends, rarely triangular in shape.
Subgroup 2 – Ascospores ovate to sub-globose, apiculate or umbonate at ends, filled with refractive globules.
Subgroup 3 – Ascospores ellipsoid brown, sometimes filled with refractive globules.
Subgroup 4 – Ascospores dark brown, almond-shaped.
Subgroup 5 – Ascospores olive brown to dark brown ellipsoid, sometimes apiculate at one or both ends.
Subgroup 6 – Perithecia small, terminal hairs branched, looped, or gnarled to fusiform.

Ascospores are very light colored (ovate to fusiform).

These two groups seldom show any overlapping of the species, which is an improvement on the results obtained by classical taxonomic methods.

In 2003, Abdel-Azeem studied the ecology and taxonomy of ascospores-producing taxa in Egypt. In order to find out whether possible phylogenetical relationships exist among isolated or recorded taxa, cladistic analysis using a number of non-ordered and non-weighted characters (Table 1.3) was carried out using Winclada Software (version 1.00.08, Nixon, 2002). *Saccharomyces* (a non-filamentous ascomycete taxon) was selected as outgroup genus. The same species has been selected by few authors earlier (Lumbsch 2000).

A cladogram representing the phylogenetic relationship among the 29 genera used in the present study is given in Fig. 1.7. *Saccharomyces* (a non-filamentous genus) was used as an outgroup in the cladistic analysis. A completely resolved phylogenetic tree was obtained in which two main clades are clearly distinguished. The first clade (designated as A) includes seven genera belonging to two orders *Eurotiales* (*Byssochlamys*, *Neosartorya*, *Eupenicillium*, *Talaromyces*, *Emericella*, and *Eurotium*) and *Gymnoascales* (*Narasimhella*). *Narasimhella* was nested as a

Table 1.3 Characters and character state of Egyptian ascomycetes after Abdel-Azeem (2003)

Character	Character state				
	0	1	2	3	4
Anamorph	Absent	Present			
Fruit body	Gymnothecium	Cleistothecium	Perithecium	Apothecium	Pseudothecium
Appendages	Absent	Present			
Ascus shape	Globose	Oval	Clavate	Cylindrical	
Ascus wall	Prototunicate	Unitunicate	Bitunicate		
Location of asci	Random	Hymenium	Locules		
Ascospore type	Unicellular	Multicellular			
Ascospore appendages	Absent	Present			
Ascospore color	Colorless	Colored			
Ascospore arrangement	Conglomerate	Uniseriate	Biseriate		
Poration of ascospore	Non-porate	Porate	Colpate		
Active spore discharge	No	Yes			
Habitat					
Dung habitat	Absent	Present			
Stored seeds	Absent	Present			
Leaf surfaces	Absent	Present			
Compost	Absent	Present			
Salt marsh soil	Absent	Present			
Cultivated soil	Absent	Present			
Desert soil	Absent	Present			

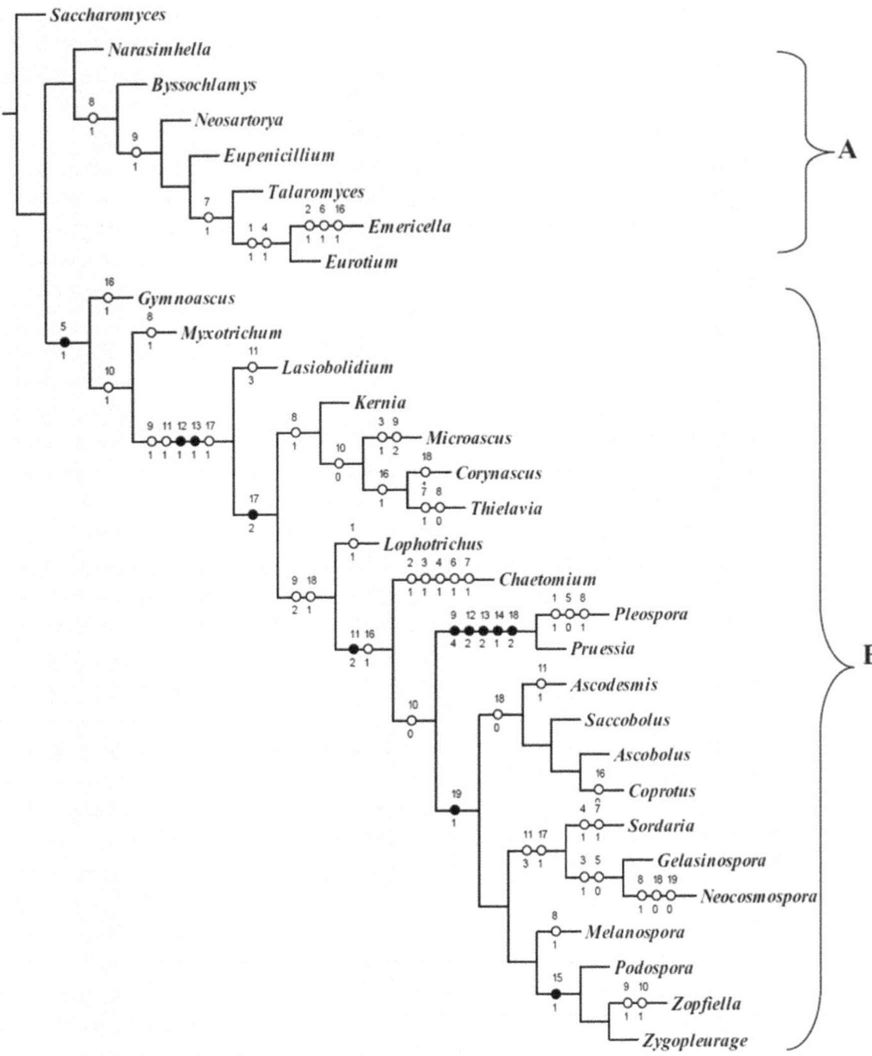

Fig. 1.7 Cladogram of examined genera with characters, character state, and synapomorphic characters indicated by black circles (Abdel-Azeem 2003)

basal taxon to the six genera belonging to the order *Eurotiales* to develop a monophyletic group. This indicates that they are closely related to each other than either one of the other genera belonging to order *Gymnoascales*.

The second clade (designated as B) includes the other 22 genera. These genera belong to five orders, namely, *Gymnoascales*, *Sphaeriales*, *Pleosporales*, *Pezizales*, and *Hypocreales*. *Pleospora* and *Preussia* from *Pleosporales* (regarded as sister taxa) are closely related to each other. Their clade is well supported by five synapomorphic evidence. Their cladogram also shows that *Pleospora* and *Preussia* are a sister-clade-group to another one comprising 11 taxa belonging to the orders *Pezizales*, *Hypocreales*, and *Sphaeriales*. The four genera *Ascodesmis*, *Saccobolus*, *Ascobolus*, and *Coprotus*, belonging to order *Pezizales*, are nested together in a separate clade which is a sister clade to another clade comprising genera of *Sphaeriales*. From the cladogram, it seems to be that order *Sphaeriales* is not a monophyletic group.

The cladogram shows also that *Lasiobolidium* is a problematic genus in the phylogenetic hypothesis presented in this study, because it is nested out the clade distinguishing the genera belonging to the order *Pezizales*; therefore, more investigations need to be done for justifying the taxonomic assignment of this genus. Similarly, *Neocosmospora*, which belongs to *Hypocreales*, is nested within the clade representing order *Sphaeriales*, and further studies are needed to clear up the ambiguity of its systematic position.

In this cladogram, a total number of 10 equally parsimonious trees were obtained with length of 68, consistency index (CI) of 41, and retention index (RI) of 77 eaC.

1.3.3 Secondary Metabolites Profiling and Chemotaxonomy

In 1981, Sekita et al. screened mycotoxin production by *Chaetomium* spp. and related fungi on rice culture. Their study was conducted by a combination of cytotoxicity tests using HeLa cells and thin-layer chromatography. Producers of sterigmatocystin, O-methylsterigmatocystin, chaetochromin, chaetocin, chctomin, cochliodinols, and mollicellin G were found, and the taxonomic significance of these findings is discussed in their study.

In a previous paper (Udagawa et al. 1979), 57 isolates of *Chaetomium* and its allied genera were screened for their ability, on rice culture, to produce chaetoglobosins, a novel class of cytochalasins (Natori 1977), and other metabolites. Five species, *C. cochliodes* Palliser, *C. globosum* Kunze ex Fr., *C. mollipilium* Ames, *C. rectum* Serg., and *C. subaflne* Serg., all belonging to the *C. globosum*-Gruppe sensu Dreyfuss (1975), were found to be producers of the cytochalasins. Unexpectedly, strains identified as *C. thielavioideum* Chen (Chen 1973) (vide infra) were found to produce sterigmatocystin (I), its 0-methyl ether (II), and a new phenolic compound, as well as a known antibiotic substance, chaetocin (111). The details of the separation of the metabolites and the structural elucidation of the new phenolic compound, designated chaetochromin (IV), were recently published (Sekita et al. 1980). Since sterigmatocystin is a well-known hepatocarcinogen, further surveys for mycotoxin production employing the same methods have been conducted on about 60 other isolates of *Chaetomium* and its allied genera.

Chaetomium and related fungi have been isolated from various agricultural commodities such as foods, feeds, and the raw materials of pharmaceutical preparations on numerous occasions, but only recently have these fungi received comprehensive attention with regard to the production of mycotoxins (Sekita et al. 1981). Udagawa et al. (1979) reviewed the literature up to the middle of 1978 concerning the production of *Chaetomium* mycotoxins and presented data on the cytotoxicity and mycotoxigenic screening of 57 cultures of *Chaetomium* and related fungi. In that screening the production of chaetoglobosins was limited to species belonging to the *C. globosum*-Gruppe sensu Dreyfuss (1975). This conclusion has been confirmed by the present screen, since there is no producer of chaetoglobosins among the 26 species of *Chaetomium* and 4 species of related genera tested.

A type strain of *A. virescens* v. Arx, ATCC 32393; an additional isolate, AC-274; and another isolate of *C. thielavioideum* produced a set of mycotoxins, i.e., sterigmatocystin (I), 0-methylsterigmatocystin (11), chaetocin (III), and chaetochromin (IV) (Udagawa et al. 1979). Reexamination of the morphology of *A. virescens* led to the conclusion that *C. thielavioideum* is identical with *A. virescens* in all macroscopic and microscopic characteristics. The genus *Achaetomiella* was erected by von Arx (1970) with *A. virescens* as the type species. The genus is characterized by very reduced terminal hairs in the perithecium, ascospores with two germ pores, and thermotolerance in growth. With the exception of the number of germ pores on the ascospores, there are only minor differences in the characteristics of the perithecia, asci, and ascospores between *Achaetomiella* and *Chaetomium*, and Udagawa (1980) has relegated *Achaetomiella* to synonymy with *Chaetomium. Chaetomium cellulolyticum* Chahal & Hawksworth (Chahal and Hawksworth 1976) is also considered to be identical to *C. thielavioideum* because of the identical pattern of its metabolites as well as its very close morphological similarities to the type strain (DAOM).

The taxonomy of *C. udagawae* Serg. ex Udagawa, another producer of sterigmatocystin (I) found in the present screening, has already been discussed (Udagawa et al. 1979). A toxic pigment, chaetochromin (IV), was originally isolated from *C. thielavioideum* cultures (Sekita et al. 1980). This pigment is of particular interest, since it has now been proven to be widely associated with the *Chaetomium* spp.: *C. caprinum* Bain. in the *C. bostrychodes*-Gruppe, *C. gracile* Udagawa in the *C. spirale*-Gruppe, and *C. tetrasporum* Hughes in the *C. crispatum-Gruppe.* The occurrence of chaetochromin from species in four different groups indicates that the toxin may be quite common and that the latter three fungi, although not closely related in their morphology, may have identical hazardous characteristics. Since oral and intraperitoneal administration of the pigment of mice results in noticeable toxicity such as selective inhibition of hematopoiesis (Sekita et al. 1981), precise examination of the toxicity of the pigment and the moldy rice is now in progress.

Chetomin (V), an antibiotic first reported from *C. cochliodes* Palliser, was subsequently discovered in a culture of *C. globosum* Kunze ex Fr., and *C. globosum* has been reported as a toxic agent of moldy corn for rats (Udagawa et al. 1979).

The production and toxicological data of chetomin from both of these species have been furnished by Brewer et al. (1972). Furthermore, in the mycotoxin survey of *Chaetomium* isolates in Canada (Brewer and Taylor 1978), the same toxin was isolated from *C. funicola* Cooke and *C. umbonatum* Brewer. A fifth chetomin-producing species, *C. subglobosum* Serg., which is taxonomically related to *C. globosum*, appeared in the study carried by Brewer et al. (1978).

A purple pigment, cochliodinol (VI), was previously isolated from liquid shake cultures of *C. cochliodes* and *C. globosum* (Brewer et al. 1968), and the antibiotic effect was reported (Brewer et al. 1970). *Chaetomium globosum* is the principal cochliodinol-producing fungus, but both species are in the *C. globosum*-Gruppe of Dreyfuss and are widely distributed in nature. In this screening, *C. elatum* Kunze ex Fr., another common species belonging to the *C. globosum*-Gruppe, also produced the compound. Two new isomers of cochliodinol, isocochliodinol and neocochliodinol, were isolated from rice cultures of *C. murorum* Corda and a new species, *C. amygdalisporum* Udagawa & Muroi (NHL 2874), respectively. The compounds were not identical with asterriquinones from *Aspergillus terreus* Thom (Yamamoto et al. 1976, 1980), and the structural elucidation will be reported in a separate paper (Sekita et al. 1981). *Chaetomium murorum*, which is assigned to a separate group (viz., the *C. murorum*-Gruppe), is also widely distributed, but it is not particularly abundant. Taxonomic consideration of *C. amygdalisporum* is also reported elsewhere (Udagawa and Muroi 1981).

Mollicellins, a series of fungal depsidones showing mutagenicity, were recently discovered in a culture of *C. mollicellum* Ames (Stark et al. 1978). The presence of mollicellin G (VII) in the *C. amygdalisporum* culture is of interest, considering the morphological resemblance between *C. mollicellum* and this fungus.

Whether mycotoxins produced by *Chaetomium* spp. are significant in outbreaks of foodborne diseases of humans and animals is still uncertain, but, because of their ubiquity in deteriorated plant materials including foods and feeds and their ability to produce a number of potentially significant mycotoxins, they should be included in the hazardous group of mycotoxigenic fungi.

1.3.4 Molecular Characterization

The ITS region was the most widely used DNA region in molecular biology for fungal identification at the species level, because of its higher degree of variation than other genic regions of rDNA. Chen et al. (2001) used polymorphic internal transcribed spacer (ITS) region 1 DNA sequences to identify medically important yeasts. Species-specific polymorphisms in the noncoding internal transcribed spacer 2 (ITS2) region of the rRNA operon have provided accurate identification of clinically significant yeasts.

Polymorphisms in the PCR product length permitted 19 species to be distinguished by ITS1 alone, compared with 16 species distinguished by using only ITS2. However, combination of both ITS alleles permitted identification of 30 species

(98% of clinical isolates). The remaining 10 species with PCR products of similar sizes contained unique ITS alleles distinguishable by restriction enzyme analysis. Further phylogenetic analyses based on ITS sequences showed a similar overall topology to 26S rRNA gene-based trees.

Pryce et al. (2003) developed a rapid identification method for fungi by sequencing the ITS 1 and ITS2 regions using an automated capillary electrophoresis system. Sequence-based identification was compared to traditional identification in a blinded manner. Of the clinical isolates tested, 88/89 had DNA sequences that were highly homologous to those of reference strains accessioned in GenBank, and 87/89 gave a sequence-based identification result that correlated with the traditional identification.

They concluded that this approach is rapid and may be a more accurate cost-effective alternative than most phenotypic methods for identification of many medically important fungi frequently encountered in a routine diagnostic microbiology laboratory.

Sharma et al. (2013) described a new endophytic species of *Chaetomium* from *Jatropha podagrica* from Maharashtra State, India. A combined sequence dataset of the ITS region, LSU rDNA, and β-tubulin genes supports recognition of this fungus as a new species that was largely concordant with morphological characters.

Rajesh et al. (2013) studied 12 isolates of *Trichoderma* obtained from soil rhizosphere of plantation crops and agricultural fields of Indian Agricultural Research Institute, New Delhi, India, and identified them based on integrated approach of morphological and molecular (ITS) characters. Similarly Prabhakaran et al. (2015) differentiated the four *Trichoderma* species, viz., *T. asperellum*, *T. harzianum*, *T. psuedokoningii*, and *T. longibrachiatum* with the help of both morphological and ITS sequence-based characterization.

ITS was a conserved rDNA sequence that has been widely used both alone and in combination with other universal sequences, such as β-tubulin and actin to identify, characterize, and perform phylogenetic analysis of fungal isolates (Balazy et al. 2008).

Aggarwal et al. (2013) studied the molecular characterization of 18 *Chaetomium* isolates collected from India based on the internal transcribed spacer (ITS) region of the rRNA gene sequences. Phylogenetic analysis of full length ITS region showed that *C. globosum* isolates (Cg1, Cg2, Cg6, Cg11, and Cg15), *Chaetomium* spp. isolates (C16, C17), and a *C. perlucidum* isolate formed a group with American isolates (SW287, SW271, and CL024) of *Chaetomium* spp., thereby supporting the close relationships among these isolates. ITS has been successfully used for species discrimination of six genera in *Zygomycetes* (Schwarz et al. 2006), *Trichoderma*, and *Hypocrea* in Ascomycota (Druzhinina et al. 2005) and *Cortinarius* section Calochroi (Frøslev et al. 2007) and *Melampsora* (Feau et al. 2009) in *Basidiomycota*.

1.3.5 DNA Barcoding

The term DNA barcoding was coined by Paul Hebert et al. (2003) in 2003 at the University of Guelph in Ontario, Canada, to create fast, simple, unified species identification system based on DNA sequence data. Barcoding uses a very short genetic sequence from a standard part of the genome the way a supermarket scanner distinguishes products using the black stripes of the Universal Product Code. As it is able to distinguish species and identify specimens (including incomplete, damaged, or immature specimens) using a very short gene sequence, it attracted many taxonomists, geneticists, and evolutionary biologists.

Leading to this an international consortium (CBOL, Consortium for the Barcode of Life) was formed in 2004 to come up with binding suggestions as what markers and protocols would be accepted for DNA barcoding and to act as mediator between the DNA barcoding community and other parties, i.e., GenBank, potential users, or technology developers. Universal markers for plants, birds, and animals were identified.

DNA barcoding shows enormous promise for the rapid identification of organisms at the species level. There has been much recent debate about the need for longer barcode sequences when these sequences were used to construct molecular phylogenies. Min and Hickey (2007) analyzed a set of fungal mitochondrial sequences of various lengths and monitored the effect of reducing sequence length on the utility of the data for both species identification and phylogenetic reconstruction. Their results demonstrated that reducing sequence length has a profound effect on the accuracy of phylogenetic trees but yields accurate species identifications. They concluded that the standard short barcode sequences (~600 bp) were not suitable for inferring accurate phylogenetic relationships, but they were sufficient for species identification among the fungi.

Molecular characters such as DNA sequence data were essentially advantageous as they offer a greater number of discrete characters, which can be analyzed statistically to infer phylogenetic relationships. The DNA sequence data have been widely used for (1) anamorph-teleomorph connections; (2) phylogenetic relationships of form-genera, of anamorph-rich clades, and of anamorphic generic complexes; and (3) species delineation of monophyletic groups of anamorphic taxa. More importantly, newly generated DNA sequence data aid to test taxonomic hypotheses at different strata of classification.

Shenoy et al. (2007) recognized the difficulty in morphology-based identification of a fungal species and in many instances may be incorrect and considered the molecular identification and phylogenetic data using multiple genes to identify fungal isolates at the species level.

In the early 1990s, most of the molecular phylogeny-related investigations focused on the establishment of taxonomic relationship at the level of class, order, or family of filamentous *Ascomycetes* fungi. In 1995, Spatafora (1995) studied phylogenetic analyses of nucleotide data from partial sequences (1150 bp) of the small subunit ribosomal DNA (SSU rDNA) of 30 taxa representing several orders of

Hymenoascomycetes and *Loculoascomycetes*. These analyses detected four major groups of filamentous ascomycetes: group 1, pyrenomycetes (*Hypocreales*, *Microascales*, *Diaporthales*, *Sordariales*) and loculoascomycetes (*Pleosporales*); group 2, operculate discomycetes (*Pezizales*); group 3, inoperculate discomycetes (*Geoglossaceae*); and group 4, plectomycetes (*Eurotiales*, *Onygenales*) and loculoascomycetes (*Chaetothyriales*). Well-supported clades, which correspond to groupings based on ascomal morphology, were obtained. However, the monophyly of the classes *Hymenoascomycetes* and *Loculoascomycetes* was rejected. Later in 1999, Liu et al. (1999) studied phylogenetic relationships among *Ascomycetes* using *rpb2* sequences. In an effort to establish a suitable alternative to the widely used 18S rRNA system for molecular systematics of fungi, they examined the nuclear gene *rpb2*, encoding the second largest subunit of RNA polymerase II. *rpb2* was a large-size single-copy gene with a modest rate of evolutionary change, provided good phylogenetic resolution of *Ascomycota*. The *rpb2* phylogeny resulted in much higher bootstrap support for all the deeper branches within the orders and for several branches between orders of the *Ascomycota*. In the same year, Lee and Richard investigated phylogenetic relationships of *Chaetomium* and related genera based on ribosomal DNA sequences. The sequence data of *rpb2* region confirmed the generic entity of *Achaetomium* and placed *Ascotricha* within the *Xylariaceae*. The analyses also defined the monophyletic lineage of the *Microascaceae* and the evolutionary linkages between the families *Sordariaceae* and *Chaetomiaceae*. The placement of *Coniochaeta* in the *Coniochaetaceae* and *Chaetomium* and *Achaetomium* in the *Chaetomiaceae*, under the order *Sordariales*, was supported by this study.

Miller and Huhndorf (2005) have indicated that multi-gene phylogenies of ascomal wall morphology was a better predictor than ascospore morphology in the *Sordariales*. Phylogenetic relationships of multiple representatives from each of several genera representing the range in ascomal wall and ascospore morphologies in the *Sordariales* were estimated using partial nuclear DNA sequences from the 28S ribosomal large subunit (LSU), β-tubulin, and ribosomal polymerase II subunit 2 (*rpb2*) genes and found that ascospore morphology was extremely homoplastic and not useful for delimiting genera. On the other hand, taxa with similar ascomal wall morphologies clustered in five well-supported clades suggesting that ascomal wall morphology was a better indicator of generic relationships in certain clades in the *Sordariales*.

Greif et al. (2009) investigated the phylogenetic affiliations of *Chaetomidium* and distribution of the cephalothecoid peridium within this genus by using sequence data of LSU, β-tubulin, and *rpb2*. The results of phylogenetic analyses showed that *Chaetomidium* is polyphyletic and should be restricted to its type, *C. fimeti* and *C. subfimeti*. The cephalothecoid species of *Chaetomidium* were distributed in three unrelated clades, suggesting that the morphological similarity among these particular species resulted from convergence instead of ancestry.

Asgari and Zare (2011) studied the phylogeny of genus *Chaetomium* in Iran. Combined sequence dataset of the ITS region, partial LSU rDNA, and β-tubulin gene sufficiently resolved five species groups of *Chaetomium* that were largely concordant with combined features of peridium structure, ascospore shape, and germ pore position.

Among the new species, *C. undulatulum* was a close relative of *C. globosum*; *C. rectangulare* was closer to *C. elatum*; *C. interruptum* and *C. grande* were closer to *C. megalocarpum*, altogether clustering with the *C. globosum* group. *C. iranianum* and *C. truncatulum* were members of the *C. carinthiacum* group, characterized by spirally coiled ascomatal hairs and fusiform ascospores.

Wang et al. (2014) studied multigene phylogenetic analyses with ribosomal ITS, partial ribosomal large subunits (28S rDNA), β-tubulin, the translation elongation factor 1α (TEF1-α), and the largest subunit of RNA polymerase II (*rpb1*) and recognized eight well-supported lineages within the monophyletic *C. indicum* group.

ITS appeared to be problematic in the identification of the blue stain fungi (Roe et al. 2010). For *Aspergillus* species diagnosis, β-tubulin gene was suggested as the potential DNA barcode locus against COI, ITS, and the intergenic spacer of the ribosomal genes (IGS) (Geiser et al. 2007). Nuclear 28S rDNA and β-tubulin genes provided a perfect capability of species separation in *Fusarium* (O'Donnell and Cigelnik 1997).

Translation elongation factor 1α gene (EF-1α) was a reliable barcode for *Fusarium* (Geiser et al. 2004).

Fungi of *Nectriaceae* were economically important and of high species diversity. For the purpose of accurate and rapid species identification, ITS, 28S rDNA, β-tubulin gene, and EF-1α gene were selected as the candidate DNA barcode markers to investigate their feasibility in identification of 28 well-circumscribed species belonging to 9 genera of the nectriaceous fungi. A total of 216 sequences of the candidate genes were analyzed.

Intra- and inter-specific variations and success rate of PCR amplification and sequencing were considered as important criteria to estimate the candidate genes. The partial β-tubulin gene met the requirements for an ideal DNA barcode and functions well for correct species delimitation. No overlapping between the intra- and inter-specific pairwise distances was found. Twenty-eight clusters were recognized in accordance with the 28 morphological species tested. In addition, it had a high PCR and sequencing success rate. Therefore, β-tubulin gene was proposed as the possible barcode for the nectriaceous fungi (Zhao et al. 2010).

To determine a suitable DNA barcode for the genus *Neonectria*, the internal transcribed spacer rDNA, β-tubulin, EF-1α, and *rpb2* genes were selected as candidate markers. A total of 205 sequences from 19 species of the genus were analyzed. Intra and inter-specific divergences and the ease of nucleotide sequence acquisition were treated as criteria to evaluate the feasibility of a DNA barcode. Results indicated that any single gene among the candidate markers failed to serve as a successful barcode, while the combination of the partial EF-1α and *rpb2* genes recognized all species tested. The combined partial EF-1α and *rpb2* genes were proposed as a DNA barcode for the genus. During this study, two new cryptic species *N. ditissimopsis* and *N. microconidia* were discovered, based on the combined data of morphology and DNA barcode information (Zhao et al. 2011).

To select an appropriate or ideal DNA barcode marker, comparison of intra- and inter-specific variations were treated as an important criterion. Barcode gap is calculated as difference between inter-specific distance and intra-specific distance. Region showing high barcode gap is considered as the best region for DNA barcoding (Suwannasai et al. 2013). Among the regions of the ribosomal cistron, the ITS region has the highest probability of successful identification (PCI) for the broadest range of fungi, with the most clearly defined barcode gap between inter- and intra-specific variation. The nuclear ribosomal large subunit, a popular phylogenetic marker in certain groups, had superior species resolution in some taxonomic groups, such as the early diverging lineages and the ascomycete yeasts, but was otherwise slightly inferior to the ITS. The nuclear ribosomal small subunit has poor species-level resolution in fungi. ITS would be formally proposed for adoption as the primary fungal barcode marker to the Consortium for the Barcode of Life, with the possibility that supplementary barcodes may be developed for particular narrowly circumscribed taxonomic groups (Schoch et al. 2012).

DNA barcoding analyses were performed by datasets from several phylogenetic lineages of the *Glomeromycota*. 1500 bp fragment spanning small subunit (SSU), ITS region, and large subunit (LSU) of nuclear ribosomal DNA were tested for species resolving power. Sub-fragments covering the complete ITS region, 800 bp of the LSU rDNA, and three 400 bp fragments spanning the ITS2, LSU-D1, or LSU-D2 domains were also analyzed. Barcode gap analyses did not resolve all species, but neighbor-joining analyses, using Kimura two-parameter (K2P) distances, resolved all species based on the 1500 bp fragment. The shorter fragments failed to separate closely related species.

The complete 1500 bp fragment was recommended as a basis for arbuscular mycorrhizal fungi DNA barcoding. This would also allow future identification of arbuscular mycorrhizal fungus (AMF) at species level based on 400 or 1000 bp amplicons in deep sequencing approaches (Stockinger et al. 2010).

Three new species of *Leohumicola* (anamorphic *Leotiomycetes*) were described using morphological characters and phylogenetic analyses of DNA barcodes.

Leohumicola levissima and *L. atra* were isolated from soils collected after forest fires in Crater Lake National Park, USA. *Leohumicola incrustata* was isolated from burned fibers from the Cape of Good Hope Nature Reserve, South Africa. The three species exhibited distinct morphology based on conidial characters. Two DNA barcode regions, the ITS nuclear rDNA region and the cytochrome oxidase subunit I (Cox1) mitochondrial gene, were sequenced. Single-gene parsimony, dual-gene parsimony, and dual-gene Bayesian inference phylogenetic analyses support *L. levissima*, *L. atra*, and *L. incrustata* as distinct phylogenetic species. Both ITS and Cox1 barcodes were effective for the molecular identification of *Leohumicola* species (Nguyen and Seifert 2008).

Despite the strength and impact of rDNA ITS as the sanctioned universal fungal DNA barcode, its resolution of higher taxonomic level relationships is inferior to many protein-coding genes such as *rpb1*, *rpb2*, or β-tubulin (Nilsson et al. 2006; Seifert 2009; Begerow et al. 2010; Schoch et al. 2014). Although many alternatives to ITS were considered by the mycological community, its sanctioning as the primary barcode marker rested on its practicality and reliability, not on the highly desired "resolution power" (Schoch et al. 2012, 2014). Many mycologists would prefer one or several universal, but phylogenetically informative, loci as barcodes, with higher species resolution power than is feasible with ITS.

Cai et al. (2009) studied for the species identification of *Colletotrichum* using ITS and other prominent regions to ascertain DNA barcode and observed that within *Colletotrichum gloeosporioides* sensu lato, GPDH, CAL, and ACT are good candidates for barcodes.

A set of six genes (ITS, ACT, CHS, GPDH, histone 3, β-tubulin) were used by Damm et al. (2009) to study *Colletotrichum* species with curved conidia from herbaceous hosts, which included six different clades. The differentiation of the species was best with both GPDH and histone 3 genes, which are superior to ITS, ACT, CHS, and β-tubulin.

Stielow et al. (2015) showed several gene sections accessible to universal primers/markers yielding a single PCR product. Barcode gap and multi-dimensional scaling analyses revealed that some of the tested candidate markers have universal properties providing adequate intra- and inter-specific variation that make them attractive secondary barcodes for species identification. Among these gene sections, a novel high-fidelity primer pair for TEF1α, already widely used as a phylogenetic marker in mycology, has potential as a supplementary DNA barcode with superior resolution to ITS.

In India, 44 isolates of *Chaetomium* representing each species were selected to establish phylogenetic relationship among *Chaetomium* species and to select the potential region for DNA barcode, using six gene regions, viz., actin, β-tubulin, calmodulin, ITS, *rpb2*, and tef-1 by Sekhar (2015). He selected ITS region as primary barcode as it resulted in best identification and grouping of *Chaetomium* species. But due to less barcode gap and probability of correct identification (PCI), it should always be supported by a secondary barcode to avoid mislead identification. Therefore, β-tubulin was proposed as secondary barcode which showed high barcode gap and PCI.

1.3.6 Evolution of the Approach: Polyphasic Taxonomy of Chaetomium

The polyphasic taxonomy takes into account all available phenotypic and genotypic data and integrates them in a consensus type of classification. Phylogenetic species recognition is increasingly being used with the internal transcribed spacers of the nrDNA (ITS) now accepted as the official DNA barcode for fungi (Schoch et al. 2012). Phylogenetic analyses by Untereiner et al. (2001) and Greif et al. (2009) provided evidence that the genus *Chaetomium* was not monophyletic.

The study carried out by de Hoog et al. (2013) demonstrated that the distinctions among *Achaetomium*, *Chaetomidium*, *Chaetomium*, and *Thielavia* were ambiguous. Taxonomic revision of *Chaetomium* is therefore needed.

Chaetomium has been connected to several anamorphic genera, such as *Acremonium*, *Botryotrichum*, *Chrysosporium*, *Histoplasma*, *Humicola*, *Phialophora*, *Scopulariopsis*, and *Scytalidium*. All these genera were described later than *Chaetomium* except for *Acremonium*, and the generic names *Botryotrichum*, *Histoplasma*, *Phialophora*, *Scopulariopsis*, and *Scytalidium* were used in much less frequencies than *Chaetomium*. Phylogenetic assessment and taxonomic revision remained to be done for these anamorphic genera, as most of them appeared to be polyphyletic.

Zhang et al. (2017) investigated 46 *Chaetomium* strains isolated from soil and compost from China. In total, 14 species were recognized, of which 7 were described as new species (i.e., *C. angulare*, *C. cirrhata*, *C. heterothallicum*, *C. laterale*, *C. longiciliata*, *C. microthecia*, *C. uniseriatum*). Growth temperature profiles of these fungi revealed that 11 species are mesophilic; *C. angulare* and *C. jodhpurense* are thermotolerant; *C. thermophilum* var. *dissitum* is thermophilic, with temperature maxima above 50 °C. Among the 46 strains, 39 were shown to be homothallic. Mating experiments were conducted for the remaining seven strains that are morphologically and phylogenetically closely related. The results of the mating experiment, together with the four-locus (ITS, LSU, RPB2, TUB) phylogeny and a pairwise homoplasy index (PHI) test, supported the division of these seven additional strains into two heterothallic species that are herein described as *C. heterothallicum* and *C. uniseriatum*. A three-locus (ITS, LSU, TUB) phylogenetic tree based on all currently accepted species that have available type-derived sequences revealed that *Chaetomium* species clearly grouped in six phylogenetic groups which showed a certain correspondence with their morphology and temperature profiles.

In their study, Zhang et al. (2017) incorporated sequences derived from 54 extype cultures and 73 authentic strains to provide a backbone tree for the genus *Chaetomium*. The result of the backbone tree showed that most species can be delimited clearly using combined ITS, LSU, and TUB dataset, which is in agreement with previous phylogenetic analysis (de Hoog et al. 2013; Wang et al. 2014). The backbone tree needs to be further expanded as there are still many species that lack type-derived sequences. Some important species (e.g., *Chaetomium globosum* and *C. jodhpurense*) need to be epitypified to facilitate future studies.

The backbone tree also recognized several major species groups in *Chaetomium* that have been traditionally recognized on the basis of morphology (Ames 1963; Seth 1970; Dreyfuss 1975). These included species groups such as *C. globosum*, *C. carinthiacum*, *C. crispatum*, *C. indicum*, *C. murorum*, *C. thermophilum*, and several distinct species.

Phylogenetically recognized *C. globosum* species group is essentially similar to the sense as defined by Dreyfuss (1975) and Asgari and Zare (2011). Four species of our isolates clustered in this species group, in which two were new species. The *C. carinthiacum* species group is a Bayesian well-supported monophyletic group, and the newly described species in this study, *C. longiciliata*, clustered in this clade, and its morphological characters also supported the inclusion of this species in this group.

The *C. murorum* species group includes two species, *C. ancistrocladum* and *C. murorum*. Both species are psychrotolerant, growing well at 15–25 °C, but are strongly reduced at 30 °C. The newly described species, *Chaetomium laterale*, grouped as a sister clade to the *C. murorum* group. *Chaetomium laterale* is similar to the species within *C. murorum* group in the ascospore shape and ascomatal hairs but is distinguished by peridium of textura angularis or irregularis and ascospores with a subapical or lateral germ pore near the attenuated end. It has a higher MinGT (15 °C in C. laterale versus below 5 °C in species within *C. murorum* group).

The *Chaetomium thermophilum* species group includes three thermophilic species. Although *C. thermophilum* var. *coprophilum* and *C. thermophilum* var. *thermophilum* were very closely related to each other, they have different MaxGT (50 °C versus 55 °C). *Chaetomium senegalense*, although suggested as thermophilic by Salar and Aneja (2007), is actually thermotolerant. Phylogenetic analysis also showed that *C. senegalense* was distinct from *C. thermophilum* species group. *C. angulare* is another thermotolerant species described in this study, which is morphologically and phylogenetically close to this species group.

Chaetomium seminis-citrulli clustered in *C. crispatum* species group. However, *C. seminis-citrulli* produces asci with ascospores uniseriate, different from other species producing asci with ascospores biseriate. Two new species, *C. heterothallicum* and *C. uniseriatum*, were described in this group. Phylogenetic analysis based on four-locus alignment clearly separated *C. seminis-citrulli* from *C. heterothallicum* and *C. uniseriatum*. *C. heterothallicum* and *C. uniseriatum* resemble to each other on the shape of ascomata, asci, ascomatal hairs, peridium, and ascospores and their heterothallic life cycle. They also have similar chlamydospores and growth temperature curve. The phylogenetic data clearly divided these isolates in two clades. PHI test also revealed that there was no significant recombination event between the two groups of strains. One clade, described as *C. uniseriatum*, contained the isolates CGMCC 3.17559, CGMCC 3.17560, and CGMCC 3.17561, which can mate with each other and produce ascomata and ascospores. The other clade, described as *C. heterothallicum*, consisted of strains CGMCC 3.17543, CGMCC 3.17544, CGMCC 3.17545, and CGMCC 3.17546. Isolates of this later group can also mate with each other with the only exception of strain CGMCC 3.17546. The isolates within the two clades could not mate with each other. In light of the phylogenetic and biological characters, Zhang et al. (2017) suggested that these two clades should be delimited to two species. Although CGMCC 3.17546 could not mate with other strains, they suggested including the isolate into species *C. heterothallicum* based on the phylogenetic analyses.

Chaetomium biapiculatum Lodha was regarded as a synonym of *C. homopilatum* Omvik by von Arx et al. (1986). However, their type strains were phylogenetically distant from each other; thus both should represent distinct species. Asgari and Zare (2011) thought MinGT and MaxGT were both important in distinguishing species within genus *Chaetomium*.

Li et al. (2012) and Wang et al. (2014) noted that MaxGT was more useful in delimiting species. They also summarized MaxGT of most species in the phylogenetic tree. The result showed that MaxGT of *Chaetomium* species is somewhat related to their phylogenetic relationships. *Chaetomium globosum* species group, *C. indicum* species group, and *C. carinthiacum* species group were located in the upper part of the phylogenetic tree comprised species with MaxGT mainly at 35–40 °C which they already have been traditionally called mesophilic. However, extremophiles such as thermophilic, thermotolerant, and psychrotolerant species all clustered in the lower part of the backbone tree. In conclusion, Zhang et al. (2017) study also suggested that morphology and growth temperature profile corresponded well with the phylogenetic relationships. For more detail, please check the original publication of Zhang et al. (2017).

1.4 *Chaetomium* Diversity in Different Habitats

1.4.1 Desert

By definition a "desert" is a region that receives extremely low rains – less than 250 mm per year – far less than the amount required to support the growth of most plants. Approximately one-third of Earth's land surface is a desert with an area more than 52,000 square kilometers (Fig. 1.8) (Abdel-Azeem et al. 2016b).

Deserts are extreme environments where intense solar radiation, limited nutrients, low organic matter content, and restricted water availability present formidable challenges for fungi inhabiting these areas. Desert soils generally are characterized by low propagule densities but high species diversity (Christensen 1981; Mouchacca

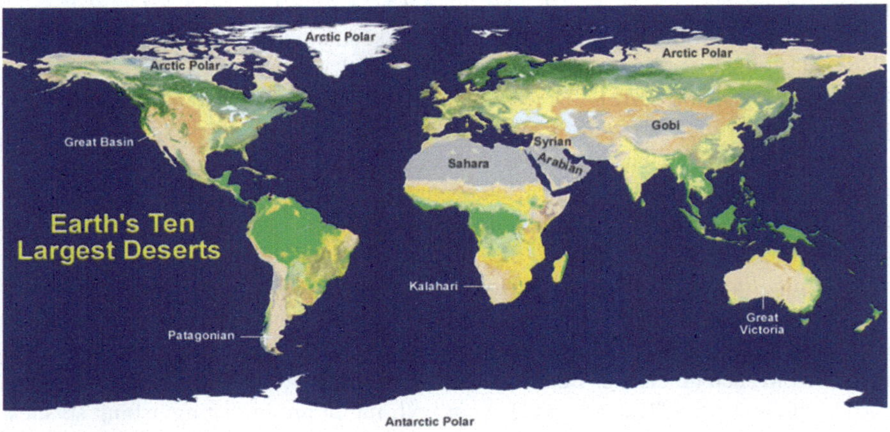

Fig. 1.8 World's desert biome

1995; Abdel-Azeem et al. 2016a). Studies on mycobiota of soils may be dated back to Adametz (1886) when Adametz started his pioneer study by isolation and naming 4 species of yeasts and 11 species of filamentous fungi including *Aspergillus* (Watanabe 2002; Abdel-Azeem et al. 2016a).

The number of mycological studies on desert soil is rather limited in comparison with other ecological habitats. Several authors assume the diversity of microbes including fungi is low compared to soil in moderate or tropical regions, and they suggest these extreme ecosystems as suitable in situ models to study the relationship between phylogenetic biodiversity and function (Adams et al., 2006).

Desert mycobiota of Egypt have been the target of many studies, viz., Montasir et al. (1956a, b), Mahmoud et al. (1964), Besada and Yusef (1968), Moubasher and Moustafa (1970), Moubasher and El-Dohlob (1970), Salama et al. (1971), Mouchacca (1971, 1973a, b, 1977, 1982), Nagiub and Mouchacca (1970–1971), Mouchacca and Nicot (1973), Mouchacca and Joly (1974, 1976), Samson and Mouchacca (1974, 1975), Moubasher et al. (1985, 1988, 1990), Nassar (1998), Abdel-Hafez et al. (1989a, b, 1990), Abdel-Sater (1990, 2000), Abdel-Hafez and El-Maghraby (1993), Abdel-Azeem and Ibrahim (2004), and Abdel-Azeem (1991, 2003, 2009).

Moubasher and Moustafa (1970) surveyed the Egyptian soil fungi with special reference to *Aspergillus*, *Penicillium*, and *Penicillium*-related genera in 32 soil samples collected from the different localities in Egypt. They met two species of *Chaetomium*, and the highest population and occurrence were recorded for *Aspergillus*. They only recorded two species of *Chaetomium*; they were *C. murorum* and *C. olivaceum*, respectively.

Abdel-Azeem in 2003 studied the ascospore-producing taxa in Egypt as the first real solid study in that country, and he recorded two species of *Chaetomium*, namely, *C. globosum* and *C. brasiliense* recovered from desert soil, respectively. The former was recorded in 22 of desert soils out of 50 and the later in 2 soils out of 50.

In their extensive survey of Sinai terricolous fungi, Abdel-Azeem and Ibrahim (2004) recorded two species of *Chaetomium*. They recorded *C. globosum* and *C. brasiliense*. They mentioned that *C. globosum* recorded 4.95% of total count of taxa.

By scanning of available sources of information concerning genus *Chaetomium* in Egypt, it was possible to determine 53 species and 1 variety isolated/reported from different substrates.

Only five taxa are introduced to the genus *Chaetomium* as novel taxa based on type materials collected from Egyptian soil, namely, *Chaetomium gelasinosporum* Aue and Muller (1967), *C. mareoticum* Besada and Yusef (1969), *C. sinaiense* Moustafa and Ezz El-Din (1989), *C. strumarium* (J.N. Rai, J.P. Tewari and Mukerji) P.F. Cannon (1986), and *C. uniporum* Aue and Muller (1967).

Few investigations have been made on soil mycobiota in Libya. Naim (1967a, b) studied rhizosphere and soil fungi of *Artemisia herba-alba* and fungi under *Citrus* trees in Tripoli, Libya. Youssef (1974) studied the fungal biota of Libyan soil. He examined 16 different localities in Libya for their fungal microbiota. Youssef recovered different species of *Chaetomium* from Libyan soils: *Chaetomium bostrychodes*,

C. irregulare, *C. mollicellum*, *Chaetomium* sp. (new species), *C. spirale*, *C. fusisporum*, and *C. jodhpurense* with low frequency of occurrence.

By available data from the checklist of Libyan fungi (El-Buni and Rattan (1981), only 10 species of *Chaetomium* were recorded.

El-Said and Saleem (2008) studied soil fungi at western region of Libya. In 2010, Mansour studied the distribution and occurrence of various groups of fungi in different kinds of soils in eastern region of Libya. A total of 63 species and 5 varieties belonging to 30 fungal genera were collected from 75 soil samples. Desert soil were represented by 22 genera and 35 species and 2 varieties and genus *Chaetomium* only represented by 1 species, namely, *C. globosum* El-Said and Saleem (2008).

Mansour (2010) recovered 59 species belonging to 23 genera obtained from 100 soil samples collected from the eastern region in Libya. *Chaetomium* were distributed among nine soil samples and represented by *C. globosum* and unknown species.

The northern part of Israel is the most mycologically investigated territory of the country, from which 307 soil microfungal species have been isolated (Volz et al. 2001). In their book, they recorded 20 species of *Chaetomium*. Grishkan et al. (2003) isolated 11 species of the cellulolytic genus *Chaetomium* easily by means of the dilution plate method. In their study, a soil microfungal community was examined over a 1-year period (1999–2000) at the western shore of the Dead Sea. A total of 78 species from 40 genera were isolated. The most prominent features of mycobiota of the territory studied were (i) the prevailing number of melanin-containing micromycetes (46 species, 65.5% of the total isolate number); (ii) a large share of teleomorphic Ascomycota (26 species, 18.5% of isolates); (iii) combination of true soil and plant surface-inhabiting species; (iv) spatial and temporal variation of the mycobiota composition; and (v) very low fungal density (nearly 500-fold lower than in the Judean Desert soil). These features are formed under the extremely stressful xeric and oligotrophic conditions in which the Dead Sea coastal micromycete community exists. Nine species (*Alternaria alternata*, *A. raphani*, *Aspergillus niger*, *Aureobasidium pullulans*, *Chaetomium globosum*, *C. murorum*, *Cladosporium cladosporioides*, *Penicillium aurantiogriseum*, and *Stachybotrys chartarum*) were considered a characteristic micromycete complex for the Dead Sea coastal habitat based on the spatial and temporal occurrence of these species. Many of the micromycetes isolated, including almost all the species listed above, are known to be distributed worldwide occurring in different soil types. This confirms the conclusion of many mycologists working in areas with saline and arid soils that there is no halo- and thermophilous mycobiota characteristic for those soils.

Grishkan and Nevo (2010) studied the microfungal composition of the Negev desert. Results showed that soil was found to be inhabited by comparatively rich microfungal diversity – more than 360 identified species belonging to three phyla: *Zygomycota* (17 species), *Ascomycota* (87 teleomorphic or sexual species and 255 anamorphic or asexual species), and *Basidiomycota* (3 species). The species represent 121 genera; the most prominent are *Penicillium* (38 species including anamorphic states of *Eupenicillium*), *Aspergillus* (34 species including anamorphic states of *Emericella*, *Eurotium*, and *Neosartorya*), *Chaetomium* (25), *Phoma* (16), and *Drechslera* (12).

Composition of thermotolerant mycobiota in the soil of Israeli deserts and northern territories was examined in spatiotemporal dynamics by Grishkan (2018). A total of 165 species from 82 genera were isolated at 37 °C using the soil dilution plate method. Aspergilli (*Aspergillus fumigatus* and *A. niger*) and teleomorphic ascomycetes (*Canariomyces notabilis*, *Chaetomium nigricolor*, and *C. strumarium*) comprised the basic part of the thermotolerant communities.

In Tunisia 16 fungal strains were isolated from arid Tunisian soils by Mtibaà et al. (2017). The most interesting strain was identified based on the analysis of the amplified nucleotide sequences of the nuclear ribosomal ITS1–5.8-ITS4 region (∗600 bp) as *Chaetomium*. Extracellular laccase was purified 15-fold from the crude culture to homogeneity with an overall yield of 50% using ultrafiltration and anion-exchange chromatography. The purified enzyme was found to be a monomeric protein with a molecular mass of 68 kDa, estimated by SDS-PAGE, and with an isoelectric point of 5.5. The optimal temperature and pH value for laccase activity toward 2,6-DMP were 60 °C and 3.0, respectively. It was stable at temperatures below 50 °C and at alkaline conditions.

In Saudi Arabia, Ali (1977) and Abdel-Hafez (1981, 1982a, b) investigated soil fungi of desert and identified 24 genera and 47 species, and 34 genera and 80 species, respectively.

In Saudi Arabia, Abdel-Hafez (1982a) isolated and identified 34 genera and 80 species in addition to 1 variety of *A. nidulans* from 40 soil samples collected from different places in desert of Saudi Arabia. More than 42 species are new records from Saudi Arabian soils. *Aspergillus* and *Penicillium* contributed the greatest number of species (15 species +1 variety and 13 species, respectively). *Chaetomium* was represented by only one species (*C. globosum*) in this study.

A total number of 75 species and 5 varieties belonging to 27 genera of cellulose-decomposing fungi were recovered during this investigation carried out by Abdel-Hafez (1982b). *Aspergillus* (12 species +4 varieties), *Alternaria* (4 species), *Stachybotrys* (3 species +1 variety), and *Penicillium* (12 species) were of high frequency of occurrence, of which *A. fumigatus*, *A. niger*, *A. terreus*, *A. alternata*, *A. tenuissima*, *S. atra*, *P. citrinum*, and *P. corylophilum* were the most common. Six genera were of moderate frequency, and these were *Botryotrichum* (two species), *Macrophomina* (one species), *Drechslera* (five species), *Ulocladium* (four species), *Chaetomium* (seven species), and *Curvularia* (four species) of which *B. atrogriseum*, *D. spicifera*, *D. sativus*, *U. botrytis*, *C. spirale*, and *C. lunata* were the most prevalent. Five and twelve genera were recovered on cellulose-Czapek's agar plates in low and rare frequency of occurrence, respectively. The number of recovered taxa of *Chaetomium* was seven namely: *C. spirale*, *C. globosum*, *C. bostrychodes*, *C. cochliodes*, *C. olivaceum*, *C. jodphurence*, and *C. subglobosum*.

Bahkali and Khiyami (1996) isolated 30 fungal species belonging to 15 genera from 30 soil samples on cellulose Czapek's agar. The highest number of fungal species was isolated from Dammam (20 species) followed by Nomas (18 species), Makkah and Riyadh (17 species each), Tabouk (16) species, and Jizan (11 species). The most frequent genera isolated were *Aspergillus*, *Penicillium*, *Alternaria*,

Ulocladium, and *Curvularia*. Throughout this study, six fungal species belonging to four genera, *Ulocladium septosporum*, *Emericella nidulans*, *Trichoderma harzianum*, *T. koningii*, *T. pseudokoningii*, and *Cochliobolus lunatus*, were found to be new records in Saudi Arabian soils. Genus *Chaetomium* were represented only by one unknown species in this study.

Saadabi (2006) studied the fungal biota (flora) of Saudi Arabian soils. Sixteen different localities in Southern area of Saudi Arabia were examined. A modified soil-dilution plate method was used for isolating the fungi. A toal of 64 species in 22 genera were reported. Of these, 5 species were *Phycomycetes*, 10 were *Ascomycetes*, and 49 were *Deuteromycetes*. Of the reported species, 55 are new records to the mycobiota of Saudi Arabia, and one of these is a new species. *Aspergillus* and *Penicillium* were the fungal genera having the highest number of isolated species. Ten species from each genus were recorded. Seven species were recovered from both *Fusarium* and *Chaetomium*, four species from *Cladosporium* and *Trichoderma*, and two species from *Alternaria*, *Pleospora*, *Rhizopus*, *Stachybotrys*, *Coniothyrium*, and *Phoma*.

The fungal community of six sand samples from Saudi Arabia and Jordan deserts was characterized by culture-independent analysis via next-generation sequencing of the 18S rRNA genes and by culture-dependent methods followed by sequencing of internal transcribed spacer (ITS) region (Murgia et al. 2018). The identified fungal phyla were *Ascomycota*, basal fungi, and *Basidiomycota*, and the most abundant detected classes were *Dothideomycetes*, *Pezizomycetes*, and *Sordariomycetes*. A total of 11 colonies of filamentous fungi were isolated and cultured from six samples, and the ITS sequencing pointed toward five different species of the class Sordariomycetes, belonging to genera *Fusarium* (*F. redolens*, *F. solani*, *F. equiseti*), *Chaetomium* (*C. madrasense*), and *Albifimbria* (*A. terrestris*). The results of this study show an unexpectedly large fungal biodiversity in the Middle East desert sand and their possible role and implications on human health. There are 327 colonies of fungi, and *Aspergillus* was one of the most common genera isolated in this study.

In Sudan, Nour (1956) made a preliminary survey of fungi from various soil types. He isolated 18 genera and 35 species; the most common species were *Rhizopus stolonifer*, *A. niger*, *A. nidulans, Curvularia lunata*, *Cladosporium sphaerospermum*, *A. terreus*, *Alternaria tenuis*, and *Cladosporium cladosporioides*. In Sudan, 120 sites, from six localities from the Sudan Gezira, were examined for soil mycobiota by Amin and Abdalla (1980). The lower fungi were mostly represented by mucoraceous genera, such as *Rhizopus*, *Mucor*, and *Cunninghamella*. Oomycetes were rare. Ascomycetes, other than the perfect states of *Aspergillus*, were not common. Only *Chaetomium globosum* was isolated from the six localities. *Monascus*, *Thermoascus*, *Neocosmospora*, and *Pyronema* were recorded for the first time from Sudanese soils. The majority of the isolates were *Hyphomycetes*. The bulk of the fungal population were species of *Aspergillus*, of which *A. niger* was the most common, followed by *A. terreus*, *A. flavus*, *A. nidulans*, and *A. fumigatus*. Other *Aspergilli* were sporadic. *A. niveus* and *A. sejunctus* were not previously reported from Sudanese soil.

The soil mycobiota of the desert of Kuwait was investigated by Halwagy et al. (1982), who identified a total of 52 genera and 130 species; *Fusarium*, *Aspergillus*, *Penicillium*, *Stachybotrys*, *Myrothecium*, *Ulocladium*, *Phoma*, and *Alternaria* were the most frequent genera. Six species of *Chaetomium* were recovered during this study; they were *Chaetomium elatum*, *C. virginicum*, *C. olivaceum*, *C. rectopilium*, *C. globosum*, and *C. spirale*.

Suleiman et al. (2019) examined the general soil fungi and AM fungal communities associated with a Lonely Tree species (*Vachellia pachyceras*) existing in the Sabah Al-Ahmad Natural Reserve located at the Kuwait desert. Their results reveal that the fungal phylotypes were classified in four major fungal phyla, namely, *Ascomycota*, *Basidiomycota*, *Chytridiomycota*, and *Zygomycota*. The largest assemblage of fungal analyses showed communities dominated by members of the phylum *Ascomycota*. The assays also revealed a wealth of incertae sedis fungi, mostly affiliated to uncultured fungi from diverse environmental conditions. They recorded only one species of *Chaetomium*, and their work thus provides a baseline of the fungal and mycorrhizal community associated with rhizosphere and non-rhizosphere soils and roots of only surviving *V. pachyceras* tree from the Kuwaiti desert and seedlings under nursery growing environments.

Maharachchikumbura et al. (2016) updated the checklist of accepted taxa in Oman. They recorded only the following taxa: *Chaetomium atrobrunneum*, *C. bostrychodes*, *C. globosum*, *C. murorum*, and *C. spirale*.

In 2017, Al-Sadi et al. (2017) studied high fungal diversity and dominance by ascomycota in dam reservoir soils of arid climates in Oman, and they recovered two species of *Chaetomium* (*C. homopilatum* and *Chaetomium* sp.).

In Qatar no previous study has been carried out on soil fungi before Moubasher and Al-Subai (1987). Their study was designed to study intensively soil fungi and their distribution in various localities and habitats. For this purpose, 42 soil samples were collected and assayed for their fungal content and for some of their physical and chemical properties. In their study there were isolated three species of *Chaetomium*.

Chaetomium thermophilum was recorded by Mandeel (2002) during his study of microfungal community associated with rhizosphere soil of *Zygophyllum qatarense* in arid habitats of Bahrain.

1.4.2 Salterns and Mangrove

When evaporation of seawater accompanied with halite (NaCl) concentrations of greater than 10% (m/w), thalassohaline hypersaline environments originated (Oren 2002) and provide some of the most extreme habitats in the world. They are common all around the globe and include, for example, marine ponds and salt marshes that are subjected to evaporation, salt or soda lakes, and sea-salt and man-made salterns (Trüper and Galinski 1986).

Life-limiting parameters in salterns are many, e.g., variable water activities (a_w), high concentrations of NaCl, low oxygen concentrations, as well as high light intensity (Brock 1979). Halotolerant and halophilic fungi were first reported as active inhabitants of solar salterns by Gunde-Cimerman et al. (2000). Later on, they were isolated by several investigators (Butinar et al. 2005a, b, c; Cantrell et al. 2006) from salterns around the world, e.g., La Trinidad in the Ebro River Delta and Santa Pola on the Mediterranean coast of Spain, Camargue in France, and the salterns on the Atlantic coast in Portugal and in Namibia, the Dominican Republic, and Puerto Rico.

Moustafa and Al-Musallam studied the fungi of salt marshes of Kuwait in Moustafa and Al-Musallam 1975. Eighty-two species and 44 genera were isolated from 40 composite soil samples representing salt marshes, salt depressions, and coastal sands. In general, the fungal population of saline soils was rather poor with a narrow spectrum of genera and species. The most frequent fungi were *Aspergillus*, *Penicillium*, *Alternaria*, *Coniothyrium*, *Stachybotrys*, *Fusarium*, *Cephalosporium*, *Ulocladium*, *Myrothecium*, and *Drechslera*. Genus *Chaetomium* was represented by three species (*C. globosum*, *C. spirals*, and *C. olivaceum*).

After a decade of research into the fungal diversity in salterns, together with new taxa, a number of fungal genera with high diversities of halotolerant and halophilic species have been described. Different species of genus *Chaetomium* are among the filamentous fungi that appear with moderate frequencies in salterns (Abdel-Azeem 2003). The group of filamentous fungi that have been isolated from different salterns around the world is mainly represented by the order *Eurotiales* by the teleomorphic genera *Eurotium* and *Emericella* and the anamorphic *Aspergillus*, *Penicillium*, and *Chaetomium* (Cantrell et al. 2006).

Abdel-Azeem (2003) recorded a list of *Chaetomium* taxa isolated from salt marches and salty soil in Egypt. He recorded seven species, namely, *C. circinatum* Chivers, *C. hamadae* (Udagawa) Arx, *C. hexagonosporum* A. Carter & Malloch, *C. homopilatum* Omvik, *C. piluliferum* J. Daniels, *C. rectopilium* Fergus & Amelung, and *C. subspirilliferum* Sergejeva.

Moustafa and Abdel-Azeem (2011) prepared an updated checklist of *Ascomycota* reported from soil and other terricolous substrates in Egypt. They mentioned that when species richness and substrate preference, as important ecological parameters, are considered, it has been noticed that Egyptian *Ascomycota* shows some interesting features noteworthy to be mentioned. At the substrate level, clay soils came first by hosting a range of 108 taxa followed by desert soils (60 taxa). At the taxonomic level, *Sordariales*, compared to other orders, accommodated the greatest number of taxa, i.e., 92 taxa followed by *Eurotiales* (61 taxa). *Chaetomiaceae* and *Trichocomaceae* are by far the richest families by housing 61 taxa. At the generic level, *Chaetomium* occupied the first place among all reported genera by including 51 species followed by *Arthroderma* (15 spp.). Provisional keys to the identification of reported taxa are given.

To conclude, in hypersaline environments, the pan-global stable taxa of genus *Chaetomium* are represented by *C. globosum*, *C. spirale*, *C. olivaceum*, *C. circinatum*, *C. hamadae*, *C. hexagonosporum*, *C. homopilatum*, *C. piluliferum*, *C. recto-*

pilium, and *C. subspirilliferum*, which are also not abundant, although more locally distributed (Abdel-Azeem 2003).

Mangroves are an assortment of tropical and subtropical trees and shrubs which have adapted to the inhospitable zone between sea and land: the typical mangrove habitat is a muddy river estuary (Kathiresan and Bingham 2001; Hogarth 2007). Mangles are considered a dynamic ecotone, and approximately 25% of the world's coastline is dominated by mangroves distributed in 112 countries encompassing an area of 18,000,000 ha (Spalding et al. 1997). Biodiversity of biota associated with mangle ecosystem is well known for animals and plants, but poorly known for fungi (Khalil et al. 2013).

Sridhar (2009) studied fungi in Pichavaram mangroves of the southeast coast of India. Damp incubation of wood, root, and leaf litter of *Avicennia marina* and *Rhizophora mucronata* yielded 15 fungi. The fungal richness was highest (nine spp.) on woody litter of both plant species and root litter of *R. mucronata*. *Aniptodera chesapeakensis*, *Halorosellinia oceanica*, *Halosarpheia marina*, *Periconia prolifica*, and *Phoma* sp. were dominant (5–6.3%). Woody litter of *A. marina* was highly colonized by *H. marina* (28%) while *R. mucronata* by *Phoma* sp. (24%). The average fungi per sample ranged between 0.3 and 0.8 with a highest on woody litter (0.7–0.8). In Pichavaram mangroves, so far about 10 niches (water, sediment, and live/dead plant parts) and 14 mangrove plant species have been surveyed for mycobiota. About 102 fungi consisting of mitosporic fungi (57 spp.), ascomycetes (37 spp.), phycomycetes (7 spp.), and basidiomycete (1 sp.) have been reported. Woody litter yielded a highest of 36 saprophytic fungi followed by 33 fungi as foliar epiphytes. *Cirrenalia pygmaea* was the most dominant fungus, followed by *H. oceanica*, *P. prolifica*, *Zalerion maritima*, and *Z. varia*. He recorded two species of *Chaetomium* (*C. globosum* and *C. olivaceum*).

1.4.3 Occurrence in Indoor Air

In 2013, a study of airborne fungus spores by viable and non-viable methods in Havana, Cuba, has been carried out by Almaguer et al. (2013). Their study on the Havana aeromycobiota diversity was extended from November 2010 to October 2011 using two complementary volumetric methods. A total of 35 fungal genera were characterized; 26 of them were recognized only by non-viable methods, six with viable methodology, and the other three with both sampling methods. Furthermore, 47 species were identified by cultivation and the spores collected with the non-viable methodology. These could not be included in a specific genus and, thus, were categorised into five fungal types. In general, the main, spread worldwide, mitosporic fungi also predominated the Havana atmosphere. The predominant species were *Cladosporium cladosporioides*, *Aspergillus flavus*, and *Penicillium citrinum*. Moreover, several *Zygomycetes* (*Syncephalastrum racemosum*, *Rhizopus stolonifer*, and *Rhizopus oryzae*), *Ascomycetes* (*Chaetomium globosum*), and *Basidiomycetes* such as *Coprinus* or *Ganoderma* were isolated.

In their extensive study, Wang et al. (2016a) recovered 145 isolates belonging to *Chaetomiaceae* which were cultured from air, swab, and dust samples from 19 countries. Countries included are Algeria, Australia, Canada, China, Cuba, Denmark, Germany, India, Indonesia, Mexico, the Netherlands, New Guinea, Solomon Islands, South Africa, Spain, Switzerland, Thailand, Uruguay, and the USA. Based on the phylogenetic analyses of DNA-directed RNA polymerase II second largest subunit (*rpb2*), β-tubulin (tub2), ITS, and 28S large subunit (LSU) nrDNA sequences, together with morphological comparisons with related genera and species, 30 indoor taxa are recognized, of which 22 represent known species, seven are described as new, and one remains to be identified to species level. In their (Wang et al. 2016a) collection, 69% of the indoor isolates with six species cluster with members of the *C. globosum* species complex, representing *Chaetomium* sensu stricto. The other indoor species fall into nine lineages that are separated from each other with several known chaetomiaceous genera occurring among them. No generic names are available for five of those lineages, and the following new genera are introduced here: *Amesia* with three indoor species, *Arcopilus* with one indoor species, *Collariella* with four indoor species, *Dichotomopilus* with seven indoor species, and Ovatospora with two indoor species. The generic concept of Botryotrichum is expanded to include *Emilmuelleria* and the *Chaetomium*-like species *B. muromum* (= *C. murorum*) in which two indoor species are included. The generic concept of Subramaniula is expanded to include several *Chaetomium*-like taxa as well as one indoor species. *Humicola* is recognized as a distinct genus including two indoor taxa. According to this study, *C. globosum* is the most abundant *Chaetomiaceae* indoor species (74/145), followed by *C. cochliodes* (17/145), *C. elatum* (6/145), and *B. piluliferum* (5/145). The morphological diversity of indoor *Chaetomiaceae* and the morphological characteristics of the new genera were described and illustrated in their study. This taxonomic study redefines the generic concept of *Chaetomium* and provides new insight into the phylogenetic relationships among different genera within *Chaetomiaceae*.

Fourteen toxigenic indoor *Chaetomium*-like isolates from buildings in Finland were investigated by Castagnoli et al. (2017). Six *C.* globosum-like strains from indoor dusts were toxic with boar sperm assay and cytotoxic to porcine kidney cells (PK-15), emitted green fluorescence, and produced chaetoglobosin inhibiting cellular glucose transport. OT7 and OT7b strains from indoor dust were cytotoxic with PK-15 cells and were non-fluorescent, and produced the extremely cytotoxic protein synthesis inhibitor, chaetomin. The six *C. globosum*-like strains were resistant to borax and very sensitive to the wetting agent genapol used in cleaning chemicals. This may indicate that indoor *Chaetomium*-like fungi occupy their own ecological niche in buildings.

Abdel-Rahim et al. (2018) evaluated the correlation between fungi causing paint deterioration and air contamination in Assiut University hospital to give a complete picture for the fungal quantity and spectrum. Seventeen fungal species were isolated from 15 samples of deteriorated water-based paint collected from the hospitals. *Chaetomium globosum* was the most common paint-deteriorating fungal species, followed by *Alternaria alternata*, *Aspergillus parasiticus*, *Penicillium oxalicum*, and *Setosphaeria rostrata*. Direct examination confirmed the ability of these fungi to colonize the paint samples producing mycelia, conidia, and fruiting bodies. In vitro, these fungi exhibited high potential to utilize the thin layer of polyacrylic paint and significant enzymatic activities of cellulase, lipase, and urease that may play a main role in paint degradation and as virulence factor of human diseases. Moreover, 27 fungal species were isolated as air-contaminating mycobiota. *Aspergillus* spp., *Cladosporium cladosporioides*, *P. oxalicum*, *A. alternata*, and *C. globosum* caused a considerable amount of indoor air contamination. The results indicated that there is a clear correlation between fungi causing paint deterioration and air contamination, whereas certain fungi were responsible for wall paint deterioration and frequently indoor air contamination. The current study suggests that improvement of antimicrobial additives of paints may be a promising approach to reduce paint biodeterioration and subsequently air contamination of indoor environments.

Key to the Most Common *Chaetomiaceae* Species (or Species Complex) from the Indoor Environment (Wang et al. 2016a)

1. Asexual morph present .... 2.

1. Asexual morph absent .... 3.

2. Ascomata absent or rare; conidiophores usually produced together with brown serile setae; conidia subglobose, usually hyaline, smooth to slightly roughened, 11–17.5 μm in diameter .... *Botryotrichum piluliferum*.

2. Ascomata numerous, terminal ascomatal hairs repeatedly dichotomously branched, usually less than 5 μm in diameter near the base; ascospores limoniform, bilaterally flattened, 11.5–14 × 8.5–10 × 7–8 μm, with an apical germ pore asexual morphs acremonium-like .... *Chaetomium elatum*.

3. Terminal ascomatal hairs repeatedly dichotomously branched or seta-like, often composed of two types of hairs different in length; ascospores ovate to elongate ovate, bilaterally flattened, usually less than 7.5 μm long (*Dichotomopilus indicus* or *D. funicola* species complex (including *D. funicola*, *D. pseudofunicola*, *D. subfunicola*, and *D. variostiolatus*)) .... 4.

3. Terminal ascomatal hairs not repeatedly dichotomously branched, ascospores limoniform, bilaterally flattened, with an apical germ pore................................ .... 5.

4. Shorter terminal hairs, if present, usually forming a net structure to enfold ascospores .... *D. funicola* species complex.

4. Shorter terminal hairs, if present, inconspicuous due to the masses of ascospores, not easy to observe .... *Dichotomopilus indicus*.

5. Ascomata subglobose to obovate with a dark collar-like apex around the ostiolar pore; terminal hairs spirally coiled, ascospores limoniform, bilaterally flattened, 6–7 × 5.5–6.5 × 4.5–5.5 μm, with an apical germ pore .... *Collariella bostrychodes*.

5. Ascomata without a dark collar-like apex around the ostiolar pore; ascospores 8.5–11 × 7–9 × 5.5–7 μm .... 5.

6. Terminal hairs flexuous, undulate to slightly coiled............................... *Chaetomium cochliodes*.

6. Terminal hairs spirally coiled, usually with coils regularly tapering in diameter .... *Chaetomium globosum*.

1.4.4 Occurrence in Water

Besides actinomycetes and cyanobacteria, *C. globosum* was identified as a producer of the volatile metabolites geosmin and 2-phenylethanol (Kikuchi et al. 1981), which contribute to earthy–musty odor and taste of public water supplies. It is probable that production of these odor metabolites by fungi takes place predominantly in soil (rather than in water) with respect to the ecology of potential producers (Kikuchi et al. 1981; Frisvad et al. 2004) and is then washed into water sources (Zaitlin and Watson 2006; Hageskal et al. 2009). However, *Chaetomium* (Nasser 2004; Gonçalves et al. 2006; Hageskal et al. 2006; Gashgari et al. 2013) and anamorphic *Humicola* (Göttlich et al. 2002; Nasser 2004) and *Botryotrichum* (Nasser 2004) were repeatedly isolated from drinking water samples, and further investigations are needed to clarify the origin of odor metabolites and the significance of the fungal contribution to production in water systems. Similarly, the potential mycotoxin production by *Chaetomium* spp. in water and their concentration has not been evaluated. *C. globosum* was isolated in some studies from relatively high number of drinking water samples (Nasser 2004), and such contamination may tend to increase in stored water. Some *Chaetomium* species have the potential to cause life-threatening infections in immunocompromised individuals, but these species have not been reported from drinking water, and other sources in the indoor environment are more likely the source of infection.

1.4.5 Occurrence in Foods

Chaetomium spp. are frequently isolated from food products, but most frequently as simple contamination, whereas only in a limited number of cases do they act as spoilage fungi. The isolates are often not identified to species level, or the identification is questionable. However, this information is valuable with regard to species-specific secondary metabolite spectra. *Chaetomium* is the only ascomycetous genus producing brown to black ascomata, which is commonly encountered in food products (Pitt and Hocking 2009) making them straightforward to identify.

Only a few *Chaetomium* species are commonly found in foods. *C. globosum* is the most frequently encountered species; additionally, *C. brasiliense* and *C. funicola* are common species in tropical commodities (Saito et al. 1976; Pitt and Hocking 2009). Relatively high infection levels by these species were encountered in soybeans, mung beans, black beans, rice, maize, barley, nuts overall, copra, and sorghum (Pitt et al. 1993, 1994, 1998; Freire et al. 1999; Pitt and Hocking 2009; Sumalan et al. 2011).

C. globosum has been isolated from a variety of commodities, particularly wheat, barley, maize (also as *C. cochliodes*), oat grains (as *C. ochraceum* and *C. cochliodes*), rice (also as *C. olivaceum*), sorghum, millet, beans (also as *C. cochliodes*), soybeans (also as *C. olivaceum* and *C. cochliodes*), mung beans, copra, peanuts, kemiri nuts, cashew nuts, walnuts, hazelnuts, pea seeds (also as *C. cochliodes*), tomato (as *C. cochliodes*), margarine, green tea, black tea, sugarcane (as *C. fibripilium*), and spices such as black pepper, white pepper, chili and hot pepper seeds (as *C cochliodes*), ginger, cinnamon, cumin, fennel, and Bishop's weed (Abdel-Azeem 2003; Hubka 2015). High level of infection by *C. globosum* was observed particularly in rice, copra, soybean, mung beans, maize, barley, cashew nuts, candle nuts, and nuts overall (Hubka 2015). This species has also been implicated in causing disease in pears in Egypt (Ismail and Abdalla 2005).

C. funicola has been recorded at relatively high frequencies from soybean samples, beans overall, and cashew nuts (Hubka 2015). The species was further isolated from rice, maize, peanuts, cashew nuts, copra, beans, mung beans, soybeans, pea seeds (also as *C. dolichotrichum*), sorghum, radish, eggplant, and spices (as *C. dolichotrichum*). *C. indicum* is phylogenetically close to *C. funicola*, but distinct. This species also resembles *C. funicola* in morphology and is distinguished solely by the absence of stiff, unbranched ascomatal hairs. *C. indicum* has been isolated from beans, soybeans, pea seeds, rice, and spices, and it is possible that it is commonly misidentified with *C. funicola*. Another species clearly distinct from *C. funicola* based on ITS and LSU rDNA data, but morphologically similar, is *C. reflexum* that has been reported from pepper and pea seeds. *C. brasiliense* was isolated from a similar spectrum of tropical commodities as *C. funicola* including soybeans, mung beans, black beans, rice, maize, cashew nuts, peanuts, sorghum, and black pepper (Hubka 2015).

Other species are isolated rarely from food products, viz., *C. aureum*, *C. atrobrunneum*, and *C. murorum* were reported from rice; *C. nigricolor*, *C. raii*, and *C. subaffine* from cereals; *C. bostrychodes*, *C. elatum*, *C. murorum*, and *C. succineum* from pea seeds; *C. aureum* and *C. bostrychodes* from oat grains; and *C. aureum* from snap beans and butter (Hubka 2015). *C. convolutum* (as *C. biapiculatum*) was reported from spices; *C. carinthiacum* from thyme; *Achaetomium globosum* from cumin; *C. bostrychodes*, *C. robustum*, and *C. aureum* from pepper; *Chaetomium* cf. *fusiforme* and *C. aureum* from tea (Hubka 2015). Webb and Mundt (1978) isolated *C. fimeti* (usually classified in the genus *Chaetomidium*) at relatively high frequencies from vegetables; *C. crispatum* was isolated from rotting potatoes, *C. elatum* from rotting onion, and *C. murorum* from Latundan banana (Hubka 2015).

Chaetomium spp. were reported from various seeds, some of which may be used as food or for oil extraction. *C. funicola* (as *C. dolichotrichum*), *C. globosum* (also as *C. cochliodes*), *C. indicum*, *C. madrasense*, and *C. murorum* were isolated from linseed (*Linum usitatissimum*); *C. bostrychodes* and *C. globosum* from okra (*Hibiscus esculentus*); *C. aureum*, *C. globosum*, *C. funicola*, and *C. murorum* from cucumber seeds (*Cucumis sativus*); *C. elatum*, *C. globosum*, and *C. murorum* from pumpkin seeds (*Cucurbita maxima*); *C. carinthiacum* from poppy seeds (*Papaver somniferum*); *C. funicola* from sesame (*Sesamum indicum*); and *C. globosum* from safflower (*Carthamus tinctorius*) and vegetable marrow (*Cucurbita pepo*, also as *C. cochliodes*) (Hubka 2015). Unidentified *Chaetomium* spp. were found on meat products, dried milk, rice, maize, wheat, sorghum, soybeans, beans, mung beans, copra, tapioca, black pepper, peanuts, cashew nuts, sorghum, sesame seeds, Bishop's weed, cumin, lotus seeds (Pitt et al. 1994), pumpkin seeds (Weidenbörner 2001), and herbal drugs (Chourasia 1995; Hubka 2015).

Key to *Chaetomium* Species Relevant to Foods After Hubka (2015)

(1a) Ascomatal hairs – dichotomously branched	2
(1∗) Ascomatal hairs – unbranched (arcuate, coiled, straight, or flexed)	5
(2) Ascospores >10 μm in long axis	*C. elatum*
(2∗) Ascospores <10 μm in long axis	3
(3) Terminal hairs – arcuate with reflexed branches	*C. reflexum*
(3∗) Branches of terminal hairs – essentially straight	4
(4) Ascomatal hairs – partly branched, partly straight, and unbranched	*C. funicola*
(4∗) All or nearly all ascomatal hairs – dichotomously branched at maturity	*C. indicum*
(5) Asci – cylindrical; ascospores uniseriate	*C. brasiliense*
(5∗) Asci – clavate	6
(6) Ascospores >13 μm in long axis	*C. murorum*
(6∗) Ascospores <13 μm in long axis	7
(7) Ascospores – unevenly fusiform to navicular	*C. aureum*
(7∗) Ascospores – symmetrical	8
(8) Ascospores – ellipsoidal, fusiform, width (shorter dimension) <6 μm	*C. carinthiacum*
(8∗) Ascospores – lemon-shaped, width > 6 μm	9
(9) Terminal hairs – undulate	*C. globosum*
(9∗) Terminal hairs – spirally coiled	*C. bostrychodes*

1.4.6 Polar

Around 2.3% of the world's fungal biota exists in the Arctic, and fungi in this region have been isolated from various substrates and habitats (Ivarson 1965; Reeve et al. 2002; Säwström et al. 2002; Callaghan et al. 2004; Ozerskaya et al. 2009; Pathan et al. 2009). More than 1000 species and over 400 genera of non-lichenized fungi are reported from Antarctic regions (including the sub-Antarctic) (Bridge and Spooner 2012; Arenz et al. 2014) including genus *Chaetomium*. Selbmann et al. (2015) published distributional records of Antarctic fungi based on strains preserved in the Culture Collection of Fungifrom Extreme Environments (CCFEE) Mycological Section associated with the Italian National Antarctic Museum (MNA) with only one record of *Chaetomium* spp.

Fungal diversity in the Arctic and Antarctic permafrost has been studied intensively over the last decade, and it has been shown to have considerable taxonomic diversity, with significant numbers of new taxa (Ruisi et al. 2007). In some permafrost regions, yeasts represented an important, or even the major (up to 100%), part of all of the fungi isolated and 20–25% of the total aerobic heterotrophs (Vorobyova et al. 1997; Steven et al. 2006).

In the cold deserts in Antarctica, like in the McMurdo Dry Valleys and the Ross Desert in southern Victoria Land, which is considered one of the harshest environments known on Earth (Nienow and Friedmann 1993), primarily xerophilic, basidiomycetous yeasts have been isolated (Vishniac and Onofri 2003; Onofri et al. 2004; Takano 2004). These were found at the highest frequencies in the youngest layers, which were less than 10,000 years old, although they have also been detected in three-million-year-old Pliocene samples (Dmitriev et al. 1997a, b; Rivkina et al. 2000).

In both the active layer and the perennially frozen Arctic sediments, a large variety of filamentous fungi have been detected, belonging to *Ascomycota*, *Basidiomycota*, *Zygomycota*, *Chytridiomycota*, and *Glomeromycota* (Wallenstein et al. 2007). The most frequently occurring genera of the filamentous fungi are *Aspergillus*, *Cladosporium*, *Geomyces*, and *Penicillium*. This last showed the highest number of species. Additionally, the following ascomycetous genera have been isolated: *Acremonium*, *Alternaria*, *Acrodontium*, *Arthrinium*, *Aureobasidium*, *Bispora*, *Beauveria*, *Botrytis*, *Camposporium*, *Chaetomium*, *Chaetophoma*, *Chalara*, *Chrysosporium*, *Diatrype*, *Engyodontium*, *Epicoccum*, *Eurotium*, *Exophiala*, *Fusarium*, *Geotrichum*, *Gliocladium*, *Gymnascella*, *Helminthosporium*, *Hormoconis*, *Lecythophora*, *Malbranchea*, *Myceliophthora*, *Monodictys*, *Nectria*, *Paecilomyces*, *Papulaspora*, *Phialophora*, *Phoma*, *Racodium*, *Rhinocladiella*, *Scolecobasidium*, *Scopulariopsis*, *Stachybotrys*, *Sepedonium*, *Sphaeronaemella*, *Sporothrix*, *Sporotrichum*, *Stephanosporium*, *Thelebolus*, *Thysanophora*, *Trichoderma*, *Trichophyton*, *Tritirachium*, *Ulocladium*, *Verticillium*, *Wardomyces*, *Xylohypha*, and mycelia sterilia (Vishniac 1993; Azmi and Seppelt 1998; Ivanushkina et al. 2005; Kurek et al. 2007; Ozerskaya et al. 2008, 2009; Stakhov et al. 2008).

1.4.7 On Herbivore Dung

Although *Chaetomium* is not a strictly coprophilous fungus genus, the diversity of representatives of the genus *Chaetomium* on dung is continuously represented in studies of dung fungi diversity worldwide. One of the most representative and recent research on the occurrence of this genus on dung, and that was precursor of others, was developed by Doveri (2004, 2011).

This author presented important contributions to the study of the genus occurring as fimicolous, with detailed diagnosis, substrate preferences, and descriptions of new species, some occurring exclusively on dung (Doveri 2008, 2013, 2016). For *Chaetomium* species recorded on dung, Doveri (2004) presented a representative selection of species reported from dung at least once until 2004. In 2008, the author presented an updated key to coprophilous species recorded from Italy (Doveri 2008), followed by another updated key to *Chaetomium* on dung, with new descriptions and records (Doveri 2013). These publications are, undoubtedly, important guides to taxonomy and biogeography of *Chaetomium* species that occur on dung. However, a most deep search in dung substrates, allied to phylogenetic approaches, can obviously improve our knowledge about this genus, including possibility of new species on this substrate. Abdel-Azeem and Salem (2015) studied the coprophilous fungi in arid Sinai of Egypt, and they recorded seven species of *Chaetomium* on camel, donkey, and goat dung. Doveri (2018) provided an updated key to sexual morph genera of *Chaetomiaceae*, where a new species of the related genus *Chaetomidium* (*C. vicugnae* Doveri) from sample of vicuña dung (*Vicugna vicugna* Molina 1782) is described. For more detail, please refer to Chap. 4.

1.4.8 Living Plants, Lichens, and Animals

Endophytes colonize symptomlessly the living, internal tissues of their host, even though the endophyte may, after an incubation or latency period, cause disease (Petrini 1991). In the literature, the term "fungal endophytes" is normally used to describe fungal organisms, which, in contrast to mycorrhizal fungi, reside entirely within the host tissues and emerge during host senescence (Rodriguez and Redman 2008).

Endophytic fungi have been classified into two groups based on differences in taxonomy, evolution, plant hosts, and ecological functions into clavicipitaceous, which are able to infect only some species of grasses, and non-clavicipitaceous, which are found in the asymptomatic tissues of bryophytes, ferns, gymnosperms, and angiosperms (Rodriguez et al. 2009).

There are 1.3 million species of endophytic fungi alone, the majority of which are likely found in tropical ecosystems (Verma et al. 2014). There has been great interest in endophytic fungi as potential producers of novel biologically active products (Schulz et al. 2002; Wildman 2003; Strobel and Daisy 2003; Tomita 2003;

Urairuj et al. 2003; Spiering et al. 2006; Manoharachary et al. 2013; Abdel-Azeem et al. 2016a, 2018).

Unique species of endophytic fungi with a wide range of potential practical applications in plant protection as repellents, insecticides, antimicrobials, anthelmintics, and vermicides have been found (Strobel et al. 2008; Vega et al. 2008). In the last 5 years, endopytic fungi as a source of new drugs have been attracted many researchers world-wide as producers of anticancer, antimicrobial, antirheumatoid, liver protection, and antioxidant compounds and also in a biotransformation process (Pimentel et al. 2011; Salem and Abdel-Azeem 2014; Abdel-Azeem et al. 2016a, 2018).

Species of *Chaetomium* as a member of non-clavicipitaceous endophytes attracted the attention of researchers as effective producers of bioactive metabolites. Such studies may result in the description of new *Chaetomium* species, e.g., Sharma et al. (2013) described *C. jatrophae* as a new species recovered from *Jatropha podagrica* in India. Also Blanchette et al. (2017) and Abdel-Azeem et al. (2018) recovered new Egyptian and African records of *Chaetomium*, namely, *C. iranianum* and *C. grande*, respectively.

In 2008, Abdel-Lateff produced chaetominedione as a new tyrosine kinase inhibitor isolated from the algicolous marine fungus *Chaetomium* sp. This fungus was recovered from seaweed *Valonia utricularis*, collected from the waters around the Azores (Atlantic Ocean).

Abdel-Azeem and Salem (2012) studied the laccase-producing fungi in Egypt and by screening sources under investigation, namely, soil, wood, seaweeds, sponge, ascidia, drifted decaying wood, plants and miscellaneous materials it was possible to encounter as many as 60 species belonging to 33 genera. *Zygomycota* represented by 6 species (10.16% of the total species number), teleomorphic *Ascomycota* by 9 species (15.25%), anamorphic *Ascomycota* by 44 species (74.57%), and *Basidiomycota* by 1 species (1.69%). Soil showed the highest Simpson's species diversity index of 0.83, while contaminated wax samples and *Adiantum capillus-veneris* showed the lowest value (0). All isolated taxa were tested for laccase production using a qualitative plate assay method by using guaiacol as color indicator. Sixteen isolates showed positive reaction indicating a lignin-degrading potentiality, and out of them, eight measured the highest zone diameter with high oxidation scale. The most promising taxa were endophytic, namely, *Chaetomium globosum*, *Phoma exigua*, *Thanatephorus cucumeris*, and *Sordaria fimicola*. pH 7, incubation temperature 30 °C, 1% maltose, and 0.3% peptone supported the highest biomass and laccase production for *Chaetomium*.

Mustafa et al. (2013) exploited some Egyptian endophytic taxa for extracellular biosynthesis of silver nanoparticles. They isolated endophytic fungi from medicinal plants in arid Sinai. Their results showed that *Zygomycota* represented by 2 species (9.5% of the total species number), teleomorphic *Ascomycota* by 3 species (14.2%), and anamorphic *Ascomycota* by 16 species (76.19%). The prevailing genera were *Aspergillus* (3 species including anamorph stages of one *Eurotium* species; 14.28% of the total isolates) and *Alternaria* (2 species, 9.5%). The remaining taxa were represented only by one species each. The most abundant species were *Alternaria alternata* (41.6%), *Nigrospora oryzae* (38.3%), and *Chaetomium globosum* (11.1%). A total 13 species belonging to 11 genera were screened for the production of AgNPs. They recorded that *Aspergillus niger* synthesized AgNPs in a moderate rate in comparison with other taxa.

In their extensive study on anticancer-producing endophytic taxa in Egypt, Salem and Abdel-Azeem (2014) were able to isolate *Chaetomium atrobrunneum*, *C. bostrychodes*, *C. brasiliense*, *C. arinthiacum*, *C. globosum*, *C. gracile*, *C. hamadae* (Udagawa) Arx, *C. iranianum*, *C. mareoticum*, *C. murorum*, *C. nigricolor*, *C. perlucidum*, *C. piluliferum*, *C. senegalense*, *C. strumarium*, and *C. subspirilliferum* from eight medicinal plants in Saint Katherine Protectorate, South Sinai, Egypt.

Jin et al. (2014) published their research articles titled "*Pezizomycotina* dominates the fungal communities of South China Sea Sponges *Theonella swinhoei* and *Xestospongia testudinaria*" and they recorded *C. globosum* in their study.

Endolichenic fungi in lichens of Champawat district, Uttarakhand, northern India, have been the target of intensive study carried out by Suryanarayanan and Thirunavukkarasu in 2017. Eleven lichen species belonging to five families (two fruticose and nine foliose) growing on the bark of *Quercus leucotrichophora* trees from four forests of Champawat district, Uttarakhand state, northern India, were studied for their endolichenic fungal assemblage. In this study, they isolated species of *Acremonium*, *Chaetomium*, and *Xylaria* as endolichenic fungi, genera which are also common endophytes of plants.

1.4.9 Human

Infections caused by fungi are a significant contributor to morbidity and mortality in humans. These infections range from comparatively innocuous superficial skin diseases caused by dermatophytes to invasive life-threatening infections. *Chaetomium* species belong to a large genus of saprobic ascomycetes found on dung, straw, paper, bird feathers, seeds, and plant debris and in soil. Although *Chaetomium* species are rarely implicated in human disease, their spectrum of mycoses includes keratitis, onychomycosis, and sinusitis in immunocompetent individuals and empyema, pneumonia, and fatal disseminated cerebral disease in immunocompromised hosts and intravenous drug users. Since many of the diseases caused by *Chaetomium* are opportunistic in nature, the pathogenesis of fungal infections is complex, and our understanding of how these taxa cause disease has lagged behind bacterial and viral pathogens. Despite being saprophytic ascomycetes with only occasional involvement in human disease processes, *Chaetomium* species are capable of inducing a broad spectrum of mycoses including onychomycosis, sinusitis, empyema, pneumonia, and fatal disseminated cerebral disease, especially in immunocompromised patients and intravenous drug users (Hoppin et al. 1983; Anandi et al. 1989; Yeghen et al. 1996; Aru et al. 1997; Thomas et al. 1999).

Chaetomium atrobrunneum is a notably invasive, neurotropic species, and its ability to grow at elevated temperatures may contribute to its neurotropism (Stiller et al. 1992; Guarro et al. 1995; Friedman 1998; Guppy et al.1998; Rock 1998; Lesire et al. 1999). Thomas et al. (1999) described a case of fatal brain abscess due to *C. atrobrunneum* in a bone marrow transplant patient. The rapid progression of cerebral infection indicates that the brain tissue provides a favorable environment for growth and proliferation of the fungus.

Chaetomium strumarium is another invasive, neurotropic species. Abbott et al. (1995) reported three *C. strumarium*-related cases of fatal cerebral mycosis in males with prior histories of intravenous drug use from the USA and Australia. *C. strumarium* was detected by histopathology and isolated from the brain tissue. First case of phaeohyphomycosis caused by *C. murorum* in China by Lin and Li (1995).

Chaetomium perlucidum is recently confirmed as a neurotropic species. Barron et al. (2003) documented the first two cases of invasive human mycoses caused by this phaeoid ascomycete. The first case concerned a 45-year-old female patient with acute myelogenous leukemia, who had an unrelated, 4/5 HLA-matched umbilical cord blood transplant. The patient became disoriented and febrile, and computed tomography of the chest revealed a 3 × 2 cm mass in the right lower lobe. After suffering a massive right-sided intraparenchymal hemorrhage, the patient died. Autopsy revealed disseminated invasive fungal infection in the lungs, brain, and myocardium; and cultures from the surgically obtained lung tissue yielded *C. perlucidum*. The second case involved a 78-year-old female with a history of asthma and chronic bronchiectasis. The patient underwent a lobectomy due to worsening symptoms, and cultures from the lung tissue grew *C. perlucidum*. The patient showed no further manifestations of disease after the lobectomy. For more detail, please refer to Chap. 5.

1.4.10 Decaying Wood

Wood decay by fungi is typically classified into three types: soft rot, brown rot, and white rot. Brown rot fungi are the most prevalent with regard to attack on coniferous, structural wood products. The wood decayed by brown rot fungi is typically brown and crumbly, and it is degraded via both non-enzymatic and enzymatic systems. A series of cellulolytic enzymes are employed in the degradation process by brown rot fungi, but no lignin-degrading enzymes are typically involved. White rot fungi are typically associated with hardwood decay, and their wood decay patterns can take on different forms. White rot fungi possess both cellulolytic and lignin-degrading enzymes, and these fungi therefore have the potential to degrade the entirety of the wood structure under the correct environmental conditions. Soft rot fungi typically attack higher moisture and lower lignin content wood and can create unique cavities in the wood cell wall. Less is known about the soft rot degradative enzyme systems, but their degradative mechanisms are reviewed along with the degradative enzymatic and non-enzymatic systems known to exist in the brown rot and white rot fungi.

In 1973, Kowalik and Sadurska (1973) studied the microbiota of papyrus from samples of Cairo museums. Different fungi imperfecti, *Ascomycetes*, and *Actinomycetes* were isolated from samples of papyrus of Cairo museums. They recovered many fungi such as *Alternaria geophila*, *Botryodiplodia theobromae*, *Emericellopsis minima*, *Fusarium lactis*, *Helminthosporium sativum*, *Spondylocladium australe*, some species of the genus *Chaetomium*, and some actinobacteria related to genus *Streptomyces* which seemed to be specific for papyrus and/or for Egyptian climatic conditions. The genus *Chaetomium* and *Emericellopsis* may play a great role in decomposition of basic polymers of papyrus. Considering the nitrogen source of microbiota, it can be observed that papyrus-destroying microorganisms preferred ammonium to nitrate ions. It was found that papyrus-decomposing microorganisms may grow equally well at 24–26 °C as at 30 °C and some fungi are growing even at 42 °C. The members of the genus *Penicillium*, which prefer rather low temperatures and are frequent inhabitants of paper, were isolated from papyrus only once. By using the method of paper sheets damped with a 10 percent ethyl alcohol solution of thymol, pentachlorophenol, dichlorophene, and p-chloro-m-cresol as microbiocides, it could be concluded that only the last fungicide may assure the protection of papyrus. They recovered 12 species of *Chaetomium*: *C. angustum*, *C. atrobrunneum*, *C. bostrychodes*, *C. cochliodes*, *C. elatum*, *C. fusiforme*, *C. globosum*, *C. indicum*, *C. ochraceum*, *C. olivaceum*, *C. trilaterale*, and *C. turgidopilosum*.

Among the fungi commonly found in environmental studies performed in Archives and Museums, some of them display cellulolytic properties such as species from the genera *Trichoderma*, *Penicillium*, *Botrytis*, *Trichothecium*, *Phoma*, *Chaetomium*, *Aspergillus*, *Cladosporium*, *Stemphylium*, *Alternaria*, *Hormodendrum*, and *Aureobasidium*. Among the proteolytic genera, one can find *Aureobasidium*, *Chaetomium*, *Cladosporium*, *Botrytis*, *Trichoderma*, *Verticillium*, *Mucor*, *Epicoccum*, and *Gymnoascus* (Pinheiro 2014). *Aureobasidium*, *Botrytis*, *Chaetomium*, *Cladosporium*, *Epicoccum*, and *Mucor* genera were identified in the present study, in both air and surface samples (Paiva de Carvalho et al. 2018). Regarding the wooden sculptures, *Chaetomium globosum* Kunze and *P. copticola* were the most frequent fungal species in the study carried out by Paiva de Carvalho et al. (2018) on fungal contamination of paintings and wooden sculptures inside the storage room of a museum. Moreover, *Aspergillus sclerotiorum* G.A. Huber and *Penicillium oxalicum* Currie & Thom were exclusively found on wooden sculptures samples and not in any of the air samples.

In 2019, Abdel-Azeem et al. studied assessment of biodegradation in ancient archaeological wood from the Middle Cemetery at Abydos, Egypt. Abydos is a large, complex archaeological site located approximately 500 km south of Cairo in Upper Egypt. The site has served as a cemetery for thousands of years and is where most of the Early Dynastic royal tombs are located. North Abydos includes the Middle Cemetery and the North Cemetery, which are separated from each other by a wadi. The Middle Cemetery was the burial ground for important Sixth Dynasty (2407–2260 BC) officials and over time for thousands of elite and non-elite individuals as well. Excavations at the core area of the Old Kingdom mortuary landscape have revealed many culturally important wooden objects, but these are often found with extensive deterioration that can compromise their preservation. The objectives of this study were to characterize the biodegradation that has taken place in excavated wooden objects, elucidate the type of wood degradation present, obtain information on soil properties at the site, and identify fungi currently associated with the wood and soils. Light and scanning electron microscopy studies were used to observe the micromorphological characteristics of the wood, and culturing on different media was done to isolate fungi. Identification of the fungi was done by examining morphological characteristics and extracting rDNA from pure cultures and sequencing the ITS region. Wooden objects, made from *Cedrus*, *Juniperus*, and *Acacia* as well as several unidentified hardwoods, were found with extensive degradation and were exceedingly fragile. Termite damage was evident, and frass from the subterranean termites along with sand particles was present in most woods. Evidence of soft rot attack was found in sections of wood that remained. Fungi isolated from wood and soils were identified as species of *Aspergillus*, *Chaetomium*, *Cladosporium*, *Fusarium*, *Penicillium*, *Stemphylium*, *Talaromyces*, and *Trichoderma*. Results provide important information on the current condition of the wood and give insights to the identity of the fungi in wood and soils at the site. These results provide needed information to help develop conservation plans to preserve these degraded and fragile wooden objects. For more detail, please refer to Chap. 15.

1.5 Conclusion

The studies discussed above reflect that the genus *Chaetomium* can be characterized with high adaptability to various ecological environments. However, it is important to mention that the results of any study aimed at the examination of *Chaetomium* biodiversity should always be evaluated in the context of the developmental stage of *Chaetomium* taxonomy and the species identification methods available at the time of the publication of the respective paper. Due to the constant development of the taxonomy of the genus and the description of new species, more recent examinations of a specific habitat may reveal higher biodiversity of the genus and refine the results of previous studies. By introducing new techniques and methods in biodiversity studies during the past two decades, the amount of information available about the distribution of *Chaetomium* taxa is constantly growing; therefore, it can be expected that the biogeography of the genus will be understood more deeply in the near future.

References

Abbott SP, Sigler L, McAleer R, McGough DA, Rinaldi MG, Mizell G. (1995) Fatal cerebral mycoses caused by the ascomycete *Chaetomium strumarium*. J Clin Microbiol 33:2692–2698.

Abdel-Azeem AM (1991) Effect of overgrazing on vegetation, microbes and soil in Ismailia-desert habitat. Biological Diversity Symposium, Madrid-Spain, pp 241–246

Abdel-Azeem AM (2003) Ecological and taxonomical studies on ascospore-producing fungi in Egypt. Ph.D Thesis, Faculty of Science, Suez Canal University
Abdel-Azeem AM (2009) Operation Wallacea in Egypt. I- A preliminary study on diversity of fungi in the world heritage site of Saint Katherine, Egypt. Assiut Univ J Bot 38(1):29–54
Abdel-Azeem AM, Ibrahim ME (2004) Diversity of terrophilous mycobiota of Sinai. Egypt J Biol 6:21–31
Abdel-Azeem AM, Salem FM (2012) Biodiversity of laccase producing fungi in Egypt. Mycosphere 3(5):900–920
Abdel-Azeem AM, Salem FM (2015) Fungi fimicola Aegyptiaci: I. Recent investigations and conservation in arid South Sinai. Mycosphere 6(2):174–194. https://doi.org/10.5943/mycosphere/6/2/8
Abdel-Azeem AM, Zaki SM, Khalil WF, Makhlouf NA, Farghaly LM (2016a) Anti-rheumatoid activity of secondary metabolites produced by endophytic *Chaetomium globosum*. Front Microbiol 7(1477):1–11
Abdel-Azeem AM, Salem FM, Abdel-Azeem MA, Nafady NA, Mohesien MT, Soliman EA (2016b) Biodiversity of the genus *Aspergillus* in different habitats. In: Gupta VK (ed) New and future developments in microbial biotechnology and bioengineering: *aspergillus* system properties and applications. Elsevier, Amsterdam, pp 3–40
Abdel-Azeem AM, Omran MA, Mohamed RA (2018a) Evaluation of the curative probability of bioactive metabolites from endophytic fungi isolated from some medicinal plants against paracetamol-induced liver injury in mice. LAP LAMBERT Academic Publishing. ISBN: 978–613–9-89820-6
Abdel-Azeem AM, Blanchette RA, Held BW (2018b) New record of *Chaetomium grande* Asgari & Zare (Chaetomiaceae) for the Egyptian and African mycobiota. Phytotaxa 343(3):283–288
Abdel-Azeem AM, Held BW, Richards JE, Davis SL, Blanchette RA (2019) Assessment of biodegradation in ancient archaeological wood from the middle cemetery at Abydos, Egypt. PLoS One 4(3):e0213753. https://doi.org/10.1371/journal.pone.0213753
Abdel-Hafez S (1981) Halophilic fungi of desert soils in Saudi Arabia. Mycopathologia 75:75e80
Abdel-Hafez SI (1982a) Survey of microflora of desert soils in Saudi Arabia. Mycopathologia 80:3–8
Abdel-Hafez SI (1982b) Osmophilic fungi of desert soils in Saudi Arabia. Mycopathologia 80:9–14
Abdel-Hafez SII, El-Maghraby OMO (1993) Thermophilic and thermotolerant fungi of Wadi-Bir-El-Ain soils. Eastern desert, Egypt. Abhath Al-Yarmouk. Pure Sci Eng 2:55–66
Abdel-Hafez AII, Mazen MB, Galal AA (1989a) Keratinophilic and cycloheximide resistant fungi in soils of Sinai Governorate, Egypt. Cryptogam Mycol 10(3):265–275
Abdel-Hafez AII, Mazen MB, Galal AA (1989b) Some ecological studies of osmophilic and halophilic soil fungi of Sinai Peninsula, Egypt. J Sohag Pure Appl Sci Bull Faculty Sci 5:67–83
Abdel-Hafez AII, Mazen MB, Galal AA (1990) Glycophilic and cellulose-decomposing fungi from soils of Sinai Peninsula, Egypt. The Arab Gulf J Sci Res 8(1):153–168
Abdel-Lateff A (2008) Chaetominedione, a new tyrosine kinase inhibitor isolated from the algicolous marine fungus *Chaetomium* sp. Tetrahedron Lett 49:6398–6400
Abdel-Rahim IR, Nafady NA, Bagy MMK, Abd-Alla MH, Abd-Alkader AM (2018) Fungi-induced paint deterioration and air contamination in the Assiut University hospital. Egypt Indoor Built Environ 28(3):384–400
Abdel-Sater MA (1990) Studies on the mycoflora of the New Valley area, Western Desert, Egypt. Ph.D. Thesis, Faculty of Science, Assiut University
Abdel-Sater MA (2000) Soil fungi of the New Valley area, Western desert, Egypt. Bulletin Faculty of Science, Assiut University 29 (2-D), pp 255–271
Abdullah, S. K. and Al-Bader, S. M. 1989. A new thermotolerant species of *Chaetomium* from Iraqi forest soil. Int. J. Mycol. Lichenol. 4:83–91
Abdullah, S. K. & Zora, S. E. (1993). *Chaetomium mesopotamicum*, a new thermophilic species from Iraqi soil. Cryptogamic Botany 3:387–389
Adams BJ, Bardgett RD, Ayres E, Wall DH, Aislabie J, Bamforth S, Bargagli R, Cary C, Cavacini P, Connell L, Convey P, Fell JW, Frat F, Hogg ID, Newsham KK, O'Donnell A, Russell N,

Seppelt RD, Stevens MI, 2006. Diversity and distribution of Victoria Land biota. Soil Biology and Biochemistry 38:3003e3018

Adametz L (1886) Untersuchungen über die niederen Pilze der Ackerkrume. Inaug Diss, 1–78. Leipzig

Aggarwal R, Tewari AK, Srivastava KD, Singh DV (2004) Role of antibiosis in the biological control of spot blotch (*Cochliobolus sativus*) of wheat by *Chaetomium globosum*. Mycopathologia 157(4):369–377

Aggarwal R, Kharbikar LL, Sharma S, Gupta S, Yadav A (2013) Phylogenetic relationships of *Chaetomium* isolates based on the internal transcribed spacer region of the rRNA gene cluster. Afr J Biotechnol 12(9):914–920

Ahammed SK, Aggarwal R, Singh DV (2005) Morphological variability in different isolates of *Chaetomium globosum*. Indian Phytopath 58(1):71–74

Ali MI (1977) Studies on the fungal flora of Saudi Arabia. 1-Wadi Hanif. Bull Fac Sci Riyadh Univ 8:7–20

Almaguer M, Aira MJ, Rodríguez-Rajo FJ, Rojas TI (2013) Study of airborne fungus spores by viable and non-viable methods in Havana. Cuba Grana 52(4):289–298

Al-Sadi AM, Al-Khatri B, Nasehi A, Al-Shihi M, Al-Mahmooli IH, Maharachchikumbura SSN (2017) High fungal diversity and dominance by ascomycota in dam reservoir soils of arid climates. Int J Agric Biol 19(4):682–688

Ames LM (1961) A monograph of the Chaetomiaceae. The United States Army research and development series, pp 125

Ames LM (1963) A monograph of the Chaetomiaceae. United States Army Research and Development Series 2, pp 1–65

Ames LM (1969) A monograph of the Chaetomiaceae. United States Army Research and Development Service. Cramer Publication, New York

Amin EE, Abdalla MH (1980) Survey of soil fungi from the Sudan Gezira. Mycopathologia 71:131–136

Anandi V et al (1989) Cerebral phaeohyphomycosis caused by *Chaetomium globosum* in a renal transplant recipient. J Clin Microbiol 27:2226–2229

Arenz BE, Blanchette RA, Farrell RL (2014) Fungal diversity in Antarctic soils. Pages 35–53. In: Cowan D (ed) In Antarctic terrestrial microbiology: physical and biological properties of Antarctic soils. Springer, Berlin

Aru A, Munk-Nielsen L, Federspiel BH (1997) The soil fungus *Chaetomium* in the human paranasal sinuses. Eur Arch Otorhinolaryngol 254:350–352

Asgari B, Zare R (2011) The genus *Chaetomium* in Iran, a phylogenetic study including six new species. Mycologia 103(4):863–882

Aue R, Muller E (1967) Vergleichende Untersuchungen an einigen *Chaetomium* arten. In Ber Schweiz bot Ges 77:187–207

Azmi OR, Seppelt RD (1998) The broadscale distribution of microfungi in the Windmill Islands region, continental Antarctica. Polar Biol 19:92–100

Bahkali AH, Khiyami MA (1996) Cellulose-decomposing fungi from Saudi Arabian soils. QatarUniv Sci J 16(1):77–80

Balazy S, Wrzosek M, Sosnowska D, Tkaczuk C, Muszewska A (2008) Laboratory trials to infect insects and nematodes by some acaropathogenic *Hirsutella* strains (Mycota: Clavicipitaceous anamorphs). J Invertebr Pathol 97:103–113

Barron MA et al (2003) Invasive mycotic infections caused by *Chaetomium perlucidum*, a new agent of cerebral phaeohyphomycosis. J Clin Microbiol 41:5302–5307

Begerow D, Nilsson H, Unterseher M et al (2010) Current state and perspectives of fungal DNA barcoding and rapid identification procedures. Appl Microbiol Biotechnol 87:99–108

Berkson BM (1966) Cytomorphological studies of the ascogenous hyphae in four species of *Chaetomium*. Mycologia 58(1):125–130

Besada WH, Yusef HM (1968) On the mycoflora of UAR soil. Proc Egypt Acad Sci 21:103–109

Besada WH and Yusef HM (1969): *Chaetomium mareoticum* sp. nov. Transactions of the British Mycological Society 52:502–504

Biswas SK, Aggarwal R, Srivastava KD, Gupta S, Dureja P (2012) Characterization of antifungal metabolites of *Chaetomium globosum* Kunze and their antagonism against fungal plant pathogens. J Biol Cont 26(1):70–74

Blanchette RA, Held BW, Abdel-Azeem AM (2017) New record of *Chaetomium iranianum* MF787598 (Chaetomiaceae) for the Egyptian and African mycobiota. Microb Biosyst J 2(2):6–9

Brewer D, Jerram WA, Taylor A (1968) The production of cochliodinol and related metabolites by *Chaetomium* species. Can. J Microbial. 14:861–866

Brewer D, Jerram WA, Meiler, Taylor A (1970) The toxicity of cochliodinol, and antibiotic metabolite of *Chaetomium* spp. Can J Microbiol 16:433–440

Brewer D, Taylor A (1978) The production of toxic metabolites by *Chaetomium* spp. isolated from soils of permanent pasture. Can J Microbiol 24:1078–1081

Brewer D, Duncan JM, Jerram WA, Leach CK, Safe S,Taylor A, Vining LC, Archibald RM, Stevenson RG, Mirocha CJ, Christensen CM (1972) Ovine ill-thrift in Nova Scotia. 5. The production and toxicology of chetomin, a metabolite of *Chaetomium* spp. Can J Microbiol 18:1129–1137

Brewer D, McInnes AG, Smith DG, Taylor A, Walter JA, Loosly HR, Kis ZL (1978) Sporidesmins. Part 16. The structure of chetomin, a toxic metabolites of *Chaetomium cochliodes*, by nitrogen-15 and carbon-13 nuclear magnetic resonance spectroscopy. J. Chem. Soc. Perkin I, 1248–1254

Bridge PD, Spooner BM (2012) Non-lichenized Antarctic fungi: transient visitors or members of a cryptic ecosystem. Fungal Ecol 5:381–394

Brock TD (1979) Ecology of saline lakes. In: Shilo M (ed) Strategies of microbial life in extreme environments. Dahlem Konferenzen, Berlin, pp 29–47

Buston HW, Moss MO, Tyrrell D (1966) The influence of carbon dioxide on growth and sporulation of *Chaetomium globosum*. Trans Brit Mycol Soc 49:387–396

Butinar L, Santos S, Spencer-Martins I, Oren A, Gunde-Cimerman N (2005a) Yeast diversity in hypersaline habitats. FEMS Microbiol Lett 244(2):229–234

Butinar L, Sonjak S, Zalar P, Plemenitasˇ A, Gunde-Cimerman N (2005b) Melanized halophilic fungi are eukaryotic members of microbial communities in hypersaline waters of solar salterns. Bot Mar 48(1):73–79

Butinar L, Zalar P, Frisvad JC, Gunde-Cimerman N (2005c) The genus *Eurotium* – members of indigenous fungal community in hypersaline waters of salterns. FEMS Microbiol Ecol 51(2):155–166

Cai L, Hyde KD, Taylor PWJ, Weir BS, Waller J, Abang MM, Zhang JZ, Yang YL, Phoulivong S, Liu ZY, Prihastuti H, Shivas RG, McKenzie EHC, Johnston PR (2009) A polyphasic approach for studying *Colletotrichum*. Fungal Divers 39:183–204

Callaghan TV, Björn LO, Chernov Y, Chapin T, Christensen TR, Huntley B, Ims RA et al (2004) Biodiversity, distributions and adaptations of Arctic species in the context of environmental change. Ambio 33:404–417

Cannon PF. 1986 – A revision of *Achaetomium*, *Achaetomiella* and *Subramaniula*, and some similar species of *Chaetomium*. Transactions of the British Mycological Society 87, 45–76. http://dx.doi.org/10.1016/S0007-1536(86)80004-3

Cantrell SA, Casillas-Martinez L, Molina M (2006) Characterization of fungi from hypersaline environments of solar salterns using morphological and molecular techniques. Mycol Res 110:962–970

Cano J, Guarro J. 1987 – Soil ascomycetes from Spain XII. *Chaetomium biporatum* sp. nov. Nova Hedwigia 44, 543–546.

Carter A (1982) A taxonomic study of the ascomycete genus *Chaetomium* Kunze [doctoral dissertation].Department of Botany, University of Toronto, Toronto

Carris, L. M. & Glawe, D. A. (1987). *Chaetomium histoplasmoides*, a new species isolated from cysts of Heterodera glycines in Illinois. Mycotaxon 29. 383-391.

Castagnoli E, Andersson MA, Mikkola R, Kredics L, Marik T, Kurnitski J, Salonen H (2017) Indoor chaetomium-like isolates: resistance to chemicals, fluorescence and mycotoxin production. In: Säteri J, Ahola M (eds) Sisäilmastoseminaari 2017, Helsinki, 15.03.2017, vol 35. (SIY raportti 35). Sisäilmayhdistys ry, pp 227–232

Chahal DS, Hawksworth DL (1976) *Chaetomium cellulolyticum*, a new thermotolerant and cellulolytic *Chaetomium*, I. isolation, description and growth rate. Mycologia 68:600–610

Chapman ES, Fergus CL (1975) Germination of ascospores of *Chaetomium globosum*. Mycologia 67(5):1048–1052

Chen KT (1973) Some new species of *Chaetomium* and *Septaria*. Acta Microbiol Sin 13:124–128

Chen Y, Eisner JD, Kattar MM, Rassoulian-barrett SL, Lafe K, Bui U, Limaye AP, Cookson BT (2001) Polymorphic internal transcribed spacer region 1 DNA sequences identify medically important yeasts. J Clin Microbiol 39(11):4042–4051

Chivers AH (1912) Preliminary diagnoses of new species of *Chaetomium*. Proc Am Acd Arts Sci 48:83–88

Chivers AH (1915) A monograph of the genera *Chaetomium* and *Ascotricha*. Memoirs Torrey Bot Club 14:155–240

Chourasia HK (1995) Mycobiota and mycotoxins in herbal drugs of Indian pharmaceutical industries. Mycol Res 99:697–703

Christensen M (1981) A synoptic key and evaluation of species in the *Aspergillus flavus* group. Mycologia 73:1056–1084

Cooke JC (1969a) Morphology of *Chaetomium erraticum*. Mycologia 56(3):335–340

Cooke JC (1969b) Morphology of *Chaetomium funicolum*. Mycologia 61(6):1060–1065

Cooke JC (1970) Morphology of *Chaetomium trilaterale*. Mycologia 62(2):282–288

Corda ACJ (1837) Icones Fungorum 1. Prague

Cullen D, Andrews JH (1984) Evidences for the role of antibiosis in the antagonism of *Chaetomium globosum* to the apple scab pathogen Venturia inaequalis. Canadian J Bot 62:1819–1823

Damm U, Woudenberg JHC, Cannon PF, Crous PW (2009) *Colletotrichum* species with curved conidia from herbaceous hosts. Fungal Divers 39:45–87

de Hoog GS, Ahmed SA, Najafzadeh MJ, Sutton DA, Keisari MS, Fahal AH, Eberhardt U, Verkleij GJ, Xin L, Stielow B, van de Sande WWJ (2013) Phylogenetic findings suggest possible new habitat and routes of infection of human eumyctoma. PLoS Negl Trop Dis 7:e2229. https://doi.org/10.1371/journal.pntd.0002229

Decock C, Hennebert GL (1997) A new species of *Chaetomium* from Ecuador. Mycol Res 101:309–310

Dhingra OD, Mizubuti ESG, Santana FM (2003) *Chaetomium globosum* for reducing primary inoculum of Diaporthe phaseolorum f. sp. meridionalis in soil surface soybean stubble in field conditions. Biol Control 26:302–310

Di Pietro A, Gut-Rella M, Pachlatko JP, Schwin FJ (1992) Role of antibiotics produced by *Chaetomium globosum* in biocontrol of *Pythium ultimum*, a causal agent of damping off. Phytopathology 82:131–135

Dmitriev VV, Gilichinsky DA, Faizutdinova RN, Shershunov IN, Golubev WI, Duda VI (1997a) Occurrence of viable yeasts in 3-million-year-old permafrost in Siberia. Mikrobiologiya 66:655–660

Dmitriev VV, Gilichinsky DA, Faizutdinova RN, Ostroumova NV, Golubev WI, Duda VI (1997b) Yeasts in late Pleistocene-early Pleistocene Siberian permafrost. Cryosphere of the Earth 1:67–70

Doveri F (2008) – Aggiornamento sul genere *Chaetomium* con descrizione di alcune specie coprofile, nuove per l'Italia – An update on the genus *Chaetomium* with descriptions of some coprophilous species, new to Italy. Pagine di Micologia 29:1–60

Doveri F (2004) Fungi Fimicoli Italici. Trento, A.M.B: Fondazione Studi Micologici. 1104 p

Doveri F (2011) Additions to "Fungi Fimicoli Italici": an update on the occurrence of coprophilous Basidiomycetes and Ascomycetes in Italy with new records and descriptions. Mycosphere 2(4):331–427

Doveri F (2013) An additional update on the genus *Chaetomium* with descriptions of two coprophilous species, new to Italy. Mycosphere 4:820–846
Doveri F (2016) Description of *Chaetomium aureum*, *Corynascus sepedonium* and *Coniochaeta hansenii* newly recorded from Italy and a key to coprophilous Chaetomiaceae and Coniochaetaceae. Ascomycete.org 8(1):7–24
Doveri F (2018) On a new species of *Chaetomidium*, *C. vicugnae*, with a cephalothecoid peridium and its relationships with Chaetomiaceae (Sordariales). Ascomycete.org 10(2):86–96. https://doi.org/10.25664/ART-0231
Dreyfuss M (1975) Taxonomische Untersuchungen innerhalb der Gattung *Chaetomium*. Sydowia 28:50–133
Druzhinina IS, Kopchinskiy AG, Komoń M, Bissett J, Szakacs G, Kubicek CP (2005) An oligonucleotide barcode for species identification in *Trichoderma* and *Hypocrea*. Fungal Genet Biol 42:813–828
El-Buni AM, Rattan SS (1981) Check list of Libyan Fungi. Department of Botany, Al Faateh University, Tripoli, Libya, 169 pages
Ellis DH (1981) Ascocarp morphology and terminal hair ornamentation in thermophilic *Chaetomium* species. Mycologia 73(4):755–773
El-Said AHM, Saleem A (2008) Ecological and physiological studies on soil fungi at western region, Libya. Mycobiology 36:109
Feau N, Vialle A, Allaire M, Tanguay P, Joly DL, Frey P, Callan BE, Hamelin RC (2009) Fungal pathogen identifications: a case study with DNA barcodes on *Melampsora* rusts of Aspen and White poplar. Mycol Res J 113:713–724
Fergus CL, Delwiche CJ (1975) The effects of nutrients and temperature on germination and viability of the ascospores of *Chaetomium rectopilium*. Mycologia 67(4):722–732
Figueras MJ, Guarro J (1988) A scanning electron microscopic study of ascoma development in *Chaetomium malaysiense*. Mycologia 80(3):298–306
Freire FCO, Kozakiewicz Z, Paterson RRM (1999) Mycoflora and mycotoxins of Brazilian cashew kernels. Mycopathologia 145:95–103
Friedman AH (1998) Cerebral fungal infections in the immunocompromised host: a literature review and a new pathogen—*Chaetomium atrobrunneum*: case report— comment. Neurosurgery 43:1469
Frisvad JC, Smedsgaard J, Larsen TO, Samson RA (2004) Mycotoxins, drugs and other extrolites produced by species in *Penicillium* subgenus *Penicillium*. Stud Mycol 2004:201–241
Frøslev TG, Jeppesen TS, Læssøe T, Kjoller R (2007) Molecular phylogenetics and delimitation of species in *Cortinarius* section Calochroi (Basidiomycota, Agaricales) in Europe. Mol Phylogenet Evol 44:217–227
Gashgari RM, Elhariry HM, Gherbawy YA (2013) Molecular detection of mycobiota in drinking water at four different sampling points of water distribution system of Jeddah City (Saudi Arabia). Geomicrobiol J 30:29–35
Geiser DM, Jiménez-Gasco M, Kang S, Makalowska I, Veeraraghavan N, Ward TJ, Zhang N, Kuldau GA, O'Donnell K (2004) *Fusarium*-ID v.1.0: a DNA sequence database for identifying *Fusarium*. Eur J Plant Pathol 110:473–479
Geiser DM, Klich MA, Frisvad JC, Peterson SW, Varga J, Samson RA (2007) The current status of species recognition and identification in *Aspergillus*. Stud Mycol 59:1–10
Gené J, Guarro J (1996) A new *Chaetomium* from Thailand. Mycol Res 100:1005–1009
Gonçalves AB, Paterson RRM, Lima N (2006) Survey and significance of filamentous fungi from tap water. Int J Hyg Environ Health 209:257–264
Göttlich E et al (2002) Fungal flora in groundwater-derived public drinking water. Int J Hyg Environ Health 205:269–279
Greif MD, Stchigel AM, Miller AN, Huhndorf SM (2009) A re-evaluation of genus *Chaetomidium* based on molecular and morphological characters. Mycologia 101:554–564
Grishkan I (2018) Thermotolerant mycobiota of Israeli soils. J Basic Microbiol 58:30–40. https://doi.org/10.1002/jobm.201700517
Grishkan I, Nevo E (2010) Spatiotemporal distribution of soil microfungi in the Makhtesh Ramon area, central Negev desert, Israel. Fungal Ecol 3:326e337

Grishkan I, Nevo E, Wasser SP (2003) Soil micromycete diversity in the hypersaline Dead Sea coastal area, Israel. Mycol Progress 2:19–28
Guarro J, Soler L, Rinaldi MG (1995) Pathogenicity and antifungal susceptibility of *Chaetomium* species. Eur J Clin Microbiol Infect Dis 14:613–618
Gunde-Cimerman N, Zalar P, de Hoog S, Plemenitasˇ A (2000) Hypersaline waters in salterns –natural ecological niches for halophilic black yeasts. FEMS Microbiol Ecol 32(3):235–240
Guppy KH, Chinnamma T, Kurian T, Douglas A (1998) Cerebral fungal infections in the immunocompromised host: a literature review and a new pathogen—*Chaetomium atrobrunneum*: case report. Neurosurgery 43:1463–1468
Hageskal G, Knutsen AK, Gaustad P, de Hoog GS, Skaar I (2006) Diversity and significance of mold species in Norwegian drinking water. Appl Environ Microbiol 72:7586–7593
Hageskal G, Lima N, Skaar I (2009) The study of fungi in drinking water. Mycol Res 113:165–172
Halwagy R, Moustafa AE, Kame SM (1982) Ecology of the soil mycoflora in the desert of Kuwait. J Arid Environ 5:109–125
Hawksworth DL, Wells H (1973) Ornamentation on the terminal hairs in *Chaetomium* Kunze ex Fr. and some allied genera. Mycol Pap 134:1–24
Hebert PDN, Ratnasingham S, de Waard JR (2003) Barcoding animal life: cytochrome c oxidase subunit 1 divergences among closely related species. *Proc R Soc B* 270:S96–S99
Hess WM, Müller E, Aue R (1967) Germ pores of Chaetomium ascospores. Naturwissenschaften 19:521–522. https://doi.org/10.1007/BF01129384
Hogarth PJ (2007) The biology of mangroves and seagrasses. Oxford University Press, Oxford, 272pp
Hoppin EC, McCoy EL, Rinaldi MG (1983) Opportunistic mycotic infection caused by *Chaetomium* in a patient with acute leukemia. Cancer 53:555–556
Horie, Y. and Udagawa, S. 1990. New or interesting *Chaetomium* species from herbal drugs. Trans. Mycol. Soc. Japan 31: 249–258.
Hubka V (2015) *Chaetomium*. In: Paterson RRM, Lima NMVS (eds) Molecular biology of food and water borne mycotoxigenic and mycotic fungi of humans. CRC Press, Boca Raton, pp 211–228
Index Fungorum Partnership – IFP (2019) Index Fungorum. Available at http://www.indexfungorum.org/. Accessed June 15, 2019
Ismail ME, Abdalla HM (2005) The fungus *Chaetomium globosum* a new pathogen to pear fruits in Egypt. Assiut J Agri Sci 36:177–188
Ivanushkina NE, Kochkina GA, Ozerskaya SM (2005) Fungi in ancient permafrost sediments of the Arctic and Antarctic regions. In: Castello JD, Rogers SO (eds) Life in ancient ice. Princeton University Press, Princeton, pp 127–139
Ivarson KC (1965) The microbiology of some permafrost soils in the McKenzie Valley, N.W.T. Arctic 18:256–260
Jin L, Liu F, Sun W, Zhang F, Karuppiah V, Li Z (2014) *Pezizomycotina* dominates the fungal communities of South China Sea Sponges *Theonella swinhoei* and *Xestospongia testudinaria*. FEMS Microbiol Ecol 90:935–945. https://doi.org/10.1111/1574-6941.12446
Kathiresan K, Bingham BL (2001) Biology of mangroves and mangrove ecosystems. Adv Mar Biol 40:81–251
Khalil AMA, El-sheikh HH, Sultan MH (2013) Distribution of Fungi in mangrove soil of coastal areas at Nabq and Ras Mohammed protectorates. Int J Curr Microbiol App Sci 2(12):264–274
Kikuchi T, Kadota S, Suehara H, Nishi A, Tsubaki K (1981) Odorous metabolites of a fungus, *Chaetomium globosum* Kunze ex Fr.: identification of geosmin, a musty-smelling compound. Chem Pharm Bull 29:1782–1784
Kowalik R, Sadurska I (1973) Microflora of papyrus from samples of Cairo Museum. Stud Conserv 18:1–24
Kunze G, Schmidt JC (1817) *Chaetomium*. Myc Heft 1:15
Kurek E, Korniłłowicz-Kowalska T, Słomka A, Melke AJ (2007) Characteristics of soil filamentous fungi communities isolated from various micro-relief forms in the high Arctic tundra (Bellsund region, Spitsbergen). Pol Polar Res 28:57–73

Legendre P, Rogers DJ (1972) Characters and clustering in taxonomy: a synthesis of two taximetric procedures. Taxon 21:567–606
Lesire V et al (1999) Possible role of *Chaetomium globosum* in infection after autologous bone marrow transplantation. Intensive Care Med 25:124–125
Li J, Zhao XM, Wang XW (2012) Growth temperature of *Chaetomium* species and its taxonomic value. Mycosystema 31:213e222
Lin Y, Li X (1995) First case of phaeohyphomycosis caused by *Chaetomium murorum* in China. Chin J Dermatol 28:367–369
Liu YJ, Whelen S, Hall BD (1999) Phylogenetic relationships among Ascomycetes: evidence from an RNA Polymerse II Subunit. Mol Biol Evol 16(12):1799–1808
Lumbsch HT (2000) Phylogeny of filamentous ascomycetes. Naturwissenschaften 87:335–342
Maharachchikumbura SSN, Al-Sadi AM, Al-Kharousi M, AlSaady NA, Hyde KD (2016) A checklist of fungi in Oman. Phytotaxa 273:219–261
Mahmoud SAZ, Abou E-FM, El-Mofty M (1964) Studied on the rhizosphere microflora of a desert plants. Folia Microbiol 9:1–8
Mandeel Q (2002) Microfungal community with rhizosphere soil of *Zygophyllum qatarense* in arid habitats of Bahrain. J Arid Environ 50:665–681
Manoharachary C, Kunwar IK, Tilak KV (2013) Diversity and characterization of fungi and its relevance. Indian Phytopath 66(1):10–13
Mansour AMA (2010) Contribution to knowledge of some soil fungi in eastern region, in Libya. J Product Develop (Agri Res) 15(3):395–404
Miller AN, Huhndorf SM (2005) Multi-gene phylogenies indicate ascomal wall morphology is a better predictor of phylogenetic relationships than ascospores morphology in the Sordariales (Ascomycota, Fungi). Mol Phylogenet Evol 35:60–75
Millner PD (1977) Radial growth responses to temperature by 58 *Chaetomium* species, and some taxonomic relationships. Mycologia 69:492–502
Millner PD, Motta JJ, Lentz PL (1977) Ascospores, germ pores, ultrastructure and thermophilism of *Chaetomium*. Mycologia 69:720–733. https://doi.org/10.2307/3758862
Min XJ, Hickey DA (2007) Assessing the effect of varying sequence length on DNA barcoding of fungi. Mol Ecol Notes 7:365–373
Montasir AH, Mostafa MA, Elwan SH (1956a) Development of soil microflora under *Zygophyllum album* L. and *Zygophyllum coccineum* L. Ain Shams Sci Bull 1:9–22
Montasir AH, Mostafa MA, Elwan SH (1956b) Development of soil microflora in relation to vegetation along a transect line at yellow hills, North Cairo. Ain Shams Sci Bull 1:23–32
Moubasher AH, Al-Subai AAT (1987) Soil fungi in State of Qatar. Publications of Sientific and Applied Research Centre, University of Qatar, Doha, Qatar
Moubasher AH, El-Dohlob SM (1970) Seasonal fluctuation of Egyptian soil fungi. Trans Br Mycol Soc 54:45–51
Moubasher AH, Moustafa AF (1970) A survey of Egyptian soil fungi with special reference to *Aspergillus*, *Penicillium* and *Penicillium* related genera. Trans Br Mycol Soc 54:35–44
Moubasher AH, Abdel-Hafez SII, El-Maghraby OMO (1985) Studies on soil mycoflora of Wadi Bir- El- Ain, Eastern Desert, Egypt. Cryptogam Mycol 6:129–143
Moubasher AH, Abdel-Hafez SII, El-Maghraby OMO (1988) Seasonal fluctuations of soil and air borne fungi of Wadi Bir- El-Ain in Eastern Desert of Egypt. Nat Monspel Ser Bot 52:57–70
Moubasher AH, Abdel-Hafez SII, Bagy MMK, Abdel-Sater MA (1990) Halophilic and halotolerant fungi in cultivated, desert and salt marsh soils from Egypt. Acta Mycol 27:65–81
Mouchacca J (1971) *Pseudeurotium desertorum* sp. nov. Revue de Mycologie 36:123–127
Mouchacca J (1973a) Deux *Alternaria* des sols arides d'Egypte: *A. chlamydospora* sp. nov. et *A. phragmospora* van Emden. Mycopathol Mycol Appl 50:217–225
Mouchacca J (1973b) Les *Thielavia* des sols arides: espèces nouvelles et analyse générique. Bulletin de la Société Mycologique de France 89:295–311
Mouchacca J (1977) Sur un nouveau Discomycetes *Ascobolus egyptiacus*. Travaux dédiès à G. Viennot-Bourgin, Société Francaise de Phytopathologoie, Paris, pp 236–267

Mouchacca J (1982) Etude analytique de la mycoflore de quelques sols de régions arides de l'Egypte. Thèse de Doctorat d'Etat, Muséum National d'Histoire Naturelle et Université Pierre et Marie Curie (Paris VI), 247 pp., 90 tabs., 30 figs
Mouchacca J (1995) Check-list of novel fungi from the Middle East described mainly from soil since 1930. Sydowia 47:240–257
Mouchacca J, Joly P (1974) Etude de la mycoflore des sols arides de l'Egypte. I. Le genre Penicillium. Revue d'Ecologie et de Biologie du Sol 11:67–88
Mouchacca J, Joly P (1976) Etude de la mycoflore des sols arides de l'Egypte. II. Le genre *Aspergillus*. Revue d'Ecologie et de Biologie du Sol 13:293–313
Mouchacca J, Nicot J (1973) Les *Fusariella* des sols arides. Revue de Mycologie 37:168–182
Moustafa AF, Abdel-Azeem AM (2005) The genus *Chaetomium* in Egypt. *El-Minia*. Sci Bull 16:235–256
Moustafa AF, Abdel-Azeem AM (2011) An annotated check-list of Ascomycota reported from soil and other terricolous substrates in Egypt. J Basic Appl Mycol 2:1–27
Moustafa AF, Al-Musallam AA (1975) Contribution to the fungul flora of Kuwait. Trans Br Mycol Soc 65(3):547–553
Moustafa AF, Ess El-Din EK (1989) *Chaetomium sinaiense* sp. nov., a new soil ascomycete from Egypt. Canadian Journal of Botany 67:3417–3419
Mtibaà R, de Eugenio L, Ghariani B et al (2017) A halotolerant laccase from *Chaetomium* strain isolated from desert soil and its ability for dye decolourization. 3 Biotech 7:329. https://doi.org/10.1007/s13205-017-0973-5
Murgia M, Fiamma M, Barac AM, Deligios M, Mazzarello V, Paglietti B, Cappuccinelli PA, Al-Qahtani A, Squartini A, Rubino S, Al-Ahdal MN (2018) Biodiversity of fungi in hot desert sands. MicrobiologyOpen. https://doi.org/10.1002/mbo3.595
Mustafa AI, Abdel-Azeem AM, Salem FM (2013) Surveying and exploitation of some taxa for extracellular biosynthesis of silver nanoparticles. Third international congress on fungal conservation, Akyaka, Mugla, Turkey, 11–15 November 2013. Abstract book: 44
Naguib AI, Mouchacca J (1970–1971) The mycoflora of Egyptian desert soils. Bulletin de l'Institut d'Egypte 52:37–61
Naim MS (1967a) Contribution to the knowledge of soil fungi in Libya. I. Rhizosphere and soil fungi of *Artemisia herba alba* in Tripoli. Mycopath Mycol Appl 31:296–299
Naim MS (1967b) Contribution to the knowledge of soil fungi in Libya. II. Fungus flora under *Citrus* trees in Libya. Mycopath Mycol Appl 31:300–304
Nassar MSM (1998) Soil mycoflora of Wadi Abu-Subayrah at Aswan region at Eastern Desert of Egypt. Egypt J Bot 38(1–2):21–46
Nasser LA (2004) Incidence of terrestrial fungi in drinking water collected from different schools in Riyadh region, Saudi Arabia. Pak J Biolog Sci 7:1927–1932
Natori S (1977) Toxic cytochalasins. In: Rodricks JV, Hesseltine CW, Mehlman MA (eds) Mycotoxins in human and animal health. Pathotox Publishers, Park Forest South, Illinois, pp 559–581
Nienow, J. A., and Friedmann, E. I. 1993. Terrestrial lithophytic (rock) communities, in Antarctic Microbiology, ed. E. I. Friedmann (New York, NY: Wiley-Liss), 343–412.
Nixon, K. 2002. WINCLADA v.1.00.08. Character analysis program. Published by the author.
Nguyen HDT, Seifert KA (2008) Description and DNA barcoding of three new species of *Leohumicola* from South Africa and the United States. Persoonia 21:57–69
Nilsson RH, Ryberg M, Kristiansson E et al (2006) Taxonomic reliability of DNA sequences in public sequence databases: a fungal perspective. PLoS One 1:e59
Nour MA (1956) A preliminary survey of fungi in some Sudan soils. Trans Br Mycol Soc 39:357–360
O'Donnell K, Cigelnik E (1997) Two divergent intragenomic rDNA ITS2 types within a monophyletic lineage of the fungus *Fusarium* are non-orthologous. Mol Phylogenet Evol 7:103–116
Onofri S, Selbmann L, Zucconi L, Pagano S (2004) Antarctic microfungi as models for exobiology. Planet Space Sci. 52, 229–237. https://doi.org/10.1016/j.pss.2003.08.019

Oren A (2002) Halophilic microorganisms and their environments. Kluwer Academic Publishers, Dordrecht

Ozerskaya SM, Kochkina GA, Ivanushkina NE, Knyazeva EV, Gilichinskii DA (2008) The structure of micromycete complexes in permafrost and cryopegs of the Arctic. Microbiology 77:482–489

Ozerskaya S, Kochkina G, Ivanushkina N, Gilichinsky DA (2009) Fungi in permafrost. In: Margesin R (ed) Permafrost soils. Soil biology, vol Vol. 16. Springer, pp 85–95

Pathan AAK, Bhadra B, Begum Z, Shivaji S (2009) Diversity of yeasts from puddles in the vicinity of Midre Lovénbreen glacier, Arctic and bioprospecting for enzymes and fatty acids. Curr Microbiol 60:307–314

Paiva de Carvalho, Hugo; Mesquita, Nuno; Trovão, João; Fernández Rodríguez, Santiago; Pinheiro, Ana Catarina; Gomes, Virgínia; Alcoforado, Ana; Gil, Francisco; Portugal, António. 2018. Fungal contamination of paintings and wooden sculptures inside the storage room of a museum: Are current norms and reference values adequate? (Acknowledged Proofreader and Translator). Journal of Cultural Heritage 34: 268-276. http://dx.doi.org/10.1016/j.culher.2018.05.001

Petrini O (1991) Fungal endophytes of tree leaves. In: Fokkema NJ, van den Heuvel J (eds) Microbial ecology of the leaves. Cambridge University Press, Cambridge, pp 185–187

Pimentel M, Lembo A, Chey W et al (2011) Rifaximin therapy for patients with irritable bowel syndrome without constipation. N Engl J Med 364:22–32

Pinheiro, A. C. M. da S. (2014) Fungal communities in archives: assessment strategies and impact on paper conservation and human health. Universidade Nova de Lisboa.

Pitt JI, Hocking AD (2009) Fungi and food spoilage, 3rd edn. Springer, Heidelberg, Germany

Pitt JI et al (1993) The normal mycoflora of commodities from Thailand. 1. Nuts and oilseeds. Int J Food Microbiol 20:211–226

Pitt JI et al (1994) The normal mycoflora of commodities from Thailand. 2. Beans, rice, small grains and other commodities. Int J Food Microbiol 23:35–53

Pitt JI et al (1998) The mycoflora of food commodities from Indonesia. J Food Mycol 1:41–60

Plomley NJB (1959) Formation of the colony in the fungus *Chaetomium*. Aust J Biol Sci 12(1):53–64

Pornsuriya C, Soytong K, Poeaim S, Kanokmedhakul S, Khumkomkhet P, Lin FC, Wang HK, Hyde KD. 2011 – *Chaetomium siamense* sp. nov., a soil isolate from Thailand, produces a new chaetoviridin, G. Mycotaxon 115, 19–27. http://dx.doi.org/10.5248/115.19

Prabhakaran N, Prameeladevi T, Sathiyabama M, Kamil D (2015) Screening of different *Trichoderma* species against agriculturally important foliar plant pathogens. J Environ Biol 36:191–198

Prokhorov VP, Linnik MA (2011) Morphological, cultural and biodestructive peculiarities of *Chaetomium* species. Mosc Univ Biol Sci Bull 66(3):95–101

Pryce TM, Palladino S, Kay ID, Coombs GW (2003) Rapid identification of fungi by sequencing the ITS 1 and ITS 2 regions using an automated capillary electrophoresis system. Med Mycol 41:369–381

Rai JN, Tewari JP, Mukerji KG. 1964 – Achaetomium, a new genus of ascomycetes. Canadian Journal of Botany 42, 693–697. http://dx.doi.org/10.1139/b64-064

Rajesh AM, Prameeladevi T, Malathi VG, Kamil D, Borah JL, Prabhakaran N, Srinivasa N (2013) A combined morpho-genetic approach towards identification of *Trichoderma* species. J Mycopathol Res 51(1):19–28

Reeve JN, Christner BC, Kvitko BH, Mosley-Thompson E, Thompson LG (2002) Life in glacial ice (Abstract). In: Rossi M, Bartolucci S, Ciaramella M, Moracci M (eds) "Extremophiles 2002," 4th International Congress on Extremophiles, 22–26 September 2002, Naples, Italy 27

Rivkina, E.N., Friedmann, E.I., McKay, C.P. and Gilichinsky, D.A. (2000) Metabolic activity of permafrost bacteria below the freezing point. Appl. Environ. Microbiol. 66, 3230–3233.

Rock JP (1998) Cerebral fungal infections in the immunocompromised host: a literature review and a new pathogen—*Chaetomium atrobrunneum*: case report—comment. Neurosurgery 43:1469

Rodriguez R, Redman R (2008) More than 400 million years of evolution and some plants still can't make it on their own: plant stress tolerance via fungal symbiosis. J Exp Bot 59(5):1109–1114
Rodríguez K, Stchigel AM, Guarro J (2002) Three new species of *Chaetomium* from soil. Mycologia 94:114–124
Rodriguez RJ, White JF Jr, Arnold AE, Redman RS (2009) Fungal endophytes: diversity and functional roles. New Phytol 182:314–330
Roe AD, Rice AV, Bromilow SE, Cooke JEK, Sperling FAH (2010) Multilocus species identification and fungal DNA barcoding: insights from blue stain fungal symbionts of the mountain pine beetle. Mol Ecol Res. https://doi.org/10.1111/j.1755-0998.2010.02844
Ruisi S, Barreca D, Selbmann L, Zucconi L, Onofri S (2007) Fungi in Antarctica. Rev Environ Sci Biotechnol 6:127–141
Saadabi AMA (2006) On the fungal flora of Saudi Arabian soils. Res J Microbiol 1:280–284
Saito M, Singh RB, Saite M (1976) Reports on study of mycotoxins in foods in relation to liver diseases in Malaysia and Thailand, p. 85 (University of Tokyo, Tokyo, Japan, 1976)
Salama AM, Elbatanoni K, Ali MI (1971) Studies on the fungal flora of Egyptian soils. I. Western Mediterranean coast and Libyan Desert. United Arab Republic J Bot 14(1):99–114
Salar RK, Aneja K (2007) Thermophilic fungi: taxonomy and biogeography. J Agric Technol 3:77e107
Salem FM, Abdel-Azeem AM (2014) Screening of anticancer metabolites produced by endophytic fungi. LAP LAMBERT Academic Publishing. ISBN 978–3–659-53697-7
Samson RA, Mouchacca J (1974) Some interesting species of *Emericella* and *Aspergillus* from Egyptian desert soil. Antonie Van Leeuwenhoek 40:121–131
Samson RA, Mouchacca J (1975) Additional notes on species of *Aspergillus*, *Eurotium* and *Emericella* from Egyptian desert soil. Antonie Van Leeuwenhoek 41:343–351
Säwström C, Mumford P, Marshall W, Hodson A, Laybourn-Parry J (2002) The microbial communities and primary productivity of cryoconite holes in an Arctic glacier (Svalbard 79°N). Polar Biol 25:591–596
Schoch CL, Seifert KA, Huhndorf A, Robert V, Spouge JL, André Levesque C, Chen W (2012) Nuclear ribosomal internal transcribed spacer (ITS) region as a universal DNA barcode marker for Fungi, and fungal barcoding consortium. Proc Natl Acad Sci 109(16):6241–6246
Schoch CL, Robbertse B, Robert V et al (2014) Finding needles in haystacks: linking scientific names, reference specimens and molecular data for fungi. Database (Oxford). https://doi.org/10.1093/database/bau061
Schulz B, Boyle C, Drager S et al (2002) Endophytic fungi; a source of novel biologically active secondary metabolites. Mycol Res 106:996–1004
Schwarz P, Bretagne S, Gantier JC, Garcia-Hermoso D, Lortholary O, Dromer F, Dannaoui E (2006) Molecular identification of Zygomycetes from culture and experimentally infected tissues. J Clin Microbiol 44:340–349
Seifert KA (2009) Progress towards DNA barcoding of fungi. Mol Ecol Resour 9:83–89
Sekhar VC (2015) Morphology and selection of potential region for DNA barcoding to identify chaetomium species. PhD thesis, Plant Pathology Section, Faculty of Post-Graduate School, Indian Agricultural Research Institute, New Delhi, India
Sekita S, Yoshihira K, Natori S (1980) Chaetochromin, a bis(naphthodihydropyran-4-one) mycotoxin from Chaetomium thielavioideum: application of 13C-IH long-range coupling to the structure elucidation. Chem Pharm Bull 28:2428–2435
Sekita S, Yoshihira K, Natori S, Udagawa S, Mouri T, Sugiyama Y, Kurata H, Umeda M (1981) Mycotoxin production by Chaetomium spp. and related fungi. Can J Microbiol 27:766–772
Selbmann L, Onofri S, Zucconi L, Isola D, Rottigni M, Ghiglione C, Piazza P, Alvaro MC, Schiaparelli S (2015) Distributional records of Antarctic fungi based on strains preserved in the Culture Collection of Fungi from Extreme Environments (CCFEE) Mycological Section associated with the Italian National Antarctic Museum (MNA). MycoKeys 9:57–71. https://doi.org/10.3897/mycokeys.10.5343
Seth HK (1970) A monograph of the genus *Chaetomium*. Beih Nova Hedwig 37:1–133
Seth HK (1972) A monograph of the genus *Chaetomium*. In Beih Nova Hedwigia, 37, pp 133

Seth HK (1983) Numerical taxonomy of the genus *Chaetomium* Kze. In: Felsenstein J (ed) Numerical taxonomy. NATO ASI series (Series G: ecological sciences), vol vol 1. Springer, Berlin, Heidelberg
Sharma G, Pandey RR (2010) Influence of culture media on growth, colony character and sporulation of fungi isolated from decaying vegetable wastes. J Yeast Fungal Res 1(8):157–164
Sharma R, Kulkarni G, Sonawane MS, Shouche YS (2013) A new endophytic species of *Chaetomium* from *Jatropha podagrica*. Mycotaxon 124:117–126
Shenoy BD, Jeewon R, Hyde KD (2007) Impact of DNA sequence data on the taxonomy of anamorphic fungi. Fungal Divers 26:1–54
Skolko AJ, Groves JW (1948) Notes on seed-borne fungi V. *Chaetomium* species with dichotomously branched hairs. Can J Res C 26:269–280
Skolko AJ, Groves JW (1953) Notes on seed-borne fungi. VII *Chaetomium*. Can J Bot 31:779–809
Sokal RR, Sneath PHA (1963) Principles of numerical taxonomy. W.H. Freeman & Co., New York.
Somrithipol S (2004) Coprophilous fungi. In: Jones EBG, Tanticharoen M, Hyde KD (eds) Thai fungal diversity. BIOTEC, Thailand, pp 119–128
Somrithipol S, Hywel-Jones NL, Jones EBG (2004) Seed fungi. In: Jones EBG, Tanticharoen M, Hyde KD (eds) Thai fungal diversity. BIOTEC, Thailand, pp 129–140
Sörgel G (1960) Zum Problem der Trennung von Arten bei Pilzen, dargestellt am Beispiel der Ascomycetengattung *Chaetomium*. Arch Mikrobiol 36:51–66. https://doi.org/10.1007/BF00405942
Sörgel G (1961) Zur Variabilitât in der Gattung *Chaetomium*, dargestellt am Beispiel der Art Ch. carinthiacumn. sp. In Arch Mikrobiol 40:383–394
Soytong K (1992a) Antagonism of *Chaetomium cupreum* to *Pyricularia oryzae*. J Plant Protect Tropics 9:17–24
Soytong K (1992b) Biological control of tomato wilt caused by *Fusarium oxysporum* f.sp. *lycopersici* using *Chaetomium cupreum*. Kasetsart J (Nat Sci) 26:310–313
Soytong K, Kanokmedhakul S, Kukongviriyapa V, Isobe M (2001) Application of *Chaetomium* species (Ketomium®) as a new broad spectrum biological fungicide for plant disease control: a review article. Fungal Divers 7:1–15
Spalding MD, Blasco F, Field CD. 1997. World mangrove atlas. International Society for Mangrove Ecosystems, Okinawa, Japan 178.
Spatafora JW (1995) Ascomal evolution of filamentous ascomycetes: evidence from molecular data. Can J Bot 73(S1):811–815
Spiering MJ, Greer DH, Schmid J (2006) Effects of the fungal endophyte, *Neotyphodium lolii*, on net photosynthesis and growth rates of perennial ryegrass (*Lolium perenne*) are independent of in plant endophyte concentration. Ann Bot 98(2):379–387
Sridhar KR (2009) Mangrove fungi of the Indian peninsula. In: Sridhar KR (ed) Frontiers in fungal ecology, diversity and metabolites. IK International Publishing House Pvt. Ltd, New Delhi, pp 28–50
Stakhov V, Gubin S, Maksimovich S, Rebrikov D, Savilova A, Kochkina G, Ozerskaya S, Ivanushkina N, Vorobyova E (2008) Microbial communities of ancient seeds derived from permanently frozen Pleistocene deposits. Microbiology 77:348–355
Stark AA, Kobbe B, Matsuo D, Buchi G, Wogan GN, Demain AL (1978) Mollicellins: Mutagenic and antibacterial mycotoxins. Appl. Environ. Microbiol. 36:412–420
Steven B, Leveille R, Pollard WH, Whyte LG (2006) Microbial ecology and biodiversity in permafrost. Extremophiles 10:259–267
Stielow JB, Lévesque CA, Seifert KA, Meyer W et al (2015) One fungus, which genes? Development and assessment of universal primers for potential secondary fungal DNA barcodes. Persoonia 35:242–263
Stiller MJ, Rosenthal S, Summerbell RC, Pollack J, Chan A (1992) Onychomycosis of the toenails caused by *Chaetomium globosum*. J Am Acad Dermatol 26:775–776
Stockinger H, Kruger M, Schubler A (2010) DNA barcoding of arbuscular mycorrhizal fungi. New Phytol 187:461–474

Strobel G, Daisy B (2003) Bioprospecting for microbial endophytes and their natural products. Microbiol Mol Biol Rev 67(04):491–502

Strobel GA, Knighton B, Ren Y et al (2008) The production of mycodiesel hydrocarbons and their derivatives by the endophytic fungus *Gliocladium roseum* (NRRL 50072). Microbiology 154(11):3319–3328

Suleiman MK, Dixon K, Commander L, Nevill P, Quoreshi AM, Bhat NR, Manuvel AJ, Sivadasan MT (2019) Assessment of the diversity of fungal community composition associated with *Vachellia pachyceras* and its rhizosphere soil from Kuwait Desert. Front Microbiol 10:63. https://doi.org/10.3389/fmicb.2019.00063. eCollection 2019

Sumalan R, Alexa E, Pop G, Dehelean C, Sumalan R (2011) The biodiversity and dissemination of mycotoxin-producing fungi in cereals and cereal products. In Proceedings of the 46th Croatian and 6th International symposium on agriculture, Opatija, Croatia, 2011, pp 770–773

Suryanarayanan TS, Thirunavukkarasu N (2017) Endolichenic fungi: the lesser known fungal associates of lichens. Mycology 8(3):189–196. https://doi.org/10.1080/21501203.2017.1352048

Suwannasai N, Martin MP, Phosri C, Sihanonth P, Whalley AJS (2013) Fungi in Thailand: a case study of the efficacy of an ITS barcode for automatically identifying species within the *Annulohypoxylon* and *Hypoxiylon* genera. PLoS One 8(2):545–529

Takano Y, Kobayashi K, Marumo K, Ishikawa Y. 2004. Biochemical indicators and enzymatic activity below permafrost environment. In: Extremophiles 2004, 5th International Conference on Extremophiles, 19–23 Sep. Cambridge, Maryland. Abstract p 84.

Thomas C, Mileusnic D, Carey RB, Kampert M, Anderson D (1999) Fatal *Chaetomium* cerebritis in a bone marrow transplant patient. Hum Pathol 30:874–879

Tomita T (2003) Amylin in pancreatic islets and pancreatic endocrine neoplasms. Pathol Int 53(9):591–595

Trüper HG, Galinski EA (1986) Concentrated brines as habitats for microorganisms. Experientia 42(11–12):1182–1187

Tschudy RH (1937a) Experimental morphology of some species of *Chaetomium*. I. Use of cultural reactions in determining species characteristics. Am J Bot 24(7):472–480

Tschudy RH (1937b) Experimental morphology of some species of *Chaetomium*. II. Reactions of species of *Chaetomium* under various conditions of cultivation. Am J Bot 24(10):657–665

Udaiyan K, Hosagoudar VS. 1991 – Some interesting fungi from the industrial water cooling towers of Madras – II. Journal of Economic and Taxonomic Botany 15, 649–666.

Udagawa S (1960a) A taxonomic study on the Japanese species of *Chaetomium*. J Gen Appl Microbiol (Tokyo) 6:223–251

Udagawa S (1960b) A taxonomic study on the Japanese species of *Chaetomium*. J Gen Appl Microbiol 6:223–251

Udagawa S (1980) New or noteworthy ascomycetes from southeast Asian soil. 1. Trans Mycol Soc Jpn 21:17–34

Udagawa S, Mouri T, Kurata H, Sekita S, Yoshihira K, Natori S, Umeda M (1979) The production of chaetoglobosins, sterigmatocystin, O-methylsterigmatocystin, and chaetocin by *Chaetomium* spp. and related fungi. Can J Microbiol 25:170–177

Udagawa S, Muroi T (1981) Notes on some Japanese Ascomycetes XVI. Transactions of the Mycological Society of Japan. 22(1):11–26

Udagawa S, Sugiyama Y (1982) New records and new species of ascomycetous fungi from Nepal, a preliminary report on the expedition of 1980. Reports on the Cryptogamic Study in Nepal, March 1982 (Miscellaneous Publication of the N ational S cience Museum, Tokyo) 11–46

Udagawa S, Uchiyama S, Kamiya S (1994) Two new species of pyrenomycetous Ascomycetes from New Caledonia. Mycoscience 35:319–325

Udagawa S, Toyazaki N, Yaguchi T (1997) A new species of *Chaetomium* from house dust. Mycoscience 38:399–402

Umikalsom MS, Ariff AB, Shamsuddin ZH, Tong CC, Hassan MA, Karim MIA (1997) Production of cellulase by a wild strain of *Chaetomium globosum* using delignified oil palm empty-fruit-bunch fibre as substrate. Appl Microbiol Biotechnol 47(5):590–595
Umikalsom MS, Ariff AB, Hassan MA, Karim MIA (1998) Kinetics of cellulase production by *Chaetomium globosum* at different levels of dissolved oxygen tension using oil palm empty fruit bunch fibre as substrate. World J Microbiol Biotechnol 14(4):491–498
Untereiner WA, Debois V, Naveau FA (2001) Molecular systematics of the ascomycete genus *Farrowia* (Chaetomiaceae). Can J Bot 79:321–333
Urairuj C, Khanongnuch C, Lumyong S (2003) Ligninolytic enzymes from tropical endophytic Xylariaceae. Fungal Divers 13:209–219
Van Teighem P (1875) Sur le developpement du fruit des *Chaetomium* et les pretendue sexualite des Ascomycetes. C R Acad Sci 81:1110–1113
Van Teighem P (1876) Nouvelles observations sur le developpement du perithece des *Chaetomium*. Bull Soc Bot 23:364–366
Vega FE, Posada F, Aime MC et al (2008) Entomopathogenic fungal endophytes. Biol Control 46:72–82
Verma A, Johri BN, Prakash A (2014) Antagonistic evaluation of bioactive metabolite from endophytic fungus, *Aspergillus flavipes* KF671231. J Mycol Article ID 371218:5 pages. https://doi.org/10.1155/2014/371218
Vishniac HS (1993) The microbiology of Antarctic soils. In: Friedmann EI (ed) Antarctic microbiology. Wiley-Liss, New York, pp 297–342
Vishniac HS, Onofri S (2003) *Cryptococcus antarcticus* var. *circumpolaris* var. nov., a basidiomycetous yeast from Antarctica Anton. Leeuw./Int. J.G. Mol. Microbiol., 83:231–233
Volz PA, Ellanskaya IA, Wasser SP, Nevo E, Grishkan I (2001) Soil microfungi of Israel. Biodiversity of Cyanoprocaryotes, Algae and Fungi of Israel. In: Subramanian CV, Wasser SP (eds). A. P. A. Gantner Verlag, Ruggel/Liechtenstein
von Arx JA, Guarro J, Figueras MJ (1986) The ascomycete genus *Chaetomium*. Beih Nova Hedwig 84:1–162
von Arx JA (1970) The genera of fungi sporulating in pure culture. 288: 134 Figs. Ed. J. Cramer, Germany.
Vorobyova E, Soina V, Gorlenko M, Minkovskaya N, Zalinova N, Mamukelashvili A, Gilichinsky D, Rivkina E, Vishnivetskaya T (1997) The deep cold biosphere: facts and hypothesis. FEMS Microbiol Rev 20:277–290
Wallenstein MW, McMahon S, Schimel J (2007) Bacterial and fungal community structure in Arctic tundra tussock and shrub soils. FEMS Microbiol Ecol 59:428–435. https://doi.org/10.1111/j.1574-6941.2006.00260.x
Walther D, Gindrat D (1988) Biological control of damping off of sugar beet and cotton with Chaetomium globosum or a fluorescent *Pseudomonas* sp. Canadian J Microbiol 34:631–637
Wang X-W, Zheng R-Y (2005a) *Chaetomium acropullum* sp. nov. (Chaetomiaceae, Ascomycota), a new psychrotolerant mesophilic species from China. Nova Hedwig 80:413–417. https://doi.org/10.1127/0029-5035/2005/0080-0413
Wang X-W, Zheng R-Y (2005b) *Chaetomium ampulliellum* sp. nov. (Chaetomiaceae, Ascomycota) and similar species from China. Nova Hedwig 81:247–255. https://doi.org/10.1127/0029-5035/2005/0081-0247
Wang XW, Wang XL, Liu FJ, Zhao XM, Li J, Cai L (2014) Phylogenetic assessment of *Chaetomium indicum* and allied species, with the introduction of three new species and epitypification of *C. funicola* and *C. indicum*. Mycol Progress 13(3):719–732. https://doi.org/10.1007/s11557-013-0955-x
Wang XW, Houbraken J, Groenewald JZ, Meijer M, Andersen B, Nielsen KF, Crous PW, Samson RA (2016a) Diversity and taxonomy of *Chaetomium* and chaetomium-like fungi from indoor environments. Stud Mycol 84:145–224. https://doi.org/10.1016/j.simyco.2016.11.005
Wang XW, Lombard L, Groenewald JZ, Lil J, Videira SIR, Samson RA, Liu XZ, Crous PW (2016b) Phylogenetic reassessment of the *Chaetomium globosum* species complex. Persoonia 36:83–133. https://doi.org/10.3767/003158516X689657

Watanabe T (2002) Pictorial atlas of soil and seed fungi, morphologies of cultured fungi and key to species, 2nd edn. CRC Press, Boca Raton.

Webb TA, Mundt JO (1978) Molds on vegetables at the time of harvest. Appl Environ Microbiol 35:655–658

Weidenbörner M (2001) Pumpkin seeds—the mycobiota and potential mycotoxins. Eur Food Res Technol 212:279–281

Whiteside WC (1957) Perithecial initials of *Chaetomium*. Mycologia 49(3):420–425

Wicklow DT (1979) Hair ornamentation and predator defence in *Chaetomium*. Trans Br Mycol Soc 72(1):107–110

Wildman HG (2003) The rise and fall of natural products screening for drug discovery. Fungal Divers 13:221–231

Yamamoto Y, Kiriyama N, Shimizu S, Koshimura S (1976) Antitumor activity of asterriquinone – metabolic product from *Aspergillus terreus*. Gann 67(4):623–4

Yamamoto H, Moriyama K, Jinnouchi H, Yagishita K (1980) Studies on terreic acid. Jpn. J. Antibiot. 33:320–328. https://doi.org/10.11553/antibiotics1968b.33.320

Yeghen T et al (1996) *Chaetomium* pneumonia in patient with acute myeloid leukaemia. J Clin Pathol 49:184–186

Youssef YA (1974) On the fungal flora of Libyan soils. Arch Microbiol 99:167–171

Zaitlin B, Watson SB (2006) Actinomycetes in relation to taste and odour in drinking water: myths, tenets and truths. Water Res 40:1741–1753

Zhang Y, Wu W, Cai L (2016) Polyphasic characterisation of *Chaetomium* species from soil and compost revealed high number of undescribed species. Fungal Biol 121(1):21–43. https://doi.org/10.1016/j.funbio.2016.08.012

Zhang Y, Wu W, Cai L (2017) Polyphasic characterisation of *Chaetomium* species from soil and compost revealed high number of undescribed species. Fungal Biol 21(1):21–43. https://doi.org/10.1016/j.funbio.2016.08.012. Epub 2016 Sep 1

Zhao P, Luo J, Zhuang GW (2010) Practice towards DNA barcoding of the nectriaceous fungi. Fungal Divers. https://doi.org/10.1007/s13225-010-0064-y

Zhao P, Luo J, Zhuang WY (2011) DNA barcoding of the fungal genus *Neonectria* and the discovery of two new species. Sci China Life Sci 54:664–674

Chapter 2
Developmental Morphology of *Chaetomium* and *Chaetomiopsis*

Ahmed M. Abdel-Azeem

2.1 Introduction

The morphology of the ascomata of Ascomycetes has been used as an important tool for the classification of these fungi at generic and suprageneric levels. Mycologists usually recognize five or six ascoma types based on their shape, the presence of an ostiole, and the peridial structure. According to Webster and Weber (2007), the following ascomata are encountered: cleistothecia, perithecia, apothecia, gymnothecia, and pseudothecia (Fig. 2.1).

In higher forms of Ascomycetes, sexual reproduction results in the formation of several types of ascocarps showing basic differences from one class to another. These differences are used as key characters in the classification of the subdivision Ascomycotina down into classes.

Luttrell (1951) erected a taxonomic system for Ascomycetes based on the structure of the asci and the ascoma centrum, which included all the structures within the peridial wall. He divided the class Ascomycetes in four subclasses: Plectoascomycetidae, Hymenoascomycetidae, Loculoascomycetidae, and Laboulbenioascomycetidae (arthropod biotrophs with ascomata lacking hyphae).

Perithecia are the typical product of sexual reproduction in the class Pyrenomycetes. The ascoma development is usually initiated with the formation of an ascogonial coil as a side branch of vegetative hyphae. The ascogonial coil (developing later into centrum) is subsequently enveloped by sterile hyphae (developing later into a multiseriate wall usually referred to as peridium), resulting in the formation of a protoperithecium. Luttrell (1951) introduced the term bitunicate for asci showing two distinct walls and also recognized eight types of centrum, viz., *Dothidea* type, *Pleospora* type, *Elsinoe* type, *Diaporthe* type, *Ophiostoma* type,

A. M. Abdel-Azeem (✉)
Botany Department, Faculty of Science, Suez Canal University, Ismailia, Egypt
e-mail: ahmed_abdelazeem@science.suez.edu.eg

A. M. Abdel-Azeem (ed.), *Recent Developments on Genus Chaetomium*, Fungal Biology, https://doi.org/10.1007/978-3-030-31612-9_2

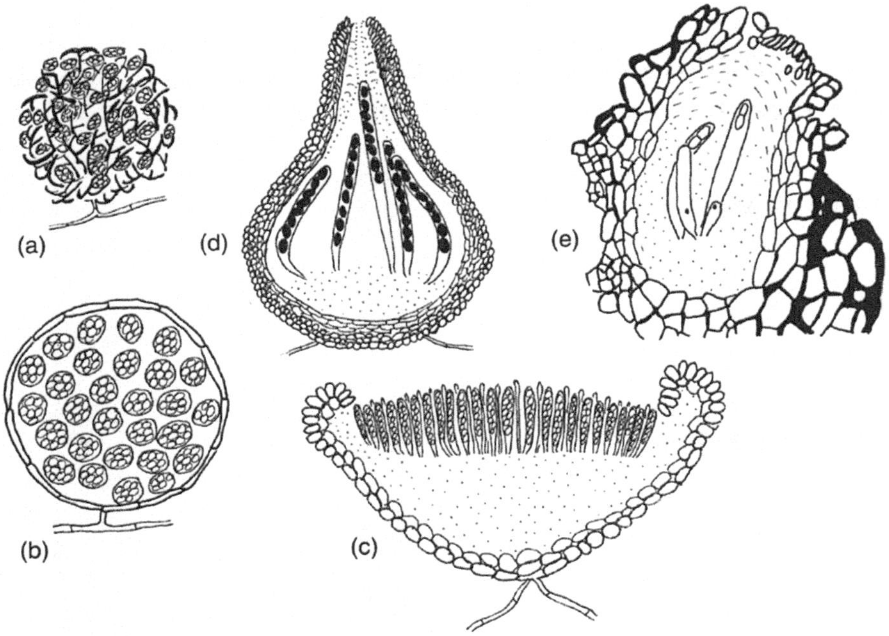

Fig. 2.1 Different types of ascocarp, diagrammatic and not to scale. (**a**) Gymnothecium made up of branched hyphae which do not completely enclose the asci. (**b**) Cleistothecium completely enclosing the asci which are formed throughout the ascocarp. There is no opening. (**c**) Apothecium, an open cup lined by a layer of asci and associated structures forming the hymenium. (**d**) Perithecium with a layer of asci at the base. It opens by a pore or ostiole. Its wall or peridium is made up of flattened cells. (**e**) Pseudothecium. The asci are formed within locules in a pseudoparenchymatous ascostroma. There is no peridium. (After Abdel-Azeem 2003)

Xylaria type, *Nectria* type, and *Phyllactinia* type, and discussed the systematic of Pyrenomycetes based on centrum types.

Genera of the family Chaetomiaceae belong to *Xylaria*-type centrum (Luttrell 1951; Parguey-Leduc and Janex-Favre 1981). In this type, ascogonia develop either freely on the mycelium or from the somatic hyphae within stroma. Branches from the ascogonial stalk cells or from subtending hyphae grow out and form an ascocarp wall. Paraphyses grow upward from the base and inward from the sides of the ascocarp wall, and the pressure they exert is thought to function in expanding the ascocarp and creating a central cavity. Growth of paraphyses in the upper part of the ascoma toward the apex creates an ostiole. The club-shaped or cylindrical asci arising from the ascogonium, now within the ascocarp at its base, grow among paraphyses and form their spores. Paraphyses may persist until the spores are expelled from the ascoma, or may gelatinize at an early stage, leaving the mature ascoma without paraphyses.

In the light of this, the presence or absence of paraphyses is not regarded as an important taxonomic character for taxa above the genus level (Guarro and Figueras 1989).

Although the internal features of perithecium development have been the object of numerous studies (Huang 1976; Mai 1976, 1977; Jensen 1981, 1982; Huang and Luttrell 1982; Sanni 1982, 1983; Read and Beckett 1985), the external features and its morphogenesis have received little attention (Seale 1973; Harris et al. 1975; Hock et al. 1978; Read 1983).

The developmental steps of external and internal structures until the formation of the mature ostiolate ascomata have also been followed in some species of *Chaetomium* using different types of microscopy including light, TEM, and SEM (Ellis 1981a; Figueras and Guarro 1988a, b). Other ultrastructural studies on *Chaetomium* have dealt with peridia surfaces, peridial hairs, septal pores, and ascospores (Hawksworth and Wells 1973; Millner et al. 1977; Ellis 1981b, 1982; Rosing 1982; Guarro and Calvo 1982; Guarro and Figueras 1985, 1989).

External and internal structures of sexual fruit bodies are of major importance in the classification of Ascomycetes into orders and families (Ainsworth 1973). Morphological criteria relate to ascocarp form (globose, ovoid, barrel, ostiolate, or non-ostiolate), texture (peridium loose or pseudoparenchymatic), glabrous (smooth), or tomentose (covered with characteristic appendages). Internal features, on the other hand, stress on centrum type, ascus form, ostiolar region, paraphyses, periphyses, and mode of ascospore discharge.

Chaetomiopsis dinae is a new Pyrenomycete genus described by Moustafa and Abdul-Wahid (1990) as a new member of the family Chaetomiaceae. In this chapter, I will discuss the results of ascocarp development and ascospore morphology of *Chaetomiopsis dinae* in order to reveal its relationship with the other members of the family for the first time.

2.2 Preparation of Samples for Light and Electron Microscopy

To throw some light on the above-mentioned criteria, *Chaetomiopsis dinae* have been grown on oatmeal agar at 30 °C as recommended by Abdel-Azeem et al. (2018). Pieces of agar bearing ascomata at different stages of development (2, 4, 6, 8, and 10 days old) were removed, fixed, and subjected to examination by light, transmission electron microscopy, and scanning electron microscopy (Abdel-Azeem et al. 2018).

Specimens, fixed in 3 absolute ethanol:1 glacial acetic acid (v/v) for 3 days, were dehydrated in Johansen's (1940) t-butanol series and embedded in paraffin and thereafter sectioned at 5 μm thick. Sections were brought down to water, stained with Mayer's hematoxylin using eosin Y as a counterstain (Sheehan and Hrapchak 1980), and finally mounted in Canada balsam.

The technique described by Figueras and Guarro (1988b) for preparation of fungal samples for SEM and TEM studies has been followed. According to this technique, pieces of agar (2 × 2 mm) bearing ascomata at different stages of development (as mentioned before) were removed and immersed in 2% glutaraldehyde solution in 0.1 M phosphate buffer over night at 5 °C and then rinsed with the same buffer solution and post fixed with 1% phosphate buffered OsO4 for 4–12 h at 5 °C in the dark. Fixed materials were again washed in the same buffer and dehydrated by 15-min changes in a graded series of ethanol. After dehydration, the material was embedded in Epon "812" resin. The ultrathin sections were transferred to grids, stained with uranyl acetate and Reynolds Pb citrate, and examined by Hitachi SL 72 P Microscope. Sections of 1 μm thick were stained with 1% aqueous toluidine blue for light microscopy.

Specimens were kept in 100% ethanol for 2 h and then successively transferred to mixtures of amyl acetate-ethanol in which concentration of the first substance (amyl acetate) was gradually increased through six steps until 100%. Specimens were then critical point dried with CO_2 using Polaron E 3000. After drying, specimens were mounted onto aluminum stubs and then coated with approximately 100–200 Å of gold-palladium in a Polaron sputter coater. Micrographs of textura, appendages, and ascospores were taken with a JEOL JSM 840 at 15 kV.

2.3 Volume Determination

To study the expansion rate of ascomata, a total number of 100 specimens, at different developmental stages, were examined according to Abdel-Azeem (1998). The volumes of each of the following elements have been considered for comparison: ascoma, centrum, and peridium (Fig. 2.2).

Ascoma Volume
Since the ascomata of *Chaetomiopsis dinae* are essentially globose, the ascoma volume is equivalent to a volume of a sphere which is normally determined by the formula:

$$V = 4/3\pi r^3.$$

where r = radius of the whole ascoma (in μm).

Centrum Volume
Exactly as ascoma volume except for the value of (r) which is equivalent to the radius of the centrum only.

Peridium Volume
Peridium volume was determined as follows:

$$\text{Peridium volume} = \text{volume of ascoma} - \text{volume of centrum}.$$

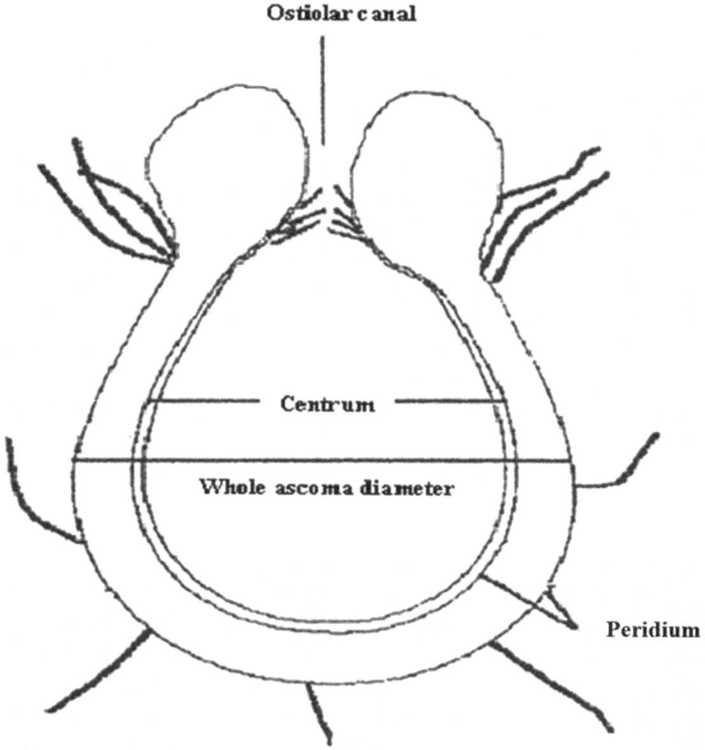

Fig. 2.2 Diagrammatic representation of vertical section through *C. dinae* mature ascoma. (© Abdel-Azeem et al. 2018)

2.4 Structure of the Ascogonium and Formation of the Ascoma

Ascogonia in this taxon are formed by specific type of initials that emerge as lateral branches from the vegetative hyphae (Fig. 2.3). These initials develop into irregular tight coils, forming sessile ascogonia (Fig. 2.4). Proliferation of initials takes place leading to the appearance of loosely packed young ascomata with hair-like appendages radiating in all directions (Fig. 2.5). With increasing size, some ascomata tend to be more ovoid and appendages became longer and denser particularly in the upper third of the ascoma (Fig. 2.6).

The underlying inner layers of hyphae below appendages flatten and cohere transforming into pseudoparenchymatous cells. At this stage, the ascoma is internally undifferentiated, just consisting of pseudoparenchymatous cells characterized by very reduced intercellular spaces (Fig. 2.7). Cells of the ascoma at

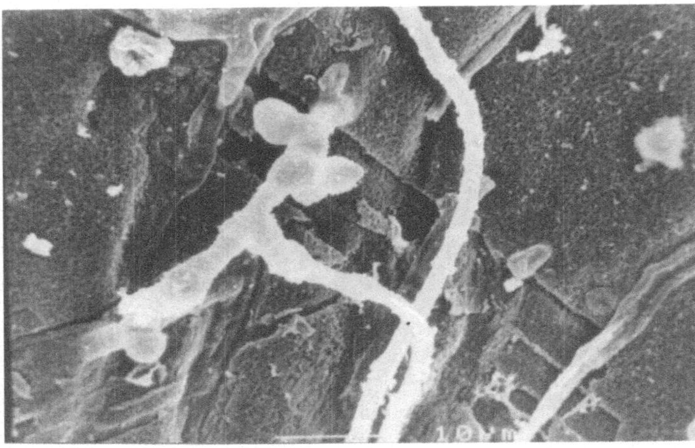

Fig. 2.3 Initials emerging as lateral branches (early phase) of *Chaetomiopsis dinae*

Fig. 2.4 Ascogonial coil irregularly curled about itself (SEM)

this stage tend to be smaller, more pigmented in the outside and larger, and less pigmented to the inside. The outermost layers of cells around the ascoma are distinctly lobed (Fig. 2.8).

By time, the central part of the ascoma starts showing differentiation. It becomes occupied by ascogenous hyphae, and at this phase an initial cavity becomes already visible (Fig. 2.9).

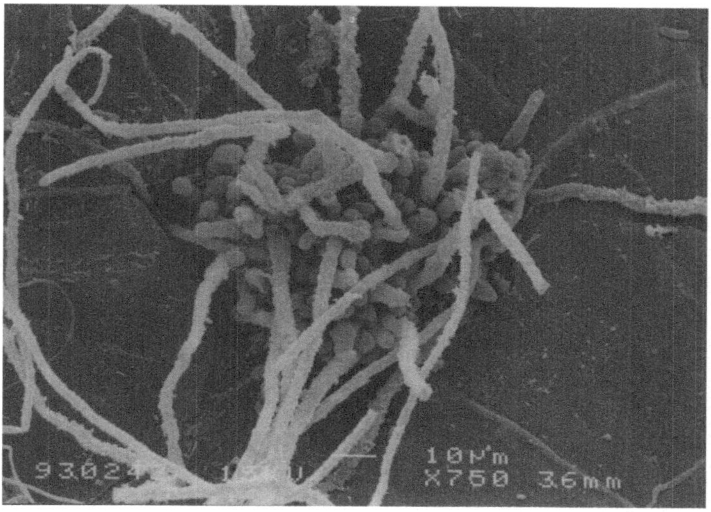

Fig. 2.5 Loosely packed young ascoma with radiating appendages of prominent roughness (SEM)

Fig. 2.6 Mature ovoid ascoma showing tuft of undulate appendages at its apex (SEM)

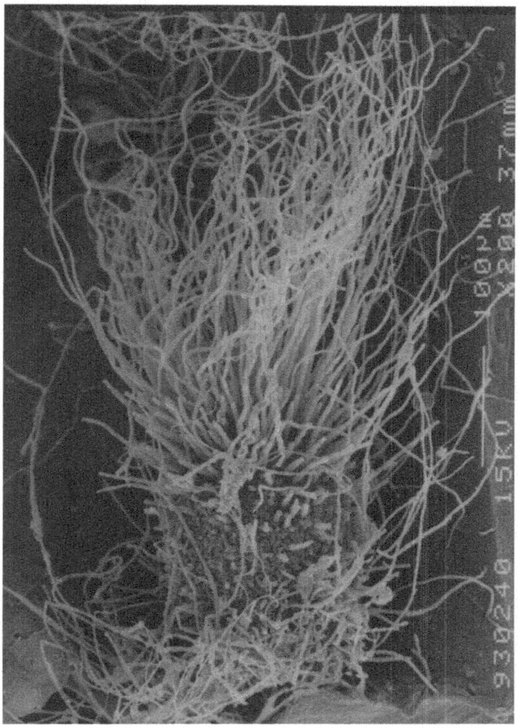

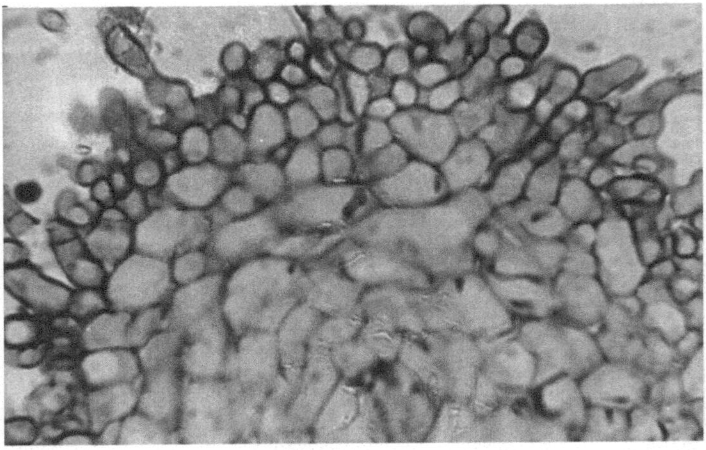

Fig. 2.7 Cross section in very young ascoma showing large-sized cells at the center and small-sized outside (light microscopy)

Fig. 2.8 SEM showing distinctly lobed peridial cells with emerging tuberculate appendages

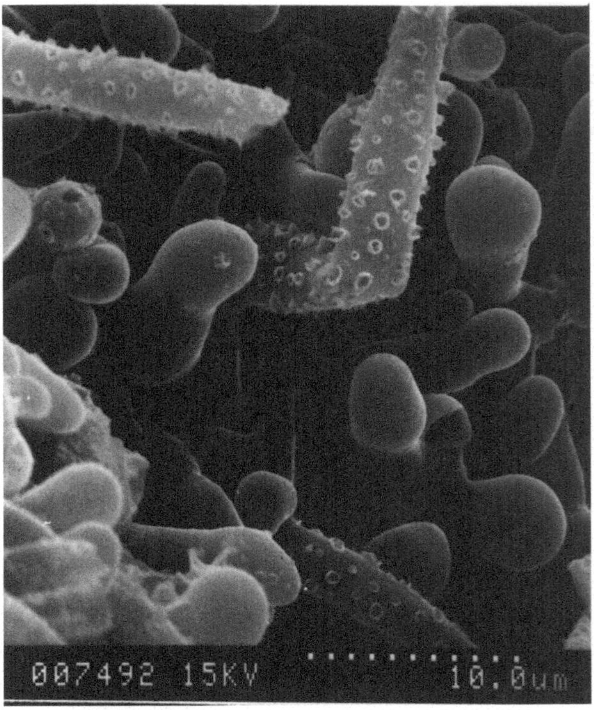

In some cases, a fine mycelium grows and interweaves between the peridial hairs of the developing ascoma (Fig. 2.10). At a later stage, the ascoma becomes more ovate and the peridial hairs of the apical part seem to proliferate more than the rest, reaching a great length (see Fig. 2.6).

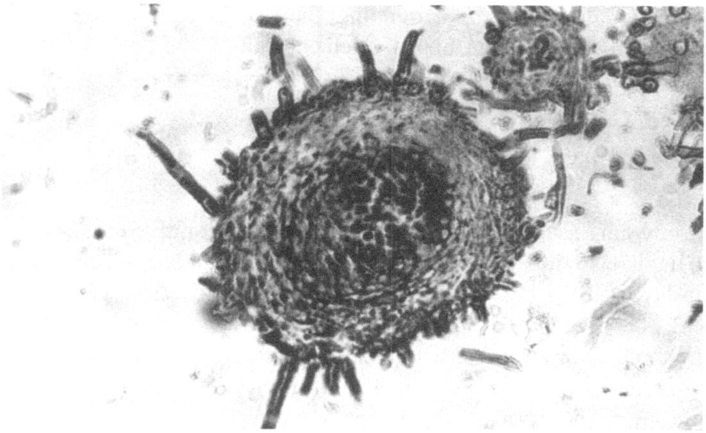

Fig. 2.9 Cross section in ascoma showing early differentiation into a fertile part constituting ascogenous hyphae and an outer sterile part developing into a peridium (light microscopy)

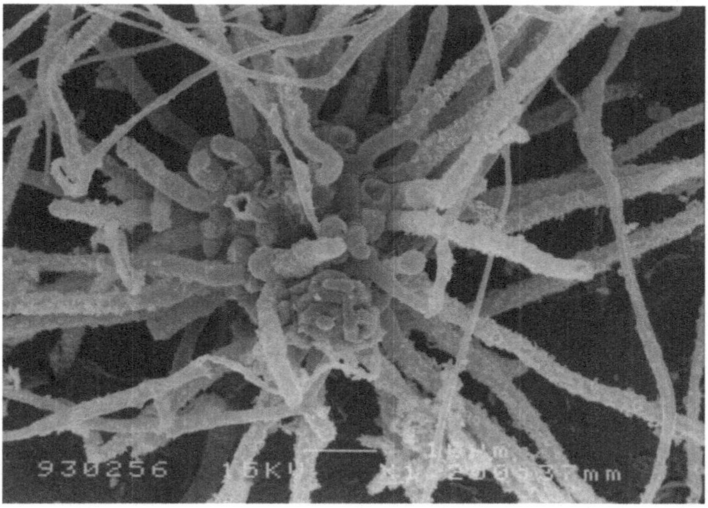

Fig. 2.10 Fine mycelia (referred to by arrows) traveling between peridial hyphae (SEM)

2.5 Structure of the Peridium

Like *Chaetomium* and other related Pyrenomycetes that have been investigated before, the peridium of *Chaetomiopsis* originates from specialized hyphal branches emerging from the ascogonium itself or from the adjacent hyphae. Growth and expansion of these enveloping hyphae (peridium) usually contribute to the formation of the ascoma in which the external cell layers normally become thicker and more pigmented. The inner layers of the peridium are consisted of thin-walled, pseudoparenchymatous cells, surrounding the ascogenous hyphae and derived structures, forming all together the so-called centrum.

2.5.1 Peridium in Surface View

Generally, in young ascomata, the external layer is a primitive network of interwoven short hyphae or lobed cells with reduced intercellular spaces, while in the underlying layers, cells become larger and intercellular spaces become more evident (Fig. 2.11). In mature ascomata, the remains of the primitive outer layer still retain a hyphoid nature but with bigger intercellular spaces, through which the underlying layers can be observed (see Fig. 2.12). It is difficult to ascertain whether the hyphoid layer is the remains of the primitive outer layer that was separated by the expansion of ascoma, or if they are the result of the growth of the hyphal tips that surround the developing ascoma.

Close-up examinations of peridial cells by SEM (Fig. 2.12) and squash preparation by light microscopy (Fig. 2.13) clearly show a peridium of textura-intricata type.

2.5.2 Peridium in Section

Cross sections in developing ascomata show typically pseudoparenchymatic peridium constituting 7–9 layers of cells arranged more or less periclinally in parallel rows. Cells of inner rows are more elongated, while those of outer rows tend to be

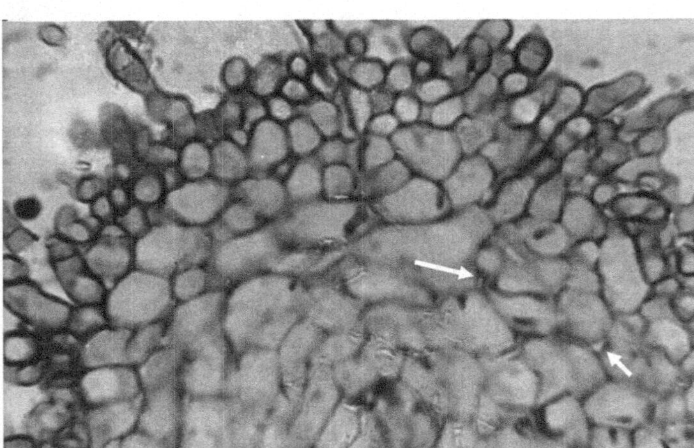

Fig. 2.11 Cross section showing outer peridium of small-sized, more pigmented cells and inner large-sized cells with developing intercellular spaces (referred by arrows)

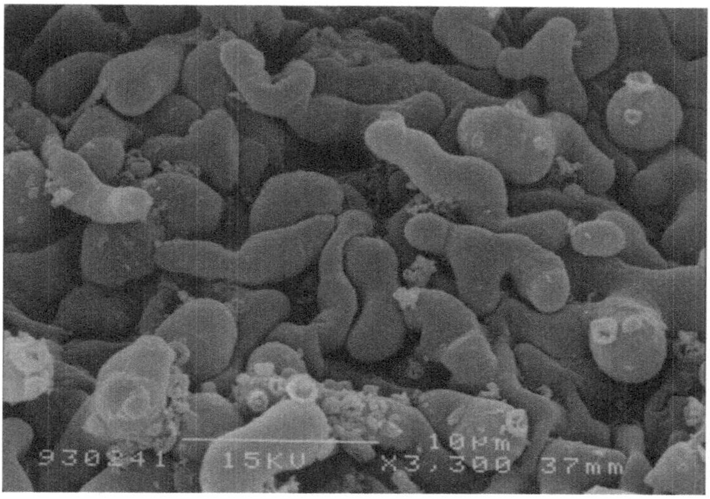

Fig. 2.12 SEM close-up showing lobed cells and intercellular spaces (textura-intricata)

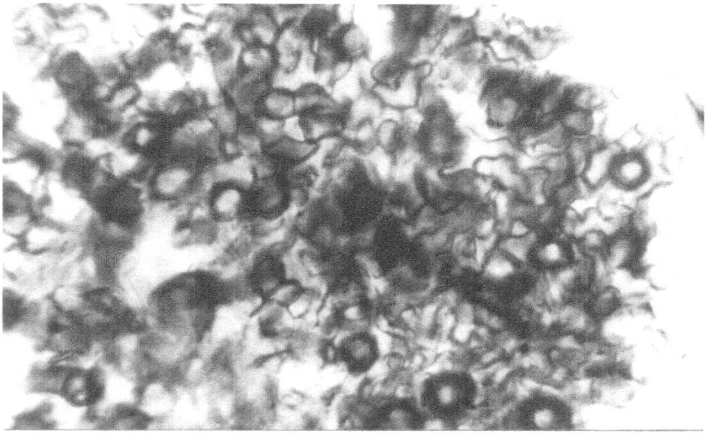

Fig. 2.13 Peridial cells in a squash preparation (light microscopy)

more isodiametric (Fig. 2.14). Cells of the external layers are irregular, more or less rounded, non-apiculate, and loosely packed. The intercellular spaces are relatively big and strongly pigmented (Fig. 2.15). The innermost cell layers of the peridium (Fig. 2.16) are obviously flattened, with reduced intercellular spaces and thin walls. The whole peridium shows an average thickness of 10–35 µm. The thickness of the peridium, however, is not usually uniform all round. It is usually broader at the base as well as the ostiolar region (Fig. 2.17).

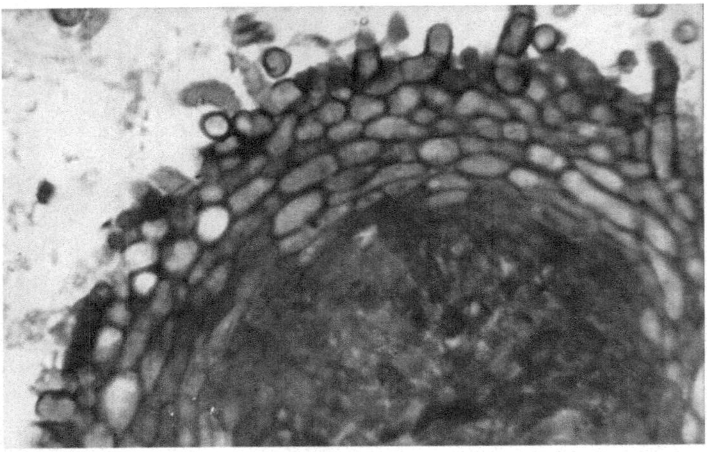

Fig. 2.14 Cross section of an ascoma showing thick peridium consisting of periclinally arranged cells

Fig. 2.15 Outermost layers of a peridium showing round cells and heavily pigmented intercellular spaces (TEM)

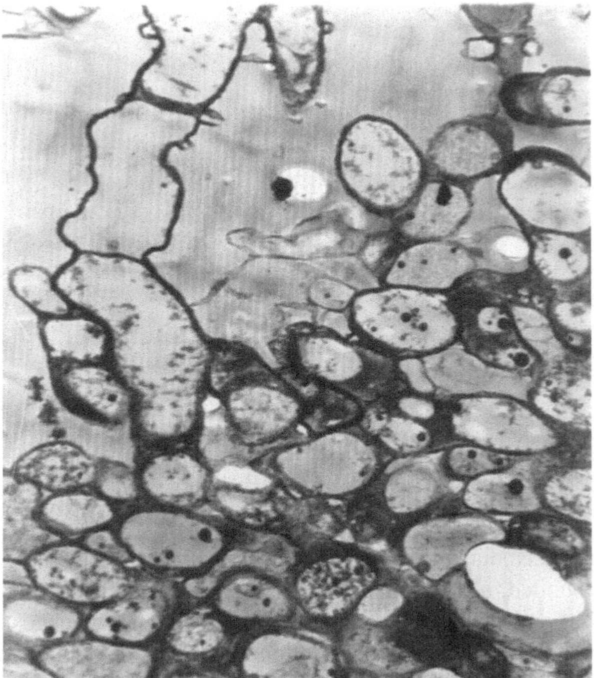

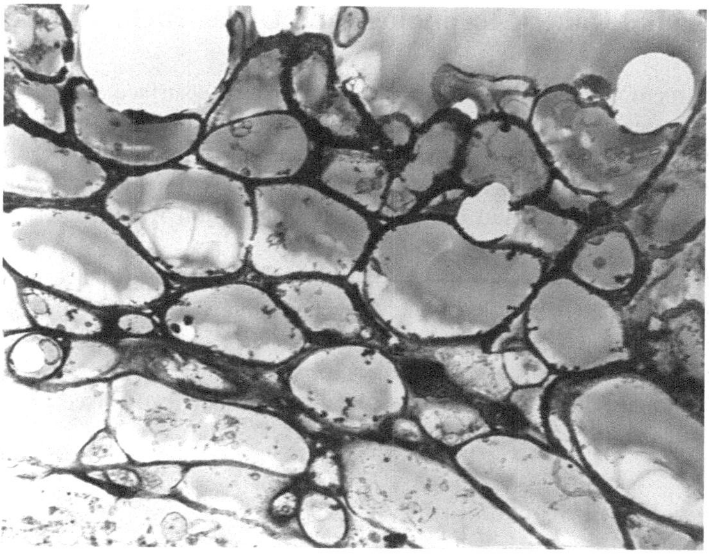

Fig. 2.16 Innermost layers of a peridium showing flattened cells of reduced intercellular spaces ad thin walls (TEM)

Fig. 2.17 Vertical section showing a peridium, thick at the base as well as at the ostiolar region (light microscopy)

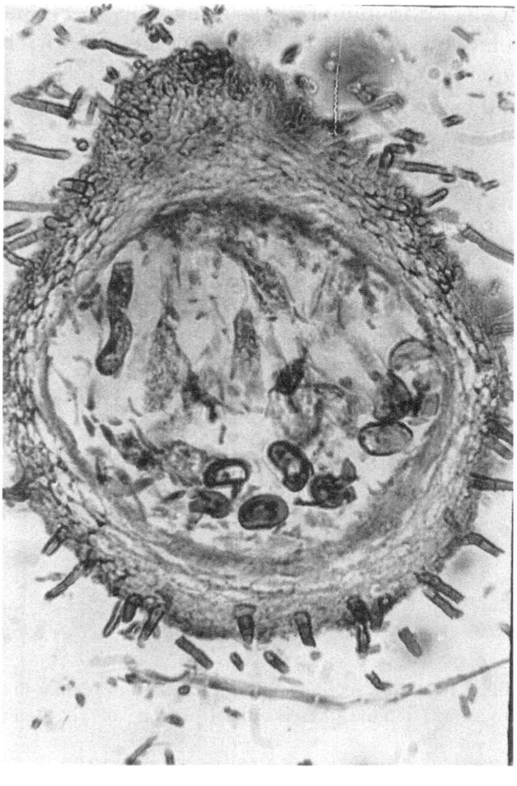

2.6 Peridial Hairs (Morphology and Ultrastructure)

Peridial hairs of *Chaetomiopsis dinae* originate from the surface and/or subsurface layers of the peridium (Fig. 2.18). They are dark-pigmented, long, dichotomously branched, mostly undulate to wavy to sinuous, and very frequently loosely coiled. Their surface texture ranges from verrucose to coarsely roughened to tuberculate. They are septate, brown, and thick at the base (4–5 μm wide), tapering toward the tip.

Wall ornamentations (tubercles) appear hemispherical when intact, turning into cup to funnel-shaped structures upon dehiscence (Figs. 2.19 and 2.20). These ornamentations are formed by the blebbing to the outside of a thin cuticle that cover the wall (Figs. 2.21 and 2.22). The cell wall comprises an inner, thin, less electron-dense zone and an outer, thick, electron-dense, heavily pigmented zone (see Fig. 2.19).

2.7 Development of Ostiolar Apparatus

The first sign of differentiation of the ostiolar apparatus starts by a prominent increase in the thickness of the peridium at the apical part of the ascoma due to meristematic activity. Continuous activity in the meristematic zone and simultaneous proliferation of resulting cells to the outside lead to the appearance of a relatively broad ostiolar region (Fig. 2.23). The opening of an ostiolar canal is soon initiated. At maturity, the ostiolar region becomes papillate to short-necked and usually surrounded by a crown of peridial hyphae (Fig. 2.24).

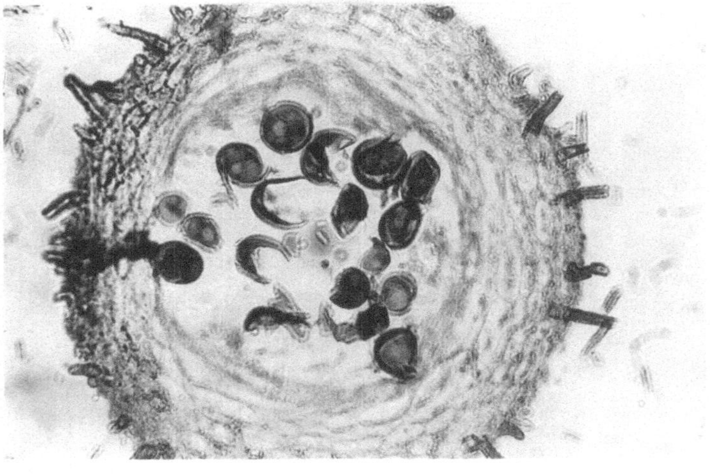

Fig. 2.18 Cross section showing peridium of regular thickness and dark-pigmented peridial hairs emerging from the surface and subsurface cells (light microscopy)

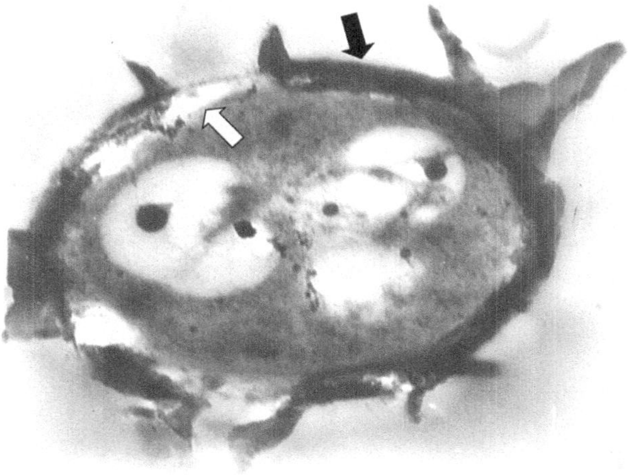

Fig. 2.19 TEM cross section showing in a peridial hair showing a cup to funnel-shaped structures (white arrow refers to the inner less electron-dense layer, while black arrow refers to the outer thick electron-dense layer)

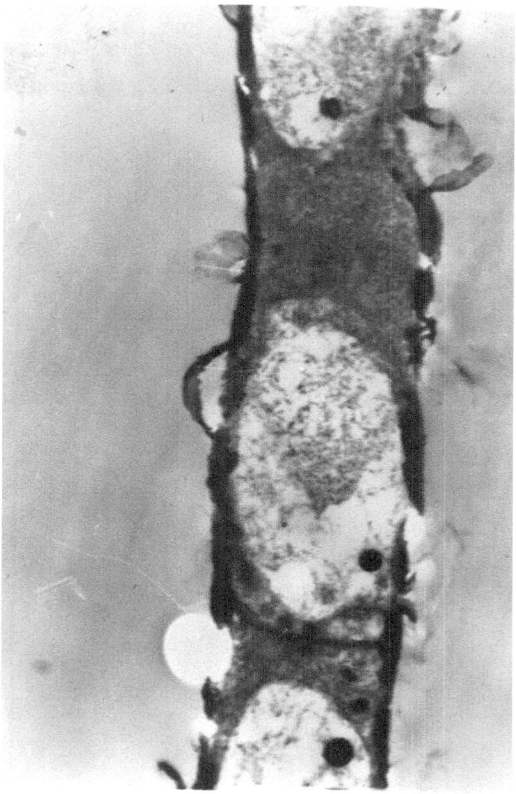

Fig. 2.20 L.S in peridial hair showing intact and ruptured tubercles (TEM)

Fig. 2.21 Peridial hair showing intact, hemispherical ornamentations of various sizes (SEM)

Fig. 2.22 Peridial hair showing dehisced ornamentations (SEM)

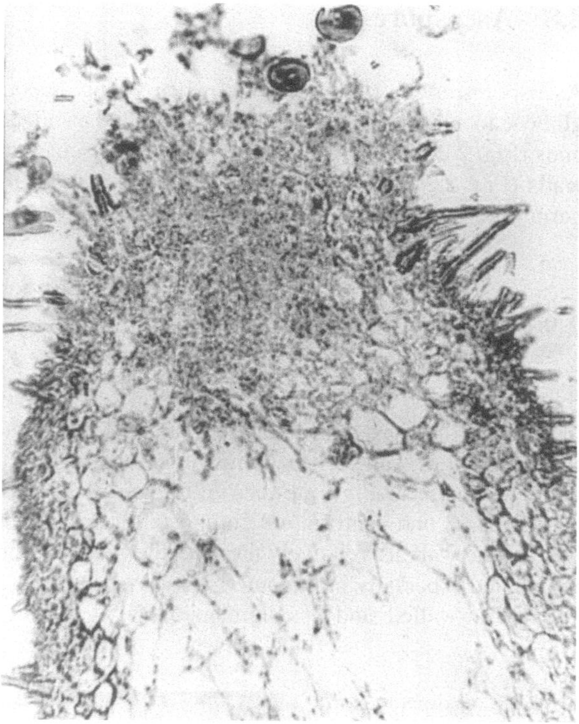

Fig. 2.23 Vertical section in an ascoma showing meristematic activity in the ostiolar region (light microscopy)

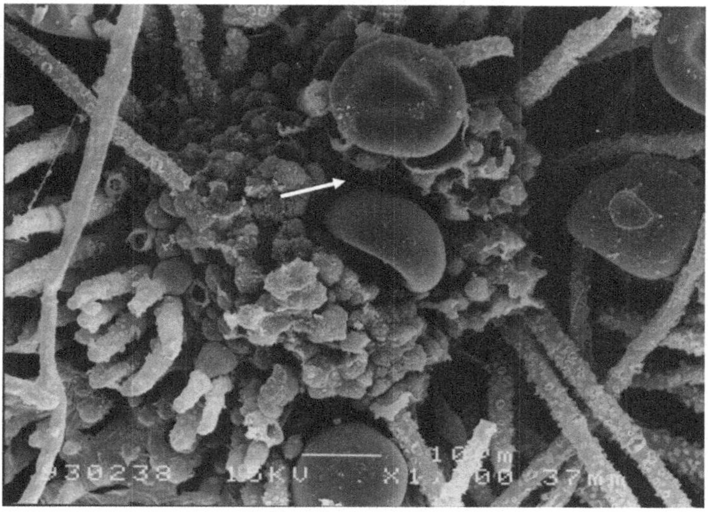

Fig. 2.24 Top view showing a crown of peridial hyphae surrounding the ostiolar region; the ostolar canal is referred to by an arrow (SEM)

2.8 Ascospores

Ascospores are dark olive colored when young, becoming dark brown at maturity, globose to subglobose or ellipsoidal to irregular, bilaterally flattened, with rounded ends (Fig. 2.25), 18–24 × 14–16 μm. They are characterized by prominently pitted walls (Fig. 2.26) and one to two (sometimes more than two) subapical to lateral germ pores, 1.5–2.0 μm wide (Figs. 2.27 and 2.28).

2.9 Expansion of the Ascoma

Expansion of the ascoma up to 60 μm (Fig. 2.29) is due to an increase in the number of cells of both wall and centrum along with some increase in their volume. Enlargement of wall cells become more evident after the ascoma reaches 60 μm in diameter and probably continues until maturity of the ascoma.

Both wall and centrum are composed of pseudoparenchymatic cells. However, cells of the wall are relatively narrow, thick-walled, flattened periclinally, and more pigmented especially in the outermost layers, while cells of the centrum tend to be wider, thin-walled, and less pigmented.

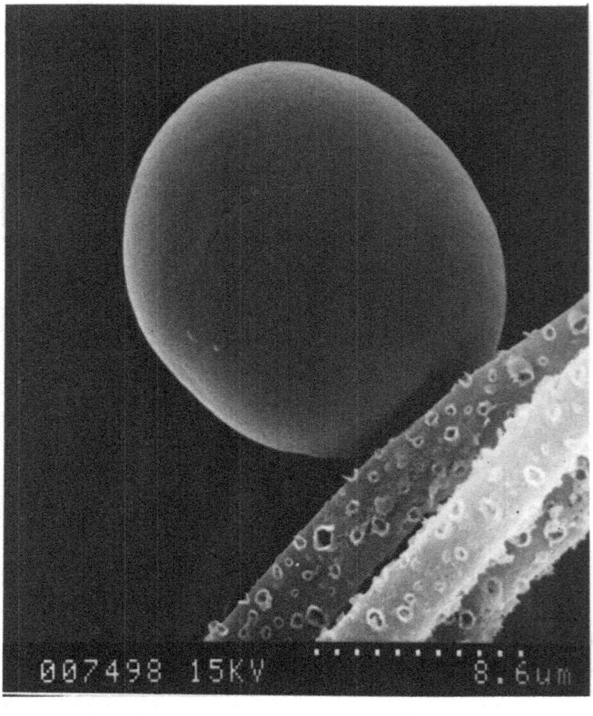

Fig. 2.25 Ascospore (SEM) showing rounded ends

Fig. 2.26 Ascospores of variable shapes showing pitted walls (light microscopy)

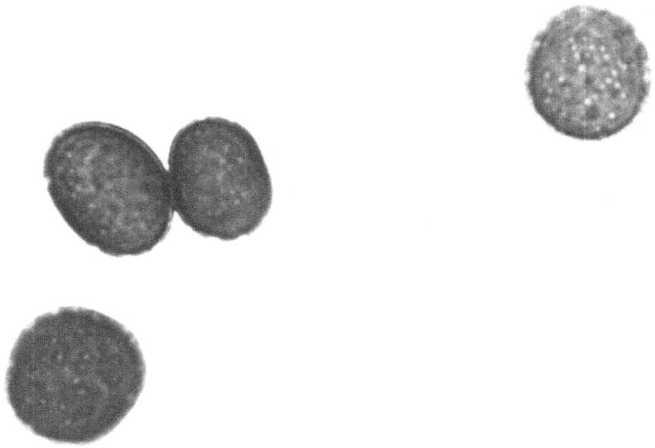

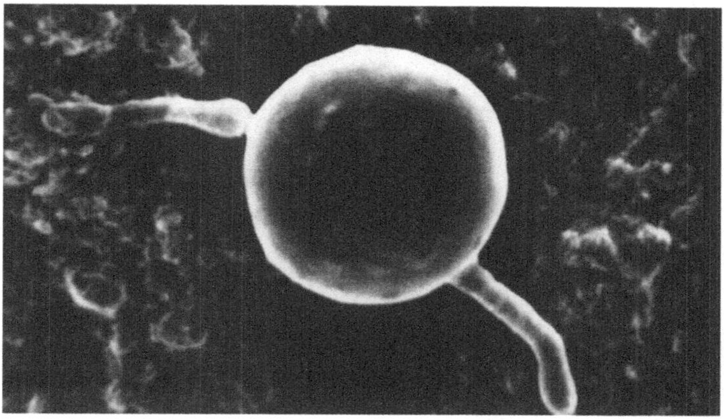

Fig. 2.27 Ascospore with two germ pores indicated by two germ tubes (SEM)

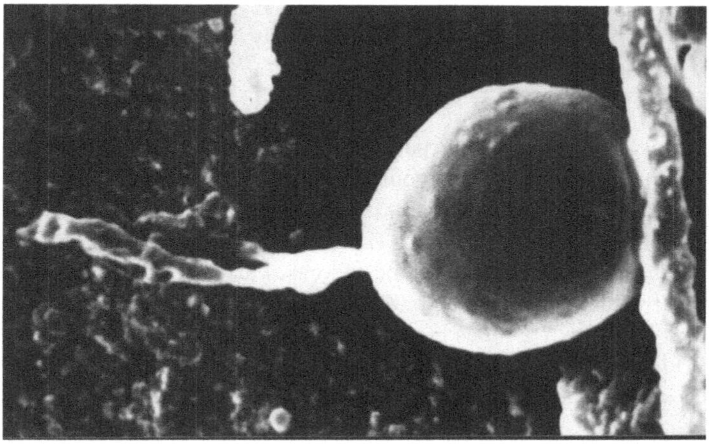

Fig. 2.28 Ascospore with single germ pores indicated by one germ tube (SEM)

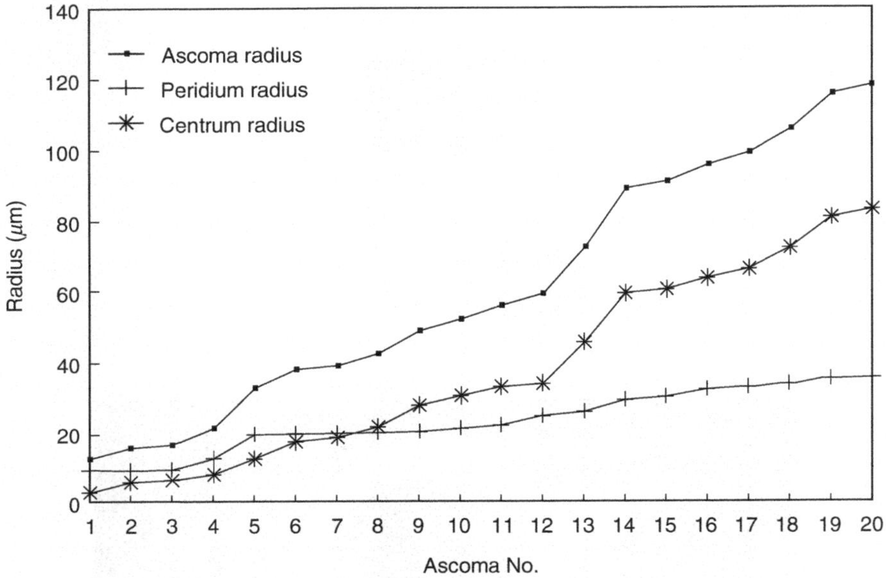

Fig. 2.29 Rate of expansion of ascoma elements in relation to time, where ascoma no: (1 to 5 = 2 days), (6 to 10 = 4 days), (11 to 20 = 6 days)

Data of Table 2.1 show clearly that the ascoma volume increases from sevenfolds to about tenfolds when the radius doubles. This is evident in the majority of cases, e.g., in ascomata numbers 8 and 4:

Volume of ascoma No. 8 = 327708.87 = 7.47-fold
Volume of ascoma No. 4 43823.95

Similarly, in ascomata 15 and 8:

Volume of ascoma No. 15 = 3126360.47 = 9.5-fold
Volume of ascoma No. 8 327708.87

Table 2.1 also indicates that the peridium volume constitutes a substantial part of the total volume of the ascoma. In young ascomata (i.e., 2 days old), it has an average ratio of 94.48%; in intermediate ascomata (i.e., 4 days old), it attains an average of 86.24%; while in mature ascomata (i.e., 6 days), it accounts for 71.65% of the ascoma volume. The peridium volume increases six- to eightfolds as the radius doubles (wall volume of ascoma No. 8 is 6.8 times greater than that of ascoma No. 4; in ascoma No. 15 is 7.2 times greater than that of ascoma No. 8).

The peridium continues to increase in thickness until it reaches a maximum of about 35.5 µm diameter forming a maximum number of eight layers of cells (see Fig. 2.14).

Table 2.1 Comparison of peridium and centrum volumes (μm^3) of 100 ascomata of representative sizes

Age (days)	Ascoma no.	Ascoma radius (μm)	Peridium radius (μm)	Centrum radius (μm)	Ascoma volume (A.V.)	Centrum volume (C. V.)	Peridium volume (P. V.)	P.V./A.V. (%)
2	1	13 ± 1.58	9.7 ± 1.2	3.3 ± 0.44	9532.26 ± 3395.2	157.43 ± 68.26	9374.8 ± 333,309	98.3
	2	16.1 ± 0.72	9.76 ± 0.72	6.34 ± 0.20	17486.04 ± 2054.3	1067.78 ± 47.15	18418.25 ± 2045.47	93.8
	3	17 ± 0.59	10 ± 0.77	7 ± 0.30	20585.47 ± 2793.2	1437.17 ± 154.98	19148.3 ± 2861.89	93
	4	21.8 ± 0.50	13.14 ± 0.44	8.72 ± 0.27	43823.95 ± 2949.8	2784.9 ± 260.74	41039.04 ± 2828.92	93.6
	5	33.16 ± 0.2	20.02 ± 0.29	13.14 ± 0.23	152796.21 ± 3375.5	9513.07 ± 505.21	143283.14 ± 3265.81	93.7
	Mean	20.21 ± 7.89	12.52 ± 4.4	7.7 ± 3.6	48844.74 ± 59494.93	2992.07 ± 3765.7	46252.71 ± 55478.53	94.48 ± 2.23
4	6	38.36 ± 0.24	20.32 ± 0.64	18.04 ± 0.38	236,629 ± 10416.2	25026.1 ± 1042.7	227751.33 ± 8468.04	96.2
	7	39.46 ± 0.43	20.24 ± 0.34	19.22 ± 0.45	257,539 ± 9530.58	29788.5 ± 2144.5	277751.33 ± 8468.04	88.4
	8	42.7 ± 0.50	20.6 ± 0.35	22.16 ± 0.24	327708.87 ± 12183.2	45608.57 ± 1492.53	282100.3 ± 10946.09	86.08
	9	49.08 ± 0.31	20.82 ± 0.49	28.26 ± 0.42	495418.05 ± 9682.04	94617.18 ± 4335.35	400800.87 ± 9997.97	80.90
	10	52.36 ± 0.39	21.56 ± 0.67	30.8 ± 0.49	601549.08 ± 13462.3	122499.45 ± 5830.99	479049.63 ± 15407.7	79.63
	Mean	44.44 ± 6.04	20.70 ± 0.52	23.6 ± 5.6	383,769 ± 158603.4	63506.25 ± 42969.4	333490.7 ± 103255.4	86.24 ± 1.75
6	11	55.49 ± 0.15	22.6 ± 0.52	33.34 ± 0.45	733481.3 ± 5954.65	155348.19 ± 6413.98	578133.13 ± 10157.32	78.8
	12	59.3 ± 0.433	25.14 ± 0.29	34.2 ± 0.29	875613.32 ± 19262.72	167636.33 ± 4299.89	707976.99 ± 16355.77	80.8
	13	72.3 ± 0.32	26.3 ± 0.54	45.9 ± 0.53	1582303.53 ± 2436.12	405317.1 ± 14108.95	1176986.35 ± 22358.93	74.3
	14	88.7 ± 0.41	29.5 ± 0.31	59.22 ± 0.37	2920542.53 ± 8000.3	870281.82 ± 16279.96	2050260.7 ± 25839.07	70.20
	15	90.70 ± 0.15	30.42 ± 0.29	60.28 ± 0.23	3126360.47 ± 16350.09	917825.58 ± 13886.12	2208534.88 ± 18508.93	70.64
	16	95.7 ± 0.16	32.32 ± 0.03	63.38 ± 0.36	3665454.98 ± 31220.9	1066853.90 ± 18309.89	2598601.07 ± 42036.72	70.8
	17	99 ± 1.58	33 ± 1.77	66 ± 0.22	4068041.67 ± 194830.47	1204641.42 ± 12243.7	2863400.24 ± 205360.64	70.3
	18	105.6 ± 0.0.42	33.8 ± 0.26	71.74 ± 0.28	4928654.15 ± 58204.01	1547087.69 ± 18589.47	3381566.45 ± 45470.39	68.6
	19	115.5 ± 0.58	35.1 ± 0.68	80.34 ± 0.51	6459693.12 ± 97408.85	2172965.14 ± 441943.53	4286727.97 ± 105504.81	66.3
	20	118.04 ± 1.1	35.5 ± 1.07	82.54 ± 0.40	6892942.23 ± 207142.14	4536632.76 ± 34472.40	4536632.76 ± 195132.05	65.8
	Mean	90.07 ± 21.6	30.36 ± 4.4	59.7 ± 17.2	3525308.73 ± 214347.85	1086426.67 ± 765206.75	2438882.05 ± 1380996.4	71.654 ± 3.4

2.10 Discussion

It is well known that the development of ascomata in the genus *Chaetomium* is variable. Two main patterns have been introduced by Whiteside (1957): one of them is designated as *Chaetomium globosum* type and the other as *Chaetomium brasiliense* type. Some intermediate patterns, however, have also been illustrated by Seth (1969).

The initials described here for *Chaetomiopsis dinae* agree to some extent with the developmental pattern characteristic of *Chaetomium globosum* type. In *Chaetomiopsis*, a sessile or slightly stalked ascogonium coils about itself in an irregular fashion. This type of perithecial initiation has also been reported in many other *Chaetomium* species such as *C. trigonosporum* (Corlett 1966), *C. erraticum* (Cooke 1969a), and *C. repens* (Figueras and Guarro 1988b). No resemblance exists between the initials of *Chaetomiopsis* and those of *C. brasiliense* type (described by Whiteside 1957) where a stalked and spirally coiled ascogonium are the most characteristic features. Nevertheless, the formation of mycelial hairs associated with the development of ascomata of *C. brasiliense* type (Whiteside 1957) have also been noticed in our taxon. Therefore, *Chaetomiopsis* from the developmental point of view represents an intermediate state between *C. globosum* type and *C. brasiliense* type. This observation has also been reported by several authors such as Seth (1969), Cooke (1972), and Figueras and Guarro (1988a, b) whom all admitted that mycelial hair production is likely a species character rather than a concrete pattern of ascogonial development.

Ascoma expansion in the genus *Chaetomium* have been interpreted, by some authors, as a multiplication and enlargement of hyphal cells that surround the ascogonium (Whiteside 1961; Corlett 1966; Cooke 1969a, b, 1970, 1972). This growth region is first localized all over the surface as well as in the central part of the ascoma but soon becoming restricted to the apical surface as well as the central part (Froeyen 1980).

In Ascomycetes, asci are usually formed as a result of plasmogamy. Asci in some members either remain singly or aggregate in clusters especially when produced by ascogenous hyphae. In the majority of Ascomycetes, however, asci are borne in ascocarps where they are protected by uni- or multiseriate peridia.

According to Talbot (1971), the ascocarp tissues may be differentiated around the sex organs after plasmogamy or the ascocarp tissues develop first, forming a stroma within which the sex organs appear later. All the tissues and structures enclosed by the peridium are collectively known as the centrum of the ascocarp; thus, the centrum includes sex organs, ascogenous hyphae, asci, and sterile tissues.

Regarding development of the central cavity within the ascoma of the genus *Chaetomium*, there are two trends, namely, either before the differentiation of asci as reported by Corlett (1966) for *C. trigonosporum* and Figueras and Guarro (1988b) for *C. repens* or after the differentiation of asci as reported for *C. globosum* (Whiteside 1961), *C. elatum* (Moreau and Moreau 1954), and *C. erraticum* (Cooke 1969a). Development of the central cavity in the new taxon (*Chaetomiopsis dinae*) belongs to the former trend, i.e., before the differentiation of asci.

The surface view of peridium (very frequently referred to as textura) has been accepted as a useful key character to discriminate between ascomata.

Textura in the family Chaetomiaceae as well as many other Pyrenomycetes have been extensively studied by Korf (1973), von Arx et al. (1986), Cannon (1986), and Guarro and Figueras (1989). They noticed that though textura is a variable character, it generally tends to be a species rather than a genus character. In the genus *Chaetomium*, for instance, textura ranges in the majority of species between intricata and epidermoidea, while other textures like angularis, prismatica, and globulosa might be present but in rare cases. In some genera like *Subramaniula*, *Achaetomium*, *Achaetomiella*, and *Thielavia*, more than one type of textura has been observed (von Arx 1975; Cannon 1986).

The ostiolar region of *Chaetomiopsis dinae* is very related to some *Chaetomium* species. According to the developmental patterns introduced by Guarro and Figueras (1989), our taxon belongs to the *Chaetomium globosum* type where the meristematic activity in the ostiolar region results in the development of an erumpent structure intermediate between papilla and short-neck. The development of ostiolar canal in *Chaetomium* and related Ascomycetes has been interpreted as usually due to a schizogenous process in the apical region resulting from internal pressure expressed by developing asci (Corlett 1966).

The morphology and ultrastructure of peridial hairs, as a taxonomic character in the family Chaetomiaceae, has been considered by Hawksworth and Wells (1973) and Ellis (1981b).

On the morphological basis of hairs, most authors divided species of the genus *Chaetomium* into six to ten groups (Skolko and Groves 1953; Ames 1963; Seth 1968a, b; Millner 1975), while on ultrastructural basis ten categories of ornamentations were noticed (Hawksworth and Wells 1973).

The formation of hair ornamentation (= excrescences) occurs in the same way as described by Guarro and Figueras (1983, 1985). These ornamentations develop as an outside blebbing of the outer layer, dragging with it the deposited dark pigment of the wall. Figueras and Guarro (1988b) designated this outer layer as a cuticle.

In the light of peridial hair morphology and ornamentation, *Chaetomiopsis dinae* is very related to *Chaetomium repens* (Figueras and Guarro 1988b) except for dichotomous branching.

Ascospores are relatively large (18–24 μm long), with prominently pitted walls, irregular in shape, with rounded surfaces (usually referred to as non-apiculate), bilaterally flattened, showing two distinct, subapical to lateral germ pores. Like all *Chaetomium* species, ascospore discharge in *C. dinae* is passive, i.e., oozing out in the form of sticky mass or cirrhus.

It should be noted finally that ascospore ornamentation represents the main element that supports the erection of the new taxon *Chaetomiopsis* as a genus. On casual examination, the new taxon may be confused with some *Chaetomium* species especially those showing related non-apiculate ascospore type. However, wall pitting might be considered as a limiting character (Moustafa and Abdul-Wahid 1990). *Chaetomium* species showing morphologically related ascospores (i.e., non-apiculate) are *C. madrasense*, *C. megalocarpum*, *C. nozdrenkoae*, and

C. varisporum. On one hand, *C. madrasense* produces relatively small ascospores (9–11 × 7–9 µm) with single germ pore, while the other three species produce also smaller ascospores never exceeding 16 µm in length, with double germ pores (von Arx et al. 1986).

References

Abdel-Azeem AM (1998) Cytogenetical and biological studies on the fungus *Chaetomiopsis*. M.Sc. thesis, Faculty of Science, Suez Canal University, Egypt

Abdel-Azeem AM (2003) Ecological and taxonomical studies on ascospore-producing fungi in Egypt. Ph.D. thesis, Faculty of Science, Suez Canal University, Egypt

Abdel-Azeem AM, Blanchette RA, Held BW (2018) New record of *Chaetomium grande* Asgari & Zare (Chaetomiaceae) for the Egyptian and African mycobiota. Phytotaxa 343(3):283–288

Ainsworth GC (1973) Introduction and keys to higher taxa. In: Ainsworth GC, Sparrow FK, Sussman AS (eds) The fungi. An advanced treatise, IVB. Academic Press, New York

Ames LM (1963) A monograph of the Chaetomiaceae. US Army Res Dev Serv 2:1–125

Cannon PF (1986) A revision of *Achaetomium*, *Achaetomiella* and *Subramaniula*, and some similar species of *Chaetomium*. Trans Br Mycol Soc 87:45–76

Cooke JC (1969a) Morphology of *Chaetomium erraticum*. Am J Bot 56:335–340

Cooke JC (1969b) Morphology of *Chaetomium funicolum*. Mycologia 61:1060–1065

Cooke JC (1970) Morphology of *Chaetomium trilaterale*. Mycologia 62:282–288

Cooke JC (1972) Perithecium development of *Chaetomium longirostre*. Can J Bot 50:2271–2274

Corlett M (1966) Perithecium development in *Chaetomium trigonosporum*. Can J Bot 44:155–162

Ellis DH (1981a) Ultrastructure of thermophilic fungi. III. Ascocarp morphology of *Chaetomium thermophile* and *C. gracile*. Trans Br Mycol Soc 77:165–177

Ellis DH (1981b) Ascocarp morphology and terminal hair ornamentation in thermophilic *Chaetomium* species. Mycologia 73:755–773

Ellis DH (1982) Ultrastructure of thermophilic fungi. IV. General discussion. Trans Br Mycol Soc 78:305–313

Figueras MJ, Guarro J (1988a) A scanning electron microscopic study of ascoma development in *Chaetomium melaysiense*. Mycologia 80:298–306

Figueras MJ, Guarro J (1988b) Developmental morphology and fine structure of the ascoma of *Chaetomium repens*. Nova Hedwigia 47:469–495

Froeyen OJ (1980) Perithecium growth and expansion in *Chaetomium globosum*. Can J Bot 58:375–383

Guarro J, Calvo MA (1982) Ornamentation de los pelos terminales de *Chaetomium montblancense* Guarro, Calvo y Ramirez y *Chaetomium teratoideum* Ames var. *minor*. Guarro, Calvo y Ramirez. Bot Soc Mycol Cast 7:41–48

Guarro J, Figueras MJ (1983) Estudios sobre la ultrastructura de hongos microscopicos del suelo. I. Los ascocarpos de *Corynascella inaequalis*. Microbiol Espan 36:93–102

Guarro J, Figueras MJ (1985) Studies on the ultrastructure of soil fungi. II. The peridium of *Chaetomium irregulare*. Int J Mycol Lichenol 2:85–94

Guarro J, Figueras MJ (1989) Morphological studies on *Chaetomium*: the peridium, the ostiolar apparatus and some considerations about the centrum. Cryptogam Bot 1:97–140

Harris JL, Howe HB, Roth IL (1975) Scanning electron microscopy of surface and internal features of developing perithecia of *Neurospora crassa*. J Bacteriol 122:1239–1246

Hawksworth DL, Wells H (1973) Ornamentation on the terminal hairs in Chaetomium Kunze ex Fr. and some allied genera. Mycol Pap 134:1–24

Hock B, Bahr M, Walk RA, Nitschke U (1978) The control of fruiting body formation in the ascomycete *Sordaria macrospora* Auersw. by regulation of hyphal development. An analysis based on scanning electron and light microscopic observations. Planta 141:93–103

Huang LH (1976) Development morphology of *Triangularia backusii* (Sordariaceae). Can J Bot 54:250–267
Huang LH, Luttrell ES (1982) Development of the perithecium in *Gnomonia comari* (Diaporthaceae). Am J Bot 69:421–431
Jensen JD (1981) The developmental morphology of *Hypoxylon serpens* in culture. Can J Bot 59:40–49
Jensen JD (1982) The development of *Gelasinospora reticulospora*. Mycologia 74:724–737
Johansen DA (1940) Plant microtechnique. McGraw-Hill Book Co., Inc., New York
Korf RP (1973) Discomycetes and tuberales. In: Ainsworth GC, Sparrow FK, Sussman AS (eds) The fungi. An advanced treatise, vol IVA. Academic Press, New York, pp 249–319
Luttrell ES (1951) Taxonomy of the pyrenomycetes. Univ Mo Stud Sci Ser 24:1–120
Mai SH (1976) Morphological studies in *Podospora anserina*. Am J Bot 63:821–825
Mai SH (1977) Morphological studies in *Sordaria fimicola* and *Gelasinospora longispora*. Am J Bot 64:489–495
Millner PD (1975) Ascomycetes of Pakistan: *Chaetomium*. Biologia (Lahore) 21:39–73
Millner PD, Motta JJ, Lentz PL (1977) Ascospores, germ pores, ultrastructure, and thermophilism of *Chaetomium*. Mycologia 69:720–733
Moreau F, Moreau V (1954) Sur Le developpment des peritheces du *Chaetomium elatum* Kunze. Rev Mycol 19:167–171
Moustafa AF, Abdul-Wahid OA (1990) *Chaetomiopsis*, a new perithecial ascomycete genus from Egyptian soils. Mycologia 82:129–131
Parguey-Leduc A, Janex-Favre MC (1981) The ascocarps of Ascohymenial Pyrenomycetes. In: Reynolds DR (ed) Ascomycete systematic: the Luttrellian concept. Springer-Verlag, New York, pp 102–123
Read ND (1983) A scanning electron microscopic study of the external features of perithecium development in *Sordaria humana*. Can J Bot 61:3217–3229
Read ND, Beckett A (1985) The anatomy of the mature perithecium in *Sordaria humana* and its significance for fungal multicellular development. Can J Bot 63:281–296
Rosing WC (1982) Ultrastructure of ascus and ascospore development in *Chaetomium brasiliense*. Mycologia 74:960–974
Sanni MO (1982) Perithecium development in *Gelasinospora tetrasperma*. Mycologia 74:320–324
Sanni MO (1983) Perithecial development of *Gelasinospora longispora*. Bot Gaz (Chicago) 143:542–545
Seale T (1973) Life cycle of *Neurospora crassa* viewed by scanning electron microscopy. J Bacteriol 113:1015–1025
Seth HK (1968a) Studies on the genus *Chaetomium* II. Taxonomy and importance of terminal hairs. Nova Hedwigia 15:345–359
Seth HK (1968b) Studies on the genus *Chaetomium* III. Ascospores. Nova Hedwigia 16:520–537
Seth HK (1969) Studies on the genus *Chaetomium*. V. Morphological studies. Ceska Mykol 23:129–134
Sheehan DC, Hrapchak BB (1980) Theory and practice of histotechnology, 2nd edn. The C.V. Mosby Co., St. Louis/Toronto/London
Skolko AJ, Groves JW (1953) Notes on seed-borne fungi VII. *Chaetomium*. Can J Bot 31:779–809
Talbot PHB (1971) Principles of fungal taxonomy. The Macmillan Press Ltd, London
von Arx JA (1975) On *Thielavia* and some similar genera of Ascomycetes. Stud Mycol 8:1–29
von Arx JA, Guarro J, Figueras MJ (1986) The ascomycete genus *Chaetomium*. Nova Hedwig Beih 84:1–62
Webster J, Weber R (2007) Introduction to fungi, 3rd edn. Cambridge University Press, Cambridge
Whiteside WC (1957) Perithecial initials of *Chaetomium*. Mycologia 49:420–425
Whiteside WC (1961) Morphological studies in Chaetomiaceae. I. Mycologia 53:512–523

Chapter 3
Molecular Approaches for Analyzing Environmental *Chaetomium* Diversity and Exploitation of *Chaetomium thermophilum* for Biochemical Analyses

Amr A. Elkelish and Ahmed M. Abdel-Azeem

3.1 Introduction

Fungi show the greatest eukaryotic diversity on the planet with their conservatively estimated 1.5 million species, and they are one of the primary decomposers in the ecosystem (Capote et al. 2012). Fungi behave as both friends and foes. Many species of fungi are beneficial to human, vegetal, animal, and environmental health because of its economic, medical, and commercial uses. However, many of them are harmful because of its ability to act as pathogen and cause diseases.

Basic methods used to detect the organism mostly rely on microscopic, cultural, and morphological approaches that require extensive time, labor, and classical taxonomy knowledge (Nilsson et al. 2011). These approaches, although the cornerstone of fungal diagnostics, can lead to the unreliable results due to problems in identification. Additionally, experts and specialist skilled in fungal identification are required for effective results (Chalupová et al. 2014).

Due to the conventional method limitations, molecular techniques came in use for the investigation of identification and classification problems. A high variety of molecular methods are increasingly becoming valuable tools in all aspects of fungal diagnostics. These techniques include immunological methods, nucleic acid-based probe technology, and polymerase chain reaction (PCR) technology. The aforementioned methods (microscopic, cultural, and morphological approaches) based on phenotypic characters, while the latter methods (immunological methods, nucleic acid-based probe technology, and polymerase chain reaction (PCR) are based on genotypic characters, which give fast, highly specific, effective, and potentially more accurate results. In contrast to the basic methods, isolation of organism does not require culturing (Badali and Nabili 2012; Spring and Thines 2010).

A. A. Elkelish (✉) · A. M. Abdel-Azeem
Botany Department, Faculty of Science, Suez Canal University, Ismailia, Egypt
e-mail: amr.elkelish@science.suez.edu.eg; ahmed_abdelazeem@science.suez.edu.eg

A. M. Abdel-Azeem (ed.), *Recent Developments on Genus Chaetomium*,
Fungal Biology, https://doi.org/10.1007/978-3-030-31612-9_3

When the morphological characters are not visible, enzyme-linked immunosorbent assay, an immunological method, was applied for the quantification and detection of fungi. These immunological methods rely on antigen–antibody reaction coupled with a fluorescent dye or enzyme and have become available for simple and rapid results (Mostafa et al. 2012; Peruski and Peruski 2003). Although the production of antibodies is generally successful for viruses, it is less well for complex organisms such as fungi. Additionally, it is difficult and expensive to produce highly specific antibodies. These methods working reliably only at the genus level is another limitation. Specificity and accuracy is most important in diagnostics applications and this may restrict the use of immunological methods (Borman et al. 2008). Nucleic acid-based methods allow the determination of closely related species and detect the minute quantity pathogen when no visible sign is present. In molecular methods, DNA/RNA probe technology includes southern hybridization, in situ hybridization, fluorescence in situ hybridization (FISH), microarray, and macroarray. Isothermal amplification technology includes loop-mediated isothermal amplification (LAMP), rolling circle amplification (RCA), and nucleic acid sequence-based amplification (NASBA). PCR technology includes multiplex PCR, nested PCR, real-time (q)PCR, and reverse transcriptase (RT)-PCR; DNA barcoding has been recently used. Each technique has its own advantages and limitations as shown in Table 3.1 and Fig. 3.1.

3.2 Molecular Methods for the Detection of Fungi

3.2.1 PCR-Based Methods

3.2.1.1 Conventional PCR

The invention of PCR brought a great revolution in the plant pathology field. This technique allows synthesizing of the specific part of DNA in million copies through alternate cycles of denaturation, annealing, and elongation by using specific primers. Initially, PCR was highly specific for the detection of diseases caused by bacteria and viruses. Now, it is widely used for the plant pathogen detection as well. PCR depends on the efficacy of DNA extraction and the concentration of deoxynucleoside triphosphate. Sometimes, the performance is affected by inhibitors present in sample assay (Fang and Ramasamy 2015). To overcome this issue in plants, the cetyltrimethylammonium bromide method is widely used with the addition of particular chemical and enzymatic treatments. Several other methods are present to reduce the effect of PCR inhibitors. In phenol–chloroform extraction method, instead of ethanol precipitation, silica matrix purification can be used to recover the DNA (Mancini et al. 2016). Moreover, for the detection of pathogens, PCR technology requires the designing of a primer to initiate the DNA replication process, which could limit the practical applicability of the technique for field sampling of disease.

Table 3.1 Molecular diagnostics techniques for the identification of fungi

Methods	Advantages	Disadvantages
Conventional PCR	Gives rapid and precise results, when using the primer of specific species.	Requires much labor and cost.
Nested PCR	Use of two sets of primer increases the yield and specificity of amplification of the target DNA.	Risk of contamination because of two cycles of amplification.
Multiplex PCR	Time and money saver by using the several pairs of primers in the same reaction.	Interference of primers and probes, reduce sensitivity.
Reverse transcriptase (RT)-PCR	Gives quantitative data about pathogens, more sensitive than conventional PCR.	Formation of each assay is time consuming and requires the expensive equipment and the reagents.
Real-time PCR (qPCR)	Automated and no need of post amplification analysis.	Cost and complexity due to simultaneous thermal cycling and fluorescence detection.
In situ hybridization	Maximum use of the short supply tissue.	Difficulty in identifying targets that have low DNA and RNA copies.
FISH	Can be used for nondividing cells.	Probe-preparing method is very difficult because it is necessary to tailor the probes to identify the particular sequences of DNA.
Microarray	Easy to use because it does not require the large-scale DNA sequencing.	Large amount of mRNA is required.
LAMP	Rapid, sensitive, and highly specific.	Primer design is complex; recognizes only one specific pathogen; risk of sample contamination.
NASBA	Expensive equipment is not needed. Better than RT-PCR.	Specificity of the reactions is dependent on thermolabile enzymes. Reaction temperature cannot be exceeded than 42 °C without compromising it.
RNA interference (RNAi)	Ability to simultaneously interrogate thousands of genes.	Variability and incompleteness of knockdowns and the potential nonspecificity of reagents.
Northern blotting	Detection of RNA size.	Applied only on a small sample of the genes.
SAGE	Prior knowledge of the subject's genome is not required.	Specificity of tag sequence.
RNA-Seq	Increased specificity and sensitivity.	Needs expensive equipment. Bioinformatics knowledge required for data analysis.

3.2.1.2 Nested PCR

Nested PCR is used to achieve a high degree of specificity and sensitivity. For example, nested PCR sensitivity is 1000 times greater than single PCR for fungal identification (Yeo and Wong 2002). In this method, two consecutive rounds occur in which a single pair of primer is used to amplify a large region of DNA, then this

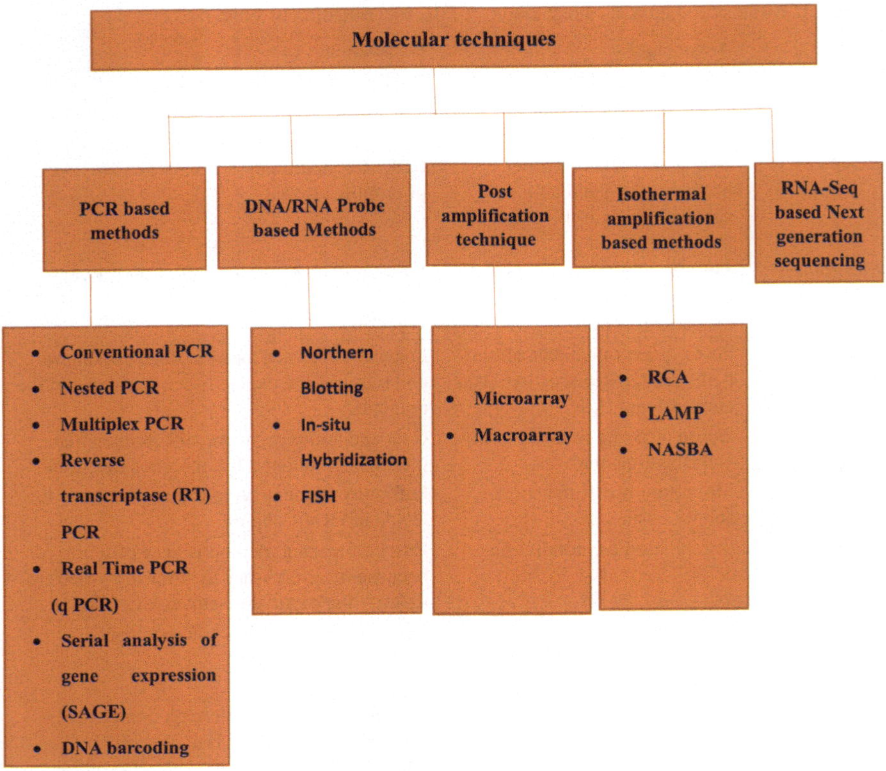

Fig. 3.1 An overview of molecular technique (Aslam et al. 2017)

amplified sequence of DNA acts as a target for the second round by using two internal primers. In this type of PCR, a considerable risk of contamination is measured because two cycles of amplification are to be performed in separate tubes. So, likely chances of false-positive results due to contamination and intense labor are the major drawbacks of this technique (Rahman et al. 2013). To overcome these limitations, laboratories must follow some strict precautions as to use the separate equipment and place for each PCR cycle (Trtkova and Raclavsky 2006).

3.2.1.3 Multiplex PCR

Multiplex PCR can be proved as time and money saver by using the several pairs of primers in the same reaction that allows the simultaneous detection of different targeted sequences of DNA. The method has so much importance in fungal ecology when taxa are mixed. Different fragments that are specific to target species were simultaneously amplified and detected on the basis of their molecular sizes on the agarose gels. DNA-synthesizing accuracy is strongly affected by amplicon size. To avoid this pitfall, primers must be designed carefully along with their relative concentration and annealing temperature (Dasmahapatra and Mallet 2006).

3.2.1.4 Reverse Transcriptase PCR

Since the discovery of standard PCR, a large number of changes have been developed in its procedure. Some of these changes have expanded the utility and diagnostics capability of PCR in many biological and medical fields (Tang et al. 1997). An important limitation of other PCR types is their inability to differentiate between dead and living fungi. This limitation is overcome by the discovery of RT-PCR. Actually, mRNA in dead cells degrades, so the detection of mRNA by RT-PCR can be done to check the cell viability (Capote et al. 2012). In this process, firstly, the RNA is reversely transcribed into cDNA by random primers and RT enzyme and then amplified by any PCR-based method. Therefore, RT-PCR is mostly used to detect and diagnose the RNA-containing viruses (such as retroviruses) infections. Diagnosis of RNA-containing viruses can be helpful in developing or checking the effectiveness of antimicrobial vaccines or therapy. RT-PCR has been used in the detection and quantification of a biocontrol agent, *Chaetomium globosum* (Aggarwal et al. 2014).

3.2.1.5 Real-Time PCR (Q PCR)

Conventional PCR methods give us only amplification of targeted DNA that is not enough for rapid taxa identification. In the past few years, a new technique, RT-PCR, has been introduced. This technique is an improved form of conventional PCR in which the DNA can be quantified along with the amplification (Mackay 2004). Monitoring of the reactions during amplification steps has been made possible by the use of fluorescent dyes such as SYBR Green I or sequence-specific fluorescent-labeled probes as the TaqMan probe (Badali and Nabili 2012). When the fluorescent dye intercalates to DNA, fluorescent signal is generated. This signal increases as the amount of targeted DNA increases after each cycle of amplification (McCartney et al. 2003). It is less costly to use the fluorescent dye as a monitoring agent but its nonspecific nature is its limitation. Actually, the binding of intercalating dye to all present DNA can produce the false results in the form of primer dimer. Then fluorogenic probes came into use due to their high specificity (Atkins and Clark 2004; Bu et al. 2005). These probes are connected with two types of fluorescent dyes: one is the reporter dye that attaches to the 5′ end and another is the quencher dye on 3′ end. Close proximity of reporter and quenching dyes prevents the emission of fluorescence. Due to the exonuclease activity of Taq polymerase, the reporter dye becomes separate from the quenching dye and starts to fluorescence (Dasmahapatra and Mallet 2006). qPCR has been used for the identification and quantification of environmental fungi, e.g., *Chaetomium* taxa (Liu et al. 2012).

3.2.1.6 Serial Analysis of Gene Expression

Serial analysis of gene expression (SAGE) is a comprehensive and sequence-based method for the quantitative gene expression profiling which allows the identification of multiple transcripts simultaneously. The method based on sequencing and quantification of the 15 bp or the longer oligonucleotides and similarity of sequences against the available genome sequences to find the corresponding expressed genes (Velculescu et al. 1995). This method uses two samples, which are ligated and labeled with separate primers and then amplified. Then, primers are removed, providing sticky ends that form the concatemers. They are cloned into vector and sequenced trailed by the wide computational analysis. SAGE has some drawbacks. First, it needs mRNA in a large quantity. Second, sometimes 15 bp tag is not enough to specifically identify the gene of origins with the more complex genomes. SAGE is used to identify environmental samples such as *C. thermophilum* (Cheng et al. 2013).

3.2.1.7 DNA Barcoding

DNA barcoding is a molecular diagnostic technique in which a small segment of DNA is used to identify the species of all domains of eukaryotic life. Barcode regions are universally found in target lineages. It exhibits adequate DNA sequence variation to distinguish different species. In DNA barcoding, standardized sequences of 500–800 base pairs are used to identify species with the markers valid for the wide range of taxonomic group (Krishnamurthy and Francis 2012). For fungi, this PCR-based method is helpful not only in the recognition of the cultured species but also in the identification of species for uncultured taxa from natural environment. Identification of species by using the barcode depends on the number of representative present in the database. By DNA barcoding, fungi can be identified in life-cycle stages not suitable for morphological identification. The efficacy of barcoding relies on the assumption that genetic variations within a species are much smaller than the variations between species. Fungi that are not suited for morphological identification and discovery can be identified and discovered at molecular level by DNA barcoding. In this technique, effective marker is very helpful to identify the poorly understood fungal species diversity in the natural environment (Roe et al. 2010). Full working of DNA barcoding technique is shown in Fig. 3.2 (Aslam et al. 2017).

There are several scientific advantages of DNA barcoding, which include (i) facilitating the identification of species at any phase of life, (ii) enabling the discovery of species based on the phylogenetic analysis of nucleic acid sequences, (iii) providing insight for diversification of life, and (iv) promoting the development of DNA sequencing tools, useful in the field of biodiversity (Savolainen et al. 2005). Reliable barcodes provide a standardized and realistic species identification tool for the evaluation of biodiversity and environmental studies. Selection of barcode region is compromised among the prospects to design ideal and universal DNA markers for the PCR amplification. Mitochondrial cytochrome c oxidase (COX I)

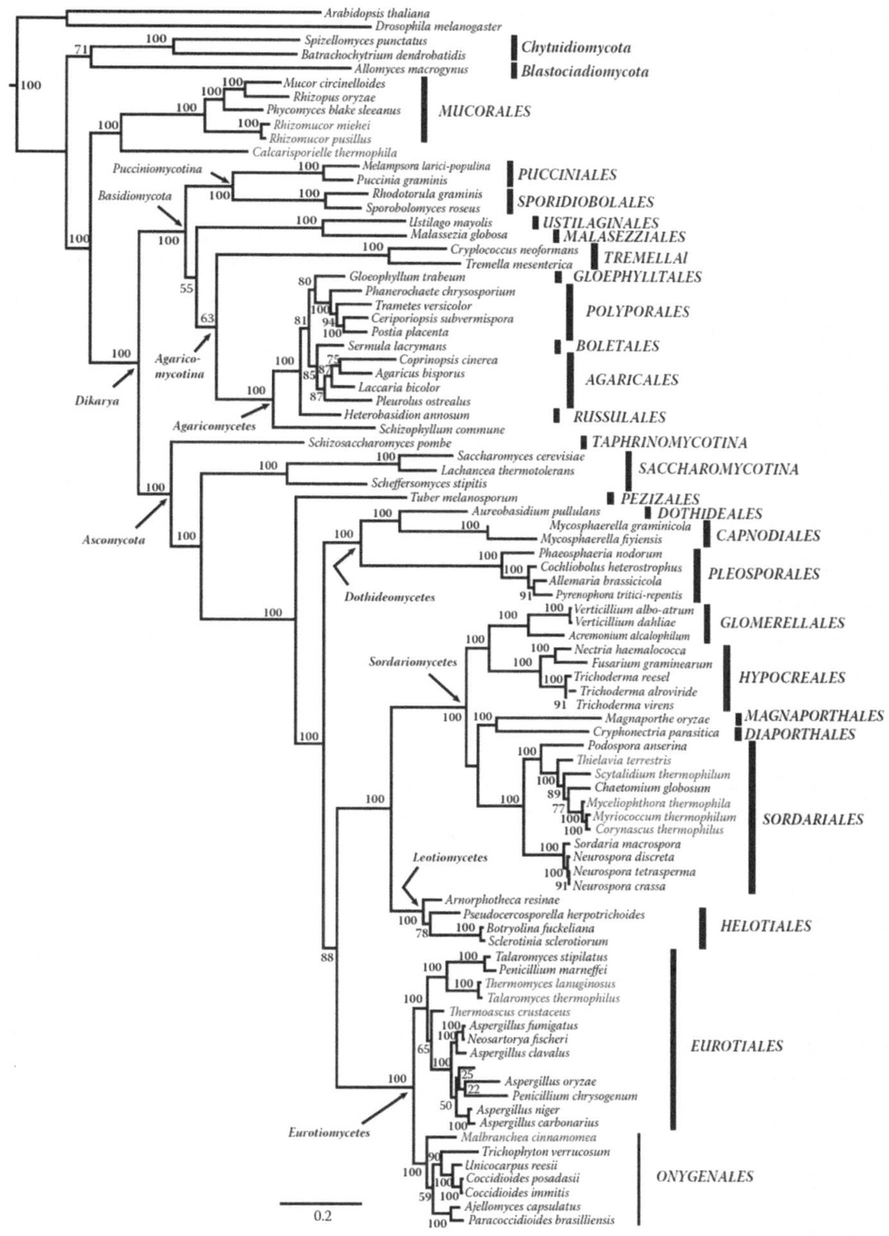

Fig. 3.2 Maximum likelihood phylogeny inferred by amino acid sequence alignments. The amino acid sequences of Mcm7, Rpb1, and Rpb2 were used for concatenated analysis. Bootstrap values are indicated at each node and are given in bold font for values above 70. Thermophilic taxa are highlighted in red font. The tree is rooted with *Arabidopsis thaliana* and *Drosophila melanogaster*. (After Morgenstern et al. 2012)

region used for animal barcoding was excluded as an ideal marker for fungi because it cannot amplify fungal DNA due to the presence of large introns. Molecular systematics of fungi are based greatly on the analysis of nuclear ribosomal RNA (rRNA) cistron, consists of small (18S) and large (28S) subunits (Hebert and Gregory 2005). Internal transcribed spacer (ITS) region can be amplified from many fungal taxa by using a limited set of primers (Schoch et al. 2012). Moreover, it has found that differences among species are often higher than those within species. So, ITS is used as a universal barcode region (Schoch et al. 2012; Li et al. 2011). This level of polymorphism makes the ITS region a strong and valid candidate for the fungal DNA barcoding. Taxa for which primary barcodes are often ineffective for identifying pathogenic fungal species, secondary barcodes are continuously being established. Simplifying ITS region to either the ITS1 spacer or ITS2 spacer only has also gained significant attention in the field of barcoding-based research. ITS2 is also not a coding region, but it has a conserved secondary structure core that helps to launch the data-handling systems. ITS2 spacer is found to be a very informative secondary barcode for plants (Xu 2016). Use of ITS2 for metabarcoding could potentially expand the comparative studies among fungi and plants. The largest subunit of ribosomal polymerase II was nominated as a representative protein-coding gene due to the good PCR success result with this gene and its function as a phylogenetic marker in the AFTOL project (Tanabe et al. 2004). Hence, ITS, SSU, LSU EF-1α, and RPB are barcoding markers that have been used for fungal DNA barcoding, as shown in Table 3.2.

3.3 DNA/RNA Probe-Based Methods (Sequencing-Independent Methods)

Nucleic acid hybridization techniques, using fluorescence tagged probes, provide high-throughput screening methods that can increase rates of sample processing and avoid some of the problems associated with the PCR-based approaches. In particular, high-density macro- and microarrays allow the rapid screening of mixed sequences to detect large numbers of specific genes, or their transcripts (Hardman 2004). These techniques were originally developed from genome sequencing projects to study gene expression (Luo et al. 2005), but they can be adopted for use in ecology.

Array-based technology has been used to identify fungal ribosomal RNA genes from soil-extracted and seawater-extracted DNA (Valinsky et al. 2002; Kiesling et al. 2002). Environmental DNA or RNA can be screened against several thousand probes fixed to a solid support (Fig. 3.3). The target probes can comprise ribosomal RNA sequences or functional genes such as those coding for metabolic enzymes, pathogenicity factors, and resistance determinants. The power of these techniques is that all of these genes can be processed in one single step. Using these systems, environmental nucleic acids can be rapidly screened and quantified for a range of

Table 3.2 Primers used for the barcoding of fungi

Name	Primer sequence (5′ → 3′)	Reference
ITS		
ITS1-F		
ITS4A	CTTGGTCATTTAGAGGAAGTA	Gardes and Bruns (1993)
ITS4B	CGCCGTTACTGGGGCAATCC	Gardes and Bruns (1993)
ITS1	CAGGAGACTTGTACACGGTCCAG	Gardes and Bruns (1993)
ITS2	TCCGTAGGTGAACCTGCGG	White et al. (1990)
ITS3	GCTGCGTTCTTCATCGATGC	White et al. (1990)
ITS4	GCATCGATGAAGAACGCAGC	White et al. (1990)
ITS5	TCCTCCGCTTATTGATATGC	White et al. (1990)
ITS9mun	GGAAGTAAAAGTCGTAACAAGG	White et al. (1990)
	TGTACACACCGCCCGTCG	Egger (1995)
SSU		
18S1F	GACTCAACACGGGGAAACTC	Vilela et al. (2005)
18S1R	AAACCTTGTTACGACTTTTA	Vilela et al. (2005)
NSA3	AAACTCTGTCGTGCTGGGGATA	Martin and Rygiewicz (2005)
NSI1	GATTGAATGGCTTAGTGAGG	Martin and Rygiewicz (2005)
NS1	GTAGTCATATGCTTGTCTC	White et al. (1990)
NS4	CTTCCGTCAATTCCTTTAAG	White et al. (1990)
NS3	GCAAGTCTGGTGCCAGCAGCC	White et al. (1990)
NS8	TCCGCAGGTTCACCTACGGA	White et al. (1990)
LSU		
LROR		
LR5	ACCCGCTGAACTTAAGC	Zhao et al. (2007)
LR16	TCCTGAGGGAAACTTCG	Zhao et al. (2007)
	TTCCACCCAAACACTCG	Vilgalys et al. (1994)
RPB		
RPB1-F		
RPB1-R	GARTGYCCDGGDCAYTTYGG	Matheny et al. (2007)
RPB2-6F	CCNGCDATNTCRTTRTCCATRTA	Matheny et al. (2007)
RPB2-6R	TGGGGKWTGGTYTGYYCCTGC	Liu et al. (1999)
	GCAGGRCARACCAWMCCCCA	Liu et al. (1999)
EF-1α		
EF1-1577F		
EF1-1567R	CARGAYGTBTACAAGATYGGTGG	Rehner (2001)
EF-3	ACHGTRCCRATACCACCRATCTT	Rehner (2001)
EF-22	GTAAGGAGGASAAGACTCACC	O'Donnell et al. (2001)
	AGGAACCCTTACCGAGCTC	O'Donnell et al. (2001)

functions. This technology is considered as the backbone to most of the current knowledge. In these methods, the probe is used for the analysis of nucleic acid without its amplification. Probes are the single-stranded shorter sequences of DNA that are labeled with the chemiluminescent reporter molecule or with radiolabeled

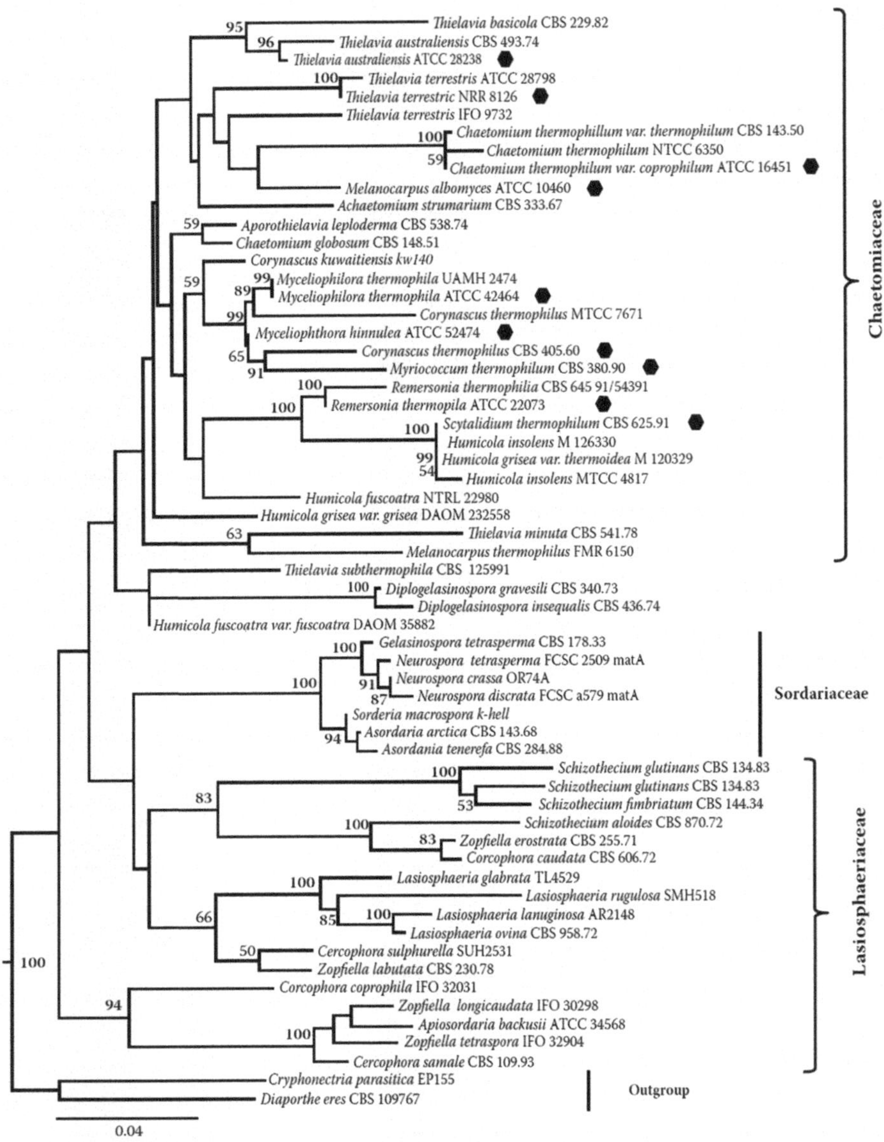

Fig. 3.3 Maximum likelihood phylogeny for the Sordariales based on concatenated small-subunit, ITS/5.8S, and large-subunit nucleotide sequence alignments. Bootstrap values of 50 and higher are given next to each node; values above 70 are indicated in bold font. Thermophilic taxa are highlighted in red font. The black hexagons indicate thermophilic strains included in the temperature-dependent growth trial. Families are indicated to the right of vertical bars (monophyletic) or curly brackets (phyletic status unresolved). The tree is rooted with *Diaporthe eres* and *Cryphonectria parasitica* belonging to the Diaporthales. (After Morgenstern et al. 2012)

isotopes, such as 32P, 33P, and 35S. These are used to identify the homologous sequence on the targeted DNA. In traditional methods, DNA probes are mostly used as an alternative to PCR for the identification of fungi. But in recent methods, these are mostly used in conjunction with PCR (McCartney et al. 2003).

3.3.1 Northern Blotting

Northern blot, which is also known as the RNA blot, is used to transfer the RNA onto a carrier for the identification of fungi. The northern blot is the same as the Southern blot except that the RNA material is used instead of the DNA. To overcome the difficulties of quantification of fungal activity in the soil, Mendoza-Mendoza et al. (2015) optimized the extraction of high-quality fungal RNA from the soil and then quantified the expression pattern of selected pattern of selected *Trichoderma* stage-specific markers in a time-course study within a soil microcosm. Firstly, the RNA should be purified so as to examine the expression of the gene of interest. The RNA material is then loaded on agarose gel. After that, the gel is passed to an electric current that migrates RNA towards the bottom of a gel. Smaller RNAs move faster compared to the larger ones (Kim et al. 2010). After that, separated RNA fragments are blotted on a special filter paper; therefore, each RNA molecule maintains its position relative to all other molecules (Berg Jeremy 2007). The filter is then exposed to radioactive probes so as to hybridize it to complementary sequences. After that, the filter is put for autoradiography so as to develop the film. Finally, a band should be detected on the autoradiograph if the probe has hybridized to a fragment of RNA on the filter. The northern blot is beneficial to study the gene expression. First, the position of bands on the blot gives the RNA size. If the size of the RNA is known, it will provide an approximation for the coding capacity of transcripts and also the size of the protein to which it encodes. Firstly, the risk of mRNA degradation during the electrophoresis is a main disadvantage of blotting technique. Secondly, detection with multiple probes is problematic. The sensitivity of northern blotting technique is relatively low compared to the RT-PCR.

3.3.2 In Situ Hybridization

In situ hybridization is also called the hybridization histochemistry. It provides invaluable information for identifying and enumerating the fungi. In this technique, single-stranded RNA probe is used, which is also called riboprobes. These probes are labeled with 35S. In situ hybridization has great similarity to northern blots. Both of these depend on the hybridization of labeled probes of DNA/RNA to the homologous sequences of mRNA. These two techniques differ in the use of starting material. In case of northern blot, sample digest is used as starting material, while in in situ hybridization histological section is used. Regardless of using the direct hybridization, the identifications using signal hybridization are most efficiently obtained after fungi growth or its biological amplification (Jensen 2014).

The main advantage of this technology is the maximum use of the short supply tissue such as clinical biopsies and cultured cells. Hundreds of different hybridizations can be carried out on the same tissue. Libraries of tissues can be prepared and stored in the freezer for future usage. There are many ways of doing in situ hybridization such as probing with synthetic oligonucleotides, cDNAs, and cRNAs. But probing with riboprobes gives the most sensitive and efficient results. Probes can be labeled either using radioactive or nonradioactive nucleotides. From these radioactive probes, 35S riboprobes are the most sensitive method for the identification of mRNA (Hayden et al. 2002).

In situ hybridization has some pitfalls. Firstly, radiolabeled probes are very costly and hazardous materials. It must be handled, transported, and disposed very carefully. Secondly, a disadvantage of using in situ hybridization technique is the difficulty in identifying targets that have low DNA and RNA copies (Qian and Lloyd 2003). In situ hybridization has been used for the identification of the pathogenic fungi (Hayden et al. 2001). ISH is a tool which is used to visualize the infection of plant tissue by the rust fungi. This method has been used to generalize the ISH protocol for the localization of rust fungi in the paraffin-embedded sections of plant tissues (Ellison et al. 2016).

3.3.3 Fluorescence In Situ Hybridization

Due to the drawback of radiolabeled probe-based hybridization, fluorescence in situ hybridization (FISH) came into development, which is used for the rapid characterization of fungi. FISH provides greater speed, resolution, and safety and paved way for the development of multiple targets detection simultaneously and quantitative analysis at phylogenetic level (Tsui et al. 2011). In addition to this, it is an important key to analyze spatial organization of fungal communities. This cytogenetic technique is used to detect specific DNA sequences on chromosomes. In FISH, a fluorescent probe is used that binds only those part of the chromosome with which it shows a greater degree of homology or complementation. Fluorescent probes are prepared by enzymatic incorporation of fluorophore-modified base throughout the probe length (Baschien et al. 2001). Traditional techniques for the identification of microorganism require that cells must be actively dividing. However, FISH can be used for the nondividing cells, which makes it a highly versatile technique. Nondividing cells can be recognized by their low level of fluorescence intensity. Different types of probes such as locus-specific probe, centromeric repeat probe, and whole chromosome probe can be used, each of which has its different applications. FISH technique has some pitfalls. In FISH, probe-preparing process is very complicated because it is necessary to tailor the probes to detect the specific sequences of DNA (Volpi and Bridger 2008).

3.4 Post-Amplification Technique

3.4.1 Microarray

The concept of microarray technology emerged from the two technical advances: firstly, the efforts on the DNA sequencing and secondly the focus on expressed component of genome. Microarray DNA chip technology allows the analysis of thousands of mRNAs simultaneously and is used to observe the changes in gene expression. This technique is different from the aforementioned techniques in the perspective that it provides the expression measurements on defined sets of genes (Eshaque and Dixon 2006). In this technology, thousands of DNA probes are arrayed within a small surface area onto a support matrix that includes glass chip or nylon filters. Location of the probe on the chip is called spot (Robinson et al. 2000). These probes are immobilized on the support matrix and targeted DNAs applied as a solution to the chip for the hybridization. Quantification of the bounded cDNA is measured by using the radiolabeled probes or fluorescent tags. Signal is produced by the hybridization of the probe to the targeted mRNA that can be detected and integrated by the dedicated software. The dedicated software generates the gene expression profile for each biological sample (Russo et al. 2003). By using this technique, comprehensive understanding of the cell of fungus can be achieved in a single array. Microarray technique is very easy to use because it does not require the large-scale DNA sequencing. Although this technique can monitor the global changes in the gene expression, significant issues are also considered. Firstly, a large amount of mRNA is required in this technique. Secondly, this study is limited by cost and access (Singh and Kumar 2013). Due to the numerous error-prone steps in the microarray experiment, replications are required to reduce the chances of error in such experiments. This technology is called destructive testing because physical disruption of cells is required to gain access to its gene expression patterns; false microarray data can be produced from the degradation of mRNA. Microarray technology has been used for the identification of the *Aspergillus candidus* species (Singh and Kumar 2013).

3.4.2 Macroarray

Macroarray is also called the DNA array hybridization, or reverse dot plot. This method uses the sensitivity of the DNA amplification and does not require the radioisotopes (Singh and Kumar 2013). Macroarray is rapidly becoming a standard molecular tool for diagnostic and epidemiological studies in an increasing number of laboratories all over the world. It works on the basis of simultaneous amplification of the related species through PCR. Along with this, it analyzes a number of amplified sequences simultaneously in one hybridization reaction. It is a more sensitive technique than PCR alone. In this assay, PCR amplification is combined with hybridization which increases sensitivity up to 1000-fold or higher than PCR only (Taoufik et al. 2004). This technique has a fast turnaround time of 1–2 days compared to radioactive culture methods that require 2–8 weeks for their completion. In this technique, the internal probes are designed to differentiate the species. These probes are fixed to the nylon membrane support. Oligose are permanently bound to the membrane by UV cross-linking. Then the PCR-amplified products are hybridized with the spotted series of species-specific probes present on the strips (Tsui et al. 2011; Leinberger et al. 2005). Macroarray in combination with PCR has been used to detect and identify the *Mycobacterium* to the species level (Leinberger et al. 2005). This assay has also been used for the identification of *Alternaria alternata*, *Aspergillus fumigatus*, *Fusarium solani*, *Candida albicans*, and *Cladosporium herbarum* (Sato et al. 2010).

3.5 Isothermal Amplification-Based Methods

3.5.1 Rolling Circle Amplification

For the identification of pathogens, single nucleotide polymorphism (SNP) detection is becoming popular. Detection of SNPs among different genotypes by the conventional real-time PCR has many challenges and drawbacks. So, to overcome these drawbacks, species-specific PLPs (circularizing oligonucleotide probes) are used. Circularizing oligonucleotide probes are the single-stranded DNA molecules having target recognition sequences of 20 nucleotides present at the both 5′ and 3′ ends, which are connected by the 40-nucleotide long linker sequence (Tsui et al. 2011). Designing of PLPs is first reported by Nilsson et al. in 1994. When hybridized to targeted region, both ends come near to each other and become circularized by the use of ligase, leaving no gaps. The circularized probe spans the entire target region in a manner similar to that of the padlocks, driven by the helical nature of the double-stranded DNA (Wang and Yang 2010). So, we can say that PLPs are useful only for detecting DNA molecules with known sequences. PLPs can be used in PCR, but such probes so far failed to yield reliable in situ single-copy gene detection. Therefore, Padlock probe is preferably suitable to template a reaction of RCA. RCA is based on rolling replication of short single-stranded DNA circular molecules. The RCA process is isothermal and also called rolling circle replication (RCR). This process requires DNA polymerase, a primer to initiate the replication, DNTPs and DNA binding, and unwinding proteins. Most of the groups are using the RCA reaction for signals amplification, where the small circular probes serve as the template. In RCA reaction, the primer has dual functions, both as RCA signal amplifier and discriminator by being complementary to the DNA-targeted sequence (Kuhn et al. 2002). In RCA, a second primer complementary to the RCA product can be used, so hyper-branched RCA (HRCA) reaction will be generated. HRCA is also used as an alternative method to PCR for DNA amplification. Recently, major molecular types of *Cryptococcus neoformans* and *C. gattii* are identified by hyperbranched rolling circle amplification (HRCA) (Trilles et al. 2014). RCA technology is a robust and simple method that can make available a universal platform for the localization of a wide variety of molecules such as nucleic acid sequence (Gusev et al. 2001).

3.5.2 Loop-Mediated Isothermal Amplification

LAMP is a robust and novel nucleic acid amplification method that is considered an alternative to the PCR. It amplifies the targeted nucleic acid under the isothermal conditions with high specificity. LAMP does not require a thermal cycler to produce the temperature changes; it instead requires single temperature for DNA amplification (Tsui et al. 2011). So, it is based on auto-cycling strand displacement amplification of DNA. In this technology, Bst DNA polymerase and a set of four primers that consist of two inner and two outer primers are used, which recognize total six unique sequences on the targeted DNA. Two inner primers are referred to as forward inner primer (FIP) and backward inner primer (BIP), while outer primers are F3 and B3 (Fakruddin 2011). Each BIP and FIP consists of two distinct sequences according to sense and antisense strand of targeted DNA. One of the inner primers is used to begin the LAMP reaction, while the other is for self-priming. LAMP reaction is processed in a heat block or water bath at 65 °C for 1 h. Then amplification product is detected by using the dye SYBR Green 1 as well as by electrophoresis. The final product has many inverted repeats of targeted sequence which exhibit cauliflower-like structure with multiple loops. LAMP is 10 times more sensitive and accurate than the conventional PCR (Ren et al. 2009). LAMP technology does not require expensive equipment to get a high level of accuracy, and it has less number of preparation steps than the conventional and qPCR. Due to high-amplification efficiency, up to 1039 copies of a target part can be attained in less than 1 h of incubation. Another advantage over the PCR in many cases is that the inhibition reactions occurred less in LAMP than that in PCR. In LAMP, no time is lost for thermal changes as it happened in PCR. For PCR, an expensive thermal cycler is required, while LAMP reaction required just a single tube (Fakruddin 2011). But several factors are present there which affect the accuracy of LAMP reaction such as the usage of specific DNA polymerase, which is critical for LAMP efficiency. Annealing of the four primers to targeted DNA is also important for the accuracy of LAMP. LAMP is useful for the detection and diagnosis but not for cloning purposes. However, a main drawback of LAMP technology is the use of indirect evaluation methods such as Mn^{2+} dye, gel electrophoresis, SYBR Green I dye, hydroxynaphthol blue dye, composite probe method, and turbidimetric method, which cannot be differentiated between desired products and nonspecifically amplified product, thus leading to false positives. The usage of molecular beacons (MBs) solves this problem by producing the fluorescence signals when it binds to target DNA. Hence, it acts as a direct detector of amplification product. Optimal conditions for the MBs (such as beacon length of 25–45 bp, reaction temperature of 60–65 °C, and beacon concentration of 0.6–1 pmol/μL) are found as an assessment tool in LAMP. A novel method based on MB-LAMP has been validated, which provides the direct detection of LAMP product. MBs are fluorescent nucleic acid probes which have a hairpin structure. The hairpin structure avoids the fluorescence because the quencher is physically closed to fluorophore. Moreover, binding to LAMP product sequences changes the MB spatial configuration, which separates the fluorophore and quencher at both ends of a single strand of nucleic acids and desorbing fluorescence (Liu et al. 2017). LAMP has been successfully used to detect the presence of thermodependent dimorphic fungus, *Paracoccidioides brasiliensis* (Endo et al. 2004). *Penicillium marneffei* has also been successfully diagnosed by using LAMP technique (Sun et al. 2010). Recently, *Ophiostoma clavatum*, a primary blue stain fungus, is identified by using LAMP (Villari et al. 2013).

3.5.3 Nucleic Acid Sequence-Based Amplification

NASBA is a very sensitive, isothermal, and transcription-based amplification system that is specially designed for RNA detection. Some NASBA systems can also amplify the DNA. It is also well known as self-sustained sequence replication (3SR). This tool is not only useful in basic research but also in the application-oriented fields, such as clinical medicine development and infectious diseases diagnosis (Sergentet-Thevenot et al. 2008).

Numerous amplification methods have already been established, such as PCR (Saiki et al. 1992), LAMP (Lee et al. 2009), and RCA (Lizardi et al. 1998). None of these methods can amplify the RNA directly with high sensitivity. NASBA provides many advantages over the other techniques of mRNA amplification. It can amplify more than 109 copies of the nucleic acid sequence in just one and a half hours by the action of three enzymes. NASBA is an isothermal reaction that is performed at 41 °C, which excludes the need for a thermal cycler, and it can facilitate the production of point-of-test devices (Fakruddin et al. 2012). An important benefit of NASBA is the production of single-stranded RNA amplicons, which can be used directly in another round of amplification or can be queried for detection without denaturation or strand separation (Chang et al. 2012). Several studies have reported that the amplification power of NASBA is comparable to, or even better than, that of the RT-PCR (Chang et al. 2012). RNA being the genomic material of many RNA viruses, an RNA-based amplification technique in contrast to the PCR keeps away from an additional reverse transcription step, thus minimizing the contamination risk and lowering hands-on time. It assists in better RT-PCR reaction as it provides the faster amplification kinetics that is especially suitable for the detection of retroviruses. It can measure the replication of DNA viruses by detecting late mRNA expression. Gene expression studies can be carried out without the intron-flanking primers or DNases. There are also some disadvantages of NASBA. First of all, RNA integrity is the foremost cause of concern for the NASBA and also for other RNA amplification techniques. Although the amplification reaction itself is isothermal at 41 °C, only a single melting step before the amplification reaction is required, which allows 416 annealing of primers to the target (Compton 1991). Moreover, due to the dependence of specificity of the reactions on thermolabile enzymes, reaction temperature cannot beat 42 °C without compromising it. Finally, the length of the targeted amplified RNA sequence should be between 120 and 250 nucleotides (Fakruddin et al. 2012).

3.6 RNA Interference

RNA interference (RNAi) is used as a latest technology to identify and control the fungi. It is an efficient technique that can be used in a highly tissue-specific way to fight mycotoxigenic fungi, which cause the infection in crop plants (Panwar et al. 2012). Positive transgenic RNAi execution depends on many factors, which include (i) designing the vectors in such a way so as to produce the double-stranded RNAs (dsRNAs) that will make the small-interfering RNA (siRNA) species for the ideal gene silencing, (ii) accessibility of plentiful target siRNAs at infection place, (iii) efficient uptake of siRNAs by fungus, (iv) siRNA half-life, and (v) amplification of the silencing effects. RNAi eliminates the negative consequences of current disease control and fights the alarming rise of the fungicide-resistant plant pathogens (Ishii and Holloman 2015). RNAi absolutely knocks down the genes by using the event of intrinsic cellular defense. Detection of the dsRNA or hairpin RNA (hpRNA) by the fungal cells leads to the targeted transcripts by using the sequence homology important for the degradation or silencing (Nakayashiki et al. 2005). To identify the unique fungal targets, cell-specific and dual RNA sequencing data should be provided. Then hpRNA or dsRNA can be modified for a definite transcript which can directly limit the fungal pathogenesis. RNAi technology poses a benefit to the cell's natural machinery, which is assisted by the short-interfering RNA molecules, to successfully knock down the expression of a gene of interest. A major disadvantage of RNAi is the possibility of an off-target effect, which leads to gene silencing, which tolerates the partial complementarity to the sense or antisense strand of targeted gene. RNAi has been used in *Zea mays* for the identification of *Aspergillus flavus* fungus (Masanga et al. 2015) and industrially relevant fungi such as *Trichoderma reesei*, *Penicillium chrysogenum*, and *Antrodia cinnamomea* (Dahlmann and Kuck 2015; Kang et al. 2013; Lin et al. 2015).

3.7 RNA-Seq-Based Next-Generation Sequencing

RNA-Seq is a newly developed deep-sequencing technology. Generally, a large population of RNA is changed to cDNA library with adaptors that linked to one or both ends. After that, each fragment with or without amplification is sequenced in a high-throughput way to get small sequences from one end as in single-end sequencing or both ends as in pair-end sequencing. Reads are usually 30–400 bp, depending on the DNA sequencing technique used. Library of the small sequences is prepared to see how closely the RNA sequencing results reveal the original RNA transcripts are mostly determined in the library preparation step. To create an RNA-Seq library, the fragmentation of either the RNA or the cDNA is required to allow the processing through next-generation sequencing. Developed mRNA should be primed for RT reaction by the use of either random primers or oligo primers. The benefit of using oligo(dT) is that the majority of cDNA produced should be poly-adenylated mRNA; hence, more of the sequences obtained should be informative (Mortazavi et al. 2008).

Three most generally used next-generation sequencing platforms for the RNA-Seq are SOLiD and Ion Torrent (Life Technologies) and HiSeq (Illumina) (Dawei and Peng 2014). Following sequencing, the resultant reads are either aligned to reference genome or assembled de novo without genomic sequence to produce a genome-scale transcription map that consists of both the transcriptional structure and level of expression for each gene. While most of RNA-Seq analysis data depend on the alignment of the reference genome sequences, new softwares such as Rnnotator and Trinity assemble the RNA-Seq data into the transcriptomes without referring a reference sequence by assembling the short adjacent reads from the RNA-sequencing data. These approaches allow the discovery of new transcripts and fair detection of the transcripts from numerous sources, permitting more effective application of the RNA-Seq for the detection of transcripts and classification of transcriptomes in the non-model organism (Grabherr et al. 2011). Through this technology, characterizations of marine fungal communities were identified by Kristiansen (2014).

3.8 Taxonomy and Molecular Phylogeny of Thermophilic *Chaetomium*

The taxonomic rearrangement of the fungi started long before molecular techniques were at hand. However, within the last four decades, molecular techniques have altered the understanding of the evolution and phylogeny of life, and more than 1000 fungal research papers reporting phylogenies have been published. The majority of these studies used 18S rDNA sequencing. Although sequences based on protein-coding genes are rarely used in fungal phylogenetics, they can resolve deep-level phylogenetic relationships (Liu et al. 1999). In fungi and for most other organisms, phylogenetic reconstruction based on a single gene is blurred, as individual genes contain a limited number of nucleotide sites and hence limited resolution. Thus, to have a robust phylogenetic analysis, multigene concatenation is performed using data from all available genes defining various characteristics of the organism. Until now, more than a hundred fungal genomes have been sequenced and are available in public databases.

3.8.1 Molecular Phylogeny of Thermophilic Fungi

Thermophilic fungi are distributed widely throughout the fungal kingdom and are not consigned to any phylum or order. The definition of thermophilic fungi has been debated and defined variously by several workers (Cooney and Emerson 1964; Crisan 1964; Maheshwari et al. 2000). Thermophilic fungi are important from a biotechnological perspective, as they produce a number of thermostable enzymes that are useful in industrial processes, such as bleaching in the paper and pulp industry, the bioremediation of polluted sites and wastewater effluents, and the production of second- and third-generation biofuels (Wesenberg et al. 2003; Gianfreda and Rao 2004; Sigoillot et al. 2005; Turner 2007). Phylogenetic analysis of these fungi may resolve the pending issues of their misidentification and taxonomic classification (Mouchacca 2000).

With the advent of polymerase chain reaction (PCR) technology, the molecular revolution of fungal taxonomy, which commenced in the early 1990s, has grown into a mature discipline where multilocus data sets, extensive taxon sampling, and rigorous analytical approaches are standard (Hibbett et al. 2007). While 16S rRNA gene-based phylogeny is the most commonly accepted taxonomic scheme in prokaryotes, its relevance to eukaryotic phylogeny is at a crossroads. Currently, there is a heated debate as to what is the best approach for reconstruction genome phylogenies, particularly in fungi. The 16S–23S internal transcribed spacer (ITS) regions of the rRNA operon might be under minimal selective pressure during evolution and therefore have more variation in the sequences than the coding regions of 16S and 23S rRNAs.

The size of the ITS varies considerably for different species and even among different operons within a single cell having multiple rRNA operons (Rajendhran and Gunasekaran 2011). To assess fungal phylogeny, single-gene phylogenies, especially those based on 18S rDNA, have established many of the accepted relationships between fungal organisms (Fitzpatrick et al. 2006). The use of the 18S rDNA approach is advantageous in view of the vertical transmission of this gene, its wide distribution, and the fact that it has slowly evolving sites. However, the phylogenetic relationships of a diverse range of fungi, including thermophilic species, must be performed using multigene concatenation, as this approach attempts to maximize the informativeness and explanatory power of the characteristic data used in the analysis (Kluge 1989). Therefore, in recent studies (Pan et al. 2010; Berka et al. 2011; Morgenstern et al. 2012; Zhou et al. 2014) on the molecular phylogeny of thermophilic fungi, the multigene concatenation approach has largely been followed to elucidate the phylogenetic relationships among the fungal orders harboring thermophilic fungi.

Considering the diversity of thermophilic fungi in various habitats, relatively few studies focusing on phylogenetic analysis have been undertaken. Pan et al. (2010) attempted to identify thermophilic fungi from geothermal sites near neutral and alkalescent thermal springs in Tengchong Rehai National Park using ITS sequencing combined with morphological analysis for the identification of thermophilic fungi to the species level. The authors isolated 102 strains belonging to *Rhizomucor miehei*, *Chaetomium* sp., *Talaromyces thermophilus*, *Talaromyces byssochlamydoides*, *Thermoascus aurantiacus* var. *levisporus*, *Thermomyces lanuginosus*, *Scytalidium thermophilum*, *Malbranchea flava*, *Myceliophthora* sp. 1, *Myceliophthora* sp. 2, *Myceliophthora* sp. 3, and *Coprinopsis* sp. Similarly, Berka et al. (2011) analyzed the genomes of two thermophilic ascomycete species, *Myceliophthora thermophila* and *Thielavia terrestris*. They reported that the two genomes are similar in organization.

The whole genome sequencing of these fungi offers new industrial applications of the enzymes from these organisms and their potential development as fungal production hosts. The authors further emphasized that with a finished genome as a scaffold and modern sequencing technologies, resequencing these strains and identifying mutations becomes relatively simple but very helpful for identifying beneficial and deleterious genetic changes.

A detailed study on the molecular phylogeny of thermophilic fungi by Morgenstern et al. (2012) investigated sequences from 86 fungal genomes to construct a robust molecular phylogeny of thermophilic fungi. Considering the criterion that a thermophilic fungus is one that optimally grows at 45 °C, their phylogenetic study suggested that the known thermophilic fungi belong to the orders *Sordariales*, *Eurotiales*, *Mucorales*, and *Onygenales*. They examined the phylogenetic relationships of a diverse range of fungi, including thermophilic and thermotolerant species, using concatenated amino acid sequences of marker genes mcm7, rpb1, and rpb2 obtained from genome sequencing projects. These markers were used to reconstruct a phylogeny of the fungi (Fig. 3.2).

The authors further observed that maximum likelihood analysis of the combined amino acid sequence alignment resulted in a tree with high bootstrap support (BSS) values even in deep internal nodes. Only two nodes received less than 50% BSS, and five nodes received support between 50% and 70%. Of the remaining 78 nodes receiving support above 70%, 61 nodes received 100% BSS (Fig. 3.2).

According to Morgenstern et al. (2012), thermophilic species in the *Sordariales* are found only in *Chaetomiaceae*. To assess this connotation, they included sequences from species belonging to the classes *Sordariaceae*, *Lasiosphaeriaceae*, and *Chaetomiaceae* of *Sordariales* (Fig. 3.3).

They reported that the concatenated data set for the *Sordariales* contains 60 operational taxonomic units (OTUs), of which 21 are represented by all three genes, 27 by two genes (mostly ITS and large subunit), and 12 by only one sequence. The tree reconstruction depicts the *Chaetomiaceae*, with the exclusion of *Thielavia subthermophila* as monophyletic and the *Lasiosphaeriaceae* as paraphyletic (Fig. 3.3). Furthermore, their study depicted that thermophilic fungi included in the order Eurotiales are not monophyletic; rather, they appear in three species pairs and as single sequences throughout the tree (Fig. 3.4).

Thermophilic fungi, as defined by their temperature preference to grow better at or above 45 °C than at 25 °C, have evolved independently in at least two lineages within the phylum *Ascomycota*, once each within the orders *Sordariales* and *Eurotiales* (Berka et al. 2011). Within the *Sordariales*, thermophily is limited to

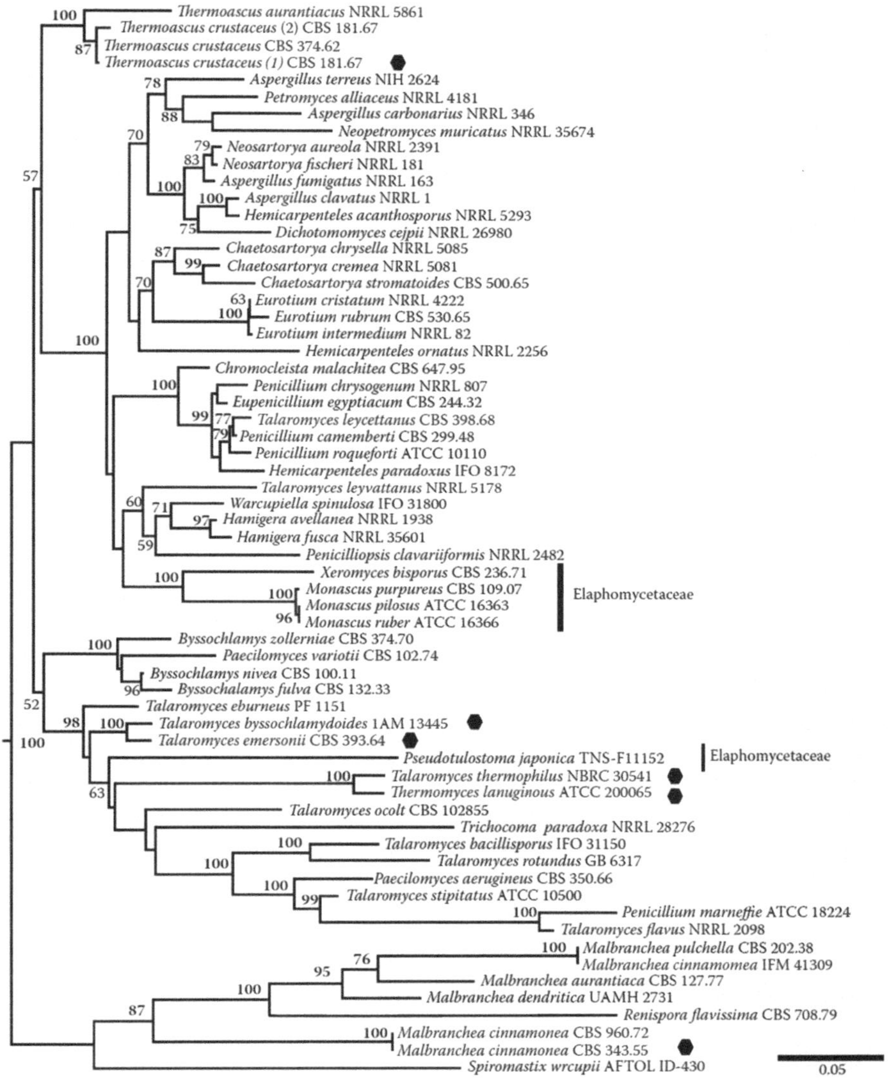

Fig. 3.4 Maximum likelihood phylogeny for the Eurotiales based on concatenated small-subunit, ITS/5.8S, and large-subunit nucleotide sequence alignments. Bootstrap values of 50 and above are given next to each node; values above 70 are indicated in bold font. Thermophilic taxa are highlighted in red font. The black hexagons indicate thermophilic strains included in the temperature-dependent growth trial. Species belonging to the Elaphomycetaceae are indicated to the right of vertical bars. The tree is rooted with species belonging to the Onygenales. (After Morgenstern et al. 2012)

subgroups of the family *Chaetomiaceae*. Among fungi more broadly, thermophily also exists in the *Zygomycota* (Zhou et al. 2014); however, it appears to be rare or altogether absent in the phyla *Basidiomycota* and *Chytridiomycota*. The evolutionary trajectory of thermophily is obscured by chaotic taxonomy. Moreover, efforts on biosystematics have lagged behind research on thermophilic fungi useful in biotechnological industries, resulting in a body of literature for these organisms that lacks accurate taxonomic treatments (Mouchacca 2000).

The recent trend in solving fungal systematic problems is to use the online fungal taxonomies that are proliferating the Internet. The following are some of the important online general classifications of fungi:

1. GenBank (www.ncbi.nlm.nih.gov/Taxonomy), a resource for a diverse community of researchers, including ecologists and molecular biologists
2. Tree of Life (www.tolweb.org/tree), which is widely used by teachers and students for basic studies and research
3. Myconet (www.fieldmuseum.org/myconet), a site on which the ascomycetous fungi are regularly updated
4. Index Fungorum (www.indexfungorum.org)
5. MycoBank (www.mycobank.org)

The integration of these and many more databases, particularly for taxonomic names, is likely to gain greater prominence for future fungal taxonomies. Further, if the classifications employed by these and other major taxonomic resources could be unified, this would promote communication and the awareness of fungal phylogeny and also provide a framework for future revisions at all taxonomic levels. Therefore, classification of fungi based on monophyletic groups that can be recommended for general use is greatly warranted. With the taxonomic advances in fungal nomenclature as supported by several phylogenetic and genome-wide studies, thermophilic fungi are described as being composed of 20 genera and 44 species (Oliveira et al. 2015).

3.8.2 Thermophilic Fungal Genomes

Until now, a few hundred fungal genomes have been sequenced, including important human pathogens, plant pathogens, and model organisms (Zhou et al. 2014). The increase in fungal genome sequencing offers an opportunity to reconstruct evolutionary events using whole genomes. The genomes of several industrially useful fungi, such as *Aspergillus niger* (Pel et al. 2007) and *Trichoderma reesei* (Martinez et al. 2008) have also been sequenced. As thermophilic fungi represent a potential reservoir of thermostable enzymes, sequencing of their genomes is advantageous from a biotechnological perspective; besides, their genomes are amenable to

Table 3.3 General features of fungal genomes of some thermophilic and thermotolerant fungi

Organism	Optimum growth temperature (°C)	Genome size (Mb)	No. of coding genes	G + C content (%)	No. of chromosomes	References
Aspergillus fumigatus	40	29.4	9926	49.9	8	Nierman et al. (2005)
Aspergillus niger	30	33.9	14,165	50.40	–	Pel et al. (2007)
Aspergillus nidulans	30	30.1	10,701	50.3	–	http://www.aspgd.org/
Myceliophthora thermophila	40–50	38.7	9110	51.4	7	Berka et al. (2011)
Rhizomucor miehei	35–45	27.6	10,345	43.83	–	Zhou et al. (2014)
Thermomyces lanuginosus	45–55	23.3	5105	52.14	–	McHunu et al. (2013)
Thielavia terrestris	45–50	36.9	9813	54.7	6	Berka et al. (2011)
Chaetomium thermophilum	45–55	28.3	7227	–	8	Amlacher et al. (2011)

manipulation using classic and molecular genetics (Berka et al. 2011). Genomes of some thermophilic fungi have been sequenced in the recent past. Characteristic features of genomes of some thermophilic fungi are presented in Table 3.3. The following section highlights the genomes of some important thermophilic fungi.

The genome of *C. thermophilum*, originally isolated by La Touche in 1948, has been sequenced by Amlacher et al. (2011). This fungus holds great promise for structural biology, has the exceptional ability to grow up to 60 °C, and has been a subject of study for its nuclear pore complex (NPC). The fungus is known to produce a variety of thermostable enzymes, such as cellulose, amylase, xylanase, and lipase, and thus is a good candidate for biotechnological explorations. Using Roche 454 FLX and XLR platforms, Amlacher et al. (2011) sequenced a nuclear genome of 28.3 Mb contained in eight chromosomes. They identified a total of 7227 protein-coding genes. Later, Bock et al. (2014) further described the genome-wide proteome and transcriptome analysis of *C. thermophilum*. The refinement of the gene structure by their transcriptomics and proteomics approach in particular ensures individual gene expression studies and subsequent experimental characterizations.

Their study implied that the individual proteins are the basis for the adaptation of a thermophilic lifestyle in thermophilic fungi. Although genome sequencing of thermophiles has been initiated only recently, several of the thermophilic fungal genomes have not been sequenced to date.

Genomic features of a few other thermotolerant and thermophilic fungi are presented in Table 3.3 with relevant references. Additionally, a variety of thermophiles may serve as model organisms that may help us to understand the possible mechanism of genomic diversification or environmental adaptation, which could be a driving force for the microbial evolution and associated thermophily.

3.8.3 Preferred Gene Sequences of Chaetomium

Nuclear ribosomal internal transcribed spacer (ITS), partial ribosomal large subunits (LSU rDNA), the second largest subunit of RNA polymerase II (RPB2), and b-tubulin (TUB) regions were recommended recently for amplification of *Chaetomium* DNA using primer sets ITS4/ITS5 (White et al. 1990), LR0R/LR5 (Vilgalys and Hester 1990), 5F-Eur/7CR-Eur (Houbraken et al. 2012), and Bt2a/Bt2b (Glass and Donaldson 1995), respectively, as recommended by recent studies carried by various investigators (Wang et al. 2015; Zhang et al. 2017).

3.9 Developing Genetic Tools to Exploit *C. thermophilum*

The use of thermostable proteins from prokaryotic thermophiles (bacteria, archaea) has been extensively exploited in the past for various biotechnological applications and in structural biology (Elleuche et al. 2015). In the eukaryotic kingdom, thermophily has been adopted in some fungi of which *C. thermophilum*, a thermophilic ascomycete species, can grow up to 60 °C with an optimum growth temperature around 50–55 °C. The thermostable nature of individual *C. thermophilum* proteins has previously been demonstrated and directly compared to their mesophilic homologues (van Noort et al. 2013). Moreover, with the recent release of a publicly available genome database with manually refined annotation (van Noort et al. 2013), *C. thermophilum* has served as a prosperous resource for thermostable eukaryotic proteins. Several proof-of-principle studies exploiting the *C. thermophilum* proteome for biotechnological applications (Li et al. 2010) or high-resolution 3D characterization of single proteins or reconstituted protein complexes (Stuwe et al. 2015) have meanwhile been reported. However, these investigations required expression of *C. thermophilum* genes in heterologous mesophilic expression systems, as *C. thermophilum* has not yet been amenable for genetic manipulation. In an effort to fully exploit the potential of this thermophilic eukaryote as a source for biochemical and structural analyses of difficult proteins and derived complexes, we developed a transformation system for *C. thermophilum*, allowing the stable integration of constructs into the genome, which renders this thermophile accessible for affinity purification of native thermostable proteins and protein complexes assembled under physiological conditions.

3.9.1 Methods (Fig. 3.5)

1. **For cultivation of *C. thermophilum***, wild type of *C. thermophilum* (La Touche) var. *thermophilum* will be obtained from any international culture collection. Mycelium will be propagated on modified CCM medium either broth or solid as recommended by Kellner et al. (2016).
2. Plates will be incubated at 52–55 °C. Submerged cultures will be inoculated with mycelium scraped off a freshly grown agar plate and typically will be grown in 250 mL baffled Erlenmeyer flasks at 50 °C and 100 rpm in a rotary shaker for approximately 24 h.
3. For large-scale growth of mycelium, submerged cultures will be shredded in a blender (4 × 20 s) and used as inoculums for 2 L CCM broth in 5 L baffled Erlenmeyer flasks incubated at 50 °C and 90 rpm for 18 h.
4. For harvesting mycelium that will be subsequently used for isolation of protein extracts or affinity purification, the cultures will be strained through a metal sieve (180 μm pore size) and washed once with sterile water, and residual liquid will be removed with a vacuum filter. Cells will be frozen in liquid nitrogen and subsequently ground to a fine powder in a cryogenic cell mill (RETSCH MM400) in two cycles of 5 min each.
5. For the generation of ascospore-producing fruiting bodies, rice extract agar (supernatant of 75 g whole grain rice, boiled for 2 h and supplemented with 15 g of agar per liter) will be inoculated with shredded mycelium and incubated at 37 °C for approximately 7 days. Spores will be harvested through scraping the agar surface, filtrated through four layers of sterile gauze, washed in 1 M sorbitol, and stored at −20 °C. Spores retained their ability to germinate on CCM agar for at least 1 year.
6. For germination of ascospores on terbinafine, the minimal inhibitory concentration will be determined and 0.1 μg/mL terbinafine will be supplemented to solid CCM medium.
7. **For cloning of *C. thermophilum* nucleoporins for native affinity purification and recombinant expression in** *S. cerevisiae* **or** *E. coli*, molecular cloning techniques will be based on protocols described by Sambrook et al. (1989). *E. coli* DH5α will be used for plasmid propagation via standard procedures.
8. Genomic DNA from *C. thermophilum* will be isolated essentially as reported by Al-Samarrai and Schmid (2000) and total RNA was isolated using the SV total RNA isolation kit (Promega).
9. cDNA will be synthesized using SuperScript III Reverse Transcriptase (Invitrogen) and an Oligo d(T) primer according to the company's instructions. cDNA will be purified with the QIAquick PCR Purification Kit (QIAGEN).
10. Construction of the terbinafine resistance marker will be achieved by cloning the open reading frame (ORF) of ctERG1 (CTHT_0032780). To obtain the actin promoter (PACTIN) and the trpC promoter (PTRPC), a 1.5 kb fragment

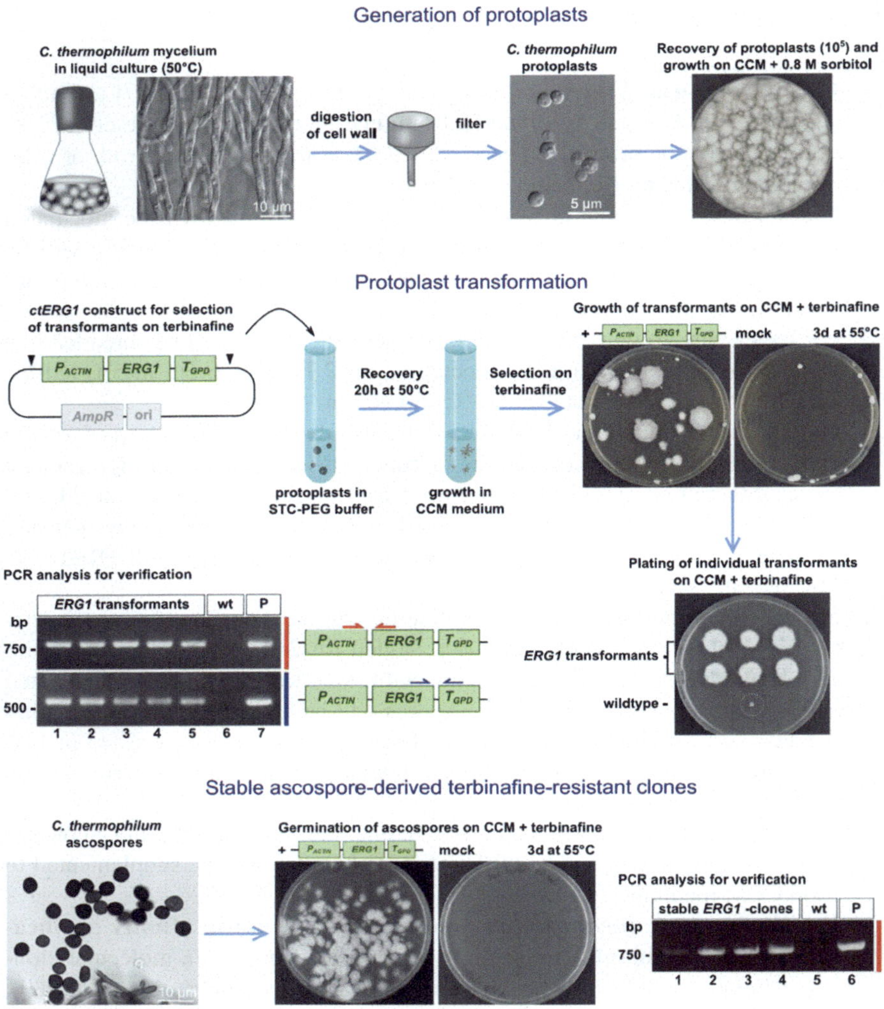

Fig. 3.5 Schematic representation of the workflow for transformation of *C. thermophilum* protoplasts and selection of positive transformants (After Kellner et al. 2016). Protoplasts are generated from *C. thermophilum* mycelium growing in submerged cultures by digestion of fungal cell walls. Upon filtration, protoplasts are plated on osmotically stabilizing medium to assess viability. For transformation, protoplasts are mixed with the *PACTIN-ERG1* plasmid for selection on terbinafine (see text for details). Arrowheads point to restriction sites for linearization. Upon recovery, mycelia are plated on solid media supplemented with terbinafine for selection of positive strains. Replating of individual transformed mycelia on selective medium confirms stable terbinafine resistance, whereas the wild type does not grow on terbinafine. Left panel: PCR analysis with oligonucleotides specific for the *PACTINERG1* construct (depicted as red and blue arrows, respectively) performed on genomic DNA (gDNA) isolated from five independent *ERG1* transformants (lanes 1–5) to verify genomic integration of the construct. The expected products of 761 nt (upper panel) and 536 nt (lower panel), respectively, could be amplified, whereas no product was amplified from

and a 1.2 kb fragment, respectively, upstream of the respective ORF (actin – CTHT_0062070; trpC – CTHT_0070860) will be amplified from genomic DNA.

11. *C. thermophilum* genes encoding nucleoporins will be previously annotated.
12. For the generation and integration of affinity-tagged fusion constructs into the *C. thermophilum* genome, nucleoporin-encoding ORFs will be PCR amplified from genomic DNA. For amplification of promoter regions, approximately 1.0–1.5 kb fragments upstream adjacent to the respective ORF will be amplified.
13. For termination of transcription, a 3′ transcription terminator DNA fragment (300 bp downstream of the glyceraldehyde-3-phosphate dehydrogenase ORF (GPD; CTHT_0004880) will be fused to the ORF of interest.
14. To clone *C. thermophilum* nucleoporins for recombinant expression, the respective gene will be PCR amplified from either cDNA or genomic DNA, when no intron was present, and inserted into appropriate expression plasmids.
15. **For transformation of *C. thermophilum* and expression tests**, protoplasts will be generated using a submerged culture of the *C. thermophilum* wild-type strain, shredded in a kitchen blender (4 × 20 s), as inoculum for 300 mL CCM medium, which was incubated at 50 °C and 90 rpm for approximately 20 h. The mycelium will be strained through a metal sieve, washed twice with PP buffer, subsequently subjected to 100 mL Erlenmeyer flasks containing 40 mL digestion solution, and incubated at 30 °C and 110 rpm for 3 h. The resulting protoplasts were filtered through a funnel (pore size 1) and collected via centrifugation (2400 rpm, 8 min, 4 °C), followed by two washing steps in PP buffer and one washing step in STC buffer.
16. The amount of protoplasts will be determined in a hemocytometer and from a standard 300 mL culture. For the determination of the recovery rate, different amounts of protoplasts will be plated on CCM medium supplemented with 0.8 M sorbitol, incubated at 50 °C for 1 day, and the resulting colonies will be counted.
17. For transformation of protoplasts with plasmid DNA, 200 μL of the protoplast solution will be mixed with 2.5 μL heparin (10 mg/mL), 2 μL spermidine trihydrochloride (50 mM), 1 μL aurintricarboxylic acid (0.4 M) as nuclease inhibitor, 40 μL STC/PEG solution (40% PEG6000 (w/vol) in STC buffer), and 5–10 μg of linearized plasmid DNA and incubated on ice for 20 min. Then, 750 μL of STC/PEG solution will be added prior to incubation at room temperature for 10 min to allow the transformation to proceed.

Fig. 3.5 (continued) wild-type gDNA (wt). The *PACTIN-ERG1* plasmid (P) was included as positive control (lane 7). Ascospores (bottom left) were generated from an *ERG1*-positive transformant to verify that the resistance trait remains stable. Upon plating of these ascospores on terbinafine, the *ERG1*-positive strain was able to germinate, whereas ascospores generated from the wild-type strain did not grow. Bottom right: PCR analysis performed on gDNA isolated from four independent ascospore-derived clones (lanes 1–4) verifies stable genomic integration of the *ERG1*-marker

18. The protoplasts will be again collected via centrifugation and recovered in 3 mL CCM medium supplemented with 0.8 M sorbitol at 50 °C and 600 rpm for approximately 20 h. After recovery, the cells will be plated on CCM agar plates supplemented with 0.8 M sorbitol as well as terbinafine hydrochloride at a final concentration of 0.5 μg/mL for selection and incubated at 50–55 °C for 2–3 days.
19. Expression of affinity-tagged fusion proteins will be tested via immunoblots of whole cell protein lysates obtained from 100 to 150 mg of mycelium, ground in 1 mL NB-HEPES buffer and 500 μL of zirconia beads in a Mini Bead Beater (5000 rpm at 4 °C, 2 × 20 s, 4 runs). The lysates will be cleared via centrifugation (14.000 rpm at 4 °C for 20 min) and supernatants will be analyzed by SDS-PAGE and Western blotting, performed using PAP and/or FLAG® antibodies according to the manufacturer's instructions.
20. **For affinity purification of native protein complexes from *C. thermophilum*** (Fig. 3.6), ground frozen mycelium will be thawed on ice in NB-HEPES buffer containing 0.1% (vol/vol) IGEPAL® CA-630 or 1% (vol/vol) Triton™ X-100, including SIGMA*FAST* complete protease inhibitor cocktail (Sigma Aldrich) at a ratio of **approximately** 1 mL buffer per gram of cells. The lysate was cleared (20.000 rpm at 4 °C for 30 min) and the FLAG-TEV-Protein A (FpA) tagged proteins were affinity-purified from the supernatant using 300 μL IgG-Sepharose suspension (IgG-Sepharose 6 Fast Flow; GE Healthcare). Upon washing, proteins were eluted by TEV protease in a 2.5 mL Mobicol column (MoBiTec) at 16 °C for 2 h and subsequently incubated with 50 μL ANTI-FLAG® M2 affinity gel (Sigma Aldrich) at 4 °C for 1 h. Bound proteins were eluted by use of 100 μg/mL FLAG peptide (Sigma Aldrich) at 4 °C for 30–60 min. Eluates were separated by SDS-PAGE and analyzed via colloidal Coomassie staining and mass spectrometry. For the detection of glycosylation of *ct*Nsp1, an O-linked N-acetylglucosamine antibody (abcam cat. no. ab2739, 1:1000 in PBS + 0.05% Tween-20/5% milk) was applied.
21. Sucrose gradient ultra-centrifugation will be performed with the FLAG eluates, which were loaded onto a continuous sucrose gradient (10–30% (w/vol) in NB-HEPES buffer), centrifuged at 26.000 rpm for 16 h in a SW60 Ti swinging bucket rotor (Beckman Coulter) and fractionated.

Fig. 3.6 (continued) lyzed by SDS-PAGE and Coomassie staining. Whole cell lysate (WCL), the complex on IgG and FLAG beads, respectively, the TEV, and final FLAG eluates are shown. Proteins were identified by mass spectrometry. (**e**) Western analysis to identify O-linked glycosylation. Wild-type and ctNup82-FpA expressing mycelium was subjected to affinity purification. WCL as well as TEV and FLAG eluates, respectively, were probed with an antibody specific for O-linked N-acetylglucosamines (O-GlcNAc). This antibody detects a band with a molecular weight corresponding to ctNsp1, both in the whole cell lysate and the purified trimeric complex, indicating O-GlcNAc modification of this nucleoporin. A molecular weight standard with molecular weights (kDa) is depicted on the left

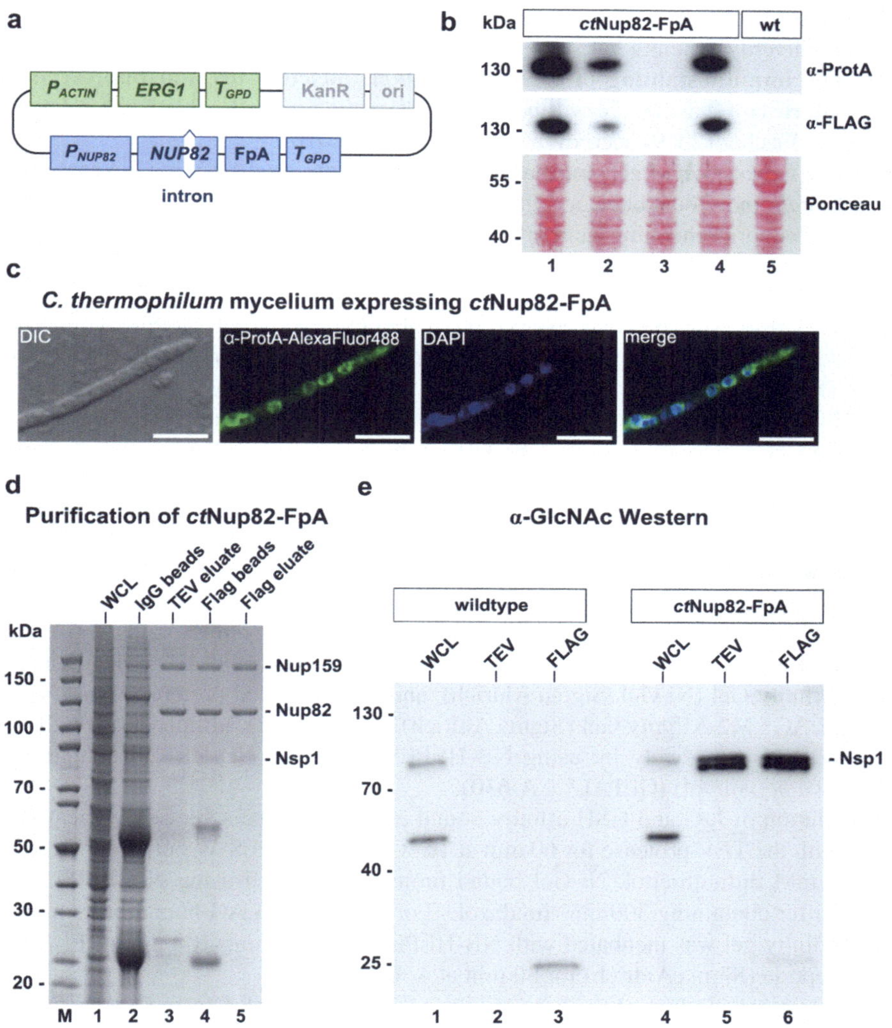

Fig. 3.6 Design and expression of constructs for affinity purification of native *C. thermophilum* proteins (After Kellner et al. 2016). (**a**) Schematic depiction of the plasmid used for transformation and integration of a C-terminally ProtA-FLAG tagged ctNup82 protein. (**b**) Western analysis of four strains positive for the PACTIN-ERG1_PNUP82-NUP82-FpA construct. Expression of the fusion protein was revealed with anti-Protein A and ANTI-FLAG antibodies on whole cell protein lysates, respectively, whereas no signal was detectable in the wild type. Equal loading was confirmed by Ponceau S staining. A molecular weight standard with molecular weights (kDa) is depicted on the left. (**c**) Indirect immunofluorescence to detect *C. thermophilum* nuclear pore complexes. An antibody directed against Protein A in combination with an Alexa Fluor 488-conjugated secondary antibody decorates nuclear rims in the strain expressing ctNup82-FpA, indicating that the fusion protein is functional and localizes to the NPC. DAPI staining was performed to visualize nuclei. DIC, GFP, and DAPI channels as well as merged fluorescent images are shown. Scale bars – 10 μm. (**d**) Tandem affinity purification of the native ctNup82 complex. Samples are ana

22. Mass spectrometric identification of proteins was carried out as described in Hakhverdyan et al. (2015).
23. **For immunostaining of *C. thermophilum* nucleoporins and fluorescence microscopy**, technique will be adapted from the protocol published by Zekert and Fischer (2009) according to Kellner et al. (2016).
24. **For expression of recombinant *C. thermophilum* nucleoporins in yeast and *E. coli***, individual nucleoporins from *C. thermophilum*, plasmids harboring the nucleoporin-encoding gene of interest under control of the *ADH1* or *GAL1/GAL1-10* promoter, respectively, will be transformed into *S. cerevisiae* strain DS1-2b43.
25. To perform simultaneous co-expression of multiple genes, the coding sequences will be inserted into appropriate yeast expression vectors, as described by Thierbach et al. (2013).
26. **For purification of recombinant *C. thermophilum* nucleoporins, reconstitution of subcomplexes, and in vitro binding assays**, cell lysis will be performed by cryogenic grinding in a cell mill. Lysed cells will be resuspended in NB-HEPES buffer containing SIGMA*FAST* complete protease inhibitor cocktail and 0.1% (vol/vol) IGEPAL® CA-630, and the lysate will be cleared by centrifugation. For purification, lysates will be incubated at 4 °C for 60 min with corresponding affinity beads: ProtA-TEV-tagged proteins will be purified by using IgG-Sepharose 6 Fast Flow, GST-TEV-tagged proteins by using Protino Glutathione Agarose 4B, His6-tagged proteins with His-Select Nickel Affinity Gel (Ni-Gel, Sigma Aldrich), and FLAG-tagged proteins with ANTI-FLAG® M2 Affinity Gel (Sigma Aldrich). After binding, affinity beads will be washed extensively by using NB-HEPES-W buffer (NB-HEPES containing 0.01% (vol/vol) IGEPAL® CA-630).
27. Elution of IgG and GSH affinity bound proteins will be achieved by cleavage with the TEV protease for 60 min at 16 °C in NB-HEPES-W buffer containing 1 mM dithiothreitol. Ni-Gel bound protein was eluted using NB-HEPES-W buffer containing 300 mM imidazole. For elution of FLAG bound protein, the affinity gel was incubated with NB-HEPES-W containing 100 μg/mL FLAG® peptide (Sigma Aldrich) for 30 min at 4 °C.
28. The reconstitution of the heterotrimeric *ct*Nup82 complex from *C. thermophilum*, consisting of *ct*Nup82, *ct*Nup159C, and *ct*Nsp1C, was performed as described. Briefly, the three constructs were co-expressed in *S. cerevisiae* and a split-tag purification of the complex was performed. In the first step, the *ct*Nup82 complex was immobilized to Ni-Gel via *ct*Nsp1C:His6, eluted with imidazole, and further purified in a second step via FLAG tagged *ct*Nup159C. The purified *ct*Nup82 complex was used as bait in the following binding experiments. As prey, individual *ct*Nup84 complex nucleoporins were purified and eluted as outlined earlier. Alternatively, the heterotrimeric *ct*Nup84 complex,

consisting of *ct*Nup120, *ct*Nup85, and *ct*Nup145C, was co-expressed in *S. cerevisiae*, purified via ProtA-TEV tagged *ct*Nup85, and eluted with TEV protease as illustrated. For binding experiments or reconstitution of higher-order complexes, the bait protein complex was incubated with tenfold excess of prey protein for 60 min at 16 °C. After extensive washing, the bait proteins were eluted as described earlier and analyzed by SDS-PAGE and Coomassie staining.

29. For gradient ultra-centrifugation, the purified and reconstituted heterotrimeric *ct*Nup82 complex and the heteroheptameric *ct*Nup82–Nup84 super-complex, respectively, were loaded onto continuous glycerol gradient (10–30% w/vol in NB-HEPES buffer) and centrifuged at 33.000 rpm for 16 h in an SW60 Ti swinging bucket rotor (Beckman Coulter). About 200 μL fractions were collected and analyzed by SDS-PAGE and Coomassie staining.

The genetic method developed by the previous method introduced by Kellner et al. (2016) is based on polyethylene glycol (PEG)-induced protoplast transformation that was successfully applied in the past for members of the Pezizomycotina clade (e.g., *Sordaria macrospora*) to which *C. thermophilum* also belongs. For the generation of protoplasts from the thermophile, a mixture of different fungal cell wall-degrading enzymes was applied to young mycelium of a wild-type *C. thermophilum* var. *thermophilum* (La Touche 1950; DSM-1495) growing as submerged cultures. The resulting protoplasts were collected by filtration and centrifugation, washed, and subsequently subjected to the transformation procedure (Fig. 3.5).

In conclusion, the genetic tools in *C. thermophilum* developed by Kellner et al. (2016), which allow applications regarding affinity purification of thermostable macromolecular complexes assembled under physiological conditions directly from the thermophile (Fig. 3.7). This method holds great potential to obtain complicated protein complexes as well as RNPs from a thermophile, including native protein and RNA modifications. Moreover, the inherent thermostability of such assemblies will facilitate the identification of transient protein–protein interactions, which may escape during isolation of complexes from mesophiles. Thus, our method could pave the way for *C. thermophilum* to further develop into a model organism for biochemical and prospective structural analyses of eukaryotic macromolecular complexes and also applications in biotechnology are expected to emerge.

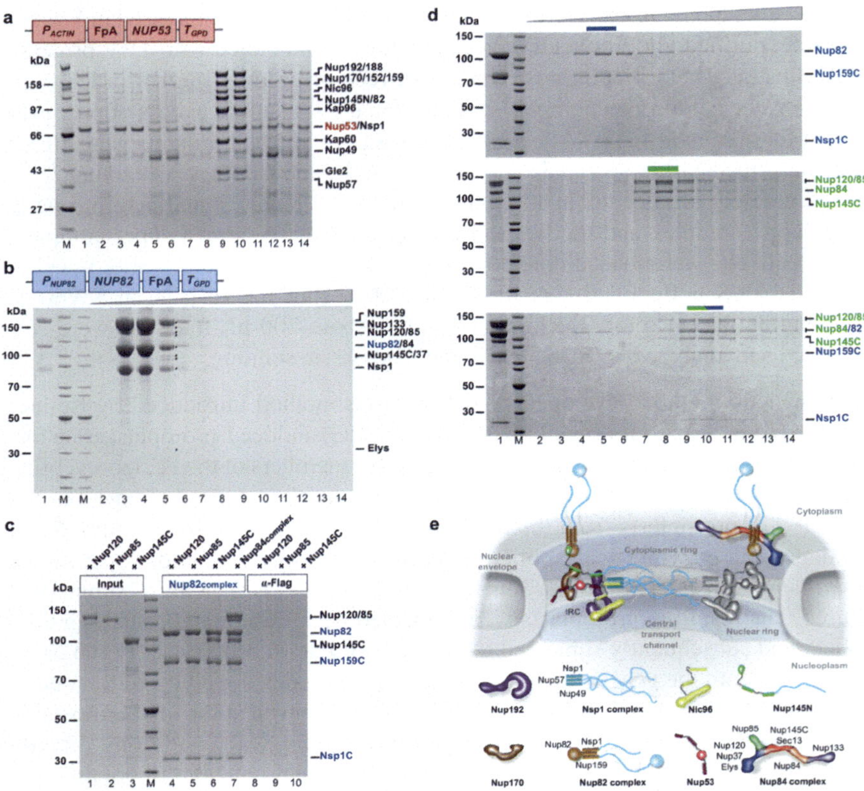

Fig. 3.7 Biochemistry of native *C. thermophilum* nuclear pore subcomplexes and reconstitution of a Nup82–Nup84 super-complex (After Kellner et al. 2016). (**a**) Affinity capture of *ct*Nup53-FpA was performed on cleared lysates under various extraction conditions. Elution of the complexes was followed by SDS-PAGE and Coomassie staining. Proteins identified by mass spectrometric analysis from the complexes isolated in condition #9 are indicated. Extraction conditions (#1–14) are presented in Supplementary Table 1. A molecular weight marker (M) is indicated on the left. (**b**) Sucrose gradient centrifugation of the *ct*Nup82 complex upon affinity capture of *ct*Nup82-FpA. Fractions were collected and analyzed by SDS-PAGE and Coomassie staining. Individual proteins were identified by mass spectrometric analysis. (**c**) In vitro binding assay with an immobilized Nup82–Nup159C1–Nsp1C complex (*ct*Nup82 complex) and full-length *ct*Nup120, *ct*Nup85, and *ct*Nup145C (*ct*Nup84 complex) (input, lanes 1–3). Samples are eluted via FLAG peptide and analyzed by SDS-PAGE and Coomassie staining (lanes 5–11). MW, molecular weight; kDa, kilodalton; α-FLAG, ANTI-FLAG M2 Affinity Gel. (**d**) Glycerol-gradient centrifugation of in vitro reconstituted complexes. Fractions were collected and analyzed by SDS-PAGE and Coomassie staining. (Top) *ct*Nup82 complex, (middle) *ct*Nup84 complex including *ct*Nup84, (bottom) *ct*Nup82–*ct*Nup84 complex including *ct*Nup84. (**e**) Schematic model for the connection of the Nup53-containing inner pore ring complex (IRC) with the Nup82 outer ring and the Nsp1 channel complex (left) and how the Nup82 complex might be linked to the Y-shaped octameric Nup84 complex within the NPC protomer

References

Aggarwal R, Gupta S, Sharma S, Shukla R (2014) Development of conventional and real time PCR assay for rapid detection and quantification of a biocontrol agent, Chaetomium globosum. J Pl Pathol 96(3):477–485

Al-Samarrai TH, Schmid J (2000) A simple method for extraction of fungal genomic DNA. Lett Appl Microbiol 30:53–56

Amlacher S et al (2011) Insight into structure and assembly of the nuclear pore complex by utilizing the genome of a eukaryotic thermophile. Cell 146:277–289

Aslam S, Tahir A, Aslam MF, Alam MW, Shedayi AA, Sadia S (2017) Recent advances in molecular techniques for the identification of phytopathogenic fungi – a mini review. J Plant Interact 12(1):493–504

Atkins SD, Clark IM (2004) Fungal molecular diagnostics: a mini review. J Appl Genet 45(1):3–15

Badali H, Nabili M (2012) Molecular tools in medical mycology; where we are! Jundishapur J Microbiol 6(1):1–3

Baschien C, Manz W, Neu TR, Szewzyk U (2001) Fluorescence in situ hybridization of freshwater fungi. Int Rev Hydrobiol 86(4–5):371–381

Berg Jeremy M (2007) Biochemistry, 6th edn. W. H. Freeman and Company, New York

Berka RM, Grigoriev IV, Otillar R, Salamov A, Grimwood J, Reid I, Ishmael N et al (2011) Comparative genomic analysis of the thermophilic biomassdegrading fungi Myceliophthora thermophila and Thielavia terrestris. Nat Biotechnol 29:922–927

Bock T et al (2014) An integrated approach for genome annotation of the eukaryotic thermophile Chaetomium thermophilum. Nucleic Acids Res 42:13525–13533

Borman AM, Linton CJ, Miles SJ, Johnson EM (2008) Molecular identification of pathogenic fungi. J Antimicrob Chemother 61(Suppl 1):i7–i12. https://doi.org/10.1093/jac/dkm425

Bu R, Sathiapalan RK, Ibrahim MM, Al-Mohsen I, Almodavar E, Gutierrez MI, Bhatia K (2005) Monochrome LightCycler PCR assay for detection and quantification of five common species of Candida and Aspergillus. J Med Microbiol 54(3):243–248

Capote N, Aguado A, Pastrana AM, Sánchez-Torres P (2012) Molecular tools for detection of plant pathogenic fungi and fungicide resistance. INTECH Open Access Publisher, Valencia

Chalupová J, Raus M, Sedlářová M, Šebela M (2014) Identification of fungal microorganisms by MALDI-TOF mass spectrometry. Biotechnol Adv 32(1):230–241

Chang C-C, Chen C-C, Wei S-C, Lu H-H, Liang Y-H, Lin C-W (2012) Diagnostic devices for isothermal nucleic acid amplification. Sensors 12(6):8319–8337

Compton J (1991) Nucleic acid sequence-based amplification. Nature 350(6313):91–92

Cheng CK, Au CH, Wilke SK, et al. (2013) 5'-serial analysis of gene expression studies reveal a transcriptomic switch during fruiting body development in Coprinopsis cinerea. BMC Genomics.;14:195. Published 2013 Mar 20. doi:https://doi.org/10.1186/1471-2164-14-195

Cooney DG, Emerson R (1964) Thermophilic Fungi: an account of their biology, activities, and classification. W.H. Freeman and Co, San Francisco

Crisan EV (1964) Isolation and culture of thermophilic fungi. Contribution from Boyce Thompson Institute Plant Research 22:291–301

Dahlmann TA, Kuck U (2015) Dicer-dependent biogenesis of small RNAs and evidence for microRNA-like RNAs in the penicillin producing fungus Penicillium chrysogenum. PLoS One 10:e0125989

Dasmahapatra K, Mallet J (2006) DNA barcodes: recent successes and future prospects. Heredity 97(4):254–255

Dawei W, Peng Y (2014) Assessing the impact of dominant sequencing-based gene expression profiling techniques (SGEPTs) on phytopathogenic fungi. Chiang Mai J Sci 41:922–9444

Egger KN (1995) Molecular analysis of ectomycorrhizal fungal communities. Can J Bot 73:1415

Elleuche S, Schafers C, Blank S, Schroder C, Antranikian G (2015) Exploration of extremophiles for high temperature biotechnological processes. Curr Opin Microbiol 25:113–119

Ellison MA, McMahon MB, Bonde MR, Palmer CL, Luster DG (2016) In situ hybridization for the detection of rust fungi in paraffin embedded plant tissue sections. Plant Methods 12(1):37

Endo S, Komori T, Ricci G, Sano A, Yokoyama K, Ohori A, Kamei K, Franco M, Miyaji M, Nishimura K (2004) Detection of gp43 of Paracoccidioides brasiliensis by the loop-mediated isothermal amplification (LAMP) method. FEMS Microbiol Lett 234(1):93–97

Eshaque B, Dixon B (2006) Technology platforms for molecular diagnosis of cystic fibrosis. Biotechnol Adv 24(1):86–93

Fakruddin M (2011) Loop mediated isothermal amplification (LAMP) – an alternative to polymerase chain reaction (PCR). Bangladesh Res Pub J 5(4):425–439

Fakruddin M, Mazumdar RM, Chowdhury A, Mannan K (2012) Nucleic acid sequence based amplification (NASBA)-prospects and applications. Int J Life Sci Pharma Res 2:106–107

Fang Y, Ramasamy RP (2015) Current and prospective methods for plant disease detection. Biosensors 5(3):537–561

Fitzpatrick DA, Logue ME, Stajich JE, Butler G (2006) A fungal phylogeny based on 42 complete genomes derived from supertree and combined gene analysis. BMC Evol Biol 6(99). https://doi.org/10.1186/1471-2148-6-99.

Gardes M, Bruns TD (1993) ITS primers with enhanced specificity for basidiomycetes – application to the identification of mycorrhizae and rusts. Mol Ecol 2:113–118. https://doi.org/10.1111/j.1365-294X.1993.tb00005.x2

Gianfreda L, Rao MA (2004) Potential of extra cellular enzymes in remediation of polluted soils: a review. Enzym Microb Technol 35:339–354

Grabherr MG, Haas BJ, Yassour M, Levin JZ, Thompson DA, Amit I, Adiconis X, Fan L, Raychowdhury R, Zeng Q (2011) Full-length transcriptome assembly from RNA-Seq data without a reference genome. Nat Biotechnol 29:644–652

Gusev Y, Sparkowski J, Raghunathan A, Ferguson H, Montano J, Bogdan N, Schweitzer B, Wiltshire S, Kingsmore SF, Maltzman W (2001) Rolling circle amplification: a new approach to increase sensitivity for immunohistochemistry and flow cytometry. Am J Pathol 159(1):63–69

Glass NL, Donaldson GC (1995) Development of primer sets designed for use with the PCR to amplify conserved genes from filamentous ascomycetes. Appl Environ Microbiol 61:1323–1330

Hakhverdyan Z et al (2015) Rapid, optimized interactomic screening. Nat Methods 12:553–560

Hardman G (2004) Microarray platforms – comparison and contrasts. Pharmacogenomics 5:487–502. http://www.futuremedicine.co.uk

Hayden R, Qian X, Roberts G, Lloyd R (2001) In situ hybridization for the identification of yeast-like organisms in tissue section. Diagn Mol Pathol 10(1):15–23

Hayden R, Qian X, Procop G, Roberts G, Lloyd R (2002) In situ hybridization for the identification of filamentous fungi in tissue section. Diagn Mol Pathol 11(2):119–126

Hebert PD, Gregory TR (2005) The promise of DNA barcoding for taxonomy. Syst Biol 54(5):852–859

Hibbett DS, Binder M, Bischoff JF, Blackwell M, Cannon PF, Eriksson OE, Huhndorf S, James T (2007) A higher-level phylogenetic classification of the fungi. Mycol Res 111:509–547

Houbraken J, Spierenburg H, Frisvad JC (2012) Rasamsonia, a new genus comprising thermotolerant and thermophilic Talaromyces and Geosmithia species. Antonie Van Leeuwenhoek 101:403–421

Ishii H, Holloman DW (2015) Fungicide resistance in plant pathogens. Springer, Tokyo

Jensen E (2014) Technical review: in situ hybridization. Anat Rec 297(8):1349–1353

Kang K, Zhong J, Jiang L, Liu G, Gou CY et al (2013) Identification of microRNA-like RNAs in the filamentous fungus Trichoderma reesei by Solexa sequencing. PLoS One 8:e76288

Kellner N, Schwarz J, Sturm M, Fernandez-Martinez J, Griesel S, Zhang W, Chait BT, Rout MP, Kück U, Hurt E (2016) Developing genetic tools to exploit Chaetomium thermophilum for biochemical analyses of eukaryotic macromolecular assemblies. Scientific Reports. 6:20937. PMID 26864114. PMCID: PMC4750058 https://doi.org/10.1038/srep20937

Kim SW, Li Z, Moore PS, Monaghan AP, Chang Y, Nichols M, John B (2010) A sensitive non-radioactive northern blot method to detect small RNAs. Nucleic Acids Res 38:e98

Krishnamurthy PK, Francis RA (2012) A critical review on the utility of DNA barcoding in biodiversity conservation. Biodivers Conserv 21(8):1901–1919
Kuhn H, Demidov VV, Frank-Kamenetskii MD (2002) Rolling-circle amplification under topological constraints. Nucleic Acids Res 30(2):574–580
Kiesling TL, Diaz MR, Statzell-Tallman A, Fell JW (2002) Identification of yeasts with DNA hybridization macroarrays. In: Hyde KD (ed) Fungi in Marine Environments. Fungal Diversity Press, Hong Kong, pp 69–80
Kristiansen HB (2014) Characterization of marine fungal communities using next generation sequencing techniques. Master thesis. University of Svalbard. https://pdfs.semanticscholar.org/209c/9b057cb7dc009fbbdf702a70dfe80d146b6d.pdf?_ga=2.15207862.1422174194.1570959968-2025412611.1565374159
Kluge AG (1989) A concern for evidence and a phylogenetic hypothesis of relationships among epicrates (Boidae, Serpentes). Syst Zool 38(1):7–25
Lee D, La Mura M, Allnutt TR, Powell W (2009) Detection of genetically modified organisms (GMOs) using isothermal amplification of target DNA sequences. BMC Biotechnol 9(1):7
Leinberger DM, Schumacher U, Autenrieth IB, Bachmann TT (2005) Development of a DNA microarray for detection and identification of fungal pathogens involved in invasive mycoses. J Clin Microbiol 43(10):4943–4953
Li AN et al (2010) Two novel thermostable chitinase genes from thermophilic fungi: cloning, expression and characterization. Bioresour Technol 101:5546–5551
Li DZ, Gao LM, Li HT, Wang H, Ge XJ, Liu JQ, Chen ZD, Zhou SL, Chen SL, Yang JB (2011) Comparative analysis of a large dataset indicates that internal transcribed spacer (ITS) should be incorporated into the core barcode for seed plants. Proc Natl Acad Sci 108(49):19641–19646
Lin YL, Ma LT, Lee YR, Lin SS, Wang S-Y et al (2015) microRNA-like small RNAs prediction in the development of Antrodia cinnamomea. PLoS One 10:e0123245
Liu CM, Kachur S, Dwan MG, Abraham AG, Aziz M, Hsueh P, Huang Y, Busch JD, Lamit LJ, Gehring CA, Keim PS, Price LB (2012) FungiQuant: a broad-coverage fungal quantitative real-time PCR assay. BMC Microbiol 12:255. https://doi.org/10.1186/1471-2180-12-255
Liu Z, Macias MJ, Bottomley MJ, Stier G, Linge JP, Nilges M, Bork P, Sattler M (1999) The three-dimensional structure of the HRDC domain and implications for the Werner and Bloom syndrome proteins. Structure 7(12):1557–1566
Liu W, Huang S, Liu N, Dong D, Yang Z, Tang Y, Ma W, He X, Ao D, Xu Y (2017) Establishment of an accurate and fast detection method using molecular beacons in loop-mediated isothermal amplification assay. Sci Rep 7:40125. https://doi.org/10.1038/srep40125
Lizardi PM, Huang X, Zhu Z, Bray-Ward P, Thomas DC, Ward DC (1998) Mutation detection and single-molecule counting using isothermal rolling-circle amplification. Nat Genet 19(3):225–232
Luo M, Liang XQ, Dang P, Holbrook CC, Bausher MG, Lee RD, Guo BZ (2005) Microarray-based screening of differentially expressed genes in peanut in response to Aspergillus parasiticus infection and drought stress. Plant Sci 169:695–703
La Touche CJ (1948) A Chaetomium-like thermophile fungus. Nature 161:320
La Touche CJ (1950) On a thermophilic species of Chaetomium. Trans Br Mycol Soc 33:94–104
Mackay IM (2004) Real-time PCR in the microbiology laboratory. Clin Microbiol Infect 10(3):190–212
Maheshwari R, Bharadwaj G, Bhat MK (2000) Thermophilic fungi: their physiology and enzymes. Microbiol Mol Biol Rev 64:461–488
Mancini V, Murolo S, Romanazzi G (2016) Diagnostic methods for detecting fungal pathogens on vegetable seeds. Plant Pathol 65:691–703
Martin KJ, Rygiewicz PT (2005) Fungal-specific PCR primers developed for analysis of the ITS region of environmental DNA extracts. BMC Microbiol 5:28
Masanga JO, Matheka JM, Omer RA, Ommeh SC, Monda EO, Alakonya AE (2015) Downregulation of transcription factor aflR in Aspergillus flavus confers reduction to aflatoxin accumulation in transgenic maize with alteration of host plant architecture. Plant Cell Rep 34:1379–1387. https://doi.org/10.1007/s00299-015-1794-9

Matheny PB, Moncalvo JM, Redhead SA (2007) Agaricales. Version 9 May 2007. http://tolweb.org/Agaricales/20551/2007.05.09 in The Tree of Life web project

McCartney HA, Foster SJ, Fraaije BA, Ward E (2003) Molecular diagnostics for fungal plant pathogens. Pest Manag Sci 59(2):129–142

Mchunu NP, Permaul K, Abdul Rahman AY, Saito JA, Singh S, Alam M, Rahman AYA. (2013) Xylanase superproducer: genome sequence of a compost-loving thermophilic fungus, Thermomyces lanuginosus strain SSBP. Genome Announc 1, e00388-13

Mendoza-Mendoza A, Steyaert J, Nieto-Jacobo MF, Holyoake A, Braithwaite M, Stewart A (2015) Identification of growth stage molecular markers in Trichoderma sp. 'atroviride type B' and their potential application in monitoring fungal growth and development in soil. Microbiology 161:2110–2126. https://doi.org/10.1099/mic.0.000167

Mortazavi A, Williams BA, McCue K, Schaeffer L, Wold B (2008) Mapping and quantifying mammalian transcriptomes by RNA-Seq. Nat Methods 5:621–628

Mostafa AT, Armin A, Hamid P, Reza AM (2012) Rapid detection methods for analysis of fungi and mycotoxins in agriculture products. Res J Recent Sci 1(7):90–98

Mouchacca J (2000) Thermophilic fungi and applied research: a synopsis of name changes and synonymies. World J Microbiol Biotechnol 16:881–888

Morgenstern I, Powlowski J, Ishmael N, Darmond C, Marqueteau S, Moisan M, Quenneville G, Tsang A (2012) A molecular phylogeny of thermophilic fungi. Fungal Biol 116:489–502

Martinez D, Berka RM, Henrissat B, Saloheimo M, Arvas M, Baker SE, Chapman J (2008) Genome sequencing and analysis of the biomass-degrading fungus Trichoderma reesei (syn. Hypocrea jecorina). Nat Biotechnol 26:553–560

Nakayashiki T, Kurtzman CP, Edskes HK, Wickner RB (2005) Yeast prions [URE3] and [PSI+] are diseases. Proc Natl Acad Sci 102(30):10575–10580

Nilsson RH, Abarenkov K, Larsson K-H, Kõljalg U (2011) Molecular identification of fungi: rationale, philosophical concerns, and the UNITE database. Open Appl Inform J 5:81–86

Nilsson M, Malmgren H, Samiotaki M, Kwiatkowski M, Chowdhary U (1994) Padlock probes: circularizing oligonucleotides for localized DNA detection. Science 265(5181):4–5

Nierman WC, Pain A, Anderson MJ, Wortman JR, Kim HS, Arroyo J, Berriman M et al (2005) Genomic sequence of the pathogenic and allergenic filamentous fungus Aspergillus fumigatus. Nature 438:1151–1156

O'Donnell K, Lutzoni FM, Ward TJ, Benny GL (2001) Evolutionary relationships among mucoralean fungi (Zygomycota): evidence for family polyploidy on a large scale. Mycologia 93:286–296

Oliveira TB, Gomes E, Rodrigues A (2015) Thermophilic fungi in the new age of fungal taxonomy. Extremophiles 19:31–37

Panwar V, McCallum B, Bakkeren G (2012) Endogenous silencing of Puccinia triticina pathogenicity genes through in planta-expressed sequences leads to the suppression of rust diseases on wheat. Plant J 73:521–532. https://doi.org/10.1111/tpj.12047

Peruski AH, Peruski LF (2003) Immunological methods for detection and identification of infectious disease and biological warfare agents. Clin Diagn Lab Immunol 10(4):506–513

Pan WZ, Huang XW, Wei KB, Zhang CM, Yang DM, Ding JM, Zhang KQ (2010) Diversity of thermophilic fungi in Tengchong Rehai National Park revealed by ITS nucleotide sequence analyses. J Microbiol 48:146–152

Pel HJ, de Winde JH, Archer DB, Dyer PS, Hofmann G, Schaap PJ, Turner G et al (2007) Genome sequencing and analysis of the versatile cell factory Aspergillus niger CBS 513.88. Nat Biotechnol 25:221–231

Qian X, Lloyd RV (2003) Recent developments in signal amplification methods for in situ hybridization. Diagn Mol Pathol 12(1):1–13

Rahman MT, Uddin MS, Sultana R, Moue A, Setu M (2013) Polymerase chain reaction (PCR): a short review. Anwer Khan Mod Med Coll J 4(1):30–36

Rajendhran J, Gunasekaran P (2011) Microbial phylogeny and diversity: small subunit ribosomal RNA sequence analysis and beyond. Microbiol Res 166:99–110

Rehner S (2001) Primers for elongation factor 1-a (EF1-a). http://ocid.NACSE.ORG/research/deephyphae/EF1primer.pdf

Ren CH, Hu CQ, Luo P, Wang QB (2009) Sensitive and rapid identification of Vibrio vulnificus by loop-mediated isothermal amplification. Microbiol Res 164(5):514–521

Robinson B, Erle D, Jones D, Shapiro S, Metzger W, Albelda S, Parks W, Boylan A (2000) Recent advances in molecular biological techniques and their relevance to pulmonary research. Thorax 55(4):329–339

Roe AD, Rice AV, Bromilow SE, Cooke JE, Sperling FA (2010) Multilocus species identification and fungal DNA barcoding: insights from blue stain fungal symbionts of the mountain pine beetle. Mol Ecol Resour 10(6):946–959

Russo G, Zegar C, Giordano A (2003) Advantages and limitations of microarray technology in human cancer. Oncogene 22(42):6497–6507

Saiki RK, Scharf S, Faloona F, Mullis K, Horn G, Erlich H, Arnheim N (1992) Enzymatic amplification of beta-globin genomic sequences and restriction site analysis for diagnosis of sickle cell anemia. Biotechnology 24:476

Sambrook J, Fritsch EF, Maniatis T (1989) Molecular cloning. A laboratory manual. Cold Spring Harbor Laboratory Press, New York

Sato T, Takayanagi A, Nagao K, Tomatsu N, Fukui T, Kawaguchi M, Kudoh J, Amagai M, Yamamoto N, Shimizu N (2010) Simple PCR-based DNA microarray system to identify human pathogenic fungi in skin. J Clin Microbiol 48(7):2357–2364

Savolainen V, Cowan RS, Vogler AP, Roderick GK, Lane R (2005) Towards writing the encyclopaedia of life: an introduction to DNA barcoding. Philos Trans R Soc B 360(1462):1805–1811

Schoch CL, Seifert KA, Huhndorf S, Robert V, Spouge JL, Levesque CA, Chen W, Bolchacova E, Voigt K, Crous PW (2012) Nuclear ribosomal internal transcribed spacer (ITS) region as a universal DNA barcode marker for fungi. Proc Natl Acad Sci U S A 109(16):6241–6246

Sergentet-Thevenot D, Montet MP, Vernozy-Rozand C (2008) Challenges to developing nucleic acid sequence based amplification technology for the detection of microbial pathogens in food. Rev Med Vet 159:514–527

Sigoillot C, Camarero S, Vidal T, Record E, Asther M, Pérez-Boada M, Martínez MJ, Sigoillot JC, Asther M, Colom JF, Martínez AT (2005) Comparison of different fungal enzymes for bleaching high-quality paper pulps. J Biotechnol 115:333–343

Singh A, Kumar N (2013) A review on DNA microarray technology. Int J Curr Res Rev 5(22):01–05

Spring O, Thines M (2010) Molecular techniques for classification and diagnosis of plant pathogenic Oomycota. In: Gherbawy Y, Voigt K (eds) Molecular identification of fungi. Springer, Berlin, Heidelberg, pp 35–50

Stuwe T et al (2015) Architecture of the fungal nuclear pore inner ring complex. Science 350:56–64

Sun J, Li X, Zeng H, Xie Z, Lu C, Xi L, De Hoog GS (2010) Development and evaluation of loop-mediated isothermal amplification (LAMP) for the rapid diagnosis of Penicillium marneffei in archived tissue samples. FEMS Immunol Med Microbiol 58(3):381–388

Tanabe Y, Saikawa M, Watanabe MM, Sugiyama J (2004) Molecular phylogeny of Zygomycota based on EF-1α and RPB1 sequences: limitations and utility of alternative markers to rDNA. Mol Phylogenet Evol 30(2):438–449

Tang Y-W, Procop GW, Persing DH (1997) Molecular diagnostics of infectious diseases. Clin Chem 43(11):2021–2038

Taoufik A, Nijhof A, Hamidjaja R, Jongejan F, Pillay V, Sonnevelt M, de Boer M (2004, Sept) Reverse line blot hybridisation in the detection of tick-borne diseases. BTi.

Thierbach K et al (2013) Protein interfaces of the conserved Nup84 complex from Chaetomium thermophilum shown by crosslinking mass spectrometry and electron microscopy. Structure 21:1672–1682

Trilles L, Wang B, Firacative C, Lazéra MDS, Wanke B, Meyer W (2014) Identification of the major molecular types of Cryptococcus neoformans and C. gattii by hyperbranched rolling circle amplification. PLoS One 9(4):e94648

Trtkova J, Raclavsky V (2006) Molecular-genetic approaches to identification and typing of pathogenic Candida yeasts. Biomed Pap 150(1):51–61

Tsui CK, Woodhall J, Chen W, Lévesque CA, Lau A, Schoen CD, Baschien C, Najafzadeh MJ, de Hoog GS (2011) Molecular techniques for pathogen identification and fungus detection in the environment. IMA Fungus 2(2):177
Turner BM (2007) Defining an epigenetic code. Nat Cell Biol 9:2–6
Valinsky L, Vedoa GD, Jiang T, Borneman J (2002) Oligonucleotide fingerprinting of rRNA genes for analysis of fungal community composition. Appl Environ Microbiol 68:5999–6004
van Noort V et al (2013) Consistent mutational paths predict eukaryotic thermostability. BMC Evol Biol 13:7
Velculescu VE et al (1995) Serial analysis of gene expression. Science 270:484–487
Vilela R, Mendoza L, Rosa S, Faria A, Madeira S, Vladimir D (2005) Molecular model for studying the uncultivated fungal pathogen Lacazia loboi. Microbiology 43:3657–36618
Vilgalys R, Hester M (1990) Rapid genetic identification and mapping of enzymatically amplified ribosomal DNA from several Cryptococcus species. J Bacteriol 172:4238–4246
Vilgalys R, Hopple JS, Hibbett DS (1994) Phylogenetic implications of generic concepts in fungal taxonomy: the impact of molecular systematic studies. Mycol Helv 6:73–91
Villari C, Tomlinson JA, Battisti A, Boonham N, Capretti P, Faccoli M (2013) Use of loop-mediated isothermal amplification for detection of Ophiostoma clavatum, the primary blue stain fungus associated with Ips acuminatus. Appl Environ Microbiol 79(8):2527–2533
Volpi EV, Bridger JM (2008) Fish glossary: an overview of the fluorescence in situ hybridization technique. BioTechniques 45(4):385–386
Wang Z, Yang B (2010) Padlock-probes and rolling-circle amplification. In: Microrna expression detection methods. Springer, Berlin, pp 241–247. https://doi.org/10.1007/978-3-642-04928-6_16
Wang J, Wu Y, Gong Y, Yu S, Liu G (2015) Enhancing xylanase production in the thermophilic fungus Myceliophthora thermophila by homologous overexpression of Mtxyr1. J Ind Microbiol Biotechnol 42:1233–1241
White TJ, Bruns T, Lee S, Taylor J (1990) Amplification and direct sequencing of fungal ribosomal RNA genes for phylogenetics. In: Innis MA, Gelfand DH, Sninsky JJ (eds) PCR protocols: a guide to methods and applications. Academic Press, New York, pp 315–322
Wesenberg D, Kyriakides I, Agathos SN (2003) White-rot fungi and their enzymes for the treatment of industrial dye effluents. Biotechnol Adv 22:161–187
Xu J (2016) Fungal DNA barcoding 1. Genome 59(11):913–932
Yeo SF, Wong B (2002) Current status of nonculture methods for diagnosis of invasive fungal infections. Clin Microbiol Rev 35(3):465–484
Zekert N, Fischer R (2009) The Aspergillus nidulans kinesin-3 UncA motor moves vesicles along a subpopulation of microtubules. Mol Biol Cell 20:673–684
Zhao R, Jeewon R, Desjardin DE, Soytong K, Hyde KD (2007) Ribosomal DNA phylogenies of Cyathus: is the current infrageneric classification appropriate? Mycologia 99:385–395
Zhou P, Zhang G, Chen S, Jiang Z, Tang Y, Henrissat B, Yan Q, Yang S, Chen CF, Zhang B, Du Z (2014) Genome sequence and transcriptome analyses of the thermophilic zygomycete fungus Rhizomucor miehei. BMC Genomics 15:294
Zhang Y, Wu W, Cai L (2017) Polyphasic characterisation of Chaetomium species from soil and compost revealed high number of undescribed species. Fungal Biol 121(1):21–43

Chapter 4
Recent Advances on Occurrence of Genus *Chaetomium* on Dung

Francisco J. Simões Calaça, Solange Xavier-Santos, and Ahmed M. Abdel-Azeem

4.1 Life Cycle of Dung Fungi and the Classification as an Obligate Dung Species

Coprophilous fungi, also known as *dung fungi*, are a broad fungal group understood in two denomination (sense), in accordance with its life history and, mainly, its substrate preference (Doveri 2004). The true dung fungi (*stricto sensu*) are all fungal species, as *Saccobolus* spp., *Pilobolus* spp., *Hypocopra* spp., that must pass through the animal's gut as a decisive step to the emergence (i.e. germination) of fungal spores on dung, after defecation, being, therefore, known as coprophilous fungi, as defined in Doveri (2004). On the other hand, there are species of dung fungi (*lato sensu*) whose dung is one of many other suitable substrates where their can complete its life cycle, for example barks, leaves, undefined standing litter or soil. These species are preferably named as fimicolous fungi (Lundqvist 1972; Abdel-Azeem and Salem 2015; Calaça et al. 2017).

Some fimicolous species can spread its spores in the matrix and reach substrates other than live vegetation, mainly when dung is not available as an "island" for colonization, as in the case of nonobligate gut passage fungi. Feces represent an immediate source of resource for the survival of a great diversity of dung-lovers or

F. J. Simões Calaça (✉) · S. Xavier-Santos
Universidade Estadual de Goiás, Campus de Ciências Exatas e Tecnológicas Henrique Santillo, Anápolis, GO, Brazil

Laboratory of Basic and Applied Mycology and Scientific Dissemination – FungiLab, Anápolis, GO, Brazil

A. M. Abdel-Azeem
University of Suez Canal, Botany Department, Faculty of Science, Ismailia, Egypt
e-mail: ahmed_abdelazeem@science.suez.edu.eg

A. M. Abdel-Azeem (ed.), *Recent Developments on Genus Chaetomium*, Fungal Biology, https://doi.org/10.1007/978-3-030-31612-9_4

dung-dependent organisms (Bell 1983; Jones 2017; Frank et al. 2017). Other substrates can include barks, standing litter, or soil and are suitable for growth of species nonspecific to dung, like those belonging to the genus *Chaetomium*, where they can complete their life cycle (Fig. 4.1).

Many environmental factors directly and indirectly influence the expression of named copromicodiversity, as recently compiled by Calaça et al. (2017). Animal community heterogeneity is one of the first drivers in modulating dung physicochemical characteristics and which species will develop for each dung type (i.e., ruminant or monogastric type). Indeed, increased grazing by cattle is positively associated with dung fungi abundance (Eldridge and Delgado-Baquerizo 2018); once more dung are produced, more spores can be spread and emerge, developing new reproductive structures, restarting the cycle continuously. However, some fimicolous species are not influenced by biotic and abiotic factors in the same way as observed for true coprophilous fungi. Species of *Chaetomium*, for example, are able to grow on various materials, and its spores can be found free in the air, being important inhalant allergens in indoor environments (Wang et al. 2016a), soils, and live or decaying plant debris (Pornsuriya et al. 2008; Asgari and Zare 2011; Blanchette et al. 2017).

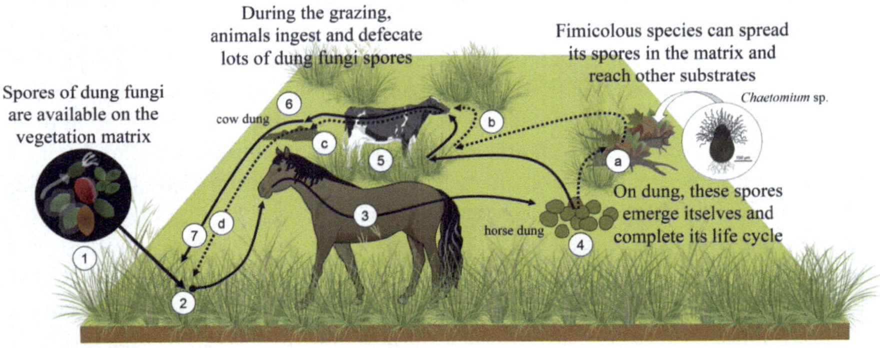

Fig. 4.1 A typical dung fungi life cycle. The entire arrows show the pathway for coprophilous (obligatory gut passengers) fungi and the dotted arrows show the pathway for fimicolous fungi. (1) A great diversity of fungal spores of both dung fungi species than other groups of fungi are dispersed on a matrix of many types of substrates especially living plants. (2) These spores are available for ingestion by animals during feeding. (3) For coprophilous fungi, as far as we know, passage through the gut is an important step to spore emergence on dung, because different phases of digestion are responsible for breaking of spore dormancy. (4) Dung fungi spores emerge and grow on dung and complete their life cycle, but some non-coprophilous species can pass through the gut without any injury (*Chaetomium* spp.) and develop on dung. These fungal species will spread its spores on the matrix (a) and develop. (5, b) Finally, the spores will be dispersed and ingested again. The life cycle is constantly restarted (6, 7 and c, d). (Figure drawn by the authors, with vectors available on "Integration and Application Network, University of Maryland Center for Environmental Science". http://ian.umces.edu/symbols/)

However, research such as those developed by mycologist Francesco Doveri (2004, 2008, 2011, 2013, 2016, 2018) and others (Piasai and Manoch 2009; Mungai et al. 2012; Altayyar et al. 2017) have shown that the genus displays a good capacity to survive on dung, considering an environment with high competition (fungi compete for space and food with other fungi, myxomycetes, bacteria, nematodes, insects), which is associated with cellulose degradation capacity and production of secondary metabolites that make the species of this genus potential competitors, frequent in most damp chamber cultures of dung fungi studies (Saito et al. 1988; Soytong et al. 2001; Doveri 2011).

4.2 Occurrence of Genus *Chaetomium* on Dung

Chaetomiaceae G. Winter is a family of saprobic or parasitic fungi, growing on plant debris, dung, soil, and other substrates (Maharachchikumbura et al. 2016) and included in class *Sordariomycetes* and order *Sordariales*. Notably, it comprises 19 genera (*Achaetomium* J.N. Rai et al., *Bommerella* Marchal, *Boothiella* Lodhi & Mirza, *Botryotrichum* Sacc. & Marchal, *Chaetomidium* (Zopf) Sacc., *Corynascella* Arx & Hodges, *Chaetomiopsis* Mustafa & Abdul-Wahid, *Chaetomium* Kunze, *Crassicarpon* Y. Marín et al., *Emilmuelleria* Arx, *Farrowia* D. Hawksw., *Guanomyces* M.C. Gonzáles et al., *Humicola* Traaen, *Madurella* Brumpt, *Myceliophthora* Costantin, *Staphylotrichum* J.A. Mey. & Nicot, *Subramaniula* Arx, *Taifanglania* Z.Q. Liang et al., *Thielavia* Zopf), where most species grow on either dung or decaying plant debris (Wijayawardene et al. 2017).

Chaetomium Kunze (species type *C. globosum* Kunze) presents saprobic species, occurring in a wide variety of substrates and biogeographic distribution (von Arx et al. 1986). As mentioned by Doveri (2004), species of this genus are not highly specialized to be considered truly coprophilous, lignicolous, or terricolous. It is characterized as one of the main genera of saprophytic Ascomycota, currently with more than 150 species described worldwide (Maharachchikumbura et al. 2016; Wijayawardene et al. 2017). *Chaetomium* produce membranaceous perithecia characteristically covered by terminal, lateral, or ascomatal hairs with different morphological structures (e.g., rigid, straight, hooked, spiral, coiled). In general, the genus is characterized by ascomata (perithecioid, rarely cleistothecioid) superficial, setulose or hairy, with ornate dark (rarely smooth hair) hairs covering the perithecium. Asci are usually clavate or cylindrical, 8-spored, very ephemeral. Spores are limoniform, ovoidal, or fusiform (sometimes triangular, spherical, or irregular) brownish to greyish in color (Moustafa and Abdel-Azeem 2005). As mentioned by Doveri (2004), spores are released as a dark, sticky mass.

According to the Index Fungorum Partnership (IFP 2018), there are 273 *Chaetomium* species accepted (see species list in Appendix). The work developed by von Arx et al. (1986) is considered as the basis for the classical taxonomic studies of the genus, which is followed in important articles and monographs on the genus (e.g., Doveri 2004). However, recent studies with molecular approaches have revealed a high number of undescribed species on this genus, especially when they are considered more potential substrates for the occurrence of a genus (Wang et al. 2014, 2016a, b; Zhang et al. 2016) (Table 4.2). Although *Chaetomium* is not a strictly coprophilous fungus genus, the diversity of representatives of the genus *Chaetomium* on dung is continuously represented in studies of dung fungi diversity worldwide. One of the most representative and recent research on the occurrence of this genus on dung, which was a precursor to others, was developed by Doveri (2004, 2011).

This author presented important contributions to the study of the genus occurring as fimicolous, with detailed diagnosis, substrate preferences, and descriptions of new species, some occurring exclusively on dung (Doveri 2008, 2013, 2016). For *Chaetomium* species recorded on dung, Doveri (2004) presented a representative selection of species reported from dung at least once until 2004. In 2008, the author presented an updated key to coprophilous species recorded from Italy (Doveri 2008), followed by another updated key to *Chaetomium* on dung, with new descriptions and records (Doveri 2013). These publications are, undoubtedly, important guides to taxonomy and biogeography of *Chaetomium* species that occur on dung. However, a most deep search in dung substrates, allied to phylogenetic approaches, can obviously improve our knowledge about this genus, including the possibility of new species on this substrate. Doveri (2018) provided an updated key to sexual morph genera of Chaetomiaceae, where a new species of the related genus *Chaetomidium* (*C. vicugnae* Doveri) is described from a sample of vicuña dung (*Vicugna vicugna* Molina 1782).

Despite the importance of the genus *Chaetomium* in several biotechnological, taxonomic, and ecological aspects, little is known about its occurrence on dung. Unfortunately, there are still few studies that focused on the biology of the group on dung (Table 4.1). Table 4.2 shows the most recent papers describing occurrences of *Chaetomium* species on dung (for more details, check the cited references).

Table 4.1 Journal articles dealing with the genus *Chaetomium* and their biology, including occurrences on dung

Keywords and Boolean indexers	Number of journal articles	Period
"Chaetomium"	701[a] and 657[b]	1945–2018
"Chaetomium" and "taxonomy"	22[a] and 38[b]	1945–2018
"Chaetomium" and "new species"	16[a] and 9[a]	1945–2018
"Chaetomium" and "coprophilous" or "dung"	3[a] and 1[b]	1945–2018

[a]Web of Science (Clarivate Analytics) database and [b]Scopus (Elsevier)

Table 4.2 Some recent reviews and papers mentioning *Chaetomium* spp. occurring on dung, published in the last 10 years. Substrates are shown in brackets in common names

Focus	Species and substrata	Reference
An update on the genus *Chaetomium* with descriptions of some coprophilous species, new to Italy	*Chaetomium carinthiacum* (bird) *C. gangligerum* (cattle, deer, goat, roe deer, and sheep) *C. homopilatum* (roe deer) *C. spinosum* (cow) *C. subaffine* (cow) *C. variostiolatum*[a] (cow) *C. bostrychodes*[a] (badger, beech marten, cattle, chamois, fallow deer, fox, goat, hare, hedgehog, hen, hooded crow, horse, lizard, snail, pigeon, porcupine, rabbit, rat, roe deer, sheep, sparrow, squirrel, toad, tortoise, unidentified bird, weasel, wild ferret, wild pig) *C. crispatum* (roe deer) *C. cuniculorum*[a] (wild rabbit) *C. elatum* (horse) *C. funicola*[a] (roe deer and horse) *C. fusisporum*[a] (sheep) *C. globosum* (cattle, dormouse, hare, hedgehog, horse, sheep, and sparrow) *C. medusarum*[a] (sparrow) *C. mollicellum*[a] (beech marten, domestic and wild pig, hedgehog, lizard, and rat) *C. oblatum* (donkey) *C. robustum*[a] (dormouse, lizard, rabbit, and roe deer) *C. seminis-citrulli* (porcupine, sparrow, and weasel) *C. trigonosporum* (wild ferret)	Doveri (2008)[1] [1]Check out substrate additions for species mentioned in this paper on Doveri (2011, 2013)
Additions to "Fungi Fimicoli Italici": An update on the occurrence of coprophilous basidiomycetes and ascomycetes in Italy with new records and descriptions	*C. ancistrocladum* (dormouse) *C. carinthiacum* (bird) *C. gangligerum* (cattle, deer, goat, roe deer, and sheep) *C. murorum* (bat, cattle, roe deer, dormouse, hare, rabbit, hedgehog, sheep, mouflon, and snail)	Doveri (2011)
Coprophilous ascomycetes of northern Thailand	*C. bostrychodes*[a] (goat)	Mungai et al. (2011)
Coprophilous ascomycetes in Kenya: *Chaetomium* species from wildlife dung	*C. convolutum* (dikdik, giraffe, and hartebeest) *C. globosum* (hartebeest and impala) *C. muelleri* (dikdik) *C. seminis-citrulli* (dikdik)	Mungai et al. (2012)
An additional update on the genus Chaetomium with descriptions of two coprophilous species, new to Italy	*C. brasiliense*[a] (fallow deer) *C. succineum* (sheep)	Doveri (2013)

(continued)

Table 4.2 (continued)

Focus	Species and substrata	Reference
Phylogenetic assessment of *Chaetomium indicum* and allied species, with the introduction of three new species and epitypification of *C. funicola* and *C. indicum*	*C. funicola*[a] (sheep) *C. indicum*[a] (dung of unidentified herbivore)	Wang et al. (2014)
A checklist of coprophilous fungi and other fungi recorded on dung from Brazil	*C. bostrychodes*[a] (camel, deer, and paca) *C. cochliodes* (domestic pig and goat) *C. globosum* (tapir) *C. funicola*[a] (paca)	Calaça et al. (2014)
Fungi fimicola Aegyptiaci: I. Recent investigations and conservation in arid South Sinai	*C. atrobrunneum*[a] (camel and donkey) *C. bostrychodes*[a] (camel, donkey, and goat) *C. globosum* (camel, donkey, and goat) *C. gracile*[a] (goat) *C. nigricolor*[a] (camel, donkey, and goat) *C. piluliferum* (camel, donkey, and goat) *C. subspirilliferum* (goat)	Abdel-Azeem and Salem (2015)
Records of coprophilous fungi – a data set	*C. bostrychodes*[a] (domestic pig, rabbit, hare, opossum, goat, squirrel, sheep, and reindeer) *C. crispatum* (rabbit, deer, squirrel, and opossum) *C. cuniculorum*[a] (hare) *C. elatum* (hare, goose, pine marten) *C. fusisporum*[a] (sheep) *C. globosum* (hare, rabbit, and sheep) *C. indicum*[a] (goat) *C. mollicellum*[a] (hare and rabbit) *C. murorum* (donkey, goat, and sheep) *C. quadrangulatum*[a] (rat)	Richardson (2015)
Description of *Chaetomium aureum*, *Corynascus sepedonium*, and *Coniochaeta hansenii* newly recorded from Italy and a key to coprophilous Chaetomiaceae and Coniochaetaceae	*C. aureum*[a] (goat)	Doveri (2016)
Diversity and taxonomy of *Chaetomium* and chaetomium-like fungi from indoor environments	*C. aureum*[a] (hyrax) *C. bostrychodes*[a] (antelope) *C. cupreum*[a] (moose) *C. fusiforme*[a] (dung of rodent) *C. quadrangulatum*[a] (rabbit) *C. mollicellum*[a] (spotted skunk) *C. senegalense*[a] (gazelle) *C. cristatum*[a] (rabbit) *C. fusisporum*[a] (marmot) *C. irregulare* (nilgai, rabbit, dung of herbivore and unidentified dung type)	Wang et al. (2016a)

(continued)

Table 4.2 (continued)

Focus	Species and substrata	Reference
Phylogenetic reassessment of the *Chaetomium globosum* species complex	*C. ascotrichoides* (sheep and goat) *C. coarctatum* (unidentified dung type) *C. cochliodes* (unidentified dung type) *C. megalocarpum* (horse) *C. olivaceum* (camel) *C. rectangulare* (unidentified dung type) *C. spirochaete* (unidentified dung type)	Wang et al. (2016b)
Potentially harmful secondary metabolites produced by indoor *Chaetomium* species on artificially and naturally contaminated building materials	*C. longicolle* (dung, type not informed)	Došen et al. (2016)

[a]Species of genus *Chaetomium* with updated nomenclature, based on the phylogenetic analyses of DNA-directed RNA polymerase II second largest subunit (rpb2), β-tubulin (tub2), ITS, and 28S large subunit (LSU) nrDNA sequences, together with morphological comparisons, as presented by Wang et al. (2016a)

4.3 Perspectives for Studies of *Chaetomium* Species on Dung

We briefly described advances and status of *Chaetomium* taxa on dung in recent years, based on both morphological and molecular characters. The tipping point to increase our understanding on *Chaetomium* species ecology and biology is to consider potential substrates where it can be found, once that genus is not highly specialized to a specific substrate. Dung, as it has been presented, is a very important source of nutrients. Therefore, mycologists should direct their sampling efforts to this type of substrate. Few researchers were responsible for the advances in knowledge of this genus in dung. However, the potential for discoveries and applications that dung-inhabiting fungi present (Bills et al. 2013; Sarrocco 2016) is good reason for exploring these other possibilities. In addition, new molecular techniques to evaluate fungal biodiversity, as done by Wang et al. (2016a, b), can save time and effort by directly evaluating DNA extracted from dung, which would avoid moist chamber use and classical taxonomy uncertainty.

An advanced knowledge on ecology of this dung-inhabiting fungi will facilitate ecological research and efforts to recognize benefits from secondary metabolites of these fungi as a better understanding of its potential as a plant or animal pathogen. Consequently, a better diagnosis of new *Chaetomium* species on dung, an extreme environment, will help to achieve maximum benefits from these microorganisms in biotechnological applications.

Appendix

List of current *Chaetomium* species, in accordance with the Index Fungorum Partnership (http://www.indexfungorum.org). For recent publications with occurrences of species on dung, see the Table 4.2.

Chaetomium abietinum Ellis & Everh.	None
C. acropullum X.Wei Wang	None
C. affine Corda	None
C. africanum L.M. Ames	None
C. afropilosum X.Wei Wang, Crous & L. Lombard	None
C. alboavenulum L.M. Ames	None
C. alchemillae Wallr.	None
C. amberpetense P.Rama Rao & Ram Reddy	None
C. amesii Sergeeva	None
C. ampullare Chivers	None
C. ampulliellum X.Wei Wang	None
C. anatolicum Karaca & Turhan	None
C. ancistrocladum Udagawa & Cain	None
C. anguipilium L.M. Ames	None
C. anguipillioides Mazzuch.	None
C. angulare Yu Zhang & L. Cai	None
C. angusti-ellipsoideum G. Malhotra & Mukerji	None
C. angustispirale Sergeeva	None
C. angustum Chivers	None
C. apiculatum Lodha	None
C. arachnoides Massee & E.S. Salmon	None
C. araliae Corda	None
C. arcuata J.N. Rai & J.P. Tewari	None
C. aterrimum Ellis & Everh.	None
C. atrosporum Skolko & J.W. Groves	None
C. aurangabadense Tilak & D.L. Ram Reddy	None
C. carinthiacum Sörgel	None
C. chiversii (J.C. Cooke) A. Carter	*C. trilaterale* var. *chiversii* J.C. Cooke
C. ciliatum Bonord.	None
C. cinnamomeum Subrahm. & Gopalkr.	None
C. circinans Wallr.	None
C. circinatum Chivers	None
C. cirrhata Yu Zhang & L. Cai	None
C. citrinum Udagawa & T. Muroi	None
C. cochliodes Palliser	None
C. comosum Bainier	None
C. concinnatum Preuss	None

C. concinnum Sörgel	None
C. congoense L.M. Ames	None
C. contagiosum X. Wei Wang, Crous & L. Lombard	None
C. contortum Peck	None
C. convolutum Chivers	None
C. coprophilum Narendra & V.G. Rao 1976	None
C. crinitum Sörgel	None
C. crispatoideum Sergeeva	None
C. crispatum (Fuckel) Fuckel	*C. anahelicinum* Udagawa & Cain
	C. simile Massee & E.S. Salmon
	C. tortile Bainier
C. cruentatum Sörgel	None
C. cruentum L.M. Ames	None
C. cucumericola X.Wei Wang, Crous & L. Lombard	None
C. cumingii (Lév.) Sacc.	None
C. cuneatum Sörgel	None
C. cuyabenoense Decock & Hennebert	None
C. cymatotrichum Cooke	None
C. deceptivum Malloch & Benny	None
C. delhianum Mukerji & N. Singh	None
C. delicatulum Roum.	None
C. depressum Wallr.	None
C. deustum Bat. & Pontual	None
C. difforme W. Gams	None
C. discolor Starbäck	None
C. distortum L.M. Ames	None
C. diversum Lodha	None
C. douglasii Schwein.	None
C. dreyfussii Arx	None
C. elatum Kunze	*C. atrum* Link
	C. atrum var. *atrum* Link
	C. atrum var. *distinctum* Roum.
	C. comatum (Tode) Fr.
	C. comatum var. *comatum* (Tode) Fr.
	C. comatum var. *helicotrichum* Sacc.
	C. lageniforme Corda
	C. pannosum Wallr.
	C. tenuissimum Sergeeva
C. elasticae Koord.	None
C. ellipsoideum K. Gopal & Lodha	None
C. elongatum Czerepan.	None
C. epiphyllum J. Kunze	None
C. fibripilium L.M. Ames	None

C. fimisedum P. Karst.	None
C. fiscicola Petr.	None
C. flavum Omvik	None
C. flexuosum Palliser	None
C. floriforme Gené & Guarro	None
C. formosum Bainier	None
C. gangligerum L.M. Ames	None
C. gelatinosum Ehrenb.	None
C. giganigrosporum Millner & S. Ahmad	None
C. glabrescens Ellis & Everh.	None
C. globisporum Lodha	None
C. globosporum Rikhy & Mukerji	None
C. globosum Kunze	*C. chlorinum* (Sacc.) Grove
	C. chlorinum var. *chlorinum* (Sacc.) Grove
	C. chlorinum var. *rufipilum* Grove
	C. coarctatum Sergeeva
	C. fieberi Corda
	C. fieberi subsp. *fieberi* Corda
	C. fieberi subsp. *saccardoanum* E. Bommer & M. Rousseau
	C. fieberi var. *chlorina* Sacc.
	C. fieberi var. *fieberi* Corda
	C. fieberi var. *macropoda* Speg.
	C. fieberi var. *rufipilum* (Grove) Sacc.
	C. globosum var. *arhizoides* Dreyfuss
	C. globosum var. *flavoviride* E.K. Novák
	C. globosum var. *globosum* Kunze
	C. globosum var. *griseum* E.K. Novák
	C. globosum var. *ochraceoides* Dreyfuss
	C. globosum var. *rectum* Dreyfuss
	C. kunzeanum Zopf
	C. kunzeanum var. *kunzeanum* Zopf
	C. olivaceum Cooke & Ellis
	C. rectum Sergeeva
C. graminiforme X.Wei Wang, Crous & L. Lombard	None
C. grande Asgari & Zare	None
C. griseum (Berk. & Broome) Cooke	None
C. guizotiae Tilak & D.L. Ram Reddy	None
C. hamadae (Udagawa) Arx	None
C. hamatum Bat. & Pontual	None
C. helicoideum H.J. Chowdhery & J.N. Rai	None
C. helicotrichum J.P. Tewari & P.D. Agrawal	None
C. heteropilum N.J. Artemczuk	None
C. heterosporum Rikhy & Mukerji	None

C. heterothallicum Yu Zhang & L. Cai	None
C. hexagonosporum A. Carter & Malloch	None
C. hispanicum Guarro & Arx	None
C. hispidum Fr.	None
C. histoplasmoides Carris & Glawe	None
C. homopilatum Omvik	None
C. humanum P. Karst.	None
C. humicola Van Warmelo	None
C. hungaricum E.K. Novák	None
C. hyaloperidium A. Carter	None
C. hyderabadense M.A. Salam & Nusrath	None
C. importatum Henn.	None
C. incomptum L.M. Ames	None
C. interruptum Asgari & Zare	None
C. iranianum Asgari & Zare	None
C. iricolor L.M. Ames	None
C. irregulare Sörgel	None
C. jabalpurense D.P. Tewari, P.D. Agrawal & Lodh	None
C. jatrophae Roh. Sharma, G. Kulk. & Shouche	None
C. jodhpurense Lodha	None
C. kauffmanianum Povah	None
C. laeliicola Henn.	None
C. lanatum Quél.	None
C. lanosum Peck	None
C. laterale Yu Zhang & L. Cai	None
C. lawransamesii Mukerji & M. Khanna	None
C. lentum Van Warmelo	None
C. leproplocinum Werner & Cain	None
C. libertiae Roum. & Pat.	None
C. longiciliata Yu Zhang & L. Cai	None
C. longicolle Krzemien. & Badura	None
C. longipilum G. Malhotra & Mukerji	None
C. longirostre (Farrow) L.M. Ames	None
C. lucknowense J.N. Rai & J.P. Tewari	None
C. ludhianense Kang	None
C. lunasporium Udaiyan & Hosag.	None
C. lusitanicum M.R.M. Gomes	None
C. luteum (J.N. Rai & J.P. Tewari) P.F. Cannon	None
C. macrosporum Sacc. & Penz.	None
C. macrostiolatum Stchigel, K. Rodr. & Guarro	None
C. madrasense Natarajan	*C. ascotrichoides* Caviello
C. malaysiense (D. Hawksw.) Arx	None
C. marchicum Lindau	None
C. mareoticum Besada & Yusef	None

C. megalocarpum Bainier	*C. megalocarpum* var. *chlorinum* Bainier
	C. megalocarpum var. *megalocarpum* Bainier
C. megasporum Sörgel	None
C. melioloides Cooke & Peck	None
C. mesopotamicum Abdullah & Zora	None
C. microascoides Guarro	None
C. microcephalum L.M. Ames	None
C. microsporum Speg.	None
C. microthecia Yu Zhang & L. Cai	None
C. minimum J.F.H. Beyma	None
C. minutispora Aruna & Gopalkr.	None
C. minutum Krzemien. & Badura	None
C. mollipilium L.M. Ames	None
C. montblancense Guarro, Calvo & C. Ramírez	None
C. montemartinii Cavara	None
C. muelleri Arx	None
C. multispirale A. Carter, R.S. Khan & P.E. Powell	None
C. myricicola Y. Horie & Udagawa	None
C. nepalense (Udagawa & Y. Sugiy.) Arx	None
C. nivale F. Strauss	None
C. nodulosum Kulshr., Raych. & A.Z.M. Khan	None
C. novae-caledoniae Udagawa, Uchiy. & Kamiya	None
C. novozelandicum X.Wei Wang, Crous & L. Lombard	None
C. nozdrenkoae Sergeeva	None
C. oblatum Dreyfuss & Arx	None
C. ochraceum Tschudy	None
C. olivicolor K. Rodr., Stchigel & Guarro	None
C. oospora Beauverie	None
C. orientale Cooke	None
C. orientum D.K. Saha	None
C. osmaniae P.Rama Rao & Ram Reddy	None
C. pachypodioides L.M. Ames	None
C. pallidum Ellis & Everh.	None
C. pampaninii Cif.	None
C. papillosum Cocc.	None
C. parvotrichum Mazzuch.	None
C. paucisetum Fuckel	None
C. perlucidum Sergeeva	None
C. phyllophora Hoffm.	None
C. pilosum (C. Booth & Shipton) X.Wei Wang, Crous & L. Lombard	None
C. piluliferoides Udagawa & Y. Horie	None

C. piluliferum J. Daniels	None
C. pimprinum Aruna & Gopalkr.	None
C. pinnatum L.M. Ames	None
C. polypori Rehm	None
C. prasinum (Berk. & M.A. Curtis) S. Hughes	None
C. pseudocochliodes X.Wei Wang, Xin Z. Liu & Crous	None
C. pseudoerraticum Lal & J.N. Kapoor	None
C. pseudoglobosum X.Wei Wang, Crous & L. Lombard	None
C. pulchellum L.M. Ames	None
C. purpurascens (Udagawa & Y. Sugiy.) Arx	None
C. putrefactum R. Gupta & Mukerji	None
C. raii G. Malhotra & Mukerji	None
C. rajasthanense Lodha	None
C. ramipilosum Schaumann	None
C. raripilum Mont.	None
C. rectangulare Asgari & Zare	None
C. rectopilium Fergus & Amelung	None
C. repandum Bat. & Pontual	None
C. repens Guarro & Figueras	None
C. retardatum A. Carter & R.S. Khan	None
C. reticulopilium L.M. Ames	None
C. rigidulum Bainier	None
C. rostratum Speg.	None
C. rufum K. Ramakr.	*C. rufum* var. *macrosporum* P.D. Agrawal & J.P. Tewari
	C. rufum var. *rufum* K. Ramakr.
C. saccardoanum (E. Bommer & M. Rousseau) Mussat	None
C. seminis-citrulli Sergeeva	None
C. semispirale Udagawa & Cain	None
C. serpentinum L.M. Ames ex A. Carter	None
C. siamense Porns. & Soytong	None
C. signatum Preuss	None
C. silvaticum Kiril.	*C. silvaticum* var. *silvaticum* Kiril.
	C. silvaticum var. *variabile* Kiril.
C. sinaiense Moustafa & Ezz-El-din	None
C. sinense K.T. Chen	None
C. soergelii Sörgel	None
C. sphaerale Chivers	None
C. sphaerospermum Cooke & Ellis	None
C. spiculipilium L.M. Ames	None
C. spinigerum Sörgel	None
C. spinosum Chivers	None

C. spinulosum Sörgel	None
C. spirale Zopf	None
C. spiralotrichum Lodha	None
C. spirilliferum Bainier	None
C. spirochaete Palliser	None
C. stercoreum Speg.	None
C. streptothrix Quél.	None
C. strigosum Wallr.	None
C. strumarium (J.N. Rai, J.P. Tewari & Mukerji) P.F. Cannon	None
C. subaffine Sergeeva	None
C. subcircinatum A. Carter & R.S. Khan	None
C. subcurvisporum Abdullah & Al-Bader	None
C. subfimeti (Seth) X.Wei Wang & Crous	None
C. subglobosum Sergeeva	None
C. subspinale Chivers	None
C. subspirale Chivers	None
C. subspirilliferum Sergeeva	None
C. subterraneum Swift & Povah	None
C. subtorulosum Sergeeva	None
C. succineum L.M. Ames	None
C. sulphureum Sörgel ex Seth	None
C. tarraconense Stchigel, K. Rodr. & Guarro	None
C. tectifimeti X.Wei Wang & Samson	None
C. telluricola X.Wei Wang, Crous & L. Lombard	None
C. tenue X.Wei Wang, Crous & L. Lombard	None
C. terrestre R.S. Dwivedi	None
C. terreum Kiril.	None
C. terricola J.C. Gilman & E.V. Abbott	None
C. tetrasporum S. Hughes	None
C. thermophilum La Touche	*C. thermophilum* var. *coprophilum* Cooney & R. Emers.
	C. thermophilum var. *dissitum* Cooney & R. Emers.
	C. thermophilum var. *thermophilum* La Touche
C. tomentosum Preuss	None
C. tortuosum Garb.	None
C. torulosum Bainier	None
C. trigonosporum (Marchal) Chivers	None
C. trilaterale Chivers	*C. rubrogenum* Van Warmelo
	C. trilaterale f. *trilaterale* Chivers
	C. trilaterale var. *diporum* J.C. Cooke
	C. trilaterale var. *pseudocupreum* Guarro

	C. trilaterale var. *trilaterale* Chivers
C. triticicola Lal & J.N. Kapoor	None
C. truncatulum Asgari & Zare	None
C. typhae Schwein.	None
C. udagawae Sergeeva	None
C. umbonatum D. Brewer	None
C. umbratile Udagawa, Toyaz. & Yaguchi	None
C. undulatulum Asgari & Zare	None
C. undulatum Bainier	*C. formosum* var. *formosum* Bainier
	C. formosum var. *neglectum* Bainier
	C. formosum var. *ovatum* Bainier
C. unguicola X.Wei Wang, Crous & L. Lombard	None
C. uniapiculatum (J.N. Rai & H.J. Chowdhery) Arx	*C. amygdalisporum* Udagawa & T. Muroi
	C. amygdalisporum var. *amygdalisporum* Udagawa & T. Muroi
	C. amygdalisporum var. *brevissimum* Wadhwani & N. Mehrotra
	C. subapiculatum K. Gopal & Lodha
C. unipapillatum Natarajan	None
C. uniseriatum Yu Zhang & L. Cai	None
C. variosporum Udagawa & Y. Horie	None
C. varium Delacr.	None
C. velutinum Ellis & Everh.	None
C. venezuelense L.M. Ames	None
C. verrucichaeta Natarajan	None
C. virgicephalum L.M. Ames	None
C. virginicum L.M. Ames	None
Chaetomium viride Lév.	None
C. wallefii J.A. Mey. & Lanneau	None
C. warcupii A.S. Saxena & Mukerji	None

References

Abdel-Azeem AM, Salem FM (2015) Fungi fimicola Aegyptiaci: I. Recent investigations and conservation in arid South Sinai. Mycosphere 6(2):174–194. https://doi.org/10.5943/mycosphere/6/2/8

Altayyar IA, Ismail AS, Abdullah SK (2017) A preliminary study of Coprophilous Fungi in North of Jordan. Int J Appl Med Biol Res 2(1):11–15

Asgari B, Zare R (2011) The genus *Chaetomium* in Iran, a phylogenetic study including six new species. Mycologia 103(4):863–882. https://doi.org/10.3852/10-349

Bell A (1983) Dung Fungi: an illustrated guide to coprophilous fungi in New Zealand. Victoria University Press, New Zealand. 88 p

Bills GF, Gloer JB, An Z (2013) Coprophilous fungi: antibiotic discovery and functions in an underexplored arena of microbial defensive mutualism. Curr Opin Microbiol 16(5):549–565. https://doi.org/10.1016/j.mib.2013.08.001

Blanchette RA, Held BW, Abdel-Azeem AM (2017) New record of *Chaetomium iranianum* MF787598 (Chaetomiaceae) for the Egyptian and African mycobiota. Microbial Biosyst 2(2):6–9

Calaça FJS, Silva NC, Xavier-Santos S (2014) A checklist of coprophilous fungi and other fungi recorded on dung from Brazil. Mycotaxon 128:205, 22 p. https://doi.org/10.5248/128.205

Calaça FJS, Araújo JC, Xavier-Santos S (2017) O status ecológico das comunidades de fungos coprófilos. Pesquisa e Ensino em Ciências Exatas e da Natureza 1(2):136–143. https://doi.org/10.29215/pecen.v1i2.452

Došen I, Nielsen KF, Clausen G, Andersen B (2016) Potentially harmful secondary metabolites produced by indoor *Chaetomium* species on artificially and naturally contaminated building materials. Indoor Air 27(1):1–13. https://doi.org/10.1111/ina.12290

Doveri F (2004) Fungi Fimicoli Italici. A.M.B: Fondazione Studi Micologici, Trento. 1104 p

Doveri F (2008) An update on the genus *Chaetomium* with descriptions of some coprophilous species, new to Italy. Pagine di Micologia 29:1–60

Doveri F (2011) Additions to "Fungi Fimicoli Italici": an update on the occurrence of coprophilous Basidiomycetes and Ascomycetes in Italy with new records and descriptions. Mycosphere 2(4):331–427

Doveri F (2013) An additional update on the genus *Chaetomium* with descriptions of two coprophilous species, new to Italy. Mycosphere 4:820–846. https://doi.org/10.5943/mycosphere/4/4/17

Doveri F (2016) Description of *Chaetomium aureum*, *Corynascus sepedonium* and *Coniochaeta hansenii* newly recorded from Italy and a key to coprophilous Chaetomiaceae and Coniochaetaceae. Ascomyceteorg 8(1):7–24

Doveri F (2018) On a new species of *Chaetomidium*, *C. vicugnae*, with a cephalothecoid peridium and its relationships with Chaetomiaceae (Sordariales). Ascomycete.org 10(2):86–96. https://doi.org/10.25664/ART-0231

Eldridge DJ, Delgado-Baquerizo M (2018) Functional groups of soil fungi decline under grazing. Plant Soil 426(1–2):51–60. https://doi.org/10.1007/s11104-018-3617-6

Frank K, Brückner A, Hilpert A, Heethof M, Blüthgen N (2017) Nutrient quality of vertebrate dung as a diet for dung beetles. Sci Rep 7:e12141. https://doi.org/10.1038/s41598-017-12265-y

Index Fungorum Partnership – IFP (2018) Index Fungorum. Available at http://www.indexfungorum.org/. Accessed June 25, 2018.

Jones R (2017) Call of nature: the secret life of dung. Pelagic Publishing, Exeter, UK. ISBN: 9781484271053. 303 p

Lundqvist N (1972) Nordic Sordariaceae s. lat. Symbolae Botanicae Upsalienses 20(1):1–314

Maharachchikumbura SS, Hyde KD, Jones EBG et al (2016) Families of Sordariomycetes. Fungal Divers 79(1):1–317. https://doi.org/10.1007/s13225-016-0369-6.

Moustafa AF, Abdel-Azeem AM (2005) The genus Chaetomium in Egypt. El-Minia Sci Bull 16(2):235–256

Mungai P, Hyde KD, Cai L, Njogu J, Chukeatirote K (2011) Coprophilous ascomycetes of northern Thailand. Curr Res Environ Appl Mycol 1(2):135–159. https://doi.org/10.5943/cream/1/2/2

Mungai PG, Chukeatirote E, Njogu JG, Hyde KD (2012) Coprophilous ascomycetes in Kenya: *Chaetomium* species from wildlife dung. Curr Res Environ Appl Mycol 2(2):113–128. https://doi.org/10.5943/cream/2/2/3.

Piasai O, Manoch L (2009) Coprophilous Ascomycetes from Phu Luang Wildlife Sanctuary and Khao Yai National Park in Thailand. Kasetsart J (Nat Sci) 43:34–40

Pornsuriya C, Lin FC, Kanokmedhakul S, Soytong K (2008) New record of *Chaetomium* species isolated from soil under pineapple plantation in Thailand. J Agric Technol 4(2):91–103

Richardson MJ (2001) Diversity and occurrence of coprophilous fungi. Mycol Res 105:387–402. https://doi.org/10.1017/S0953756201003884

Richardson MJ (2015) Records of coprophilous fungi: a data set. Mycotaxon 130: 925–926. https://doi.org/10.5248/130.925

Saito T, Yasushi S, Kiyotaka K, Shinsaku N, Yoichi I, Takeshi K (1988) Chetracin A and Chaetocins B and C, Three New Epipolythiodioxo-piperazines from *Chaetomium* spp. Chem Pharm Bull 36(6):1942–1956. https://doi.org/10.1248/cpb.36.1942

Sarrocco S (2016) Dung-inhabiting fungi: a potential reservoir of novel secondary metabolites for the control of plant pathogens. Pest Manag Sci 72(4):643–652. https://doi.org/10.1002/ps.4206

Soytong K, Kanokmedhakul S, Kukongviriyapa V, Isobe M (2001) Application of *Chaetomium* species (Ketomium®) as a new broad spectrum biological fungicide for plant disease control: a review article. Fungal Divers 7:1–15

von Arx JA, Guarro J, Figueras MJ (1986) The ascomycete genus *Chaetomium*. Beihefte zur Nova Hedwigia 84:1–162

Wang XW, Wang XL, Liu FJ, Zhao XM, Li J, Cai L (2014) Phylogenetic assessment of *Chaetomium indicum* and allied species, with the introduction of three new species and epitypification of *C. funicola* and *C. indicum*. Mycol Prog 13(3):719–732. https://doi.org/10.1007/s11557-013-0955-x

Wang XW, Houbraken J, Groenewald JZ, Meijer M, Andersen B, Nielsen KF, Crous PW, Samson RA (2016a) Diversity and taxonomy of *Chaetomium* and chaetomium-like fungi from indoor environments. Stud Mycol 84:145–224. https://doi.org/10.1016/j.simyco.2016.11.005

Wang XW, Lombard L, Groenewald JZ, Lil J, Videira SIR, Samson RA, Liu XZ, Crous PW (2016b) Phylogenetic reassessment of the *Chaetomium globosum* species complex. Persoonia 36:83–133. https://doi.org/10.3767/003158516X689657

Wijayawardene NN, Hyde KD, Rajeshkumar KC et al (2017) Notes for genera: Ascomycota. Fungal Divers 86(1):1–594. https://doi.org/10.1007/s13225-017-0386-0.

Zhang Y, Wu W, Cai L (2016) Polyphasic characterisation of *Chaetomium* species from soil and compost revealed high number of undescribed species. Fungal Biol 121(1):21–43. https://doi.org/10.1016/j.funbio.2016.08.012

Chapter 5
Chaetomium in Indoor Environment and Medically Important Species of *Chaetomium*

Eman A. Attia and Ahmed M. Abdel-Azeem

5.1 Introduction

Fungal contamination in damp or water-damaged buildings has become an increasing problem worldwide (Andersen et al. 2011). After water damage (e.g., leaking water pipes, flooding, faulty building constructions, or severe and prolonged condensation), many building materials become good substrates for certain fungi. These growing fungi can cause adverse effects not only on the buildings but also to their occupants (WHO 2009; Samson et al. 1994; Samson et al. 2010; Flannigan and Miller 2011; Andersen et al. 2011; Miller and McMullin 2014). Members of the genus *Chaetomium* are capable of colonizing various substrates and are well-known for their ability to degrade cellulose and produce a variety of bioactive metabolites. More than 400 species have been described in *Chaetomium*. Some of these species have been reported to be important inhalant allergens. They contribute to the development of the symptoms of both rhinitis and asthma due to the production of mycotoxins and microbial volatile organic compounds as well as the liberation of ascospores and hyphal fragments in the indoor environment (Gonianakis et al. 2005; Apetrei et al. 2009; Polizzi et al. 2009; Mason et al. 2010; Andersen et al. 2011; Miller and McMullin 2014). *Chaetomium globosum* is the most common species of the *Chaetomiaceae* in the indoor environment (Vesper et al. 2007; Ayanbimpe et al. 2010; Straus 2011; McMullin et al. 2013; Miller and McMullin 2014), and this species can already be present in new gypsum wallboard (Andersen et al. 2017). *Chaetomium globosum* has been reported to produce a variety of toxic metabolites, such as chaetoglobosins, chaetomugilins, and chaetoviridins (Andersen et al. 2011; McMullin et al. 2013; Miller and McMullin 2014), while both *C. elatum* and *C. globosum* were able to produce cochliodones in pure cultures as well as on naturally contaminated building materials (Došen et al. 2017). Little is known about the other

E. A. Attia (✉) · A. M. Abdel-Azeem
Botany Department, Faculty of Science, University of Suez Canal, Ismailia, Egypt
e-mail: ahmed_abdelazeem@science.suez.edu.eg

A. M. Abdel-Azeem (ed.), *Recent Developments on Genus Chaetomium*,
Fungal Biology, https://doi.org/10.1007/978-3-030-31612-9_5

indoor *Chaetomium* species and their potential hazard to humans and buildings. Furthermore, *C. globosum* and several other *Chaetomium* species are reported as causal agents of onychomycosis or superficial infections (Naidu et al. 1991; Koch and Haneke 1965; Hubka et al. 2011; Aspiroz et al. 2007; de Hoog et al. 2013a), and some of them are capable of opportunistically causing deep or systemic infections (Hoppin et al. 1983; Guppy et al. 1998; Barron et al. 2003; Ahmed et al. 2016).

The genus *Chaetomium* is commonly recognized by having ostiolate ascomata with a membranaceous perithecial wall covered by relatively well-developed hairs, producing fasciculate and evanescent asci and single-celled, smooth, and pigmented ascospores with germ pores (Ames 1963; von Arx et al. 1986). *Chaetomium globosum*, the type species of the genus, was first described by Kunze (Kunze and Schmidt 1817). The taxonomy of *Chaetomium* has been studied by several authors (Corda 1840; Zopf 1881; Chivers 1915; Skolko and Groves 1948, 1953; Sörgel 1960; Ames 1963; Mazzucchetti 1965; Seth 1970; Dreyfuss 1976; Millner 1977; Millner et al. 1977; von Arx et al. 1984). von Arx et al. (1986) re-defined the taxonomic concept of *C. globosum*. They included species that produce globose to ovate or obovate ascomata with a wall consisting of textura intricata, covered by a diverse morphology of ascomatal hairs ranging from erect, flexuous to regularly coiled. The ascomata contain clavate (or slightly fusiform), evanescent asci, and the ascospores are limoniform and bilaterally flattened shaped and have an apical germ pore. Following this concept, 28 species were reduced to synonymy with *C. globosum*. The species concept of *C. globosum* sensu von Arx was not supported by a recent study (Asgari and Zare 2011). On the basis of phylogenetic inference of six loci and morphological characters, *C. globosum* was again revised by Wang et al. (2016), and six species that were treated as synonyms of *C. globosum* by von Arx et al. (1986) were resurrected. Furthermore, the non-ostiolate genus *Chaetomidium* was also synonymized with *Chaetomium* (Wang et al. 2016).

Melanized fungi are important causes of human infection, and about 70 genera representing hundreds of species have been implicated in human disease (de Hoog et al. 2013b; Guppy et al. 1998; Revankar and Sutton 2010; Revankar et al. 2002). Several main ecological groups can be distinguished. Members of *Chaetothyriales* exhibit pronounced virulence and cause deep and systemic infections in immunocompetent humans, e.g., chromoblastomycosis or brain infection (Badali et al. 2009; Revankar and Sutton 2010). Members of Pleosporales are preponderantly found as degraders of plants debris or as mild opportunistic pathogens; human infections mostly comprise traumatic inoculation of contaminated materials (Revankar and Sutton 2010). Recently, the significance of *Sordariales* was underlined (Badali et al. 2011; de Hoog et al. 2013a), particularly the *Chaetomiaceae*, whose prevalence has been underestimated because of diagnostic problems.

Phenotypically, the identification of clinical *Chaetomium*-like fungi has been difficult as a large proportion of them fail to produce typical diagnostic structures in culture (Najafzadeh et al. 2014; Vinod Mootha et al. 2012). Until recently, fungi lacking propagation in the form of conidia were treated in the clinical laboratory as unidentifiable 'mycelia sterilia'(Pounder et al. 2007; Santos et al. 2013; Vinod Mootha et al. 2012). Isolates forming clumps or bulbils were referred to as

Papulaspora, whereas filamentous strains from subcutaneous infection were known as *Madurella* (de Hoog et al. 2013a). With the application of molecular phylogenetic multi-gene DNA sequence analyses, such isolates were found to be of high diversity being distributed to many genera and families of ascomycetes. In addition, the species proved to comprise several sibling species that were previously thought to represent a single taxon.

Recently, more attention has been paid to the non-sporulating *Chaetomium*-like isolates that cause human infection (Najafzadeh et al. 2014; Vinod Mootha et al. 2012), such as keratitis or subcutaneous infection after trauma. Despite application of molecular phylogenetic methods, researchers were unable to identify these isolates down to species level in the genus *Chaetomium* because of the present state of morphological confusion and high phylogenetic divergence in the genus.

Chaetomium contains more than 300 described species and are generally cosmopolitan and reside in soil on cellulose-rich materials or on dung (Bell 2005; Carter and Khan 1982; Doveri 2008; von Arx et al. 1986). A certain prevalence of *Chaetomium*-like species was noted in desert soil subjected to conditions of dryness and extremely variable temperatures (Rodríguez et al. 2004). Members of the *Madurella* clade, phylogenetically located inside the genus *Chaetomium*, are typically confined to areas with arid climates. *Madurella* species are consistent agents of human subcutaneous mycoses, and the arid areas of northeastern Africa are endemic for human mycetoma (Ahmed et al. 2002, 2016). Most human infections by *Chaetomium*-like species concern traumatic inoculations into otherwise healthy humans and rarely occur as deep infections in severely immunocompromised hosts (Al-Aidaroos et al. 2007; Guppy et al. 1998; Hubka et al. 2011).

5.2 Clinical Features

Despite being saprophytic ascomycetes with only occasional involvement in human disease processes, *Chaetomium* are capable of inducing a broad spectrum of mycoses including onychomycosis, sinusitis, empyema, pneumonia, and fatal disseminated cerebral disease, especially in immunocompromised patients and intravenous drug users (Hoppin et al. 1983; Anandi et al. 1989; Yeghen et al. 1996; Aru et al. 1997; Thomas et al. 1999).

Chaetomium atrobrunneum is a notably invasive, neurotropic species, and its ability to grow at elevated temperatures may contribute to its neurotropism (Stiller et al. 1992; Guarro et al. 1995; Friedman 1998; Guppy et al. 1998; Rock 1998; Lesire et al. 1999).Thomas et al. (1999) described a case of fatal brain abscess due to *C. atrobrunneum* in a bone marrow transplant patient. The rapid progression of cerebral infection indicates that the brain tissue provides a favorable environment for growth and proliferation of the fungus. In 2019, Mhmoud et al. reported *C. atrobrunneum* for the first time causing human eumycetoma (Fig. 5.1).

Chaetomium strumarium is another invasive, neurotropic species. Abbott et al. (1995) reported three *C. strumarium*-related cases of fatal cerebral mycosis in males

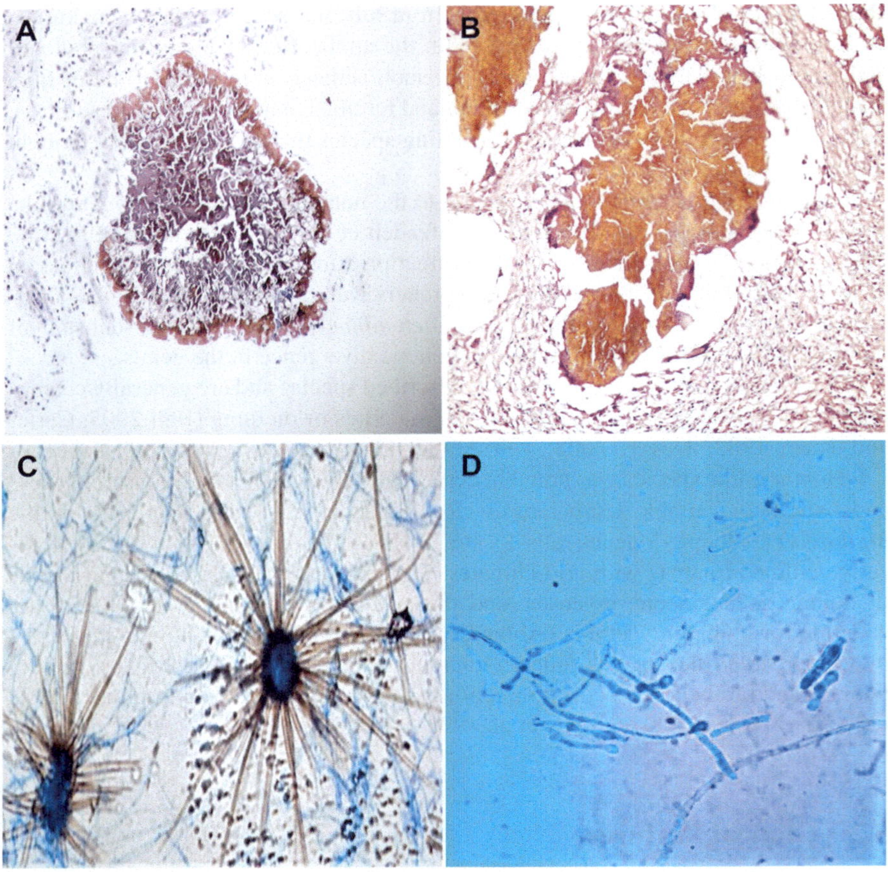

Fig. 5.1 Microphotograph showing multiple black grains surrounded by granulation tissue with marked histiocytic and mixed inflammatory cellular infiltrates (HE 10×). (**a**) Grains of *Chaetomium* spp. showed abundant extracellular matrix, which was yellow to brown in colour, and the fungus hyphae are located at the periphery of the matrix with short filamentous structure. (**b**) The filamentous pattern of *M. mycetomatis* grains consists of brown septate and branched hyphae at the centre and periphery with long filament. (**c, d**) Microphotograph of LPCB mount showing ascoma and ascospore cells resembling the typical *Chaetomium* spp. cells (**c**) and conidia of *M. Mycetomatis* (**d**). HE, haematoxylin–eosin; LPCB, lactophenol cotton blue (after Mhmoud et al. 2019)

with prior histories of intravenous drug use from the United States and Australia. *C. strumarium* was detected by histopathology and isolated from the brain tissue.

Chaetomium perlucidum is recently confirmed as a neurotropic species. Barron et al. (2003) documented the first two cases of invasive human mycoses caused by this phaeoid ascomycete. The first case concerned a 45-year-old female patient with acute myelogenous leukemia, who had an unrelated, 4/5 HLA-matched umbilical cord blood transplant. The patient became disoriented and febrile; computed tomography of the chest revealed a 3 × 2 cm mass in the right lower lobe. After suffering a massive right-sided intraparenchymal hemorrhage, the patient died. Autopsy revealed disseminated invasive fungal infection in the lungs, brain, and myocardium; and cultures from the surgically obtained lung tissue yielded *C. perlucidum*. The second case involved a 78-year-old female with a history of asthma and chronic bronchiectasis. The patient underwent a lobectomy due to worsening symptoms, and cultures from the lung tissue grew *C. perlucidum*. The patient showed no further manifestations of disease after the lobectomy.

Chaetomium globosum is an occasional agent of onychomycosis (Stiller et al. 1992; Lesire et al. 1999). In addition, Teixeira et al. (2003) reported that *C. globosum* was responsible for a systemic infection with enlargement of the axillary and cervical lymph nodes in a chronic myeloid leukemia patient, who underwent an allogeneic sibling-matched bone marrow transplant. Culture of the aspirates from both lymph nodes resulted in the growth of *C. globosum*. The infection was successfully treated with amphotericin B.

5.3 Diagnosis

Considerable similarities exist between mycosis caused by *Chaetomium* and *Aspergillus* from radiographical and histopathological standpoints. A melanin-specific stain (i.e., the Masson-Fontana stain) is helpful reputedly for distinguishing the melanin-containing *Chaetomium* from most *Aspergillus* species. However, this is contradicted by reports of melanin from *Aspergillus* (e.g., *Aspergillus fumigatus*) (Youngchim et al. 2004), and so the information obtained may have limited value in differentiating the causal fungi.

In vitro culture techniques offer a slow but valuable way to isolate and propagate *Chaetomium* organisms for subsequent macroscopic and microscopic characterization. Inoculation of sterilized plant material with ascospore suspension may enhance induction of mature perithecia, leading to the production of well-developed ascomata on the surface of the substrate.

Molecular methods such as (i) PCRand (ii) sequencing analysis of the rRNA and internal transcribed spacer (ITS) regions provide an approach for the rapid and accurate identification of *Chaetomium* species from other fungi.

5.4 Methods

5.4.1 Sample Preparation

Clinical specimens are examined by microscopy with a melanin-specific stain (i.e., the Masson-Fontana stain). However, as mentioned above, a melanin stain may not be particularly useful. Portions of the samples are inoculated on Sabouraud glucose agar with or without antibiotics. The resulting isolates are identified on the basis of macroscopical and microscopical features.

Genomic DNA of fungal pellets was extracted using a cetyltrimethylammonium bromide (CTAB) method described previously by Möller et al. (1992). Amplification and sequencing were performed for the internal transcribed spacer (ITS) and D1/D2 domains of the 28S rRNA gene, partial translation elongation factor 1-α (TEF1), β-tubulin (Btub), and DNA-dependent RNA polymerase II largest subunit (RPB1) and second largest subunit (RPB2). Primers used for amplification and sequencing are according to de Hoog et al. (2013b).

DNA sequences were assembled and edited using SEQMAN from the Lasergene package (DNASTAR, Madison, WI, U.S.A.). Sequences will be deposited in GenBank. To study the phylogenetic position of the unknown species, two alignments will be generated. Sequences will be aligned with the online version of MAFFT v.7 (http://mafft.cbrc.jp) and manually adjusted using BIOEDIT v.7.1.3 software (Hall 1999). Each gene will be aligned independently and concatenated matrices will be prepared using DATACONVERT v. 1.0. The first alignment consisted of ribosomal ITS and LSU sequences of representative species of *Chaetomium*, *Chaetomidium*, *Thielavia*, *Papulaspora*, *Subramaniula*, and *Madurella*. The second alignment consisted of the protein coding loci TEF1, Btub, RPB1 and RPB2 sequences of a selected number of strains. Alignments and trees will be deposited in TreeBASE database.

Phylogenetic analyses using maximum likelihood were performed in RAxML v. 8.0.24 (Stamatakis 2014). Bayesian analyses with default priori of MRBAYES v. 3.1.2 were conducted using the CIPRES Science Gateway server. Two simultaneous Markov chain Monte Carlo samplings were performed with four chains of which one was cold and three were heated.

The run was conducted for 30,000,000 generations with sampling every 100 generations and the 'burn in' was set at 25% of resulting trees. Convergence was evaluated from the two independent runs using AWTY and TRACER v. 1.5 (Nylander et al. 2008; Rambaut and Drummond 2007).

Phaeohyphomycosis and onychomycosis due to *Chaetomium* spp., including the first report of *Chaetomium brasiliense* infection, have been studied by Hubka et al. (2011). *Chaetomium* species have been rarely described as etiological agents of invasive and dermatomycotic infections in humans. The majority of cases have been reported within the last two decades. Treatment failed in most of these cases. In their study, they presented two cases in which *Chaetomium spp.* can be clearly identified as an etiological agent in pathological conditions. In the first report, they described a new etiological agent, *Chaetomium brasiliense*, which was implicated in a case of otitis externa in a patient with spinocellular carcinoma basis cranii. The patient had been repeatedly treated for relapsing otitis externa and had previously undergone surgery several times for otitis media. The fungal etiology was confirmed by repeated positive culture and histologic studies. The second case involved onychomycosis with strikingly brown nail discoloration due to *Chaetomium globosum* in an otherwise healthy patient. The nail lesion was successfully cured by oral terbinafine. The determination of both species was supported by sequencing of rDNA regions.

Eumycetoma is a traumatic fungal infection in tropical and subtropical areas that may lead to severe disability. *Madurella mycetomatis* is one of the prevalent etiologic agents in arid Northeastern Africa. The source of infection has not been clarified. Subcutaneous inoculation from plant thorns has been hypothesized, but attempts to detect the fungus in relevant material have remained unsuccessful. de Hoog et al. (2013a) tried to find clues to reveal the natural habitat of *Madurella* species using a phylogenetic approach, i.e., by comparison of neighboring taxa with known ecology. They found four species of *Madurella* were included in a large data set of species of *Chaetomium, Chaetomidium, Thielavia*, and *Papulaspora* ($n = 128$) using sequences of the universal fungal barcode gene rDNA ITS and the partial LSU gene sequence. Their study demonstrates that *Madurella* species are nested within the *Chaetomiaceae*, a family of fungi that mainly inhabit animal dung, enriched soil, and indoor environments. They hypothesized that cattle dung, ubiquitously present in rural East Africa, plays a significant role in the ecology of *Madurella*. If cow dung is an essential factor in inoculation by *Madurella*, preventative measures may involve the use of appropriate footwear in addition to restructuring of villages to reduce the frequency of contact with etiologic agents of mycetoma. On the other hand, they mentioned that *Chaetomiaceae* possess a hidden clinical potential which needs to be explored.

A consistent human pathogen is thus introduced in the family *Chaetomiaceae*. Traditionally, most species of the family were considered to be insignificant as agents of human disease. The majority of *Chaetomium* clinical strains analyzed in de Hoog et al. (2013a) study were probably transient colonizers or agents of mild superficial disorders. Twenty-seven were involved in onychomycosis or cutaneous and eye infections in otherwise healthy individuals. This matches with literature data (Hattori et al. 2000; Hubka et al. 2011). Data of de Hoog et al. (2013a) illustrated that *Chaetomium globosum* showed a definite bias toward superficial infection, with 17 out of 29 strains analyzed. The species is able to degrade keratin by production of extracellular keratinases (Kaul and Sumbali 1999). Fatal, disseminated, and cerebral infections by *Chaetomiaceae* have also been reported. In the literature, about 20 deep and disseminated cases were described, nearly all in immunocompromised and severely debilitated patients (Guppy et al. 1998; Badali et al. 2011). Several *Chaetomium*-like fungi thus show rather pronounced pathology, sometimes with species-specific predilections.

Grain formation in tissue by *Chaetomiaceae* other than *Madurella* is not known. A single case of chromoblastomycosis by *Chaetomium funicola* was reported by Piepenbring et al. (2007). The few subcutaneous cases (Lin et al. 1995) all showed hyphae in tissue rather than the compact grains of *Madurella mycetomatis*. In contrast to *Madurella*, none of the infecting *Chaetomiaceae* was exclusively clinical; all contained environmental strains as well. If agents of black-grain mycetoma have a relatively limited distribution in the phylogeny of *Sordariales*, i.e., are clustered within a single family, *Chaetomiaceae*, one may hypothesize that these fungi are predisposed to human infection and thus are likely to share a set of fundamental virulence factors. Many members of *Chaetomiaceae* have their natural habitat in soil or on mammal dung. A possible explanation of their recurrent virulence may lie in physiological properties such as growth at the human body temperature of 37 °C and the production of secondary metabolites such as inhibitors of chemokines and TNF-a (Rether et al. 2004; Chan and Chu 2007). Particularly the fatal brain infections, which were repeatedly reported in *Achaetomium strumarium* (synonym of *Chaetomium strumarium*) (Abbott et al. 1995; Aribandi et al. 2005), in *C. atrobrunneum* (Hubka et al. 2011), and in *Thielavia subthermophila* (Badali et al. 2011), all belonging to the *Chaetomiaceae*, are remarkable. The hidden clinical diversity of *Chaetomiaceae* urgently needs to be explored. The role of mammal dung and dung-enriched soil is one of the prime ecological niches in the order *Sordariales*, and this also holds true for *Chaetomium* (Zhang et al. 2006). Some species in the current study exclusively grow in dung, such as *Chaetomium homopilatum*. Multiple *Chaetomium* and *Thielavia* species have been isolated in East Africa from different kinds of dung, ranging from cow and horse to more exotic types of dung such as that of elephant and wildebeest (Carter and Khan 1982). Conversely, the position of *Madurella* in *Chaetomiaceae* is informative for the natural habitat of this pathogen. In the highly endemic area in Sudan, *M. mycetomatis* has as yet not been cultured, whereas the isolation of other causative agents of mycetoma, *Nocardia brasiliensis, Actinomadura madurae*, and *Streptomyces somaliensis*, has been successful (Aghamirian and Ghiasian 2009).

In Sugiyama et al. (2008) described a case of erythematous epilation of a dog caused by *C. globosum*. A mixed-breed young dog, a 4-month-old male, weighing 7.25 kg, showed depilation, scales, and dermatitis with slight itchiness on his skin. The main symptom was an erythematous epilation on the left subocular skin, 7.5 cm in diameter, accompanied by elephantiasis-like hyperplasia and scales. Similar lesions were observed on the skin on both sides of the ear lobes, the heels, tail, and left angulus oris. The scales from the crusted lesion were cultured on chloramphenicol-added potato dextrose agar plates at the first visit; this was followed by ambulatory practices. The isolates at the first visit, 1 and 3 weeks after treatment, were identified as *C. globosum* by mycological study and the D1/D2 domain of the large subunit rRNA gene sequence.

Fungal keratitis is a common cause of corneal ulcers in developing nations, accounting for 44% of corneal ulcers in South India (Reddy et al. 2017). Fungi are opportunistic in the eye since they rarely infect healthy, intact ocular tissues. *Chaetomium* species is an uncommon etiological agent when it comes to causing keratitis in humans. A global review of the literature with search words "*Chaetomium*" and "keratitis" reveals only four reported cases of *Chaetomium* keratomycosis (Vinod Mootha et al. 2012; Kaliamurthy et al. 2011; Balne et al. 2012; Ghosh et al. 2016; Reddy et al. 2017). Two cases were identified by ITS sequencing to *Chaetomium*-like species (non-sporulating), *Chaetomium atrobrunneum*, and yet another morphologically as *Chaetomium* spp. *Chaetomium globosum* has also been reported as a causative agent in <1% cases of fungal keratitis in a series from North India (Ghosh et al. 2016).

In Plumlee et al. (2017) investigated equine hyphalmycotic encephalitis, characterize key histopathologic features, and classify causative organisms with molecular diagnostic techniques. Seven cases were evaluated by histopathology. Panfungal PCR targeting the ribosomal RNA large subunit coding region and the noncoding internal transcribed spacer 2 region was performed on DNA extracted from formalin-fixed, paraffin-embedded sections of affected brain, and the resulting sequences were queried against published fungal genomes. Affected animals ranged from 8 to 22 years of age and presented with neurologic signs. Macroscopic lesions within affected brains included multifocal hemorrhage, focal swelling of the thalamus with red and yellow discoloration, and focal cerebral malacia. Major histologic findings included multifocal discrete foci of necrosis, neutrophilic to granulomatous inflammation, vasculitis, and intralesional fungal hyphae variably affecting the cerebrum, thalamus, and brainstem. DNA sequences in four cases showed >98% homology with species within the *Chaetomiaceae* family, including *Acrophialophora fusispora, Acrophialophora levis*, and *Chaetomium strumarium*. Histomorphologically, *Chaetomiaceae* fungi were 7–10 mm wide, septate, parallel walled, and nonpigmented, with dichotomous branching in affected horses. This case series is the first report of equine mycotic encephalitis caused by members of the *Chaetomiaceae* family, previously reported as rare emerging pathogens in humans.

5.5 Conclusion

The role of genus *Chaetomium* in human and animal disease has increased significantly in the last decade. *Chaetomium* infections and infections by species clustering in the *Chaetomium* phylogenetic tree, such as *Chaetomidium* and *Thielavia*, have been reported from the skin, hair, and nails. Moreover, several species of the *Chaetomiaceae* have been reported to cause serious opportunistic infections in immunocompromised patients. It seems that members of *Chaetomiaceae* indeed have an underestimated clinical potential, and re-evaluation of the role of the genus in human pathology is urgently required. The natural habitat of many species in arid climates and their survival at high temperatures probably enhance their survival in mammalian tissue. Due to identification difficulties on the basis of phenotypic criteria, some older cases of *Chaetomium* or *Subramaniula* species might have been erroneously disregarded or reported as cases of *Madurella* or *Papulaspora* infection. Moreover, antifungal susceptibility studies are scant, and treatment protocols are urgently needed.

References

Abbott SP, Sigler L, McAleer R et al (1995) Fatal cerebral mycoses caused by the ascomycete Chaetomium strumarium. J Clin Microbiol 33(10):2692–2698

Aghamirian MR, Ghiasian SA (2009) Isolation and characterization of medically important aerobic actinomycetes in soil of Iran (2006–2007). Open Microbiol J 3:53–57

Ahmed A, Adelmann D, Fahal A, Verbrugh H, van Belkum A, de Hoog GS (2002) Environmental occurrence of Madurella mycetomatis, the major agent of human eumycetoma in Sudan. J Clin Microbiol 40:1031–1036
Ahmed SA, Khan Z, Wang XW et al (2016) Chaetomium-like fungi causing opportunistic infections in humans: a possible role for extremotolerance. Fungal Divers 76:11–26
Al-Aidaroos A, Bin-Hussain I, El Solh H, Kofide A, Thawadi S, Belgaumi A, Al Ahmari A (2007) Invasive Chaetomium infection in two immunocompromised pediatric patients. Pediatr Infect Dis J 26:456–458
Ames LM (1963) A monograph of the Chaetomiaceae. U.S. Army Research and Development, Series 2
Anandi V et al (1989) Cerebral phaeohyphomycosis caused by *Chaetomium globosum* in a renal transplant recipient. J Clin Microbiol 27:2226–2229
Andersen B, Frisvad JC, Søndergaard I et al (2011) Associations between fungal species and water-damaged building materials. Appl Environ Microbiol 77:4180–4188
Andersen B, Dosen I, Lewinska AM, Nielsen KF (2017) Pre-contamination of new gypsum wallboard with potentially harmful fungal species. Indoor Air 27(1):6–12. https://doi.org/10.1111/ina.12298
Apetrei IC, Draganesc GE, Popescu IT et al (2009) Possible cause of allergy for the librarians: books manipulation and ventilation as sources of fungus spores spreading. Aerobiologia 25:159–166
Aribandi M, Bazan Iii C, Rinaldi MG (2005) Magnetic resonance imaging findings in fatal primary cerebral infection due to Chaetomium strumarium. Australas Radiol 49(2):166–169
Aru A, Munk-Nielsen L, Federspiel BH (1997) The soil fungus *Chaetomium* in the human paranasal sinuses. Eur Arch Otorhinolaryngol 254:350–352
Asgari B, Zare R (2011) The genus Chaetomium in Iran, a phylogenetic study including six new species. Mycologia 103:863–882
Aspiroz C, Gene J, Rezusta A et al (2007) First Spanish case of onychomycosis caused by Chaetomium globosum. Med Mycol 45:279–282
Ayanbimpe GM, Wapwera SD, Kuchin D (2010) Indoor air mycoflora of residential dwellings in Jos metropolis. Afr Health Sci 10:172–176
Badali H, Carvalho VO, Vicente V, Attili-Angelis D, Kwiatkowski IB, Gerrits van den Ende AHG, de Hoog GS (2009) Cladophialophora saturnica sp. nov., a new opportunistic species of Chaetothyriales revealed using molecular data. Med Mycol 47:51–62. https://doi.org/10.1080/13693780802291452
Badali H, Chander J, Gupta A, Rani H, Punia RS, de Hoog GS, Meis JF (2011) Fatal cerebral phaeohyphomycosis in an immunocompetent individual due to Thielaviasubthermophila. J Clin Microbiol 49:2336–2341. https://doi.org/10.1128/JCM.02648-10
Bakhiet S M, Rubino S, Faha A H (2019). Chaetomium atrobrunneum causing human eumycetoma: The first report. PLOS Neglected Tropical Diseases 13(5): e0007276. https://doi.org/10.1371/journal.pntd.0007276
Balne PK, Nalamada S, Kodiganti M, Taneja M (2012) Fungal keratitis caused by *Chaetomium atrobrunneum*. Cornea 31:94–95
Barron MA, Sutton DA, Veve R et al (2003) Invasive mycotic infections caused by Chaetomium perlucidum, a new agent of cerebral phaeohyphomycosis. J Clin Microbiol 41(11):5302–5307
Bell A (2005) An illustrated guide to the coprophilous Ascomycetes of Australia. CBS Fungal Biodivers Ser 3:1–172
Carter A, Khan RS (1982) New and interesting Chaetomium species from East Africa. Can J Bot 60:1253–1262
Chan T-M, Chu M (2007) Sch 213766, a novel chemokine receptor CCR-5 inhibitor from Chaetomium globosum. J Antibiot 60:524–528
Chivers AH (1915) A monograph of the genera Chaetomium and Ascotricha. Mem Torrey Bot Club 14:155–240
Corda ACJ (1840) Icones fungorum hucusque cognitorum, vol 4, Praha

de Hoog GS, Ahmed SA, Najafzadeh MJ et al (2013a) Phylogenetic findings suggest possible new habitat and routes of infection of human eumyctoma. PLoS Negl Trop Dis 7(5):e2229
de Hoog G, Guarro J, Gené J, Figueras MJ (2013b) Atlas of clinical fungi, 3rd edn. (e-version). Centraalbureau voor Schimmelcultures, Utrecht, The Netherlands/Universitat Rovirai Virgili, Reus, Spain
Došen I, Nielsen KF, Clausen G et al (2017) Potentially harmful secondary metabolites produced by indoor Chaetomium species on artificially and naturally contaminated building materials. Indoor Air 27(1):34–46. https://doi.org/10.1111/ina.12290
Doveri F (2008) An update on the genus Chaetomium with descriptions of some coprophilous species, new to Italy. Pagine Micol 29:1–60
Dreyfuss M (1976) Taxonomische Untersuchungen innerhalb der Gattung Chaetomium. Sydowia 28:50–133
Flannigan B, Miller JD (2011) Microbial growth in indoor environments. In: Flannigan B, Samson RA, Miller JD (eds) Microorganisms in home and indoor work environments. Diversity, health impacts, investigation and control, 2nd edn. CRC Press, Boca Raton, FL, pp 57–107
Friedman AH (1998) Cerebral fungal infections in the immunocompromised host: a literature review and a new pathogen—*Chaetomium atrobrunneum*: case report— comment. Neurosurgery 43:1469
Ghosh AK, Gupta A, Rudramurthy SM, Paul S, Hallur VK, Chakrabarti A (2016) Fungal keratitis in North India: spectrum of agents, risk factors and treatment. Mycopathologia 181:843–850
Gonianakis M, Neonakis I, Darivianaki E et al (2005) Airborne Ascomycotina on the island of Crete: seasonal patterns based on an 8-year volumetricsurvey. Aerobiologia 21:69–74
Guarro J, Soler L, Rinaldi MG (1995) Pathogenicity and antifungal susceptibility of Chaetomium species. Eur J Clin Microbiol Infect Dis 14:613–618
Guppy KH et al (1998) Cerebral fungal infections in the immunocompromised host: a literature review and a new pathogen—*Chaetomium atrobrunneum*: case report. Neurosurgery 43:1463–1469
Hall TA (1999) BioEdit: a user-friendly biological sequence alignment editor and analysis program for Windows 95/98/NT. Nucleic Acids Symp Ser 41:95–98
Hattori N, Adachi M, Kaneko T, Shimozuma M, Ichinohe M, Iozumi K (2000) Case report. Onychomycosis due to Chaetomium globosum successfully treated with itraconazole. Mycoses 43(1–2):89–92
Hoppin EC, McCoy EL, Rinaldi MG (1983) Opportunistic mycotic infection caused by Chaetomium in a patient with acute leukemia. Cancer 52:555–556
Hubka V, Mencl K, Skorepova M et al (2011) Phaeohyphomycosis and onychomycosis due to Chaetomium spp., including the first report of Chaetomium brasiliense infection. Med Mycol 49:724–733
Kaliamurthy J, Kalavathy CM, Nelson Jesudasan CA, Thomas PA (2011) Keratitis due to Chaetomium sp. Case Rep Ophthalmol Med 2011:696145
Kaul S, Sumbali G (1999) Production of extracellular keratinases by keratinophilic fungal species inhabiting feathers of living poultry birds (Gallus domesticus). Mycopathologia 146:6–19
Koch HA, Haneke H (1965) Chaetomium funicolum Cooke alsmoglicher Erreger einer tiefen Mykose. Mkosen 9:23–28
Kunze G, Schmidt JK (1817) Mykologische Hefte: 1. Germany, Leipzig
Lesire V et al (1999) Possible role of *Chaetomium globosum* in infection after autologous bone marrow transplantation. Intensive Care Med 25:124–125
Lin Y, Li X et al (1995) First case of phaeohyphomycosis caused by Chaetomium murorum in China. Chin J Dermatol 28:367–369
Mason S, Cortes D, Horner WE (2010) Detection of gaseous effluents and by products of fungal growth that affect environments (RP-1243). HVAC&R Res 16:109–121
Mazzucchetti G (1965) Microfungi della cellulosa eddla carta attivita's e inquadramento sistematico – Il genere "Chaetomium". Pubblicazioni Dell'entenazionale Per La Cellulosa e Per La Carta, Roma

McMullin DR, Sumarah MW, Miller JD (2013) Chaetoglobosins and azaphilones produced by Canadian strains of Chaetomium globosum isolated from the indoor environment. Mycotoxin Res 29:47–54
Mhmoud NA, Santona A, Fiamma M, Siddig EE, Deligios M, Bakhiet SM, Rubino S, Faha AH (2019). Chaetomium atrobrunneum causing human eumycetoma: The first report. PLoS Negl Trop Dis 13(5):e0007276. https://doi.org/10.1371/journal.pntd.0007276
Miller JD, McMullin DR (2014) Fungal secondary metabolites as harmful indoor air contaminants: 10 years on. Appl Microbiol Biotechnol 98:9953–9966
Millner PD (1977) Radial growth responses to temperature by 58 Chaetomium species, and some taxonomic relationships. Mycologia 69:492–502
Millner PD, Motta JJ, Lentz PL (1977) Ascospores, germ pores, ultrastructure, and thermophilism of Chaetomium. Mycologia 69:720–733
Möller EM, Bahnweg G, Sandermann H, Geiger HH (1992) A simple and efficient protocol for isolation of high molecular weight DNA from filamentous fungi, fruit bodies, and infected plant tissues. Nucleic Acids Res 20:6115–6116
Naidu J, Singh SM, Pouranik M (1991) Onychomycosis caused by Chaetomium globosum Kunze. Mycopathologia 113:31–34
Najafzadeh MJ, Fata A, Naseri A, Keisari MS, Farahyar S, Ganjbakhsh M, Ziaee M, Dolatabadi S, de Hoog GS (2014) Implantation phaeohyphomycosis caused by a non-sporulating Chaetomium species. J Mycol Med 24:161–165. https://doi.org/10.1016/j.mycmed.2013.09.007
Nylander JA, Wilgenbusch JC, Warren DL, Swofford DL (2008) AWTY: a system for graphical exploration of MCMC convergence in Bayesian phylogenetic inference. Bioinformatics 24:581–583
Piepenbring M, Caceres MOA, Espino EAA, Kirschner R, Schofer H (2007) Chromoblastomycosis caused by Chaetomium funicola: a case report from Western Panama. Br J Dermatol 157:1025–1029
Plumlee Q, Meason-Smith C, Dieterly A, Gomez G, Porter BF, Rodrigues HA (2017) Chaetomiaceae fungi, novel pathogens of equine neurotropic phaeohyphomycosis. Vet Pathol 54(5):813–819
Polizzi V, Delmulle B, Adams A et al (2009) JEM spotlight: fungi, mycotoxins and microbial volatile organic compounds in mouldy interiors from water-damaged buildings. J Environ Monit 11:1849–1858
Pounder J, Simmon KE, Barton CA, Hohmann SL, Brandt ME, Petti CA (2007) Discovering potential pathogens among fungi identified as non-sporulating molds. J Clin Microbiol 45:568–571
Rambaut A, Drummond AJ (2007) Tracer v. 1.4. Available from http://beast.bio.ed.ac.uk/tracer. Accessed 29 Sept 2014
Reddy M, Venugopal R, Prakash PY, Kamath YS (2017) Corneal ulcer due to a rare coelomycetes fungus *Chaetomium strumarium*: case report and global review of *Chaetomium* keratomycosis. Indian J Ophthalmol 65:871–874
Rether J, Erkel G, Anke T, Sterner O (2004) Inhibition of inducible TNF-alpha expression by oxaspirodion, a novel spiro-compound from the ascomycete *Chaetomium subspirale*. Biol Chem 385:829–834
Revankar SG, Sutton DA (2010) Melanized fungi in human disease. Clin Microbiol Rev 23:884–928. https://doi.org/10.1128/CMR.00019-10
Revankar SG, Patterson JE, Sutton DA, Pullen R, Rinaldi MG (2002) Disseminated phaeohyphomycosis: review of an emerging mycosis. Clin Infect Dis 34:467–476
Rodríguez K, Stchigel AM, Cano JF, Guarro J (2004) A new species of Achaetomium from Indian soil. Stud Mycol 50:77–82
Rock JP (1998) Cerebral fungal infections in the immuno compromised host: a literature review and a new patho- gen—*Chaetomium atrobrunneum*: case report—comment. Neurosurgery 43:1469
Samson RA, Flannigan B, Flannigan ME et al (1994) Health implications offungi in indoor environments. Elsevier, Amsterdam

Samson RA, Houbraken J, Thrane U et al (2010) Food and indoor fungi. In: CBS Laboratory Manual Series 2. CBS-KNAW Fungal Biodiversity Centre, Utrecht, The Netherland

Santos DW, Padovan AC, Melo AS, Gonçalves SS, Azevedo VR, Ogawa MM, Freitas TV, Colombo AL (2013) Molecular identification of melanised non-sporulating moulds: a useful tool for studying the epidemiology of phaeohyphomycosis. Mycopathologia 175:445–454. https://doi.org/10.1007/s11046-012-9608-x

Seth HK (1970) A monograph of the genus Chaetomium. Beiheftezur Nova Hedwigia 37:1–133

Skolko AJ, Groves JW (1948) Notes on seed-borne fungi V. Chaetomium species with dichotomously branched hairs. Can J Res 26:269–280

Skolko AJ, Groves JW (1953) Notes on seed-borne fungi VII. Chaetomium. Can J Bot 31:779–809

Sörgel G (1960) Zum problem der trennung von arten bei pilzen, dargestellt am beispiel der ascomycetengattung Chaetomium. Arch Microbiol 36:51–66

Stamatakis A (2014) RAxML Version 8: a tool for phylogenetic analysis and post-analysis of large phylogenies. Bioinformatics 30:1312–1313. https://doi.org/10.1093/bioinformatics/btu033

Stiller MJ, Rosenthal S, Summerbell RC et al (1992) Onychomycosis of the toenails caused by Chaetomium globosum. J Am Acad Dermatol 26:775–776

Straus DC (2011) The possible role of fungal contamination in sick building syndrome. Front Biosci 3:562–580

Sugiyama K, Sano A, Murakami M et al (2008) Three isolations of Chaetomium globosum from erythematous epilation of canine skin. Med Mycol 46(5):505–510

Teixeira AB et al (2003) Phaeohyphomycosis caused by *Chaetomium globosum* in an allogeneic bone marrow trans- plant recipient. Mycopathologia 156(4):309–312

Thomas C et al (1999) Fatal *Chaetomium* cerebritis in a bone marrow transplant patient. Hum Pathol 30:874–879

Vesper S, McKinstry C, Ashley P et al (2007) Quantitative PCR analysis ofmolds in the dust from homes of asthmatic children in North Carolina. J Environ Monit 9:826–830

Vinod Mootha V, Shahinpoor P, Sutton DA, Xin L, Najafzadeh MJ, de Hoog GS (2012) Identification problems with sterile fungi, illustrated by a keratitis due to a non-sporulating chaetomium-like species. Med Mycol 50:361–367. https://doi.org/10.3109/13693786.2011.611179

von Arx JA, Dreyfuss M, Müller E (1984) A reevaluation of Chaetomium and Chaetomiaceae. Persoonia 12:169–179

von Arx JA, Guarro J, Figueras MJ (1986) The Ascomycete genus Chaetomium. Beiheftezur Nova Hedwigia 84:1–162

Wang XW, Lombard L, Groenewald JZ et al (2016) Phylogenetic reassessment of the Chaetomium globosum species complex. Persoonia 36:83–133

WHO (2009) WHO guidelines for indoor air quality: dampness and mould. World Health Organization Regional Office for Europe, Copenhagen

Yeghen T et al (1996) *Chaetomium* pneumonia in a patient with acute myeloid leukaemia. J Clin Pathol 49:184–186

Youngchim S et al (2004) Production of melanin by *Aspergillus fumigatus*. J Med Microbiol 53:175–181

Zhang N, Castlebury LA, Miller AN, Huhndorf SM, Schoch CL et al (2006) An overview of the systematics of the Sordariomycetes based on a four-gene phylogeny. Mycologia 98:1076–1087

Zopf W (1881) Zur Entwicklungsgeschichte der Ascomyceten: Chaetomium. Nova Acta der Kaiserlich Leopoldinisch-Carolinisch Deutschen Akademieder Naturforscher 42:199–292

Part II
Metabolites of *Chaetomium*: The Good, the Bad and the Ugly

Chapter 6
Recent Advancements on the Role of Biologically Active Secondary Metabolites from *Chaetomium*

Shimaa Ragab Hamed, Ahmed M. Abdel-Azeem, and Parsa Mahmood Dar

6.1 Introduction

Natural products have served as an important source of drugs since ancient times and about half of the useful drugs today are derived from natural sources, for example, plants, microorganisms, and marine organisms. The natural world still offers a valuable source for novel drug discovery, but rapid identification of the bioactive compounds of natural product mixtures remains a critical factor to ensure that this gift from nature can compete with state-of-the-art developed technologies such as chemical compound libraries and high-throughput screening of combinatorial synthetic efforts. Fungi are of growing importance as a promising source of bioactive natural products (Cutler 1984; Stadler and Keller 2008). The fungal kingdom contains numerous organisms, including mushrooms, rusts, smuts, puffballs, mold or yeast that play remarkable economic and ecological roles (Gurnani et al. 2014). Mushrooms have been used for the production of food and alcoholic beverages (yeast), for medication in traditional medicine, as well as for cultural purposes (Dias et al. 2012). Among the more fascinating properties of fungi is their ability to synthesize a wide variety of secondary metabolites typically defined as small molecules that are not necessary for normal growth or development (Fox and Howlett 2008). The study of micro fungi has led to the discovery of many compounds that are used now as antibiotics, antifungal, immunosuppressive, or cholesterol lowering agents.

S. R. Hamed (✉)
Microbial Biotechnology Department, National Research Centre, Giza, Egypt

A. M. Abdel-Azeem
Botany Department, Faculty of Science, University of Suez Canal, Ismailia, Egypt
e-mail: ahmed_abdelazeem@science.suez.edu.eg

P. M. Dar
Institute of Chinese Medical Sciences, Avenida da Universidade, Taipa, Macau, China

A. M. Abdel-Azeem (ed.), *Recent Developments on Genus Chaetomium*,
Fungal Biology, https://doi.org/10.1007/978-3-030-31612-9_6

6.2 *Chaetomium*

The genus *Cheatomium*, comprising more than 350 species, has been widely studied to isolate more and more bioactive secondary metabolites with interesting structures (Itoh et al. 1980; Jiao et al. 2004; Zhang et al. 2010) Up to recent times, more than 200 compounds of diverse structural types have been obtained from the genus *Cheatomium* (Zhang et al. 2012), with diverse bioactivities such as cytotoxicity, antimicrobial activity, and the ability to inhibit root elongation (Xu et al. 2012). *Chaetomium* is a genus of fungi in the Chaetomiaceae family. It is a dematiaceous (dark-walled) mold normally found in soil, air, and plant debris and also consider as soil contaminant. *Chaetomium* colonies grow rapidly. They appear cottony and white in color initially. Mature colonies become grey to olive in color. From the rear, the color is tan to red or brown to black. Fruiting structures of this mold look like olive-green cockleburs—1.588 mm in diameter—that develop on compost; they frequently emit a musty odor. Lack of oxygen, when the temperatures are greater than 61 °C, permits formation of compounds produced in anaerobic conditions. Most species are prolific producers of the enzyme cellulase that breaks down cellulose. Destruction of paper and other materials containing cellulose (including foods, feeds, paper, textile, bird feathers, seeds and military equipment) by species of this mold is well documented. Due to their strong ability to destroy material, *Chaetomium* is perhaps the third most common indoor fungal contaminant of moldy damp buildings. Therefore, *Chaetomium* species are often used in testing materials for resistance to mold growth. The most widespread species in this genus are *Chaetomium atrobrunneum*, *C. funicola*, *C. globosum*, and *C. strumarium*. The most widespread and common species is *Chaetomium globosum*. This species causes many problems of biodeterioration of paper and other cellulose containing material. It is considered a "weed" of mushroom beds, where it inhibits the growth of cultivated mushrooms. *Chaetomium globosum* is of particular interest because it appears to produce mycotoxins, compounds which are detrimental to human health. These fungi have been definitively linked with allergies in people who are sensitive to molds; they appear to produce mutagenic mycotoxins which interfere with DNA damage in humans and other animals (Sutton et al. 1998; Nielsen et al. 1999).

6.3 Extraction, Isolation, and Structure Elucidation

Conforming to earlier and recent reports concerning the production of natural products from *Chaetomium* sp., extraction and isolation of natural products from *Chaetomium* sp. were performed as follows: the fungal culture was extracted with organic solvents; the resultant extract was subjected to column chromatography on silica gel or RP-C18, and Sephadex LH-20; and the obtained fractions were further subjected to preparative HPLC to get pure compounds. The structures of the metabolites which were isolated from the *Chaetomium* genus were determined through extensive spectroscopic analysis (Jiao et al. 2004; Ding et al. 2006) including UV, IR, MS, 1D (1H and 13C NMR and DEPT) and 2D NMR (COSY, NOESY, ROESY, HMQC or HSQC, and HMBC). The absolute configurations of several compounds have been established by X-ray analysis, CD spectra, and modified Mosher's NMR (nuclear magnetic resonance), in conjunction with computational chemistry.

6.4 Secondary Metabolites

Fungi produce biomass in the form of mycelium and spores, both of which contain an array of secondary metabolites and bioactive compounds. Demain and Fang (2000) noted that these metabolites function (i) as competitive weapons used against other bacteria, fungi, amoebae, plants, insects, and large animals; (ii) as metal transporting agents; (iii) as agents of symbiosis between microbes and plants, nematodes, insects, and higher animals or other possible ecological functions. There is now strong evidence supporting the fact that secondary metabolites allow organisms, particularly microorganisms, to carve out an ecological niche (Keller et al. 2005; Gloer 2007). The production of toxic secondary metabolites by an organism would directly affect a competitor's ability to access the shared resource, increasing the organism's own chance for survival. Fungi may colonize a resource and produce secondary metabolites that render the nutrient substrate unappealing to animals or other microorganisms. Currently there are more than 200 known bioactive metabolites produced by different *Chaetomium* spp.; more bioactive secondary metabolites might be found in this member of fungi (Zhang et al. 2012). There are many distinguished secondary metabolites which are synthesized through many and different pathways as shown in Fig. 6.1.

6.4.1 Chaetochromones

Chaetochromones A (**1**) and B (**2**), two novel polyketides, were isolated from the crude extract of fungus *Chaetomium indicum* (CBS.860.68) together with three known analogs PI-3 (**3**), PI-4 (**4**), and SB236050 (**5**) as shown in Fig. 6.2. The structures of these compounds were determined by HRESI-MS (high resolution electrospray ionization mass spectroscopy) and NMR experiments. Chaetochromones A (**1**) and B (**2**) are members of the polyketides family; they may have originated from a similar biogenetic pathway as the known compounds PI-3 (**3**), PI-4 (**4**), and SB236050 (**5**). The biological activities of these secondary metabolites were evaluated against eight plant pathogens, including *Alternaria alternata*, *Ilyonectria radicicola*, *Trichoderma viride*, *Aspergillus niger*, *Fusarium verticillioide*, *Irpex lacteus*, *Poria placenta*, and *Coriolus versicolor*. Compound **1**

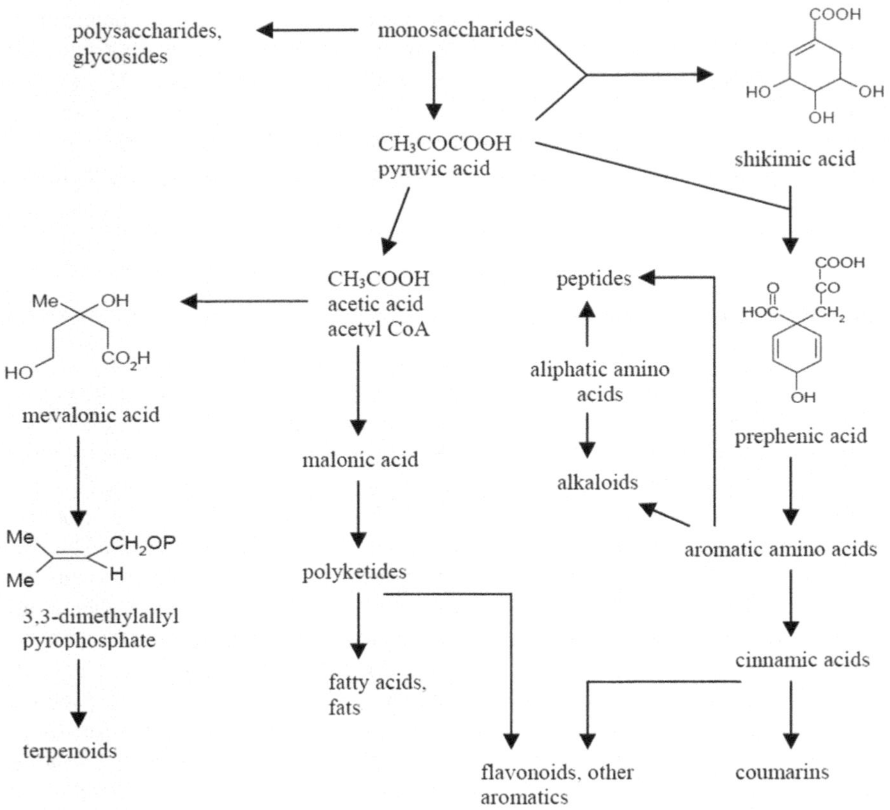

Fig. 6.1 Scheme of secondary metabolism

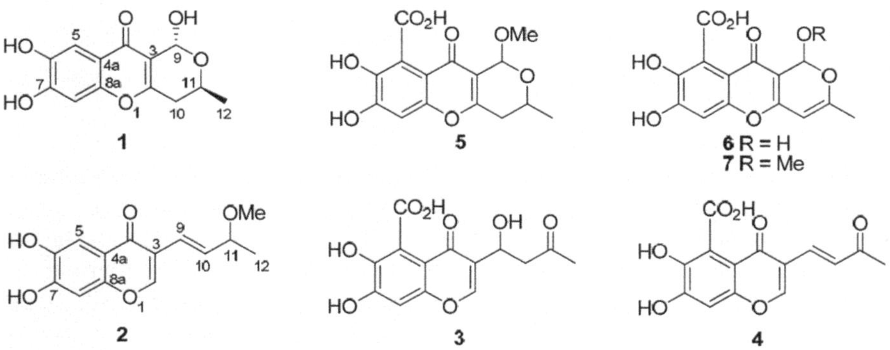

Fig. 6.2 Structure of compounds **1**–**6**

displayed moderate inhibitory rate (>60%) against the brown rot fungus *Poria placenta* that causes significant wood decay. In addition, the cytotoxic activities against three cancer cell lines A549, MDA-MB-231, PANC-1 were also tested, without any inhibitory activities being detected (Keyang et al. 2013) (Fig. 6.2).

6.4.2 Mollipilin

Chaetomium mollipilium cultivated with nicotinamide (100 μM) showed a notable change in the secondary metabolites (Fig. 6.3). Five new C_{13}-polyketides, mollipilin A–E (1–5), as well as known spiroketal 6 (named as mollipilin F), and (_)-aureonitol (7) were isolated. The cell growth inhibitory activities of polyketides 1–7 were evaluated in human HCT 116 cells. Mollipilin A (1) and B (2) exhibited moderate inhibitory effects on the cell growth with GI_{50} (concentration required to inhibit growth by 50%) values of 1.8 and 3.7 μM, respectively. The Western blot with an antibody against acetylated histone H4 showed that nicotinamide treatment of *C. mollipilium* induced hyperacetylated histone H4 in a dose dependent manner (Teigo et al. 2012).

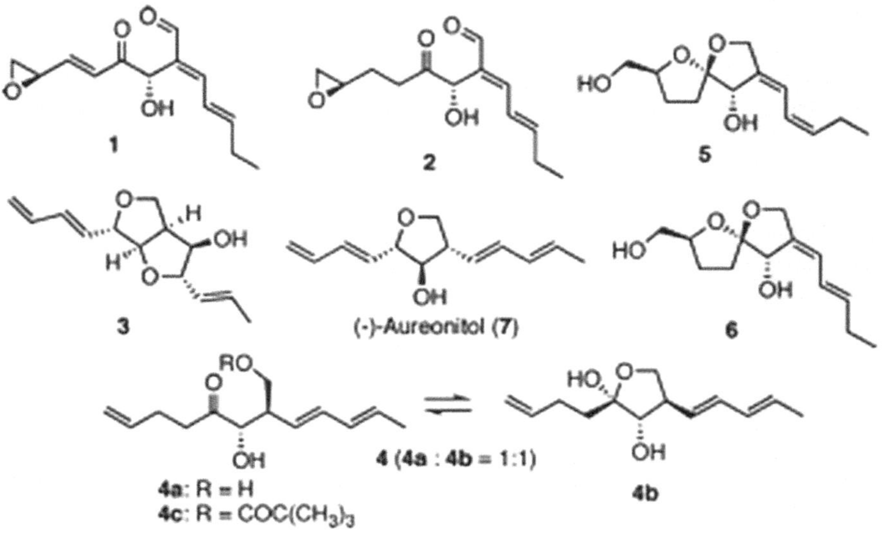

Fig. 6.3 Structure of new C_{13}-polyketides from *Chaetomium mollipilium*

6.4.3 Chaetosemins

He et al. (2018) isolated 15 polyketides including four new ones (Fig. 6.4), chaetosemins G–J (1–4), along with 11 known ones (5–15) from the culture of *C. seminudum* (C208) and *Chaetomium* sp. (C521). These metabolites were evaluated in vitro for antifungal, antioxidant, toxicity, and α-glucosidase inhibitory activities. Chaetosemin J (4) and monaschromone (5) significantly inhibited the growth of four plant pathogenic fungi *Botrytis cinerea, Alternaria solani, Magnaporthe oryzae,* and *Gibberella saubinettii* with minimum inhibitory concentrations (MIC) values ranging from 6.25 to 25.0 μM. Moreover, both epicoccone B (11) and flavipin (14) exhibited DPPH (2,2-diphenyl-1-picrylhydrazyl) free radical scavenging ability with IC_{50} (half maximal inhibitory concentration) values of 10.8 and 7.2 μM, respectively; they had more potent α-glucosidase inhibition than the drug acarbose with IC_{50} values of 27.3 and 33.8 μM, respectively. Monaschromone (5) may be used as the lead compound of pesticides (He et al. 2018).

6.4.4 Chaetominine

Chaetominine (Fig. 6.5) is tripeptide-derived alkaloidal metabolite with a new framework obtained from the culture of *Chaetomium* sp. (IFB-E015), an endophytic fungus colonizing inside normal *A. axilliflora* leaves (Rui et al. 2006). It is more cytotoxic to the human leukemia K562 and colon cancer SW1116 cell lines than the co-assayed positive reference, 5-fluorouracil, which is currently one of most frequently prescribed anticancer drugs. Chaetominine seems biogenetically related to the tryptoquivaline-like ç-lactonic metabolites isolated previously from the cultures of *Aspergillus* (Yamazaki et al. 1978) and *Penicillium* species (Ariza et al. 2002).

Chaetominine was the first tryptoquivaline-related metabolite from the *Chaetomium* species (also an endophyte) with the ç-lactone moiety as in tryptoquivalines rearranged to form a unique a-lactam ring at the center of the molecule. The tetracyclic system (rings a–d) of chaetominine was also possessed by other metabolites such as kapakahines (Yeung et al. 1996; Nakao et al. 2003).

6.4.5 Cochliodones

Cochliodones (1) (Fig. 6.6) is a purple pigment isolated from *C. cochliodes*, *C. globosum* (Brewer et al. 1968; Brewer et al. 1970) and also, isolated from the culture of *C. elatum.*, *C. murorum*, and *C. amygdalisporum* (Sekita et al. 1981). Cochliodones have an antimicrobial activity against several genera of microfungi such as *Botrytis allii* and *Fusarium moniliforme* at low concentrations (1–10 μg/ml); toxicity tests

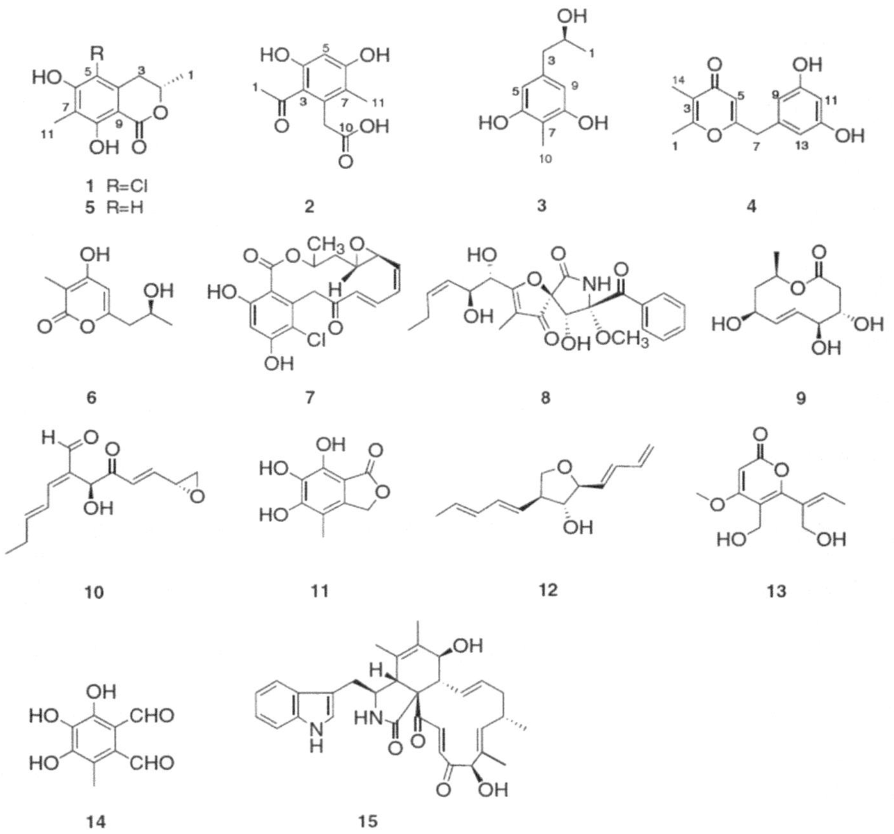

Fig. 6.4 Structures of compounds **1–9** from *C. seminudum* (C208) and **10–15** from *Chaetomium* sp. (C521)

Fig. 6.5 Structure of chaetominine

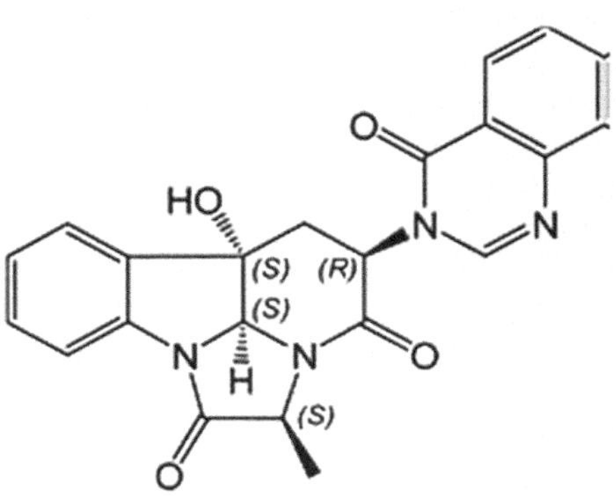

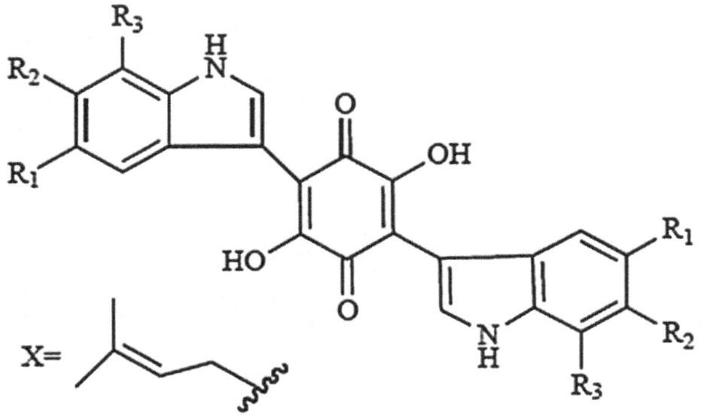

1 R1=X, R2=R3=H, **2** R1=R3=H, R2=X, **3** R1=R2=H, R3=X

Fig. 6.6 Structures of cochliodones, isocochliodinol, and neocochliodinol

on young turkeys and rats showed that the pigment does not cause acute toxicity (Meiler and Taylor 1971). Isocochliodinol (**2**), isolated from *Chaetomium murorum*, had a suppression effect against HIV-1 protease with an IC_{50} value as low as 0.17 μM. Neocochliodinol (**3**), isolated from *C. amygdalisporum*, also exhibited cytotoxic activity that depends upon the position of prenyl substituents at the indole rings (Sekita 1983).

6.4.6 Chaetoviridins

Chaetoviridin (Fig. 6.7) is a type of azaphilone. Takahashi et al. isolated new chaetoviridins A–D from *Chaetomium globosum* var. *flavoviride* (Takahashi et al. 1990). Chaetoviridins E and B, with antibiotic activity, were isolated from *Coprophilous* sp. and *Chaetomium* sp. obtained from the scat of an emu (Kingsland and Barrow 2009). From both *C. cochliodes* VTh01 and *C. cochliodes* CTh05, three new azaphliones, chaetoviridines E+ and F and *epi*-chaetoviridin A+, together with the known chaetoviridin A, were isolated with antimycobacterial activity against *M. tuberculosis*; chaetoviridine E+ also displayed an antimalarial activity against *P. falciparum* (Phonkerd et al. 2008). Moreover, both chaetoviridine E+ and F were also found to be cytotoxic against KB, BC1, and NCI-H187 cell lines. *C. globosum* (F0142) has bio-control activity against rice blast (*Magnaporthe grisea*) and wheat leaf rust (*Puccinia recondita*) due to two major antifungal products, chaetoviridins A and B. Chaetoviridin A exerts a higher antifungal efficacy against rice blast both in vitro and in vivo than chaetoviridin B (Takahashi et al. 1990).

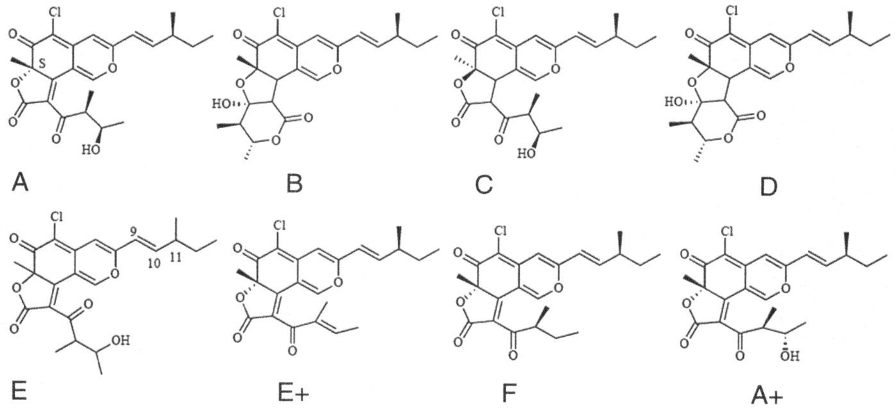

Fig. 6.7 Structure of chaetoviridins

6.4.7 Chaetomugilins

A strain of *Chaetomium globosum* was initially isolated from the marine fish *Mugil cephalus* collected in Katsuura Bay in Japan in October 2000, from which chaetomugilins I–O (1–7) (Fig. 6.8) were isolated (Muroga et al. 2009). As a primary screen for antitumor activity, the cancer cell growth inhibitory properties of the new chaetomugilins (1–7) were examined using themurine P388 leukemia cell line, the human HL-60 leukemia cell line, the murine L1210 leukemia cell line, and the human KB epidermoid carcinoma cell line. All compounds except chaetomugilin 5 exhibited significant cytotoxic activity against the cancer cell lines; chaetomugilin 1 in particular showed more potent cytotoxic activity against human cancer cell lines (HL-60 and KB cells) than 5-FU, the positive control. In addition, chaetomugilin 1 was examined using a disease-oriented panel of 39 human cell lines. Furthermore, evaluation of the pattern of differential cytotoxicity using the COMPARE program (Yamori et al. 1999) suggested that the mode of action of chaetomugilin 1 might be different from that of any other anticancer drugs developed to date. Two new products, 11-epichaetomugilin A and 4′-epichaetomugilin A, were also isolated from the strain of *C. globosum* present in the fish *Mugil cephalus*. 11-Epichaetomugilin A is the first azaphilone that has an opposite absolute configuration at C-11 to that of other natural compounds to date. It exhibited weak cytotoxic activity against P388 cells and HL-60 cells. Two more new azaphilones, seco-chaetomugilins A and D were produced from the same strain. Seco-chaetomugilin D showed growth inhibitory activity against P388, HL-60, L1210, and KB cells (Yamada et al. 2009).

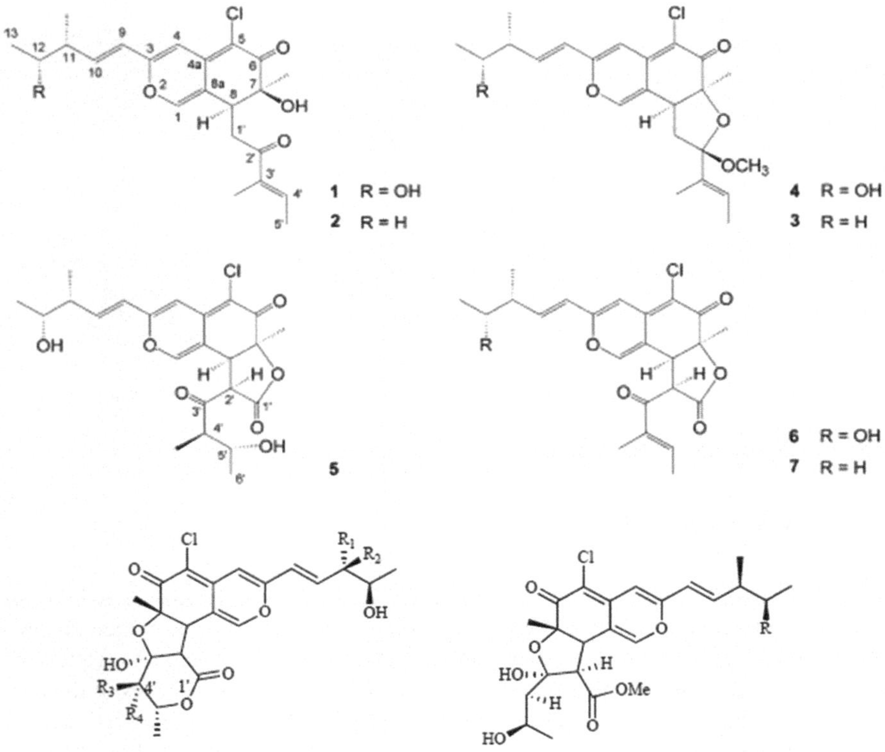

Fig. 6.8 Different structures of chaetomugilins

6.4.8 *Chaetoglobosins*

Chaetoglobosins A and B (Fig. 6.9), two azaphilone alkaloid dimers with an unprecedented skeleton, were isolated from the extract of *C. globosum* (IFB-E019), an endophytic isolate from *Imperata cylindrica* stem (Ge et al. 2008). Chaetoglobosin A can significantly inhibit the proliferation ability of MCF-7 and SW1116 cell lines. It also inhibits expressions of tumor-related genes bcl-2, c-myc, and β-catenin. Isolated an endophytic strain of *Chaetomium globosum* from the leaves of *Viguiera robusta* (Asteraceae); it was found to produce cytotoxic chaetoglobosins in a Czapek medium.

Fig. 6.9 Structure of chaetoglobosins

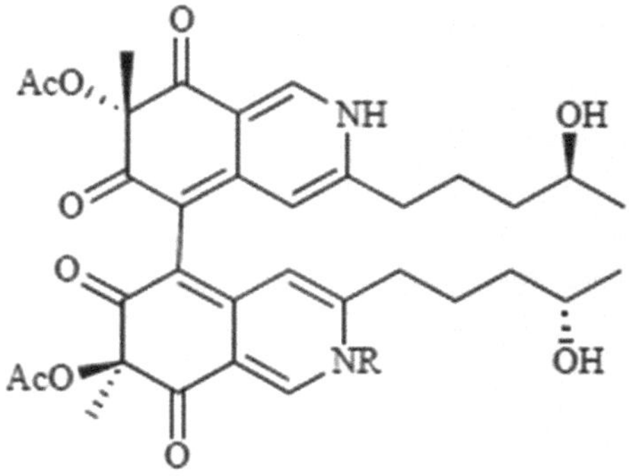

Chaetoglobins A R=H
Chaetoglobins B R=CH2CH2OH

6.4.9 Diketopiperazines

Diketopiperazines are well-known toxic products produced by a wide variety of fungi and also plants (Müllbacher et al. 1986; Dong et al. 2005). There are 24 compounds extracted from this class in the *Chaetomium* genus until now (Fig. 6.9). Gliotoxin is an example of these compounds, isolated from *Gliocladium fimbriatum* (Weindling and Emerson 1936; Gardiner et al. 2005) and also from *C. globosum* found in *Ginkgo biloba* (Qin et al. 2009). Different research have characterized three diketopiperazines: gliotoxin (**1**), methylthiogliotoxin (**2**), and fumitremorgin C (**3**). Only gliotoxin had good antifungal activity against plant pathogenic fungi including *Fusarium oxysporum* f. sp. *vasinfectum*, *F. graminearum*, *F. sulphureum*, *Cercospora sorghi*, and *Alternaria alternate*. Echinuline (**4**), isolated from *C. globosum* (KMITL-N0802), exhibited inhibitory activity on *M. tuberculosis* (Kanokmedhakul et al. 2002), while neoechinulin A (**5**) isolated from endophyte *C. globosum* of the marine red alga *Polysiphonia urceolata* produced (Wang et al. 2006). The fungus *Chaetomium* sp., isolated from the medicinal plant *Otanthus maritimus* (Asteraceae), yielded cyclo-(alanyl-tryptophane) (**6**) (Marwah et al. 2007). Chaetocin (**7**) is a metabolite of the fungus *C. minutum* with antibacterial and cytostatic activities (Hauser et al. 1970); it is isolated from a culture of *C. thielavioideum* (Udagawa et al. 1979). Tibodeau et al. (2009) proved that chaetocin has the ability to induce cellular oxidative stress, thereby selectively killing cancer cells. Chaetocin is also considered to be a selective inhibitor substrate of thioredoxin reductase (Tibodeau et al. 2009). All these observations have significant calibration with the mechanism of action and clinical development of chaetocin and related thiodioxopiperazines. Chetracin A (**8**), a dimeric diketopiperazine, is isolated from *Chaetomium nigricolor* and *C. retardatum*. Chaetocins B (**9**) and C (**10**), which are isolated from *C. virescens* var. *thielabioideum*, are two analogs of chaetocin (Saito et al. 1985, 1988). None of these compounds showed any antimicrobial activities toward the Gram-negative bacteria or fungus; however, they did exhibit a marked activity against the Gram-positive bacteria *S. aureus*. The cytotoxicities of chetomin, chaetocin, chetracin A, chaetocins B, and C to HeLa cells were highly noticeable (IC_{50} = 0.02–0.07 μg/ml), and similar structure-activity relationships were found for the antibacterial activity. Chetomin (**11**) was isolated from *Chaetomium cochliodes* with antibacterial activity (Geiger 1949). Later, it was also isolated from *C. cochliodes* and *C. globosum* present in soil samples collected in Nova Scotia as a toxic metabolite (Safe and Taylor 1972; McInnes et al. 1976). Three new chetomin-related metabolites, chetoseminudins A–C (**12–14**) and chetomin were produced by immunomodulatory extraction of *C. seminudum* (Fujimoto et al. 2004). Only chetomin and chetoseminudin A displayed high immunosuppressive activities against concanavalin A and lipopolysaccharide (LPS)-induced (B-cells) proliferations of mouse splenic lymphocytes. The activity of chetomin and chetoseminudin A might be due to the diketopiperazines moiety in their molecules. Chaetocochins A–C (**15–17**) are three new diketopiperazines derivatives which are isolated from the solid-state fermented rice culture of *Chaetomium cochliodes* (Li et al. 2006). Chaetocochins A and B, having a 14-membered ring, represent a novel type of diketopiperazines. Chaetocochins A and C have significant cytotoxicity in vitro against cancer cell lines and their activities appear to be correlated with the number of sulfur-methyls in their ring (Wu et al. 2008) (Fig. 6.10).

6.4.10 Chromones

Eleven novel unique chromones, termed chaetoquadrins A–E and F–K (Fig. 6.11) have been isolated from the EtOAc (ethyl acetate) extract of *C. quadrangulatum* (strain 71-NG-22); their monoamine oxidase inhibitory (MAO) features and structures including absolute configuration were elucidated. Successive studies on the extract of *C. quadrangulatum* guided by MAO inhibitory activity afforded six new constituents tentatively termed CQ6–11 in addition to 1–5. When the EtOAc extract was partitioned into n-hexane, EtOAc, and water layers, the EtOAc layer showed higher MAO inhibitory activity compared with the other two layers. Chromatographic fractionation of the EtOAc layer, focused on MAO inhibitory activity, gave the six novel constituents CQ6–11 (Fujimoto et al. 2003). Chaetoquadrin A, B, and C are tetracyclic spiroketals while chaetoquadrin D has a sulfonyl group. Chaetoquadrin G and H have appreciable monoamine oxidase inhibitory activity. Another novel chromone chaetoaurin and six known analogs eugenetin, eugenitol, chaetoquadrins A, B, G, and H were isolated from a solid culture of *C. aureus* (Li et al. 2010).

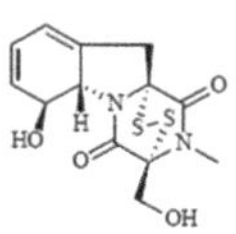

1

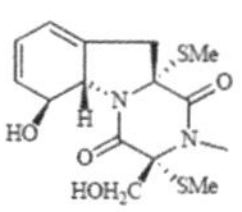

2

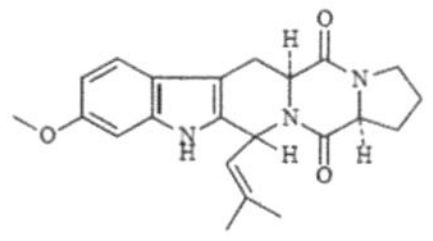

3

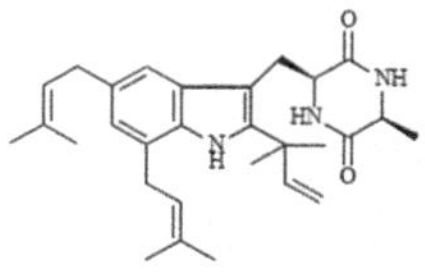

4

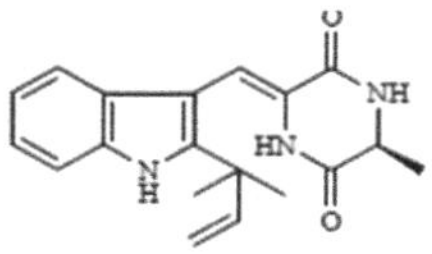

5

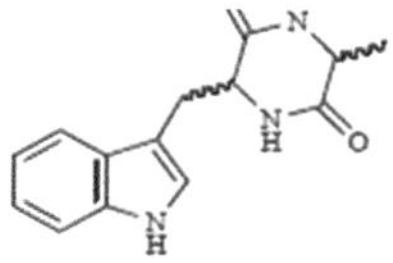

6

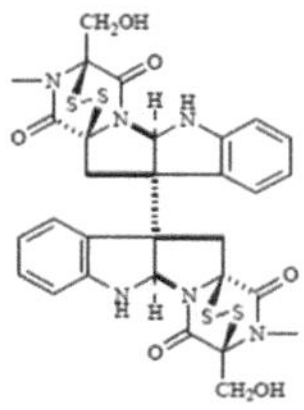

7

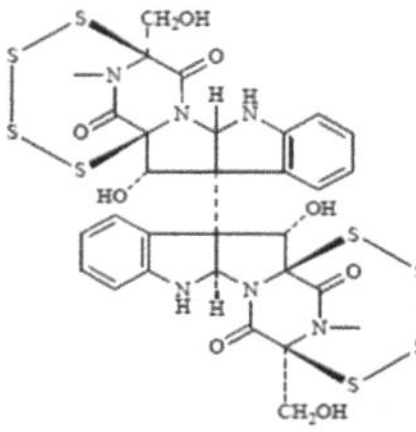

8

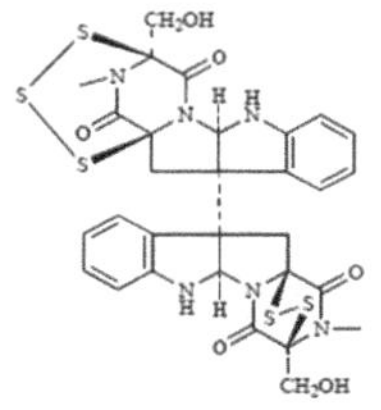

9

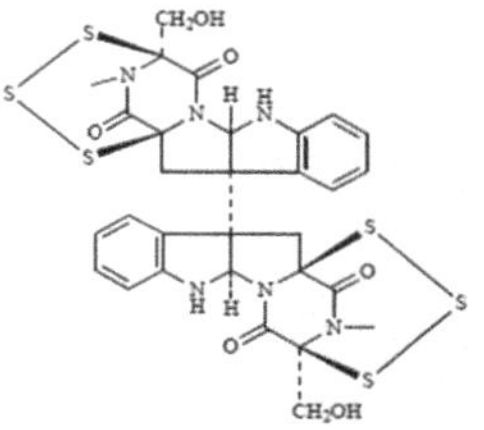

10

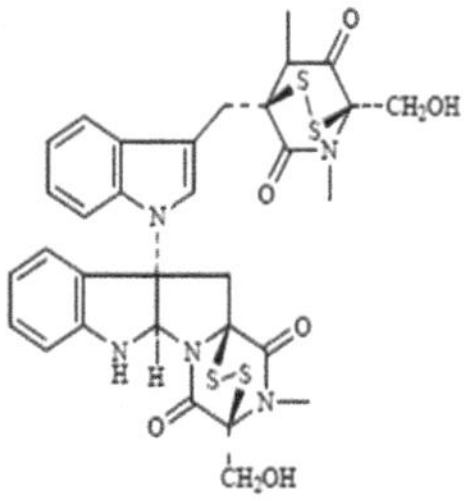

11

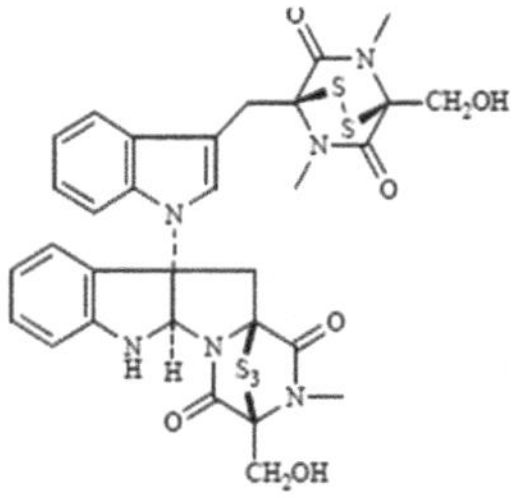

12

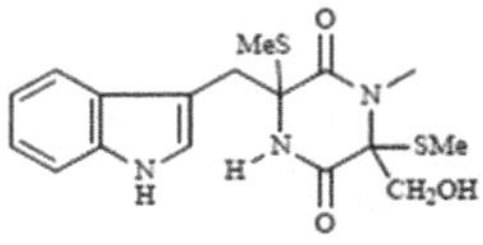

13

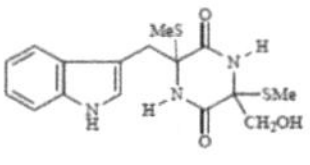

14

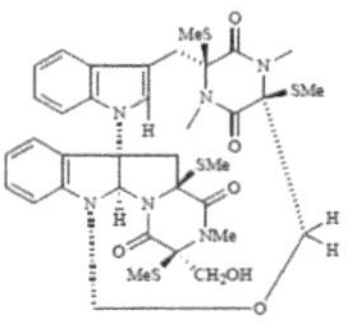

15

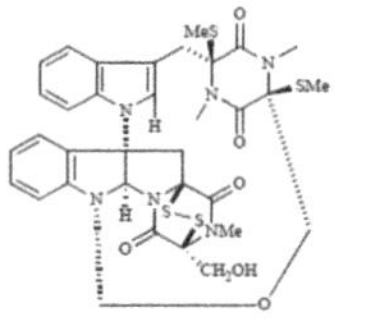

16

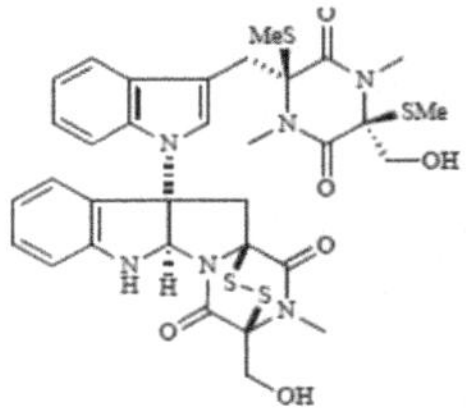

17

Fig. 6.10 Structures of diketopiperazine compounds

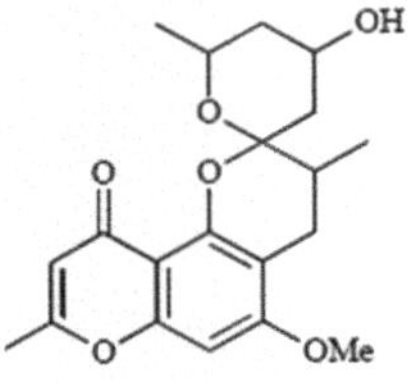

A

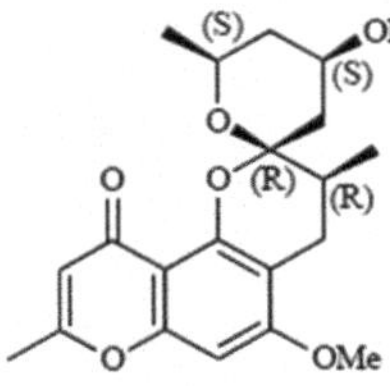

B

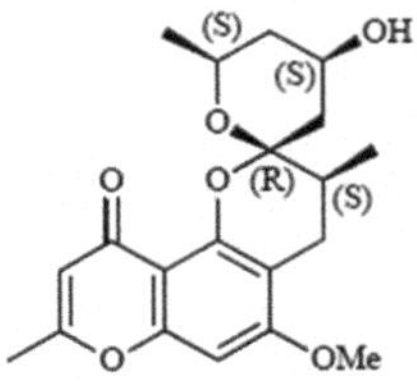

C

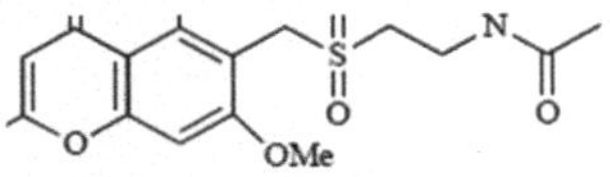

D

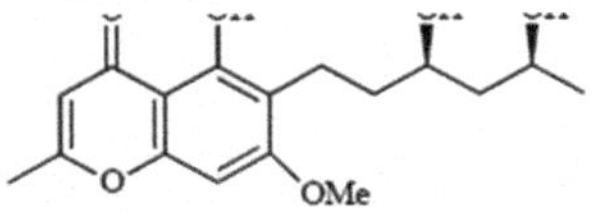

E

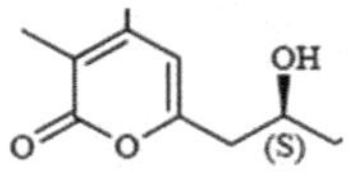

F

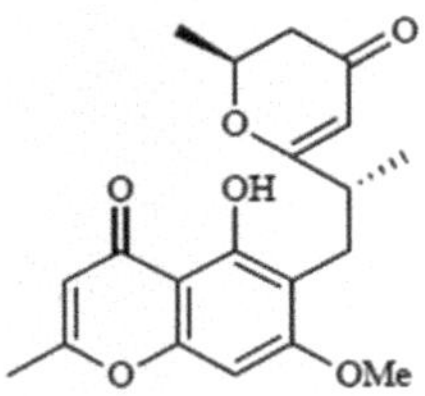

G

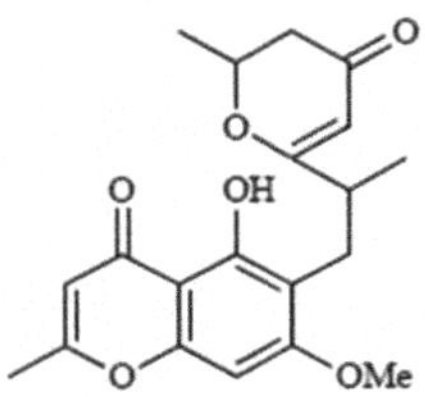

H

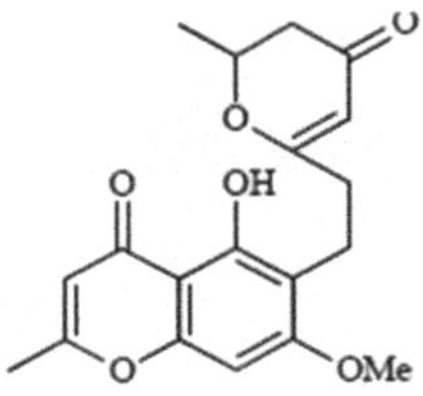

I

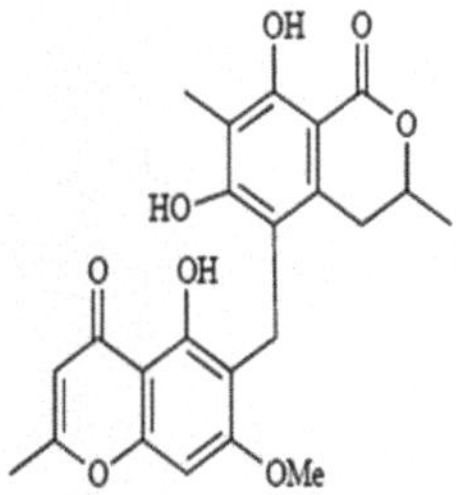

J

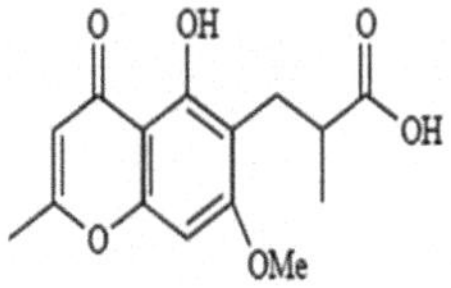

K

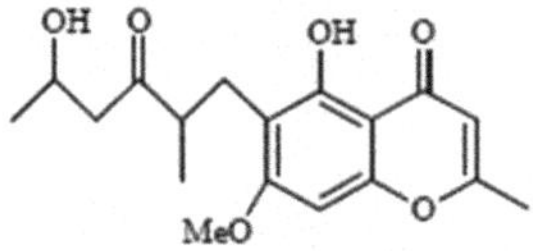

chaetoaurin

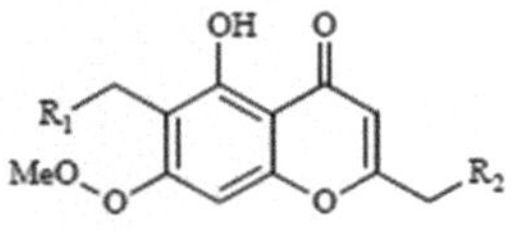

eugenetin R1= H, R2=H
eugenitol R1= OMe, R2=H

Fig. 6.11 Structures of chromones

Fig. 6.12 Structures of chaetoglocins A and B

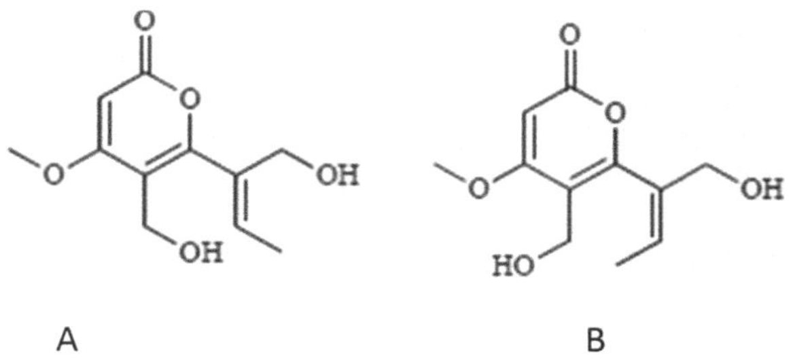

6.4.11 *Pyranone Compounds*

Two new pyranones, chaetoglocins A and B (Fig. 6.12) were isolated from the endophytic fungus *C. globosum* (IFB-E036) with antimicrobial activity against Gram-positive bacteria (Ge et al. 2011). In 1972, a new chromone metabolite was isolated from an extract of *Chaetomium minutum*, and this isolated metabolite was recognized as 6-hydroxymethyleugenin (Hauser and Zardin 1972). All these compounds have the same important structural motif, tetrahydropyran-4-one, which is characterized in many bioactive natural products (Baliah et al. 1983).

6.4.12 *Orsellides*

Orsellides (Fig. 6.13) are rare examples of fungal metabolites containing deoxyhexose building blocks that are normally characteristic of bacterial secondary metabolites. Thus, it was quite a surprise that *Chaetomium sp.* (strain Gö 100/9), an endophyte from marine algae, produces a couple of secondary metabolites which were identified as deoxy sugar derivatives. The antibacterial orsellides A–E (1–5), novel esters consisting of orsellinic acid (6) and a 6-deoxyhexose, were isolated from *Chaetomium sp.* (strain Gö 100/9) together with the known metabolites globosumones A (7) and B (8) (Schlörke and Zeeck 2006).

6.4.13 *Xanthones and Anthraquinones*

Xanthones (Fig. 6.14) are a group of natural products, extracted from sources such as plants, lichens, and fungi, with different biological activities. O-methylsterigmatocystin (**1**) and sterigmatocystin (**2**) were the first xanthones isolated in 1979 from cultures of *Chaetomium thielavioideum* (Sekita et al. 1981). Five new xanthone derivatives, globosuxanthones A–D (**3**–**6**), 2-hydroxyvertixanthone (**7**), together with two known anthraquinones chrysazin (**8**) and 1,3,6,8-tetra hydroxyanthraquinone (**9**) were isolated from an extract of the fungal strain,

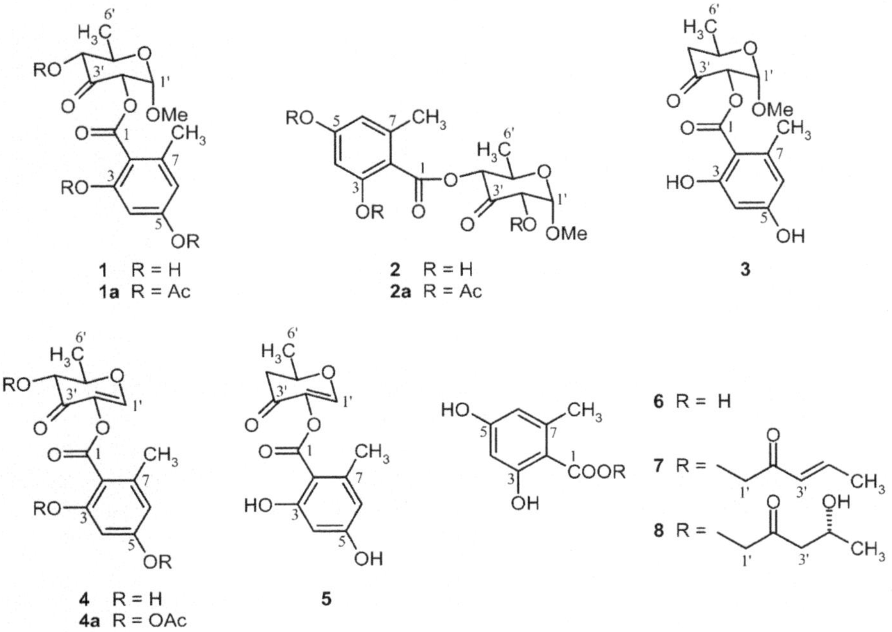

Fig. 6.13 Structure of orsellides

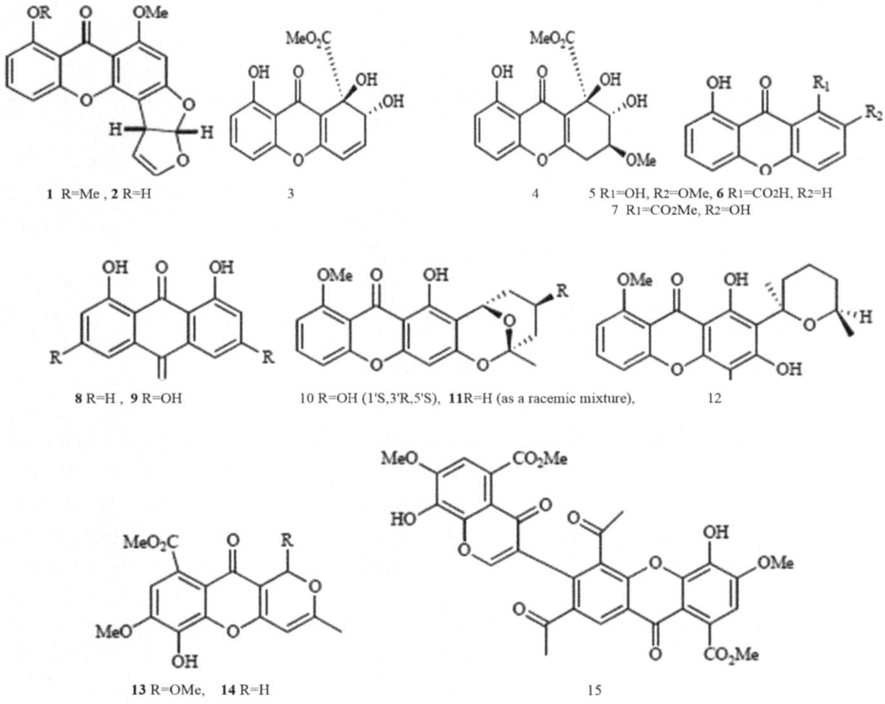

Fig. 6.14 Different structures of xanthones and anthraquinones

C. globosum Ames, inhabiting the rhizosphere of the Christmas cactus, *Opuntia leptocaulis* DC (Cactaceae). Globosuxanthone A had a strong cytotoxicity against a panel of seven human solid tumor cell lines (NCIH460, MCF-7, SF-268, PC-3, PC-3M, LNCaP, and DU-145), disrupted the cell cycle leading to accumulation of cells in either G2/M or S phase, and induced signs of apoptosis (Wijeratne et al. 2006). Three new xanthones, chaetoxanthones A–C (**10–12**), were also isolated from a marine-derived *Chaetomium* sp. (Pontius et al. 2008). Compounds **10** and **11** possess a rare dioxane/tetrahydropyran moiety and compound **12** is a chlorinated xanthone with a tetrahydropyran ring. Compound **11** had activity against *P. falciparum* without any cytotoxicity, while compound **12** showed a moderate activity against *Trypanosoma cruzi*. Chaetocyclinones A–C (**13–15**) were produced from the culture of an endosymbiotic fungus *Chaetomium* sp. (strain Gö 100/2) which was isolated from marine algae. Metabolite **13** had a fungistatic activity against a phytopathogenic fungus *Phytophthora infestans* in agar plate diffusion assay. There were no cytotoxic activities against the tumor cell lines HM02, HepG2, and MCF7 (Lösgen et al. 2007).

6.4.14 Steroids and Terpenoids

Ergosterol (**1**) and ergosterol peroxide (**2**) (Thohinung et al. 2010), ergosterol-β-D-glucoside (**3**), and ergosteryl palmitate (**4**) (Fig. 6.15) were isolated from *C. cochliodes* VTh01 and *C. globosum* KMITL-N0802, respectively. Ergosterol peroxide has an antimalarial activity against *P. falciparum* and an antimycobacterial activity against *M. tuberculosis*. A novel polyhydroxylated C29-sterol, 25β-methyl-22-homo-5α-cholest-7,22-diene-3β, 6β, 9α-triol, designated as globosterol (**5**), together with three known ergosterol derivatives, 9 hydroxycerevisterol (**6**), 9(11)-dehyoergosterol peroxide (**7**), and ergosta-4,6,8,22-tetraen-3-one (**8**), has been isolated from the cultures of an endophytic fungus *C. globosum* (ZY-22) taken from *Ginkgo biloba*. 9-Hydroxycerevisterol showed cytotoxic activity against HeLa cells in vitro and was found to be antinociceptive by using the acetic acid-induced writhing method (Kawagishi et al. 1988). Ergosterol peroxide and 9(11)-dehydroergosterol peroxide had a suppressive effect on the proliferation of mouse lymphocytes stimulated with mitogens such as Con A and LPS (Koyama et al. 2002). Heptelidic acid with antibiotic activity (**9**) was extracted from a culture of *C. globosum*, which was isolated from soil. Heptelidic acid was active against anaerobic bacteria, especially against *Bacteroides fragilis* (Kim and Lee 2009 In addition, methylglyoxal may be considered an important cause of diabetic complications. Its primary source is dihydroxyacetone phosphate whose levels are in part controlled by glyceraldehyde-3-phosphate dehydrogenase (GAPDH). The potent and specific inhibition of GAPDH of various origins by heptelidic acid was demonstrated in a concentration-dependent way by using a human red blood cell culture (Beisswenger et al. 2003). Kumagai et al. proved that heptelidic acid had high selective glycolytic cells through glucose-

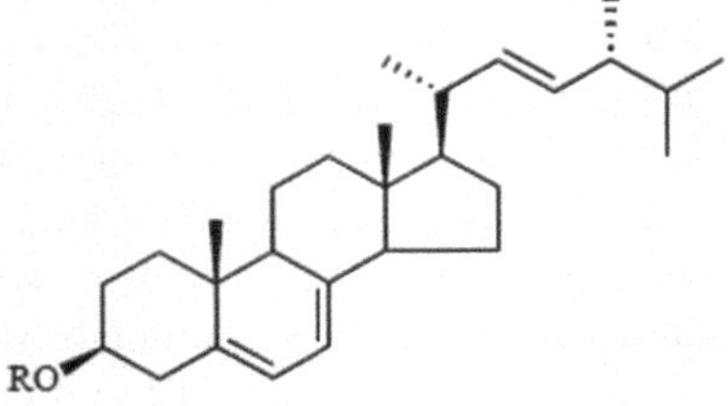

1 R=H, **3** R=glc, **4** R=CO(CH2)12CH3

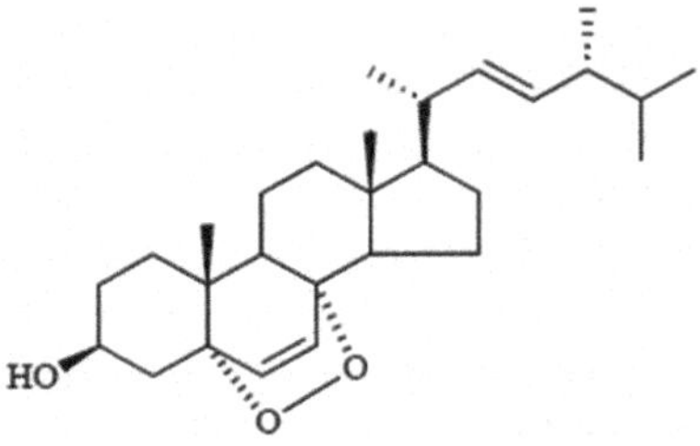

2

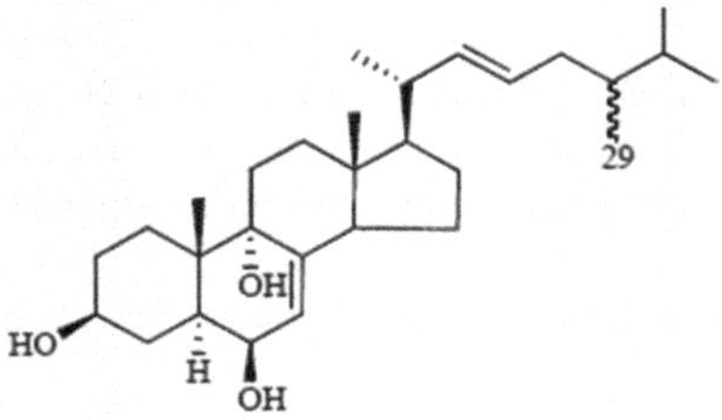

5

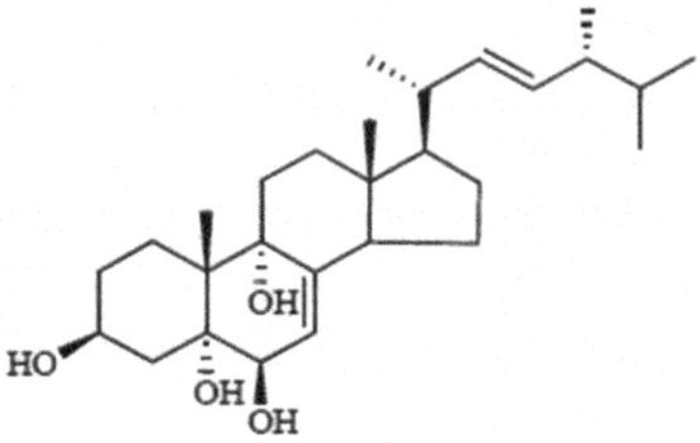

6

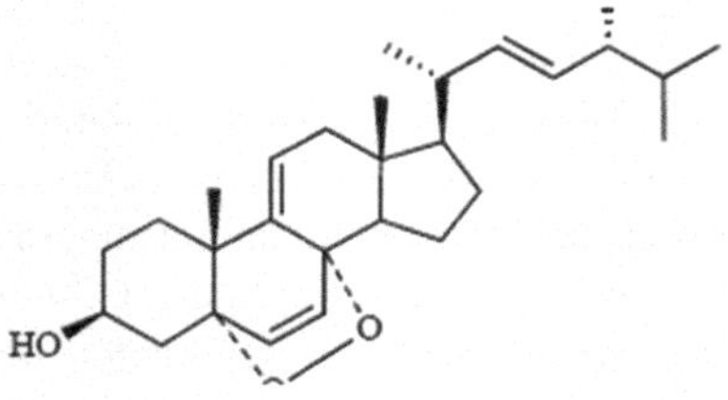

7

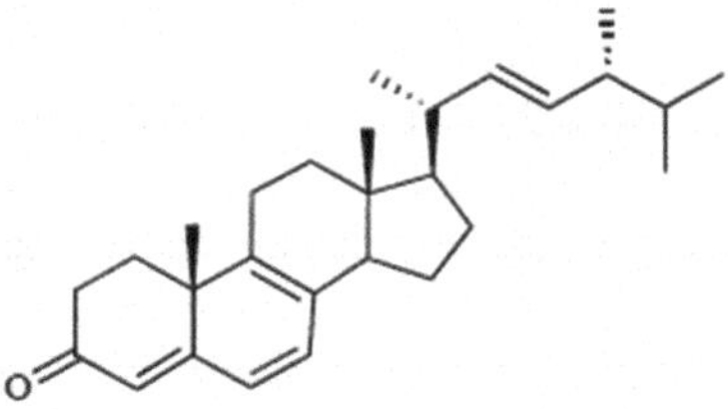

8

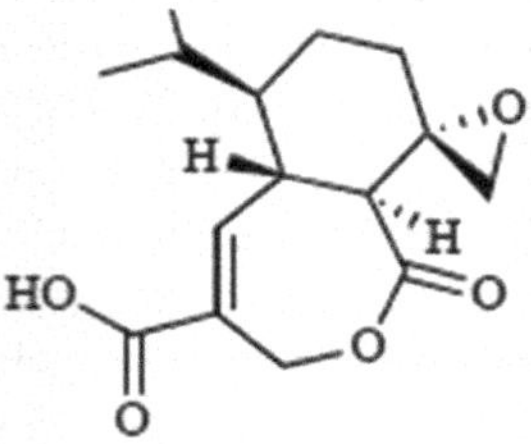

9

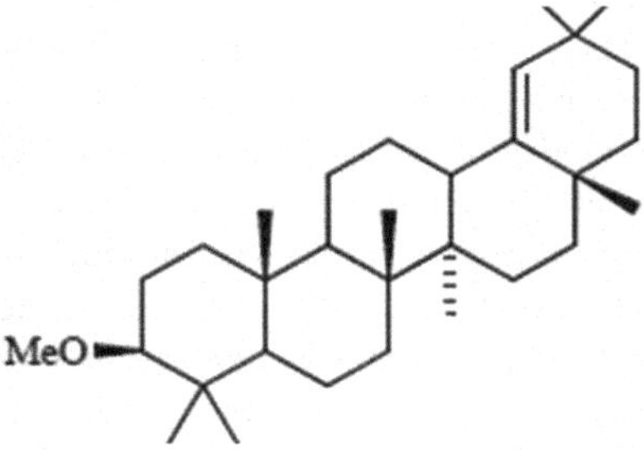

10

Fig. 6.15 Different structure of steroids and terpenoids dependent active ATP depletion, which occurred by progression of ATP dependent glucose phosphorylation, in spite of the blockade of the pathway at GAPDH. This mechanism may give effective methods of treatment of cancers that aggressively consume glucose (Kumagai et al. 2008). Furthermore, heptelidic acid induces nucleosomal DNA fragmentation in NG108-15 cells and exhibits antimalarial activity in vitro against the human malaria parasite *P. falciparum* (Nakazawa et al. 1997; Tanaka et al. 1998). Atriterpenoid, 3β-methoxyolean-18-ene, named miliacin (**10**), was isolated from the marine fungus *C. olivaceum* (Smetanina et al. 2001).

6.4.15 Aureonitols

Burrows extracted a tetrahydrofuran (THF) (**1a**) (Fig. 6.16) as secondary metabolite from *Chaetomium coarctatum* (Burrows 1967). Mason and Vane also published their work on the same sample and confirmed the previous results (Mason and Vane 1967). In 1992, Abraham extracted THF molecule from *C. cochlioides* (Abraham and Arfmann 1992) which was identical to those reported by Bohlmann for aureonitol (**1b**) from the 1H NMR data (Bohlmann and Ziesche 1979). Later, Seto et al. studied the biosynthesis of tetrahydrofuran in *C. coarctatum* using labeling experiments and reported that THF was a polyketide metabolite (Seto et al. 1979; Saito et al. 1983). The endophytic fungus *Chaetomium* sp., isolated from the plant *Otanthus maritimus*, yielded a new THF derivative, aureonitolic acid (**2**), together with cochliodinol (**3**), isocochliodinol (**4**), and orsellinic acid (Aly et al. 2009). Cochliodinol and orsellinic acid showed pronounced inhibiting activity against L5178Y mouse lymphoma cells while aureonitolic acid was inactive. Two antibacterial furano-polyenes, (−)-musanahol (**5**) and 3-epi-aureonitol (**6**), were isolated from *Chaetomium* sp.

6.4.16 Tetramic Acids

Tetramic acid (2,4-pyrrolidine-dione) is a kind of antibiotic with antimicrobial activity against different resistant microbial pathogens (Royles 1995). Chaetochalasin A (Fig. 6.17), a novel decalin derivative with an unprecedented ring system, was isolated from *Chaetomium brasiliense* (Oh et al. 1998). It displayed cytotoxicity in the NCI's 60-cell line human tumor cell panel. The derivative was also obtained from *C. cochliodes* (VTh01) and exhibited antimalarial activity against *Plasmodium falciparum* and weak antimycobacterial activity against *Mycobacterium tuberculosis* (Phonkerd et al. 2008).

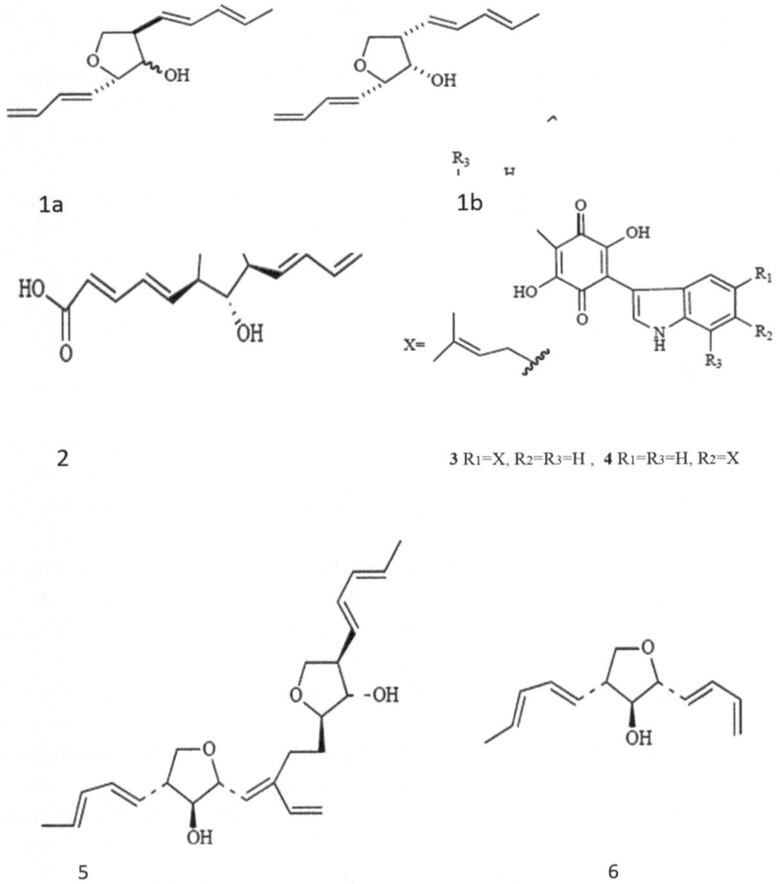

Fig. 6.16 Structures of aureonitols

6.4.17 *Depsidones*

Depsidones (Fig. 6.17) have been previously isolated from lichens, fungi, and plants [84], but now there are two new depsidones, mollicellins I (**1**) and J (**2**), and a new chromone, 2-(hydroxymethyl)-6-methylmethyleugenin (**3**), along with six known compounds, mollicellin D (**4**), mollicellin H (**5**), eugenetin (**6**), O-methylsterigmatocystin (**7**), sterigmatocystin (**8**), and chaetocin (**9**), isolated from a solid culture of *Chaetomium brasiliense*. Mollicellins I and H have anti-activity against three human tumor cell lines, breast cancer, lung, and neuroma (Li et al. 2008). Four new depsidones, mollicellins K-N (**10–13**), and their analogs, mollicellins B (**14**), C (**15**), E (**16**), F (**17**), H, and J, with two sterols, ergosterol and

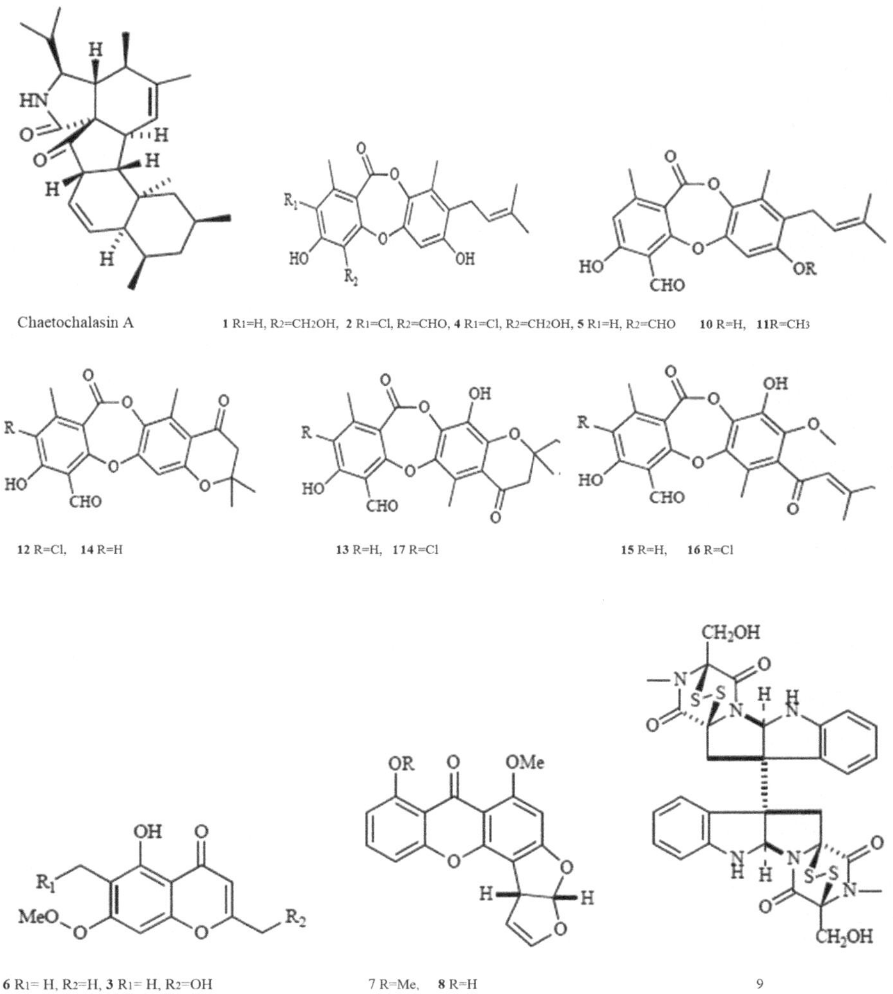

Fig. 6.17 Chaetochalasin A and depsidones

ergosterol peroxide, were isolated from *C. brasiliense*. Among these isolates, mollicellins K-M, mollicellins B, C, E, and J have antimalarial activity against *P. falciparum*. Only mollicellin K has moderate activity against *M. tuberculosis* and potent property against *Candida albicans*. Mollicellins K, N, C, E, and H have significant cytotoxicity against NCI-H187 cell lines. The aforementioned metabolites were found to have cytotoxic effect against five cholangiocarcinoma cell lines (Khumkomkhet et al. 2009).

6.4.18 *Resorcylic Acid Lactone*

In the field of treatment of cancer, and neurodegenerative diseases, fungal polyketides with resorcylic acid lactone play a very interesting role (Yang et al. 2011). During the fractionation bioassay of a Sonoran desert plant-associated fungus *C. chiversii*, the heat shock protein (Hsp90) inhibitor radicicol (**1**) (Fig. 6.18) was extracted (Turbyville et al. 2006). It is a nanomolar inhibitor of the chaperone Hsp90, whose repression leads to a combinatorial blockade of cancer-causing pathways. Recently, clustered genes for radicicol biosynthesis were identified and functionally characterized from the endophyte *C. chiversii* (Wang et al. 2008). Chaetoatrosin A (**2**), a novel chitin synthase II inhibitor, was extracted from *C. atrobrunneum* (F449). This metabolite has antifungal activities against various pathogenic fungi such as *Rhizoctonia solani*, *Pyricularia oryzae*, *Botrytis cinerea*, *Cryptococcus neoformans* and *Trichophyton mentagrophytes* (Hwang et al. 2000).

6.4.19 *Oxaspirodion*

Oxaspirodion is isolated from *Chaetomium subspirale*. It inhibited the TNF-α driven luciferase reporter gene expression with an IC_{50}-value of 2.5 μg/ml (10 μM) in TPA-/ionomycin-stimulated Jurkat T-cells by interfering with signal transduction

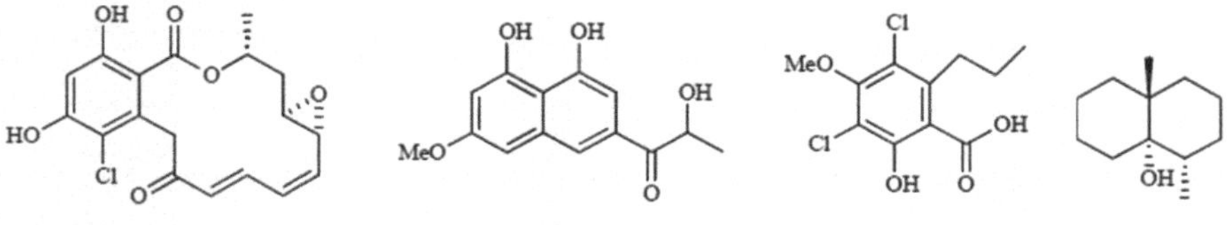

1 2 3 5

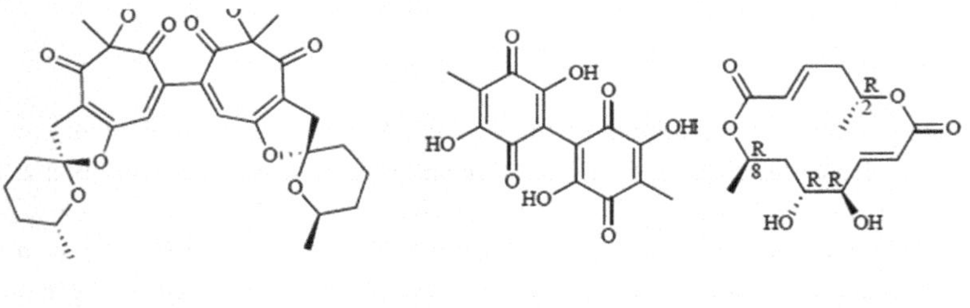

4 6 7

Fig. 6.18 Structures of resorcylic acid, lactone, naphthnol, and colletodiol pathways involved in the inducible expression of many pro-inflammatory genes. TNF-α is the main pro-inflammatory cytokine in inflammatory diseases like septic shock, rheumatoid arthritis, and Crohn's disease. Moreover, oxaspirodion also inactivated the transcription factor NF-κB, which is involved in the inducible expression of many proinflammatory genes (Rether et al. 2004).

6.4.20 Naphthnol and Colletodiol

Differanisole A (**3**) (Fig. 6.18), a new metabolite with activity against mouse leukemia cells, was extracted from *Chaetomium* (RB-001) (Oka et al. 1985). It also had activity against mouse neuroblastoma C-1300 cells in vivo (Suzuki et al. 1986) and growth of human myeloid leukemia cells. In addition, it induced G1 arrest and granulocytic differentiation of HL-60 cells (Kubohara et al. 1993). Chaetospiron (**4**), a metabolite extracted from *Chaetomium* sp. (BS 6556), has a dimeric structure containing two units related to maletinnin A (Bitzer 2005). Geosmin (**5**), an odorous metabolite of *C. globosum* Kinze ex Fr. from a soil sample, has a musty-smelling property (Kikuchi et al. 1981). In 1955, oosporein (**6**) was first isolated as a dibenzoquinone of *C. aureum* chivers (Lloyd et al. 1955) and later from a culture of *Chaetomium* sp. as a toxin to plants and poultry (Cole et al. 1974; Manning and Wyatt 1984). Oosporein inhibited herpes simplex virus type 1 DNA polymerase (Terry et al. 1992) and produced a strong antagonistic effect on late blight of tomato *Phytophthora infestans* in comparison with *Alternaria solani* and *Fusarium oxysporum* (Nagaoka et al. 2004). Colletodiol (**7**), a 14-membered bismacrolactone, was extracted from *C. funicola* (Powell and Whalley 1969).

6.5 Conclusion

In this chapter, many of the chemical compounds of different sorts have been reported from many species belonging to the genus *Chaetomium*. The variation of structures have distinct effects on biological activities including the cytotoxicity, antimalarial and antibacterial activity, and phytotoxicity, although there are many compounds which have not yet been discussed and also tested for their biological activities. Advanced studies on the biosynthetic pathway, pharmacokinetic property, structure-activity relationships, mechanism of action, and ecological roles of new isolated metabolites have great importance in their practical applications in medicine and agriculture.

References

Abraham WR, Arfmann HA (1992) Rearranged tetrahydrofurans from Chaetomium cochlioides. Phytochemistry 31:2405–2408

Aly AH, Debbab A, Edrada-Ebel R, Wray V, Müller WE, Lin W, Ebel R, Proksch P (2009) A new tetrahydrofuran derivative from the endophytic fungus Chaetomium sp. isolated from Otanthus maritimus. Z Naturforsch C 64:350–354

Ariza MR, Larsen TO, Peterson BO, Duus JO, Barrero AF (2002) Penicillium digitatum metabolites on synthetic media and citrus fruits. J Agric Food Chem 50:6361–6365

Baliah V, Jeyaraman R, Chandrasekaran L (1983) Synthesis of 2,6-disubstituted piperidines, oxanes, and thianes. Chem Rev 83:379–423

Beisswenger PJ, Howell SK, Smith K, Szwergold BS (2003) Glyceraldehyde-3-phosphate dehydrogenase activity as an independent modifier of methylglyoxal levels in diabetes. Biochem Biophys Acta 1637:98–106

Bitzer J (2005) Isolierung und Strukturaufklärung neuer Naturstoffe aus Bakterien und endophytischen Pilzen durch chemisches screening. Dissertation, University of Göttingen

Bohlmann F, Ziesche J (1979) Ein ungewöhnliches tetrahydrofuranderivat aus Helichrysum aureonitens. Phytochemistry 18:664–665

Brewer D, Jerram W, Taylor A (1968) The production of cochliodinol and a related metabolite by Chaetomium species. Can J Microbiol 14:861–866

Brewer D, Jerram WA, Meiler D, Taylor A (1970) The toxicity of cochliodinol, an antibiotic metabolite of Chaetomium spp. Can J Microbiol 16:433–440

Burrows BF (1967) A new fungal tetrahydrofuran. J Chem Soc Chem Commun 12:597–598

Cole RJ, Kirksey JW, Cutler HG, Davis EE (1974) Toxic effects of oosporein from Chaetomium trilaterale. J Agric Food Chem 22:517–520

Cutler HG (1984) Biologically active natural products from fungi: templates for tomorrow's pesticides. In: Ory RL, Rittig FR (eds) Bioregulators, chemistry and uses. American Chemical Society, Washington, DC, pp 153–170

Demain AL, Fang A (2000) The natural functions of secondary metabolites. In: Fiechter A (ed) History of modern biotechnology I, advances in biochemical engineering/biotechnology, vol 69. Springer, Berlin, pp 1–39. https://doi.org/10.1007/3-540-44964-7_1

Dias DA, Urban S, Roessner U (2012) A historical overview of natural products in drug discovery. Meta 2:303–336

Ding G, Song YC, Chen JR, Xu C, Ge HM, Wang XT, Tan RX (2006) Chaetoglobosin U, a cytochalasan alkaloid from endophytic Chaetomium globosum IFB-E019. J Nat Prod 69:302–304

Dong JY, He HP, Shen YM, Zhang KQ (2005) Nematicidal epipolysulfanyldioxopiperazines from Gliocladium roseum. J Nat Prod 68:1510–1513

Fox EM, Howlett BJ (2008) Secondary metabolism: regulation and role in fungal biology. Curr Opin Microbiol 11:481–487

Fujimoto H, Nozawa M, Okuyama E, Ishibashi M (2003) Six new constituents from an ascomycete, Chaetomium quadrangulatum, found in a screening study focused on monoamine oxidase inhibitory activity. Chem Pharm Bull 51(3):247–251

Fujimoto H, Sumino M, Okuyama E, Ishibashi M (2004) Immunomodulatory constituents from an ascomycete, Chaetomium seminudum. J Nat Prod 67:98–102

Gardiner DM, Waring P, Howlett B (2005) The epipolythiodioxopiperazine (ETP) class of fungal toxins: distribution, mode of action, functions and biosynthesis. J Microbiol 151:1021–1032

Ge HM, Zhang WY, Ding G, Saparpakorn P, Song YC, Hannongbua S, Tan RX (2008) Chaetoglobins A and B, two unusual alkaloids from endophytic Chaetomium globosum culture. Chem Commun:5978–5980

Ge HM, Zhang Q, Xu SH, Guo ZK, Song YC, Huang WY, Tan RX (2011) Chaetoglocins A-D, four new metabolites from the endophytic fungus Chaetomium globosum. Planta Med 77:277–280

Geiger WB (1949) Chetomin, an antibiotic substance from Chaetomium cochliodes III. Composition and functional groups. Arch Biochem 21:125–131
Gloer JB (2007) Applications of fungal ecology in the search for new bioactive natural products. In: Kubicek CP, Druzhinina IS (eds) The mycota, vol IV, 2nd edn. Springer-Verlag, New York
Gurnani N, Mehta D, Gupta M, Mehta BK (2014) Natural products: source of potential drugs. Afr J Basic Appl Sci 6:171–186
Hauser D, Zardin T (1972) Isolation of 6-hydroxymethyleugenin from Chaetomium minutum. Experientia 28:1114
Hauser D, Weber HP, Sigg HP (1970) Isolation and configuration of chaetocin. Helv Chim Acta 53:1061–1073
He L, Zhong-Bin L, Dan T, Wen-Bo H, Qiang Z, Jin-Ming G (2018) Polyketides from two Chaetomium species and their biological functions. J Antibiot 71:677–681. https://doi.org/10.1038/s41429-018-0047-x
Hwang EI, Yun BS, Kim YK, Kwon BM, Kim HG, Lee HB, Bae KS, Kim SU (2000) Chaetoatrosin A, a novel chitin synthase II inhibitor produced by Chaetomium atrobrunneum F449. J Antibiot 53:248–255
Itoh Y, Kodama K, Furuya K, Takahashi S, Haneishi T, Takiguchi Y, Arai M (1980) A new sesquiterpene antibiotic, heptelidic acid producing organisms, fermentation, isolation and characterization. J Antibiot 33(5):468–473
Jiao W, Feng Y, Blunt JW, Cole AL, Munro MH (2004) Chaetoglobosins Q, R, and T, three further new metabolites from Chaetomium globosum. J Nat Prod 67:1722–1725
Kanokmedhakul S, Kanokmedhakul K, Phonkerd N, Soytong K, Kongsaeree P, Suksamrarn A (2002) Antimycobacterial anthraquinone-chromanone compound and diketopiperazine alkaloid from the fungus Chaetomium globosum KMITL-N0802. Planta Med 68:834–836
Kawagishi H, Katsumi R, Sazawa T, Mizuno T, Hagiwara T, Nakamura T (1988) Cytotoxic steroids from the mushroom Agaricus blazei. Phytochemistry 27:2777–2779
Keller NP, Turner G, Bennett JW (2005) Fungal secondary metabolism – from biochemistry to genomics. Nat Rev Microbiol 3:937–947. https://doi.org/10.1038/nrmicro1286
Keyang L, Yisheng Z, Li L, Xuewei W, Gang D (2013) Chaetochromones A and B, two new polyketides from the fungus Chaetomium indicum (CBS.860.68). Molecules 18:10944–10952. https://doi.org/10.3390/molecules180910944
Khumkomkhet P, Kanokmedhakul S, Kanokmedhakul K, Hahnvajanawong C, Soytong K (2009) Antimalarial and cytotoxic depsidones from the fungus Chaetomium brasiliense. J Nat Prod 72:1487–1491
Kikuchi T, Kadota S, Suehara H, Nishi A, Tubaki K (1981) Odorous metabolites of a fungus, Chaetomium globosum Kinze ex Fr. Identification of geosmin, a musty-smelling compound. Chem Pharm Bull 29:1782–1784
Kim JH, Lee CH (2009) Heptelidic acid, a sesquiterpene lactone, inhibits etoposide-induced apoptosis in human leukemia U937 cells. J Microbiol Biotechnol 19:787–791
Kingsland SR, Barrow RA (2009) Identification of chaetoviridin E from a cultured microfungus, Chaetomium sp. and structural reassignment of chaetoviridins B and D. Aust J Chem 62:269–274
Koyama K, Akiba M, Imaizumi T, Kinoshita K, Takahashi K, Suzuki A, Yano S, Horie S, Watanabe K (2002) Antinociceptive constituents of Auricularia polytricha. Planta Med 68:284–285
Kubohara Y, Okamoto K, Tanaka Y, Asahi K, Sakurai A, Takahashi N (1993) Differanisole A, an inducer of the differentiation of friend leukemic cells, induces stalk cell differentiation in Dictyostelium discoideum. FEBS Lett 322:73–75
Kumagai S, Narasaki R, Hasumi K (2008) Glucose-dependent active ATP depletion by koningic acid kills high-glycolytic cells. Biochem Biophys Res Commun 365:362–368
Li GY, Li BG, Yang T, Yan JF, Liu GY, Zhang GL (2006) Chaetocochins A-C, epipolythiodioxopiperazines from Chaetomium cochliodes. J Nat Prod 69:1374–1376
Li GY, Li BG, Yang T, Liu GY, Zhang GL (2008) Secondary metabolites from the fungus Chaetomium brasiliense. Helv Chim Acta 91:124–129

Li LM, Zou Q, Li GY (2010) Chromones from an ascomycete, Chaetomium aureus. Chin Chem Lett 21:1203–1205
Lloyd G, Robertson A, Sankey GB, Whalley WB (1955) The chemistry of fungi. Part XXV. Oosporein, a metabolite of Chaetomium aureum chivers. J Chem Soc:2163–2165
Lösgen S, Schlörke O, Meindl K, Herbst-Irmer R, Zeeck A (2007) Structure and biosynthesis of chaetocyclinones, new polyketides produced by an endosymbiotic fungus. Eur J Org Chem 2007(13):2191–2196
Manning RO, Wyatt RD (1984) Comparative toxicity of Chaetomium contaminated corn and various chemical forms of oosporein in broiler chicks. Poult Sci 63:251–259
Marwah RG, Fatope MO, Deadman ML, Al-Maqbali YM, Husband J (2007) Musanahol: a new aureonitol-related metabolite from a Chaetomium sp. Tetrahedron 63:8174–8180
Mason SF, Vane GW (1967) The circular dichroisrn and the stereochemistry of a new fungal tetrahydrofuran. J Chem Soc Chem Commun (12):598
McInnes AG, Taylor A, Walter JA (1976) The structure of chetomin. J Am Chem Soc 98:6741–6741
Meiler D, Taylor A (1971) The effect of cochliodinol, a metabolite of Chaetomium cochliodes, on the respiration of microsopores of Fusarium oxysporum. Can J Microbiol 17:83–86
Müllbacher A, Waring P, Tiwari-Palni U, Eichner RD (1986) Structural relationship of epipolythiodioxopiperazines and their immunomodulating activity. Mol Immunol 23:231–235
Muroga Y, Yamada T, Numata A, Tanaka R (2009) Chaetomugilins I-O, new potent cytotoxic metabolites from a marine-fish derived Chaetomium species. Stereochemistry and biological activities. Tetrahedron 65:7580–7586
Nagaoka T, Nakata K, Kouno K, Ando T (2004) Antifungal activity of oosporein from an antagonistic fungus against Phytophthora infestans. Z Naturforsch C 59:302–304
Nakao Y, Kuo J, Yoshida WY, Kelly M, Scheuer PJ (2003) More kapakahines from the marine sponge Cribrochalina olemda. Org Lett 5:1387–1390
Nakazawa M, Uehara T, Nomura Y (1997) Koningic acid (a potent glyceraldehyde-3-phosphate dehydrogenase inhibitor)-induced fragmentation and condensation of DNA in NG108-15 cells. J Neurochem 68:2493–2499
Nielsen KF, Gravesen S, Nielsen PA, Andersen B, Thrane U, Frisvad JC (1999) Production of mycotoxins on artificially and naturally infested building materials. Mycopathologia 145:43–56
Oh H, Swenson DC, Gloer JB, Wicklow DT, Dowd PF (1998) Chaetochalasin A: a new bioactive metabolite from Chaetomium brasiliense. Tetrahedron Lett 39:7633–7636
Oka H, Asahi K, Morishima H, Sanada M, Shiratori K, Iimura Y, Sakurai T, Uzawa J, Iwadare S, Takahashi N (1985) Differanisole A, a new differentiation inducing substance. J Antibiot 38:1100–1102
Phonkerd N, Kanokmedhakul S, Kanokmedhakul K, Soytong K, Prabpai S, Kongsearee P (2008) Bis-spiro-azaphilones and azaphilones from the fungi Chaetomium cochliodes VTh01 and C. cochliodes CTh05. Tetrahedron 64:9636–9645
Pontius A, Krick A, Kehraus S, Brun R, König GM (2008) Antiprotozoal activities of heterocyclic-substituted xanthones from the marine-derived fungus Chaetomium sp. J Nat Prod 71:1579–1584
Powell JW, Whalley WB (1969) The chemistry of fungi. 58. Structures of colletodiol, a metabolite of Chaetomium funicola. J Chem Soc Perkin Trans 1:911–912
Qin JC, Gao JM, Zhang YM, Yang SX, Bai MS, Ma YT, Laatsch H (2009) Polyhydroxylated steroids from an endophytic fungus, Chaetomium globosum ZY-22 isolated from Ginkgo biloba. Steroids 74:786–790
Rether J, Erkel G, Anke T, Sterner O (2004) Inhibition of inducible TNF-alpha expression by oxaspirodion, a novel spiro-compound from the ascomycete Chaetomium subspirale. Biol Chem 385:829–834
Royles BJL (1995) Naturally occurring tetramic acids: structure, isolation, and synthesis. Chem Rev 95:1981–2001

Rui HJ, Shu X, Jun YL, Hui MG, Hui D, Chen X, Hai LZ, Ren XT (2006) Chaetominine, a cytotoxic alkaloid produced by endophytic Chaetomium sp. IFB-E015. Org Lett 8(25):5709–5712

Safe S, Taylor A (1972) Sporidesmins. Part XIII. Ovine 111-thrift in Nova Scotia. Part 11. The characterisation of chetomin, a toxic metabolite of Chaetomium cochliodes and Chaetomium globosum. J Chem Soc Perkin Trans 1:472–479

Saito M, Seto H, Yonehara H (1983) Biosynthesis of 2-(but-l,3-dienyl)-3-hydro-4-(penta-l,3-dienyl)-tetrahydrofuran, a metabolite of Chaetomium coarctatum. Agric Biol Chem 47:2935–2937

Saito T, Suzuki Y, Koyama K, Natori S, Iitaka Y, Kinoshita T (1985) Chetracin A and chaetocins B and C, three new epipolythiodioxopiperazines from Chaetomium spp. Chem 26:4731–4734

Saito T, Koyama K, Natori S, Iitaka Y (1988) Chetracin A, a new epipolythiodioxopiperazine having a tetrasulfide bridge from Chaetomium abuens and C. retardatum. Tetrahedron Lett Pharm Bull 36:1942–1956

Schlörke O, Zeeck A (2006) Orsellides A–E: an example for 6-deoxyhexose derivatives produced by fungi. Eur J Org Chem 2006(4):1043–1049. https://doi.org/10.1002/ejoc.200500793

Sekita S (1983) Isocochliodinol and neocochliodinol, bis(3-indolyl)-benzoquinones from Chaetomium spp. Chem Pharm Bull 31:2998–3001

Sekita S, Yoshihira K, Natori S, Udagawa S, Muroi T, Sugiyama Y, Kurata H, Umeda M (1981) Mycotoxin production by Chaetomium spp. and related fungi. Can J Microbiol 27:766–772

Seto H, Saito M, Uzawa J, Yonehara H (1979) Utilization of 13C -13C coupling in structural and biosynthetic studies. XII. biosynthesis of 2-(but-1,3-dienyl)-3-hydro-4-(penta-1,3-dienyl)-tetrahydrofuran, A metabolite of Chaetomium coarctaum. Heterocycles 13:247–253

Smetanina OF, Kuznetzova TA, Denisenko VA, Pivkin MV, Khudyakova YV, Gerasimenko AV, Popov DY, Il'in SG, Elyakov GB (2001) 3β-Methoxyolean-18-ene (miliacin) from the marine fungus Chaetomium olivaceum. Russ Chem Bull 50:2463–2465

Stadler M, Keller NP (2008) Paradigm shifts in fungal secondary metabolite research. Mycol Res 112(2):127–130

Sutton DA, Fothergill A, Rinaldi MG (1998) Guide to clinically significant fungi. Williams and Wilkins, Baltimore

Suzuki T, Oka H, Okura A, Asahi K, Takahashi N (1986) In vitro and in vivo effects of differanisole A on some tumor cells. J Antibiot 39:869–871

Takahashi M, Koyama K, Natori S (1990) Four new azaphilones from *Chaetomium globosum* var. *flavo-viridae*. Chem Pharm Bull 38:625–628

Tanaka Y, Shiomi K, Kamei K, Sugoh-Hagino M, Enomoto Y, Fang F, Yamaguchi Y, Masuma R, Zhang CG, Zhang XW, Omura S (1998) Antimalarial activity of radicicol, heptelidic acid and other fungal metabolites. J Antibiot 51:153–160

Teigo A, Shuntaro M, Naoki S, Tohru T, Kenji M, Hiroaki S, Tomoji O, Yoshiteru O (2012) Structural diversity of new c13-polyketides produced by Chaetomium mollipilium cultivated in the presence of a nad þ-dependent histone deacetylase inhibitor. Org Lett 14(21):5456–5459

Terry BJ, Liu WC, Cianci CW, Proszynski E, Fernandes P, Bush K, Meyers E (1992) Inhibition of herpes simplex virus type 1 DNA polymerase by the natural product oosporein. J Antibiot 45:286–288

Thohinung S, Kanokmedhakul S, Kanokmedhakul K, Kukongviriyapan V, Tusskorn O, Soytong K (2010) Cytotoxic 10-(indol-3-yl)-[13]cytochalasans from the fungus Chaetomium elatum ChE01. Arch Pharm Res 33:1135–1141

Tibodeau JD, Benson LM, Isham CR, Owen WG, Bible KC (2009) The anticancer agent chaetocin is a competitive substrate and inhibitor of thioredoxin reductase. Antioxid Redox Signal 11:1097–1106

Turbyville TJ, Wijeratne EM, Liu MX, Burns AM, Seliga CJ, Luevano LA, David CL, Faeth SH, Whitesell L, Gunatilaka AAL (2006) Search for Hsp90 inhibitors with potential anticancer

activity: isolation and SAR studies of radicicol and monocillin I from two plant-associated fungi of the Sonoran desert. J Nat Prod 69:178–184
Udagawa S, Muroi T, Kurata H, Sekita S, Yoshihira K, Natori S, Umeda M (1979) The production of chaetoglobosins, sterigmatocystin, O-methylsterigmatocystin, and chaetocin by Chaetomium spp. and related fungi. Can J Microbiol 25:170–177
Wang S, Li XM, Teuscher F, Li DL, Diesel A, Ebel R, Proksch P, Wang BG (2006) Chaetopyranin, a benzaldehyde derivative, and other related metabolites from Chaetomium globosum, an endophytic fungus derived from the marine red alga Polysiphonia urceolata. J Nat Prod 69:1622–1625
Wang S, Xu Y, Maine EA, Wijeratne EM, Espinosa-Artiles P, Gunatilaka AA, Molnár I (2008) Functional characterization of the biosynthesis of radicicol, an Hsp90 inhibitor resorcylic acid lactone from Chaetomium chiversii. Chem Biol 15:1328–1338
Weindling R, Emerson OH (1936) The isolation of a toxic substance from the culture filtrate of Trichoderma. Phytopathology 26:1068–1070
Wijeratne EMK, Turbyville TJ, Fritz A, Whitesell L, Gunatilaka AAL (2006) A new dihydroxanthenone from a plant associated strain of the fungus Chaetomium globosum demonstrates anticancer activity. Bioorg Med Chem 14:7917–7923
Wu ZJ, Li GY, Fang DM, Qi HY, Ren WJ, Zhang GL (2008) Analysis of epipolythiodioxopiperazines in fungus Chaetomium cochliodes using HPLC-ESI-MS/MS/MS. Anal Chem 80:217–226
Xu GB, Li LM, Yang T, Zhang GL, Li GY (2012) Chaetoconvosins A and B, alkaloids with new skeleton from fungus Chaetomium convolutum. Org Lett 14(23):6025–6055
Yamada T, Muroga Y, Tanaka R (2009) New azaphilones, secochaetomugilins A and D, produced by a marine-fish-derived Chaetomium globosum. Mar Drugs 7:249–257
Yamazaki M, Fujimoto H, Okuyama E (1978) Structure determination of six fungal metabolites, tryptoquivaline E, F, G, H, I and J from Aspergillus fumigatus. Chem Pharm Bull 26:111–117
Yamori T, Matsunaga A, Sato S, Yamazaki K, Komi A, Ishizu K, Mita I, Edatsugi H, Matsuba Y, Takezawa K, Nakanishi O, Kohno H, Nakajima Y, Komatsu H, Andoh T, Tsuruo T (1999) Potent antitumor activity of MS-247, a novel DNA minor groove binder, evaluated by an in vitro and in vivo human cancer cell line panel. Cancer Res 59:4042–4049
Yang SX, Gao JM, Zhang Q, Laatsch H (2011) Toxic polyketides produced by Fusarium sp., an endophytic fungus isolated from Melia azedarach. Bioorg Med Chem Lett 21:1887–1889
Yeung BKS, Nakao Y, Kinnel RB, Carney JR, Yoshida WY, Scheuer PJ, Kelly-Borges M (1996) The kapakahines, cyclic peptides from the marine sponge Cribrochalina olemda. J Org Chem 61:7168–7173
Zhang J, Ge HM, Jiao RH, Li J, Peng H, Wang YR, Wu JH, Song YC, Tan RX (2010) Cytotoxic chaetoglobosins from the endophyte Chaetomium globosum. Planta Med 76:1910–1914
Zhang Q, Li HQ, Zong SC, Gao JM, Zhang AL (2012) Chemical and bioactive diversities of the genus Chaetomium secondary metabolites. Mini Rev Med Chem 12(2):127–148

Chapter 7
Chaetomium's Alkaloids

Parsa Mahmood Dar, Ahmed M. Abdel-Azeem, and Shimaa Ragab Hamed

7.1 Introduction

Fungi, like higher plants and bacteria, are able to produce metabolites, including alkaloids. Alkaloids, along with other metabolites, are the most important fungal metabolites from the pharmaceutical and industrial point of view. Alkaloids have been reported as a group of basic organic substances of plant and microbial origin, containing at least one nitrogen atom in a ring structure in the molecule. The first microbial alkaloids to be recognized and studied were those of *Claviceps purpurea*, the agent causing ergot of rye (Mahmood et al. 2010). Fungi of the genus *Chaetomium* are a rich source of novel and bioactive, very important, secondary metabolites. The genus *Chaetomium* is commonly recognized as having ostiolate ascomata with a membranaceous perithecial wall covered by relatively well-developed hair, producing fasciculate and evanescent asci and single-celled, smooth, and pigmented ascospores with germ pores (Ames 1963; von Arx et al. 1986). In 1817, Gustav Kunze established the genus *Chaetomium* (Ames 1961) to classify the species *C. globosum* and *C. elatum*. No further contributions to the genus were made until 1837 when the publication of Corda described its characteristic asci in his work, *Icones Fungorum Hucusque Cognitorum* (Ames 1949). In 1915, Arthur Houston Chivers produced a complete monographic treatment of the genus, recognizing only 28 of the described 114 species (Chivers Arthur Houston 1915).

P. M. Dar (✉)
Institute of Chinese Medical Sciences, Avenida da Universidade, Taipa, Macau, China

A. M. Abdel-Azeem
Botany Department, Faculty of Science, University of Suez Canal, Ismailia, Egypt
e-mail: ahmed_abdelazeem@science.suez.edu.eg

S. R. Hamed
Microbial Biotechnology Department, National Research Centre, Giza, Egypt

A. M. Abdel-Azeem (ed.), *Recent Developments on Genus Chaetomium*,
Fungal Biology, https://doi.org/10.1007/978-3-030-31612-9_7

Chaetomium is a genus of fungi in the *Chaetomiaceae* family. It is a dematiaceous (dark-walled) mold normally found in soil, air, cellulose, and plant debris. According to the *Dictionary of the Fungi* (10th edition, 2008), there are about 95 species in the widespread genus (Kirk et al. 2008).

Till 2016, a variety of more than 200 secondary metabolites belonging to diverse structural types, such as chaetoglobosins, epipolythiodioxopiperazines, azaphilones, xanthones, anthraquinones, chromones, depsidones, terpenoids, and steroids, have been discovered. Most of these fungal metabolites exhibit antitumor, cytotoxic, antimalarial, enzyme inhibitory, antibiotic, and other activities (Zhang et al. 2012). Some important and explored species of the genus *Chaetomium* are *C. atrobrunneum*, *C. cupreum*, *C. globosum*, *C. grande*, *C. interruptum*, *C. iranianum*, *C. rectangulare*, *C. thermophilum*, *C. truncatulum*, and *C. undulatulum*.

Alkaloidal compounds of genus *Chaetomium* are:

1. Chaetoglobosins
2. Diketopiperazines
3. Isoquinolines
4. Bis(3-indolyl)-benzoquinone
5. Tetramic acids
6. Tripeptide alkaloids
7. Benzonaphthyridinedione
8. Azaphilones

7.2 Chaetoglobosins

Chaetoglobosins (Fig. 7.1) are a family of macrocyclic polyketide alkaloids belonging to the general group of cytochalasans with a wide range of biological activities targeting cytoskeletal processes (Binder and Tamm 1973; Scherlach et al. 2010). Chaetoglobosins possess an indolyl group in place of the phenyl group present in cytochalasans, namely, 10-(indol-3-yl)-[12]cytochalasans. Cytochalasans are an important class of natural products; they have attracted considerable attention from synthetic organic chemists, biologists, and phytochemists for decades because of their complex polycyclic skeletons and varied and often potent bioactivities, including antitumor, anti-HIV, and immunomodulatory activities (Chen et al. 2016). To date, 62 chaetoglobosins have been produced by fungi of the genus *Chaetomium*. *Chaetomium globosum* (Sekita et al. 1973), the most common species within this genus, is a remarkable producer of chaetoglobosins. Chaetoglobosins A (1) and B (2), two cytotoxic cytochalasan derivatives, were first isolated from *Chaetomium globosum* (Silverton et al. 1976). Later, the structure of chaetoglobosin A was established by X-ray analysis. Chaetoglobosin A is the most widely studied metabolite in the chaetoglobosin series. Meanwhile, six further metabolites, chaetoglobosins

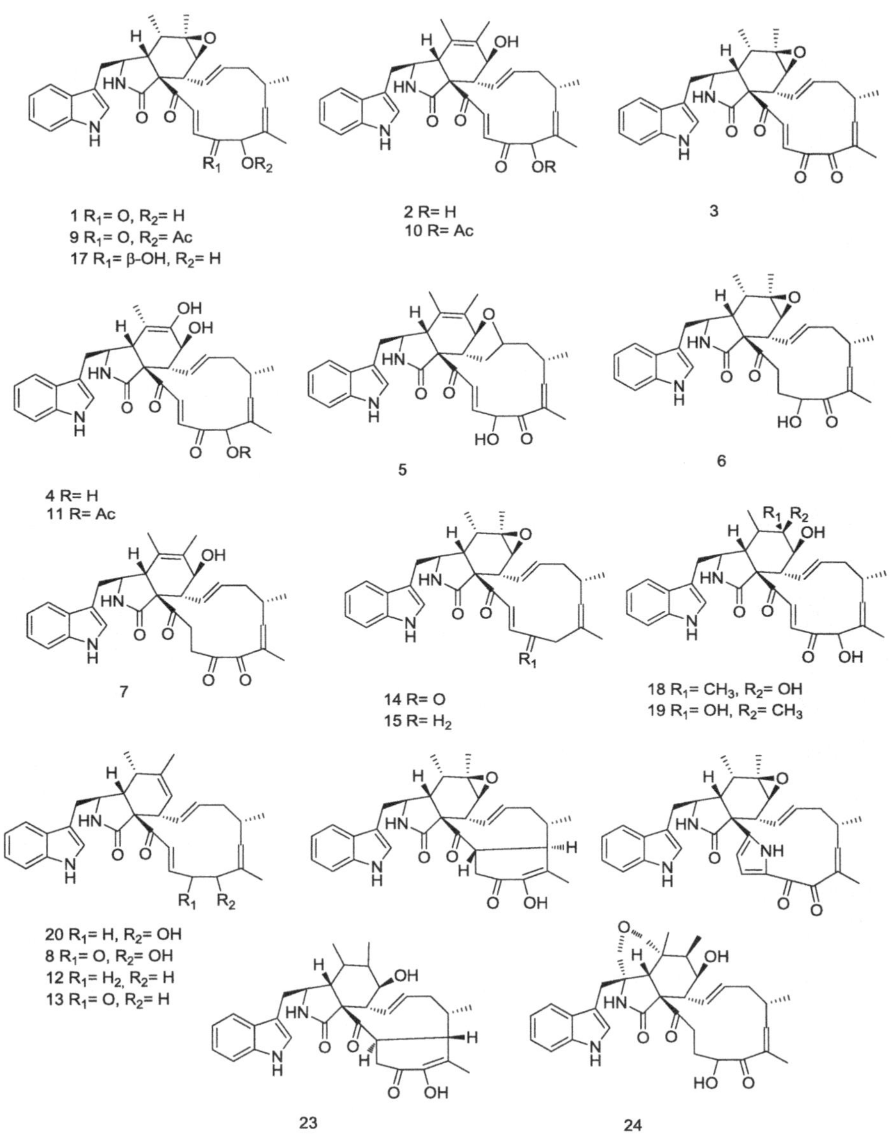

Fig. 7.1 Chaetoglobosins from Chaetomium 1–62

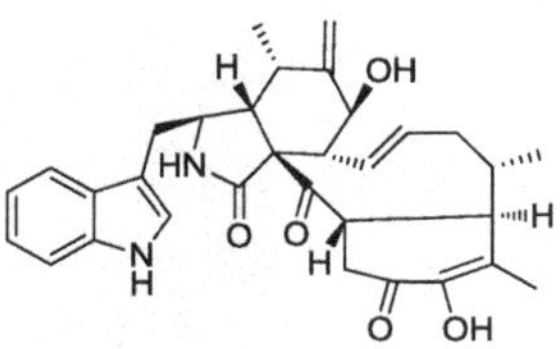

25

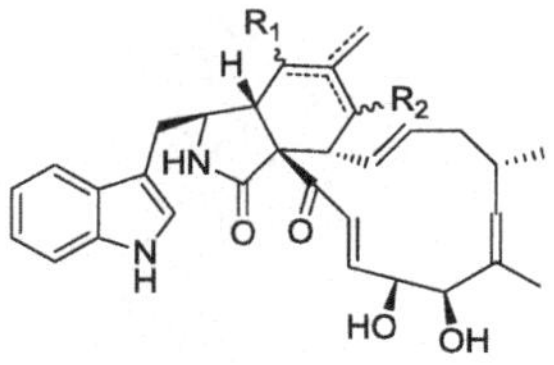

26 R_1= OMe, R_2= βOH
27 R_1=Me, R_2= βOH
28 R_1= αMe, R_2= H

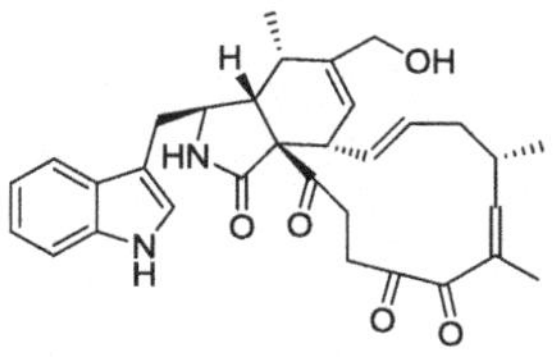

29

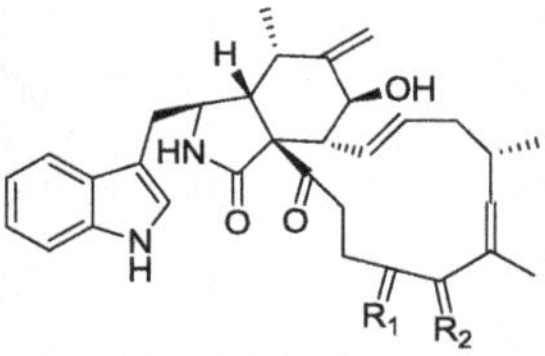

30 R_1= O, R_2= H, βOH
31 R_1= R_2= βOH
32 R_1= R_2= O
16 R_1= H, βOH, R_2= O

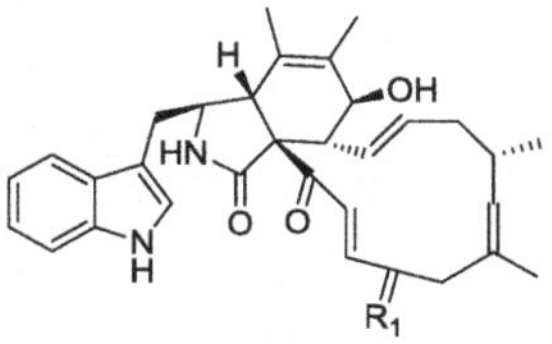

33 R_1= H_2
34 R_1= O

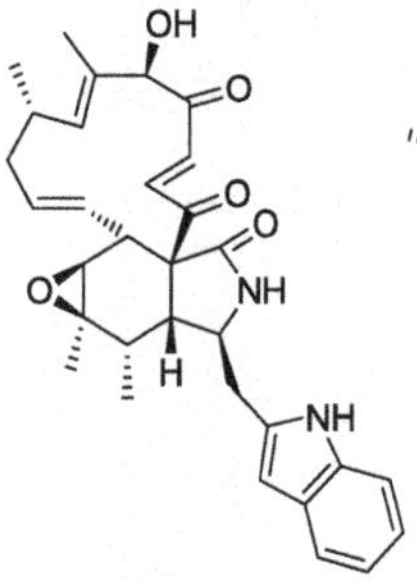

35

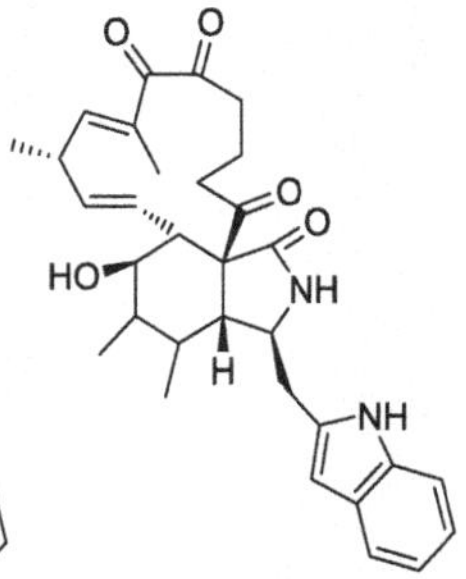

36

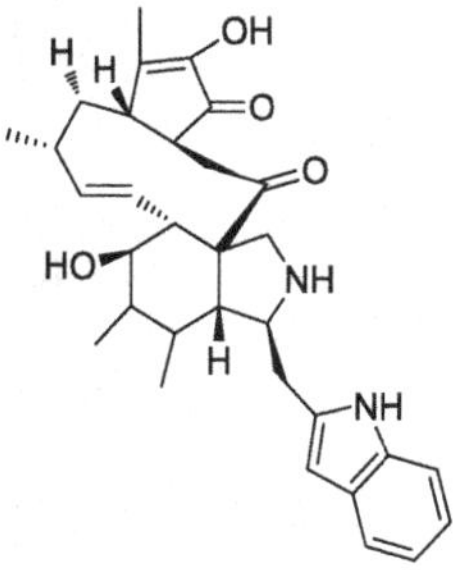

37

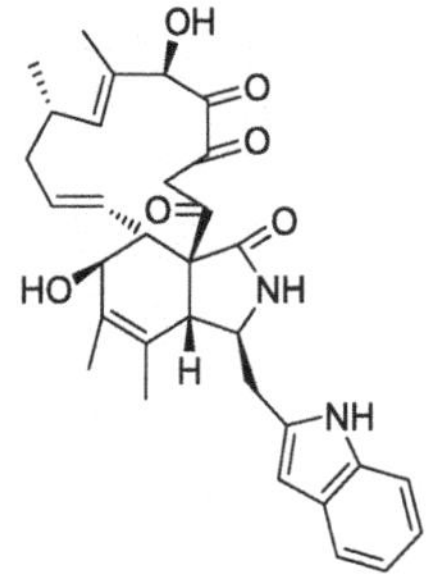

38

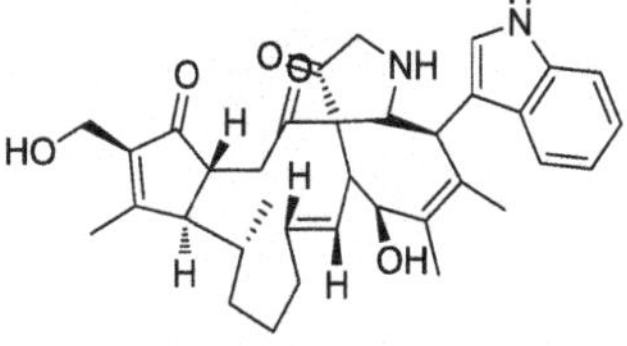

39

Fig. 7.1 (continued)

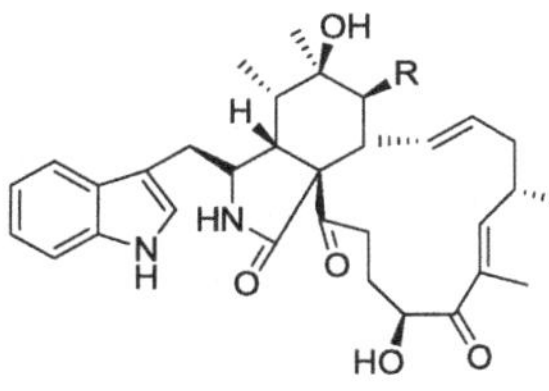

40 R= OH
41 R= OAc

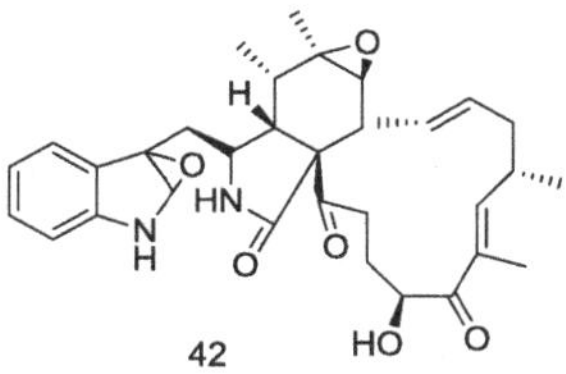

42

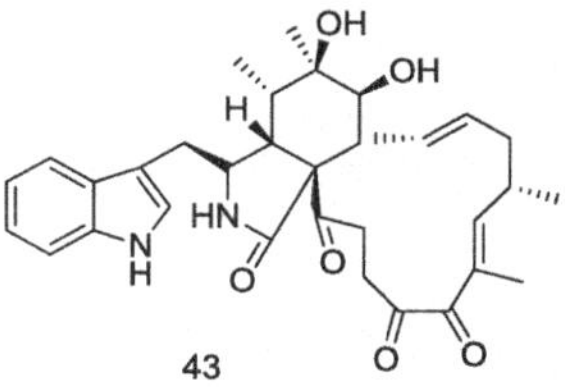

43

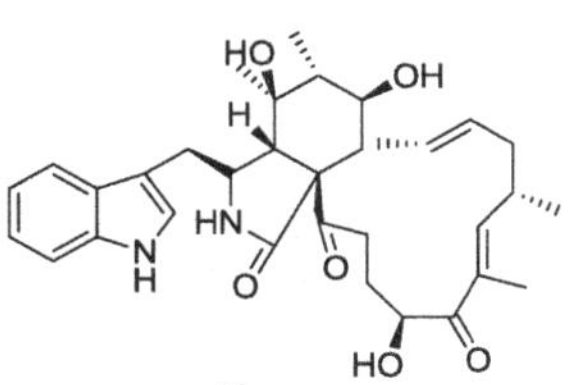

44

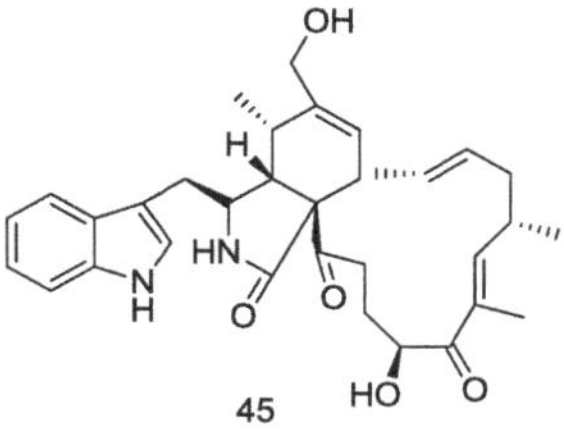

45

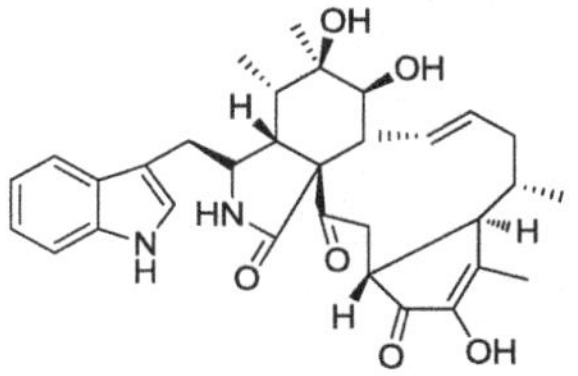

46

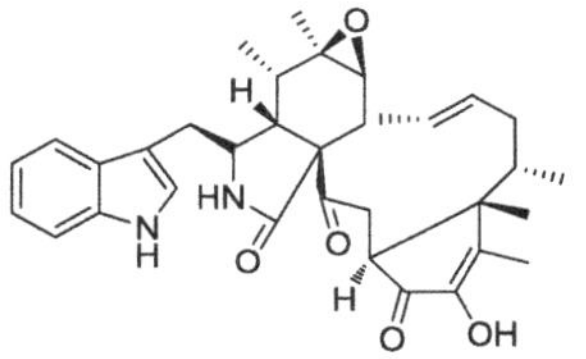

47

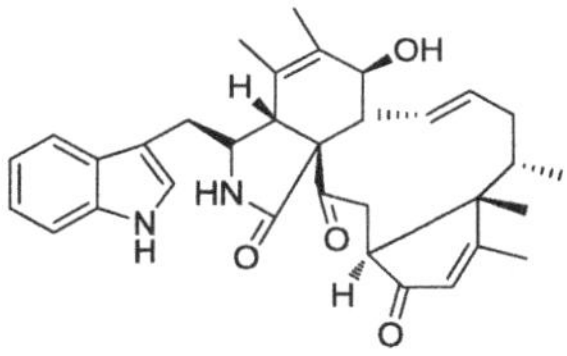

48

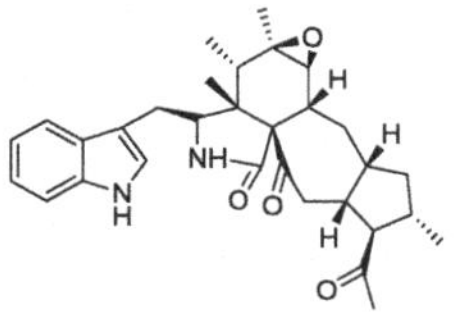

49

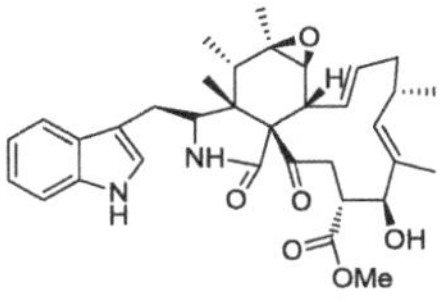

50

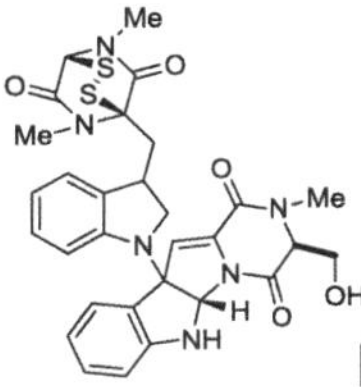

51

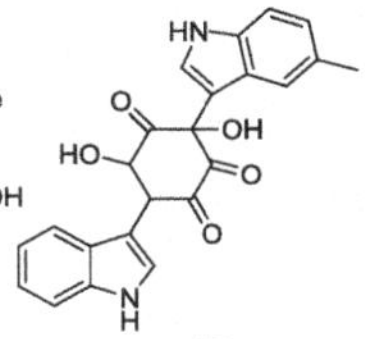

52

Fig. 7.1 (continued)

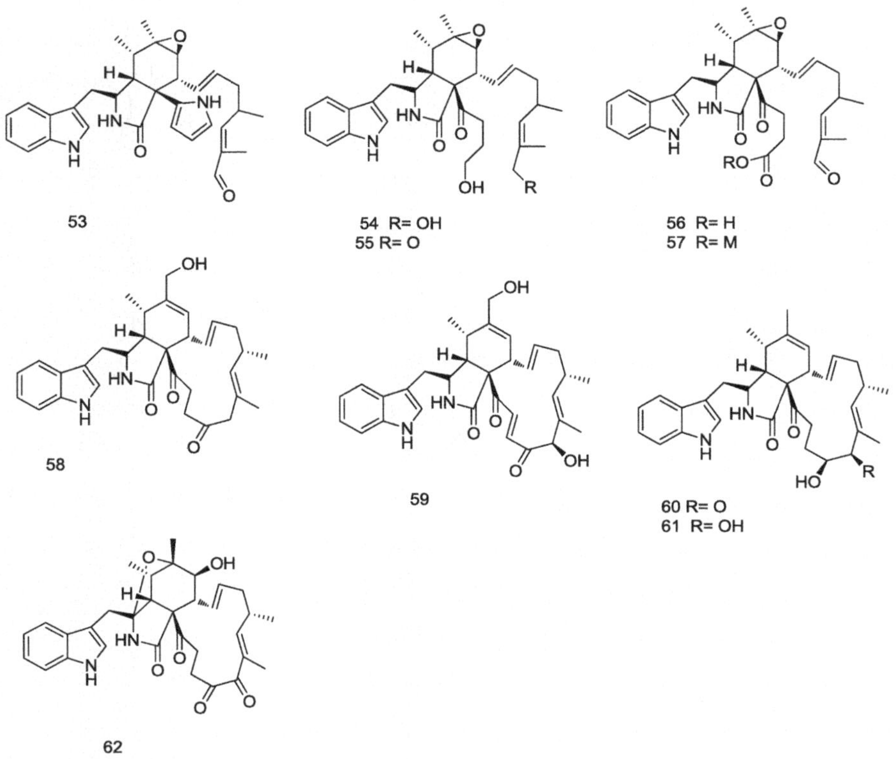

Fig. 7.1 (continued)

C–G (37) and J (8), were isolated and identified from the same fungus *C. globosum* (Silverton et al. 1976; Umeda et al. 1975; Sekita et al. 1976, 1977, 1982a, b, c). Indirect evidence for the Diels–Alder-mediated biosynthesis of cytochalasans was obtained by feeding (Sekita et al. 1983) and inhibition (Oikawa et al. 1992a, b) experiments with *Chaetomium subaffine*, which produces chaetoglobosin A (1), 19-O-acetylchaetoglobosin A (9), 19-acetylchaetoglobosin B (10), and 19-acetylchaetoglobosin D (11) (Oikawa et al. 1992a; Probst and Tamm 1981a, b). Interestingly, treatment of *C. subaffine* with specific inhibitors of cytochrome P-450 gave four new plausible chaetoglobosin A precursors, prochaetoglobosins I (12), II (13), III (14), and IV (15) (Oikawa et al. 1992a, b; Probst and Tamm 1981a, b). Two more new metabolites, chaetoglobosin Fex (16) and 20-dihydrochaetoglobosin A (17), were isolated from *C. subaffine* (ATCC 22132) (Oikawa et al. 1991), in which 20-ketoreductase activity of 1 and prochaetoglobosins was found. Nine cytotoxic metabolites, including three novel compounds, chaetoglobosins Q (18), R (19), and T (20), have been isolated from cultures of *C. globosum* (CANU N60) (Oikawa et al. 1993) together with six known chaetoglobosins A (1), B (2), D (4), and J (8) and prochaetoglobosins I (12) and II (13). The structures were elucidated primarily from NMR spectroscopic data. Chaetoglobosins A (1), B (2), D (4), J (8), Q (18), T (20) and prochaetoglobosins I (12) and II (13) displayed significant cytotoxicity against the P388 murine leukemia cell line (IC_{50} = 1.58–4.90 μg/ml). Chaetoglobosins A, B, D, and J and prochaetoglobosins I and II were also shown to be antimicrobial against *Bacillus subtilis*, *Cladosporium resinae*, and *Trichophyton mentagrophytes*. A new cytotoxic alkaloid, chaetoglobosin U (21), along with four known chaetoglobosins C (3), E (5), and F (6) and penochalasin A (22), has been characterized from a solid culture of C. *globosum* IFB-E019, an endophytic fungus residing inside the stem of *Imperata cylindrica* (Gramineae). Chaetoglobosin U (21) was shown to exhibit cytotoxic activity against the human nasopharyngeal epidermoid tumor KB cell line (IC_{50} = 16.0 μM), comparable to that of 5-fluorouracil (IC_{50} = 14.0 μM) as a positive reference. The metabolites 3, 5, 6, and 22 were moderately active to the KB cell line (Jiao et al. 2004). Bioassay-guided fractionation of a solid culture of the endophyte *C. globosum* IFB-E041 isolated from the herb *Artemisia annua* stems afforded two new alkaloids, chaetoglobosins V (23) and W (24) and known congeners chaetoglobosins A (1), C (3), E–G (5–7), and Fex (16). The structures were elucidated by 1D and 2D NMR techniques. Chaetoglobosin W (24) is unique in its possession of an oxolane ring formed via an oxygen bridge at C-3 and C-6. The metabolites exhibited moderate cytotoxic activities against human cancer cell lines (KB, K562, MCF-7, and HepG2) (IC_{50} = 18–30 μg/ml) (Ding et al. 2006). Cultures of *C. globosum* QEN-14, an endophyte derived from the marine green alga *Ulva pertusa*, produced a series of new metabolites cytoglobosins A–G (25–31), together with chaetoglobosin Fex (16) and isochaetoglobosin D (32). Cytoglobosins C (27) and D (28) displayed cytotoxic activity against the tumor cell line A-549 (IC_{50} = 2.26 and 2.55 μM, respectively) (Zhang et al. 2010). Nine cytochalasans, including a new chaetoglobosin V (33), two new natural products, prochaetoglobosins III (14) and IIIed (34), and six known chaetoglobosins B–D (2–4), F (6), and G (7) and isochaetoglobosin D (32), were isolated from *C. elatum* ChE01, a strain isolated from Thai soil. The nine compounds showed cytotoxicity against the human breast cancer BC1 cell lines (IC_{50} = 2.54–21.29 μM) and cholangiocarcinoma cell lines (IC_{50} = 3.41–86.95 μM). Among these compounds, 2 and 34 exhibited strong cytotoxicity to the BC1 cell line (IC_{50} = 2.54 and 3.03 μM, respectively) (Cui et al. 2010). Chaetoglobosins A–G (1–7) and J (8) were cytotoxic against HeLa cells (IC_{50} = 3.2–20 μg/ml) (17), and most of them induced polynucleation in HeLa cells. Chaetoglobosin A (1) exhibited strong cytotoxicity against various human cancer cell lines (Thohinung et al. 2010) and showed a strong inhibitory activity (IC_{50} = 60 pM) against bleb formation on K562 cells induced by phorbol 12,13-dibutyrate, but it did not inhibit protein kinase C activity in vitro. Its inhibitory activity might be involved in the modulation of actin filaments on the cell membrane. Of chaetoglobosins B (2), D (4), and E (5), produced by *C. globosum*, an endophytic fungus associated with *Viguiera robusta* Gardn (Asteraceae), only chaetoglobosin B (2) was active against Jurkat (leukemia) and B16F10 (melanoma) tumoral cells (Ko et al. 1998), with 89.5% and 57.1% inhibition at 0.1 mg/ml, respectively. The growth of four *Helicobacter pylori* strains and three *Staphylococcus* strains was only inhibited by 1 (Momesso et al. 2008). Chaetoglobosin A (1) was found to enhance fibrinolytic activity of bovine aortic endothelial cells at 3–100 μM via its induction of urokinase (Namgoong et al. 1998). Apart from cytotoxicity, chaetoglobosins A (1) and C (3) also showed potent phytotoxicity against alfalfa seedlings (Shinohara et al. 2000). Compound 1 and its acetyl derivative (9) were shown to be phytotoxic in a bioassay of cresson seedlings (Ichihara et al. 1996). Compound 1, identified from an antagonistic fungus, *C. cochliodes*, found in greenhouse soil, exhibited antifungal activity toward several plant pathogens (Von Wallbrunn et al. 2001): *Pythium ultimum*, *Phytophthora capsici*, *Rhizoctonia solani*, *Botrytis cinerea*, and *Fusarium oxysporum*, with ED_{50} values of 1.98, 4.01, 4.16, 2.67, and 35.14 ppm, respectively. Among seven chaetoglobosins A–F (1–6) and J (8) that were produced by *Chaetomium sp.*, a strain obtained from the green alga *Halimeda discoidea*, only 1 showed the bead shape deformation on mycelia of rice plant pathogenic fungus, *Pyricularia oryzae* at a dose as low as 0.5 μg/ml; compound 8 inhibited conidia germination at 5 μg/ml and weak hypertrophy of mycelia with growth inhibition at 0.25 μg/ml. In addition, 1 was active to three phytopathogenic fungi, *Colletotrichum lindemuthianum*, *Fusarium nivale*, and *Pythium aphanidermatum*, at less than 50 μg/ml (Kang et al. 1999). Both 1 and 3, isolated from the cultures of an endophyte, *C. globosum*, residing in the herbal plant *Ginkgo biloba* leaves, significantly inhibited the growth of the brine shrimp (*Artemia salina*) larvae and *Mucor miehei* (Kobayashi et al. 1996). Cultivation of *Chaetomium globosum*, an endophytic fungus, in *Ginkgo biloba*, produced five cytochalasan mycotoxins, chaetoglobosins A, G, V, Vb, and C (35–39), in three media. Some known chaetoglobosins were also produced. These alkaloids possess good phytoxic and cytotoxic activities (Qin et al. 2009a, b).

Chemical investigation on the methanol extract of *Chaetomium globosum* (TW1-1), a fungus isolated from the common pillbug (*Armadillidium vulgare*), resulted in the isolation of nine new highly oxygenated cytochalasan alkaloids, armochaetoglobins S–Z (40-48). Their structures and absolute configurations were elucidated by spectroscopic analyses. Among them, compound 41 appears to be the first member of the chaetoglobosin family with an acetyl group and compound 42 the first chaetoglobosin characterized by an 2′,3′-epoxy-indole moiety. The discovery of these new compounds revealed the largely untapped chemical diversity of cytochalasans and enriched their chemical research (Scherlach et al. 2010).

Armochaeglobines A (51) and B (52), two indole-based cytochalasan alkaloids with new carbon skeletons, were obtained from the fungus *Chaetomium globosum* TW1-1, which was first isolated from the arthropod *Armadillidium vulgare*. Their structures were elucidated by extensive spectroscopic analyses, ECD calculation, and single-crystal X-ray diffraction analysis. Interestingly, compound 51 featured a unique tetracyclic 5/6/7/5 fused ring system and 52 possessed a rare 12-membered carbon scaffold (42). Two new indole alkaloids chaetocochin J (49) and chaetoglobinol A (50), along with chetomin, chetoseminudin A, cochliodinol, and semicochliodinol, were isolated from the rice culture of the fungus *Chaetomium globosum*. Their structures were elucidated by spectral analysis (Chen et al. 2015a, b). Ten new cytochalasan alkaloids, termed armochaetoglobins A–J (53–62), and four known chaetoglobosins were isolated from a methanol extract of *Chaetomium globosum* TW1-1, a fungus isolated from the medicinal terrestrial arthropod *Armadillidium vulgare*. Their structures were elucidated by a combination of spectroscopy, single-crystal X-ray crystallography, and ECD calculations. Armochaetoglobins A–E (53–57) represented the first examples of seco-chaetoglobosins arising from an oxidative cleavage of C-19 and C-20. Among these compounds, armochaetoglobin A (1) features an unusual pyrrole ring. The cytotoxic activities of 54–62 were evaluated, and armochaetoglobin H (8) showed moderate inhibitory activities against five human cancer cell lines, with IC_{50} values ranging from 3.31 to 9.83 μM (Xu et al. 2015).

7.3 Diketopiperazines

Diketopiperazines (Fig. 7.2) are well-known secondary metabolites produced by fungi and plants. Epipolythiodioxopiperazines (ETPs) are an important group of bioactive compounds widely found as toxic substances of mold (Chen et al. 2015a, b; Waring et al. 1988; Dong et al. 2005). They feature either thiomethyl groups or disulfide bridges. To date, 24 compounds of this class have been discovered in the *Chaetomium* genus. In this series of compounds, gliotoxin (63) is the first and best-known ETP isolated from *Gliocladium fimbriatum* (Müllbacher et al. 1986; Weindling and Emerson 1936; Weindling 1941). Recently, from the cultures of *C. globosum* found in *Ginkgo biloba* (Kobayashi et al. 1996), our research group characterized three diketopiperazines, gliotoxin (63), methylthiogliotoxin (64), and fumitremorgin C (65). Only compound 35 had good antifungal activity against plant pathogenic fungi, including *Fusarium oxysporum f. sp. vasinfectum*, *F. graminearum*, *F. sulphureum*, *Cercospora sorghi*, and *Alternaria alternate*. The fungus *C. globosum* KMITL-N0802 contained echinuline (66), which exhibited inhibitory activity on *M. tuberculosis* (Gardiner et al. 2005). Cultivation of an endophyte *C. globosum* from the marine red alga *Polysiphonia urceolata* produced neoechinulin A (67) (Kanokmedhakul et al. 2002). The fungus *Chaetomium sp.*, isolated from the medicinal plant *Otanthus maritimus* (Asteraceae), yielded cyclo-(alanyl-tryptophane) (68) (Wang et al. 2006). In 1944, Walksman and Bugie and Geiger et al. (1944) isolated a strain of *Chaetomium cochliodes* which, in culture, produced an antibacterial chetomin (69), a metabolite containing hexahydropyrrolo[2,3-b] indole (Geiger et al. 1944). Later, in 1972, Safe and Taylor (Geiger 1949; Safe and Taylor 1972) carried out the detailed characterization of this toxic metabolite of *C. cochliodes* and *C. globosum* from soil samples collected in Nova Scotia. Subsequently, analysis of the 13C and 15N NMR spectra of [15N6] chetomin (69) established the connectivity of the sporidesmin-like and 3-(ω-skatyl)-3,6-epidithiopiperazine-2,5-dione fragments (McInnes et al. 1976). The hypoxic response in humans is regulated by the hypoxia-inducible transcription factor system; inhibition of hypoxia-inducible factor (HIF) activity has potential for the treatment of cancer. Recently, chetomin (69) was found to block the interaction between

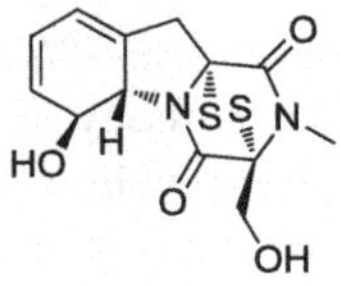

63

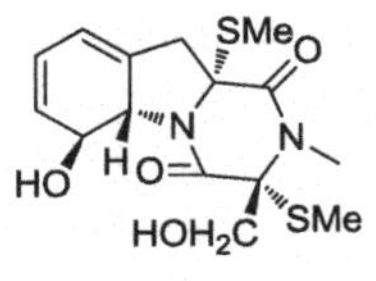

64

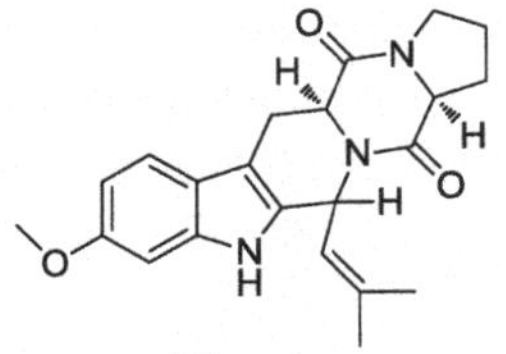

65

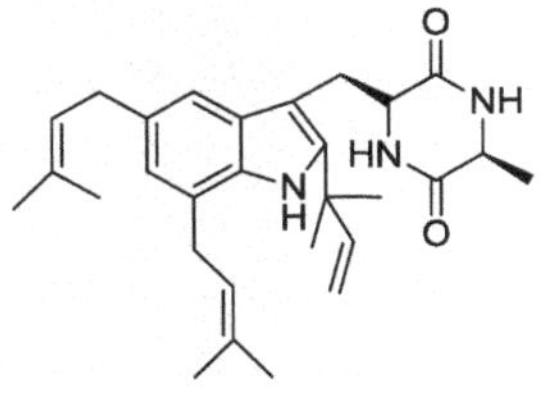

66

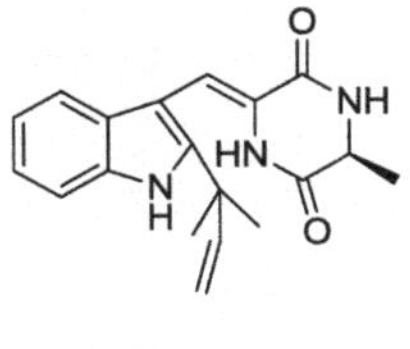

67

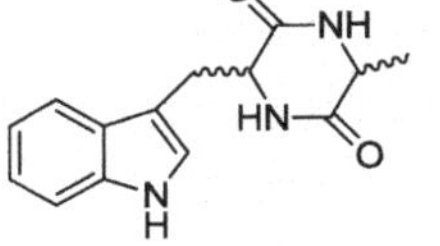

68

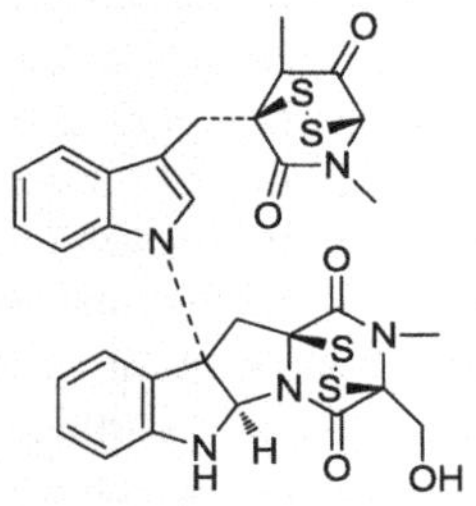

69

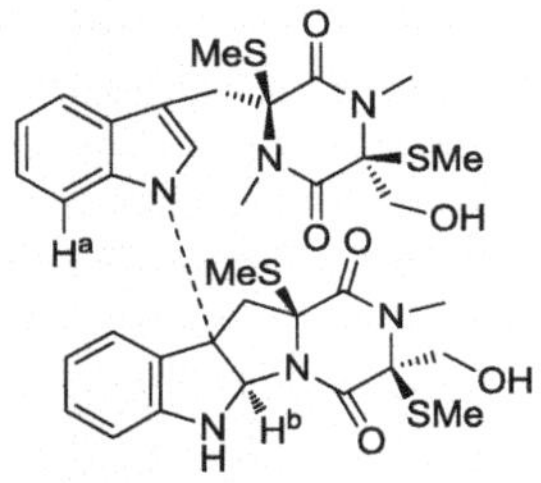

70

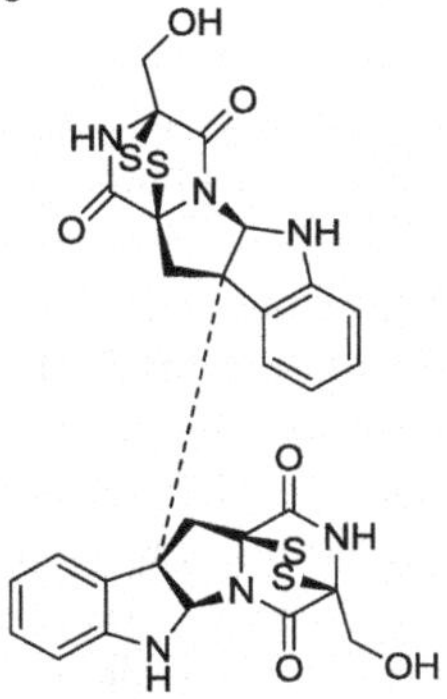

71

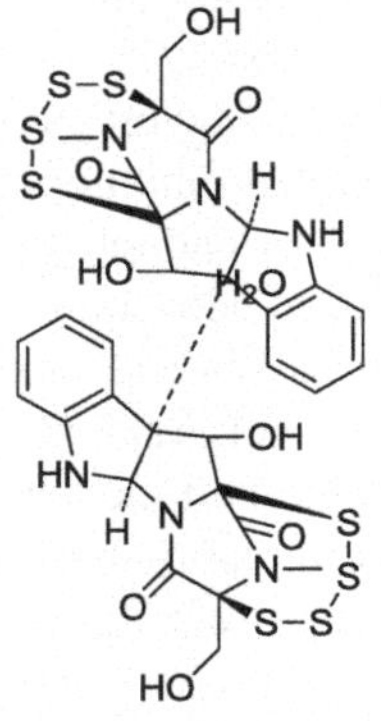

72

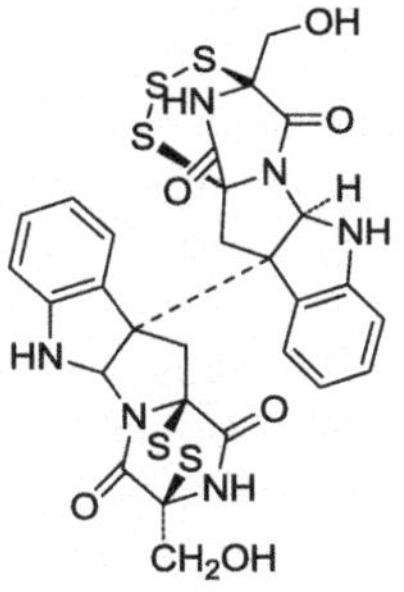

73

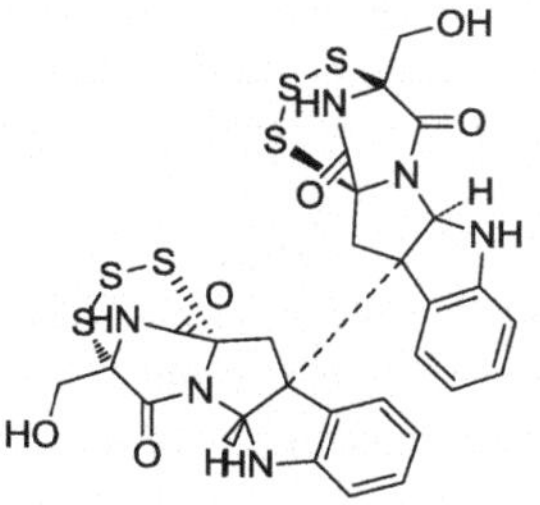

74

Fig. 7.2 Diketopiperazine alkaloids of *Chaetomium*

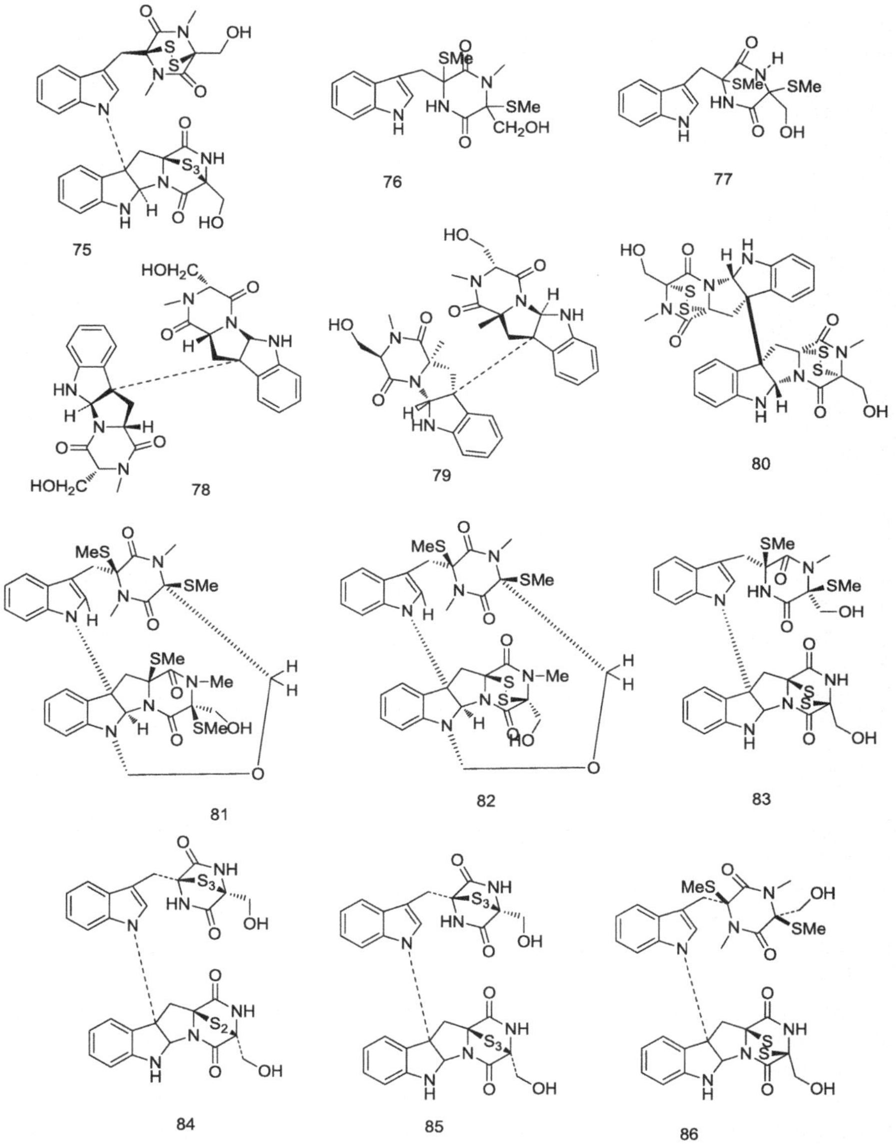

Fig. 7.2 (continued)

HIF- and the transcriptional coactivator p300 by a zinc ejection mechanism (Brewer et al. 1976). Structure-activity studies with both natural and synthetic ETP derivatives reveal that only the ETP core is required, and sufficient, to block the interaction of HIF-1 and p300. HIF-1 inhibition by chetomin (69) effectively reduces hypoxia-dependent transcription and radiosensitizes hypoxic HT 1080 human fibrosarcoma cells in vitro (Cook et al. 1996) The structure of dethiotetra(methylthio) chetomin (70), a new antimicrobial metabolite isolated from *C. globosum* Kinze ex Fr. along with chetomin (69), was determined by X-ray crystallographic analysis. Chetomin (69) and its derivative (70) showed antimicrobial activity against *Escherichia coli* and *Staphylococcus aureus* (Staab et al. 2007). The absolute configuration of chaetocin (71), a metabolite of the fungus *C. minutum* with antibacterial and cytostatic activity, was in 1970 elucidated by chemical and X-ray methods (Kikuchi et al. 1982). From the culture of *C. thielavioideum*, 71 was also obtained by Udagawa et al. (1979). A new dimeric ETP, chetracin A (72) with two tetrasulfide bridges, was isolated from *Chaetomium nigricolor* and *C. retardatum*, and its triacetate structure was established by X-ray analysis. Two analogues of compound 71, chaetocins B (73) and C (74), isolated from *C. virescens* var. *thielabioideum*, were proved to be homologs with one disulfide and one trisulfide bridge and with two trisulfide bridges (Saito et al. 1985, 1988), respectively. The absolute configurations of both 73 and 74 were determined by CD spectra. None of these compounds showed any antimicrobial activities toward Gram-negative bacteria or fungus; however, they did exhibit a marked activity against Gram-positive bacteria *S. aureus* FDA 209P. Chaetocins B (73) (MIC = 0.05 μg/ml) and C (74, MIC = 0.025 μg/ml), having trisulfide bridges, displayed potent properties, followed by chaetocin (43, MIC = 0.1 μg/ml), and chetracin A (72) (MIC = 0.39 μg/ml), having disulfide or tetrasulfide bridges, whereas their analogues without bridges did not show such activity. The cytotoxicities of chetomin (69), chaetocin (71), chetracin A (72), chaetocins B (73), and C (74) to HeLa cells were highly noticeable (IC_{50} = 0.02–0.07 μg/ml), and similar structure-activity relationships were found to the antibacterial activity. Three new chetomin-related metabolites chetoseminudins A–C (75–77) and 68 were produced by immunomodulatory extract of *C. seminudum* 72-S-204-1 (79). Only chetomin (69) and chetoseminudin A (75) displayed high immunosuppressive activities (IC_{50} = 0.17 and 0.09 and 0.18 and 0.13 μg/ml, respectively) against concanavalin A (Con A) (T-cells) and lipopolysaccharide (LPS)-induced (B-cells) proliferations of mouse splenic lymphocytes. The activity of compounds 69 and 75 might be due to their ETP moiety in their molecules. Moreover, histone methylation plays a key role in establishing and maintaining stable gene expression patterns during cellular differentiation and embryonic development (67). Chaetocin (71) was characterized as the first inhibitor of a lysine-specific histone methyltransferase (HMT), a key enzyme in the epigenetic control of gene expression (68). Furthermore, during examination of the inhibitory activity against H3K9 HMT G9a, which is a potential therapeutic target in human cancer, sulfur-deficient chaetocin (78) and its enantiomer (ent-79), recently synthesized by Iwasa et al. (2010) were found to be devoid of activity, while chaetocin (43) and its enantiomer (ent-80) showed comparable activities. These results disclose that the disulfide unit is crucial for the HMT inhibitory activity and that the chirality of 43 is insensitive to HMT. Additionally, ent-chaetocin (ent-79), the unnatural enantiomer of chaetocin (71), was more potent than the parent compound 71 in human leukemia HL-60 cells at 0.3 μM and was found to induce apoptosis through the caspase 8/caspase 3 activation pathway (Teng et al. 2010). Mainly regulated at the transcriptional level, the cellular cyclin-dependent kinase inhibitor, CDKN1A/p21(WAF1) (p21), is a major cell cycle regulator of the response to DNA damage, senescence and tumor suppression. Importantly, treatment with the specific SUV39H1 inhibitor, chaetocin (71), repressed histone H3 lysine 9 trimethylation at the p21 gene promoter, stimulated p21 gene expression, and induced cell cycle arrest (Cherrier et al. 2009). Recently, Tibodeau et al. reported that chaetocin (71) potently induces cellular oxidative stress, thus selectively killing cancer cells, and that 71 is a competitive and selective substrate and inhibitor of thioredoxin reductase (Tibodeau et al. 2009). These observations have important implications about the mechanism of action and clinical development of 71 and related thiodioxopiperazines. Three new ETPs chaetocochins A–C (81–83) and their congeners 69 and 70 were isolated from the solid-state fermented rice culture of *Chaetomium cochliodes* (Li et al. 2006). Chaetocochins A (81) and B (82) having a 14-membered ring represent a novel type of ETP. Chaetocochins A (81) and C (83) and 42 exhibited significant cytotoxicity in vitro against cancer cell lines Bre-04 (MDA-MB-231), Lu-04 (NCIH460), and N-04 (SF-268); compound 70 displayed the strongest cytotoxicity (GI_{50} = 0.06, 0.05, and 0.2 g/ml, respectively). Their activities appear to be correlated with the number of sulfur-methyls. It has been known that some ETPs are difficult to be identified even using NMR, since the numbers of S atoms in each piperazine ring are difficult to clarify. HPLC-ESI-QTOF-MS/MS/MS has proved to be a powerful tool for compound identification, especially for quite minor natural products in complex extracts. More recently, Zhang and coworkers employed low-energy collision-induced electrospray ionization tandem mass spectrometry (in positive ion mode) to characterize five ETPs: dethiotetra(methylthio)chemotin (70), chaetocochins A–C (81–83), and chemotin (69) isolated from the same fungus (Wu et al. 2007). A likely known compound, chetoseminudin A (77), and three ETPs, chaetocochins D–F (84–86), were identified based on tandem mass spectra of known ETPs (Li et al. 2006).

7.4 Isoquinolines

Chaetoindicins A–C (87–89), three novel isoquinolines (Fig. 7.3) with an unprecedented skeleton, were isolated from the solid-state fermented culture of *Chaetomium indicum*. X-ray crystallographic analysis confirmed the structure of 88 (Li et al. 2006).

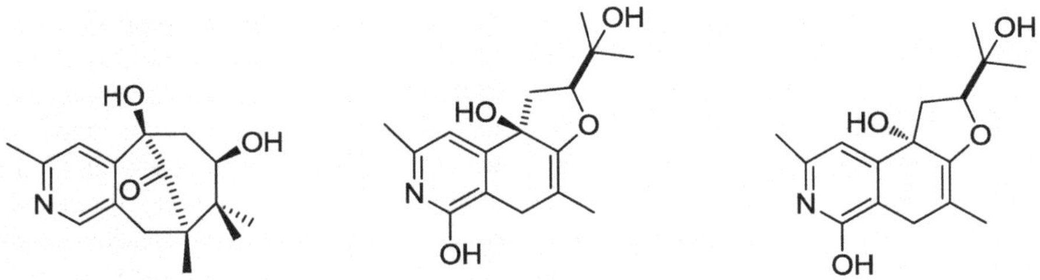

Fig. 7.3 Isoquinoline alkaloids of *Chaetomium*

7.5 Bis(3-indolyl)-benzoquinones

Cochliodinol (90) was first reported in 1968 as a purple pigment from various strains of *C. cochliodes* Palliser and *C. globosum* Kunze ex Fr. (Brewer et al. 1968, 1970). It was subsequently discovered in the cultures of *C. elatum* Kunze ex Fr., *C. murorum* Corda, and *C. amygdalisporum* Udagawa & Muroi (Sekita et al. 1981). It inhibited not only the growth of some species of several genera of microfungi but also the germination of spores of *Botrytis allii* and *Fusarium moniliforme* at 1–10 μg/ml. The growth of several strains of *Pseudomonas aeruginosa* was inhibited by 90 at 30 μg/ml. Symptoms of acute toxicity were not revealed in toxicity tests with young turkeys and rats (Brewer et al. 1970). Meiler and Taylor disclosed that 90 at 15 μg/ml decreased the respiration rate of microspores of *Fusarium oxysporum*, using glucose or succinate as substrate, by about 70%, and inhibited their germination (Meiler and Taylor 1971). Brewer et al. reported that 90 may be bactericidal for *Butyrivibrio fibrisolvens* and *Bacillus subtilis* (MIC = 1 and 5 μg/ml, respectively) (Brewer et al. 1984). Isocochliodinol (91) and neocochliodinol (92) (Sekita et al. 1983), were isolated from *Chaetomium murorum* and *C. amygdalisporum*, respectively. Isocochliodinol (91) has been reported to suppress HIV-1 protease with an IC_{50} value as low as 0.17 μM and the epidermal growth factor receptor protein tyrosine kinase at 15–60 μM and thus become lead molecules for these targets. Molecular modeling of the HIV1-protease-inhibitor complexes showed hydrogen bonding between the dihydroxybenzoquinone unit of isocochliodinol to both active-site aspartic acids (Asp25/Asp25′) of the protease and the indole moieties of the inhibitors filling the P2 and P2′ pockets of the protease (Fredenhagen et al. 1997). More recently, beside isocochliodinol (91), 90 is the main component obtained from the endophytic *Chaetomium sp.* isolated from the medicinal plant *Salvia officinalis* (Labiatae), proved to be potential in vitro cytotoxic against L5178Y mouse lymphoma cell line (EC_{50} = 7.0 μg/ml), whereas 91 (Fig. 7.4) showed a weak activity (Debbab et al. 2009). Their cytotoxic activity was in relation to the position of prenyl substituents at the indole rings.

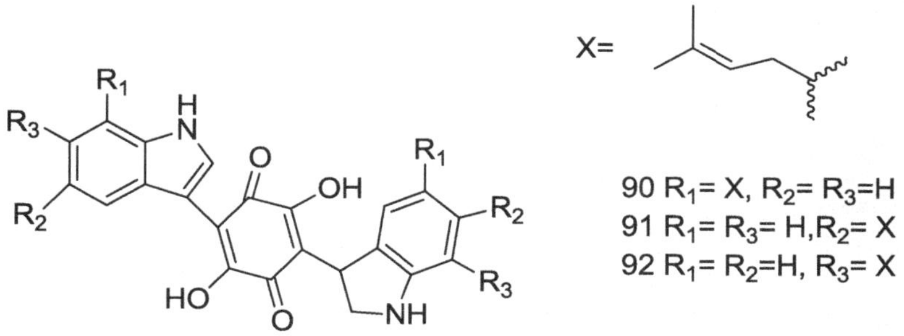

Fig. 7.4 Bis(3-indolyl)-benzoquinones

Fig. 7.5 Tripeptide alkaloid of *Chaetomium*

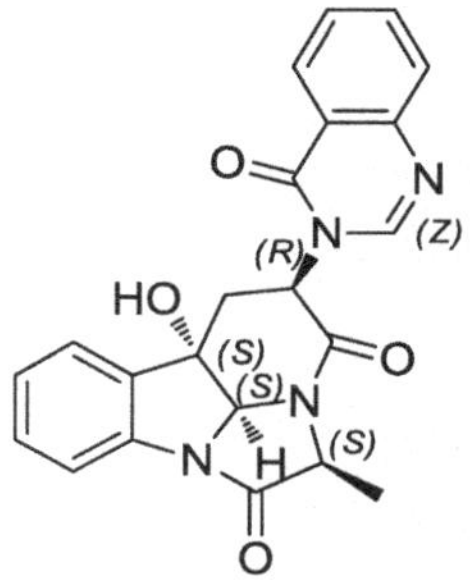

93

7.6 Tripeptide Alkaloid

Chaetominine (93) (Fig. 7.5), a tripeptide alkaloid with a new framework, was characterized from the solid substrate culture of *Chaetomium* sp. (IFB-E015), an endophytic fungus on the *Adenophora axilliflora* leaves (Jiao et al. 2006). Its structure was determined by a combination of its spectral data and X-ray diffraction analysis, with its absolute configuration elucidated by Marfey's method. Chaetominine (93) was more cytotoxic than 5-fluorouracil against the human leukemia K562 (21 nM) and colon cancer SW1116 (28 nM) cell lines. A plausible biosynthetic pathway for (−)-93 was proposed by Tan et al. (Abdel-Lateff 2008), involving L-alanine, anthranilic acid, and Dtryptophan, the last two being closely linked through the shikimate pathway.

7.7 Benzonaphthyridinedione

A marine fungal isolate *Chaetomium* sp., isolated from the alga *Valomia utricularis*, gave a novel benzonaphthyridinedione derivative, chaetominedione (94) (Fig. 7.6). It has a significant inhibitory activity toward p56lck tyrosine kinase with 93.6% enzyme inhibition at 200 μg/ml (Breinholt et al. 1996).

7.8 Other N-Containing Compounds

A new isatin derivative prenisatin (95) along with cochliodinol (90) was isolated from submerged fermentations of *Chaetomium globosum* Kunze: Fr., isolated from a soil sample. Compound 95 inhibited the in vitro growth of *Botrytis cinereal* (Qin et al. 2009a, b). Three metabolites, cerebrosides C (96) and B (97) and allantoin (98), were isolated from the mycelia of *C. globosum* ZY-22, an endophyte found in the leaves of *Ginkgo biloba*. Cerebroside C (96) has elicitor activity in plants. Two new lactams, chaetoglocins C (99) and D (100) (Fig. 7.7), were isolated from the solid culture of *C. globosum* IFB-E036, an endophytic fungus found in the root of *Cynodon dactylon* (Poaceae) and displayed no antimicrobial activity against Gram-positive bacteria (Ge et al. 2011).

Fig. 7.6 Benzonaphthyridinedione alkaloid of genus *Chaetomium*

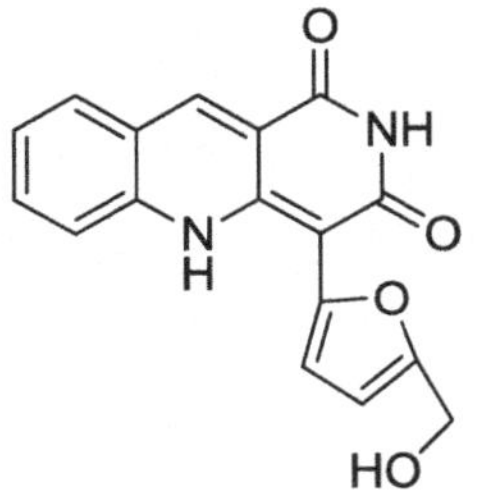

94

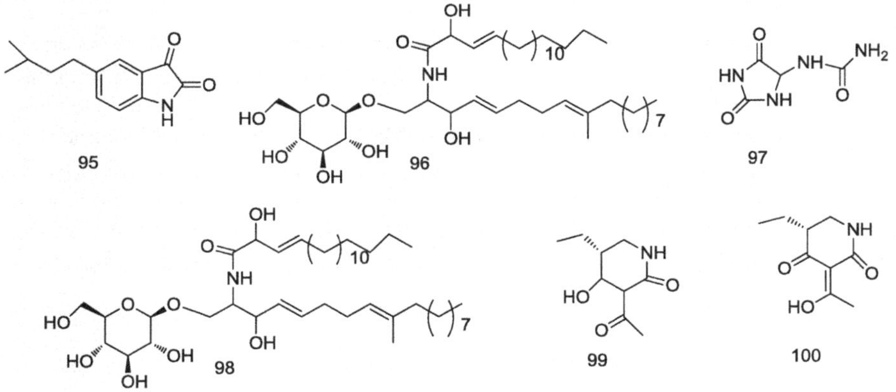

95 96 97 98 99 100

Fig. 7.7 Some N-containing compounds of *Chaetomium*

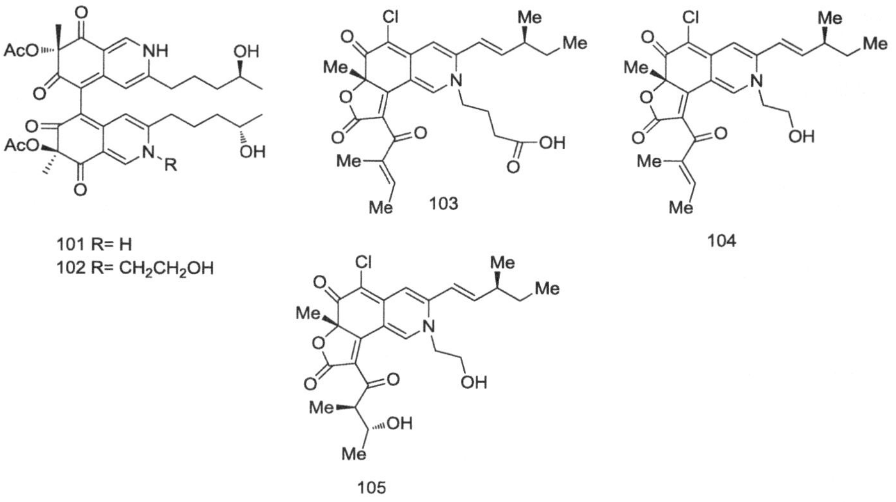

Fig. 7.8 Azophilones of *Chaetomium*

7.9 Azaphilones

Chaetoglobins A (101) and B (102), two azaphilone alkaloid dimers with an unprecedented skeleton, were isolated from the extract of *C. globosum* IFB-E019, an endophytic isolate from *Imperata cylindrica* stem. Their structure elucidation was accomplished by a combination of Mosher's reaction and CD and 2D NMR techniques. Chaetoglobin A (101) can significantly inhibit the proliferation ability of MCF-7 and SW1116 cell lines (IC_{50} = 26.8 and 35.4 μg/ml, respectively). It also inhibited expressions of tumor-related genes bcl-2, c-myc and catenin. (Ge et al. 2008). Three novel azaphilone alkaloids, namely, chaetomugilides A–C (103–105) (Fig. 7.8), together with three related compounds were isolated from the methanol extract of *Chaetomium globosum* TY1, an endophytic fungus isolated from *Ginkgo biloba*. Their structures were elucidated based on extensive 1D and 2D NMR as well as HRESI-MS spectroscopic data analysis. The isolated compounds exhibited highly cytotoxic activities against human cancer cell line HePG2 with the IC_{50} values ranging from 1.7 to 53.4 μM (Li et al. 2013).

References

Abdel-Lateff A (2008) Chaetominedione, a new tyrosine kinase inhibitor isolated from the algicolous marine fungus Chaetomium sp. Tetrahedron Lett 49:6398–6400

Ames LM (1949) New cellulose destroying fungi isolated from military material and equipment. Mycologia 41(6):637–648. https://doi.org/10.2307/3755020. JSTOR 3755020

Ames LM (1961) A monograph of the *Chaetomiaceae*. Verlag von J Cramer, Lehre, p 2, 3, 9, 21
Ames LM (1963) A monograph of the Chaetomiaceae. U.S. Army Research and Development, Series 2
Binder M, Tamm C (1973) The cytochalasans: a new class of biologically active microbial metabolites. Angew Chem Int Ed Engl 12:370–380
Breinholt J, Demuth H, Heide M, Jensen GW, Moller IL, Nielsen RI, Olsen CE, Rosendahl CN (1996) Prenisatin (5-(3-methyl2-butenyl)-indole-2,3-dione): antifungal isatin derivative from Chaetomium globosum. Acta Chem Scand 50:443–445
Brewer D, Jerram WA, Taylor A (1968) The production of cochliodinol and a related metabolite by Chaetomium species. Can J Microbiol 14:861–866
Brewer D, Jerram WA, Taylor A (1970) The toxicity of cochliodinol, an antibiotic metabolite of Chaetomium spp. Can J Microbiol 16:433–440
Brewer D, Mcinnes AG, Smith DG, Taylor A, Walter JA, Loosli HR, Kis ZL (1976) Sporidesmins. Part 16. The structure of chetomin, a toxic metabolite of Chaetomium cochliodes, by nitrogen-15 and carbon-13 nuclear magnetic resonance spectroscopy. J Chem Soc Perkin Trans 1:1248–1251
Brewer D, Jen WC, Jones GA, Taylor A (1984) The antibacterial activity of some naturally occurring 2,5-dihydroxy-l,4-benzoquinones. Canadian Journal of Microbiology 30(8):1068–1072
Chen C, Wang J, Liu J, Zhu H, Sun B, Wang J, Zhang J, Luo Z, Yao G, Xue Y, Zhang Y (2015a) Armochaetoglobins A–J: Cytochalasan alkaloids from Chaetomium globosum TW1-1, a fungus derived from the terrestrial arthropod Armadillidium vulgar. J Nat Prod 78(6):1193–1201. https://doi.org/10.1021/np5006
Chen C, Zhu H, Li X-N, Yang J, Wang J, Li G, Li Y, Tong Q, Yao G, Luo Z, Xue Y, Zhang Y (2015b) Armochaeglobines A and B, two new indole-based alkaloids from the arthropod-derived fungus Chaetomium globosum. Org Lett 17(3):644–647. https://doi.org/10.1021/ol503666b
Chen C, Tong Q, Zhu H, Tan D, Zhang J, Xue Y, Yao G, Luo Z, Wang J, Wang Y, Zhang Y (2016) Nine new cytochalasan alkaloids from Chaetomium globosum TW1-1 (Ascomycota, Sordariales). Sci Rep 6:18711
Chivers AH (1915) Details – a monograph of the genera Chaetomium and Ascotricha. Biodiversity Heritage Library. https://doi.org/10.5962/bhl.title.97551
Cherrier T, Suzanne S, Redel L, Calao M, Marban C, Samah B, Mukerjee R, Schwartz C, Gras G, Sawaya BE, Zeichner SL, Aunis D, Van Lint C, Rohr O (2009) p21WAF1 gene promoter is epigenetically silenced by CTIP2 and SUV39H1. Oncogene 28(38):3380–3389
Cui CM, Li XM, Li CS, Proksch P, Wang BG (2010) Cytoglobosins A-G, cytochalasans from a marine-derived endophytic fungus, Chaetomium globosum QEN-14. J Nat Prod 73:729–733
Cook KM, Hilton ST, Meeinovie J, Motherwell WB, Figg WD, Schofield CJ (1996) Epidithiodiketopiperazines block the interaction between hypoxia inducible factor-α (HIF- α) and p300 by a zinc ejection mechanism. J Biol Chem 284:2681–26838
Debbab A, Aly AH, Edrada-Ebel RA, Müller WEG, Mosaddak M, Hakiki A, Ebel R, Proksch P (2009) Bioactive secondary metabolites from the endophytic fungus Chaetomium sp. isolated from Salvia officinalis growing in Morocco. Biotechnol Agron Soc Environ 13:229–234
Ding G, Song YC, Chen JR, Xu C, Ge HM, Wang XT, Tan RX (2006) Chaetoglobosin U, a cytochalasan alkaloid from endophytic Chaetomium globosum IFB-E019. J Nat Prod 69:302–304
Dong JY, He HP, Shen YM, Zhang KQ (2005) Nematicidal epipolysulfanyldioxopiperazines from Gliocladium roseum. J Nat Prod 68:1510–1513
Fredenhagen A, Petersen F, Tintelnot-Blomley M, Rösel J, Mett H, Hug P (1997) Semicochliodinol A and B: inhibitors of HIV-1 protease and EGF-R protein tyrosine kinase related to asterriquinones produced by the fungus Chrysosporium merdarium. J Antibiot 50:395–401
Gardiner DM, Waring P, Howlett B (2005) The epipolythiodioxopiperazine (ETP) class of fungal toxins: distribution, mode of action, functions and biosynthesis. J Microbiol 151:1021–1032
Ge HM, Zhang WY, Ding G, Saparpakorn P, Song YC, Hannongbua S, Tan RX (2008) Chaetoglobins A and B, two unusual alkaloids from endophytic Chaetomium globosum culture. Chem Commun (45):5978–5980

Ge HM, Zhang Q, Xu SH, Guo ZK, Song YC, Huang WY, Tan RX (2011) Chaetoglocins A-D, four new metabolites from the endophytic fungus Chaetomium globosum. Planta Med 77:277–280

Geiger WB (1949) Chetomin, an antibiotic substance from Chaetomium cochliodes III. Composition and functional groups. Arch Biochem 21:125–131

Geiger WB, Conn JE, Waksman SA (1944) Chetomin, a new antibiotic substance produced by Chaetomium cochliodes I. Formation and properties. II. Isolation and concentration. J Bacteriol 48:531–536

Ichihara, A., Katayama, K., Teshima, H., Oikawa, H., and Sakamura, S. (1996) Chaetoglobosin O and other phytotoxic metabolites from Cylindrocladium floridanum, a causal fungus of alfalfa black rot disease. Biosci. Biotech. Biochem., 60, 360–361

Iwasa E, Hamashima Y, Fujishiro S, Higuchi E, Ito A, Yoshida M, Sodeoka M (2010) Total synthesis of (+)-chaetocin and its analogues: their histone methyltransferase G9a inhibitory activity. J Am Chem Soc 132:4078–4079

Jiao W, Feng Y, Blunt JW, Cole AL, Munro MH (2004) Chaetoglobosins Q, R, and T, three further new metabolites from Chaetomium globosum. J Nat Prod 67:1722–1725

Jiao RH, Xu S, Liu JY, Ge HM, Ding H, Xu C, Zhu HL, Tan RX (2006) Chaetominine, a cytotoxic alkaloid produced by endophytic Chaetomium sp. IFB-E015. Org Lett 8:57095712

Kang JG, Kim KK, Kang KY (1999) Antagonism and structural identification of antifungal compound from Chaetomium cochliodes against phytopathogenic fungi. Agric Chem Biotechnol 42:146–150

Kanokmedhakul S, Kanokmedhakul K, Phonkerd N, Soytong K, Kongsaeree P, Suksamrarn A (2002) Antimycobacterial anthraquinone-chromanone compound and diketopiperazine alkaloid from the fungus Chaetomium globosum KMITL-N0802. Planta Med 68:834–836

Kikuchi T, Kadota S, Nakamura K, Nishi A, Taga T, Kaji T, Osaki K, Tubaki K (1982) Dethiotetra(methylthio) chetomin, a new antimicrobial metabolite of chaetomium globosum Kinze ex Fr. Structure and partial synthesis from chetomin. Chem Pharm Bull 30:3846–3848

Kirk PM, Cannon PF, Minter DW, Stalpers JA (2008) Dictionary of the Fungi, 10th edn. CABI, Wallingford, p 131. ISBN 978-0-85199-826-8

Ko HR, Kim BY, Ahn SC, Oh WK, Kim JH, Lee HS, Kim HM, Han SB, Mheen TI, Ahn JS (1998) Chaetoglobosin A, an inhibitor of bleb formation on K562 cells induced by phorbol 12,13-dibutyrate. J Microbiol Biotechnol 8:705–709

Kobayashi H, Namikoshi M, Yoshimoto T, Yokochi T (1996) A screening method for antimitotic and antifungal substances using conidia of Pyricularia oryzae, modification and application to tropical marine fungi. J Antibiot 49:873–879

Li GY, Li BG, Yang T, Yan JF, Liu GY, Zhang GL (2006) Chaetoindicines A—C, Three Isoquinoline Alkaloids from the Fungus Chaetomi-um indicum. Org. Lett.8 16:3613-3615.

Li X, Tia Y, Yang S-X, Zhang Y-M, Qin J-C (2013) Cytotoxic azaphilone alkaloids from Chaetomium globosum TY1. Bioorg Med Chem Lett 23(10):2945–2947

Mahmood ZA, Ahmed SW, Azhar I, Sualeh M, Baig MT, Zoha S (2010) Bioactive alkaloids produced by fungi I. Updates on alkaloids from the species of the genera boletus, fusarium and psilocybe. Pak J Pharm Sci 23(3):349–357

McInnes AG, Taylor A, Walter JA (1976) The structure of chetomin. J Am Chem Soc 98:6741–6741

Meiler D, Taylor A (1971) The effect of cochliodinol, a metabolite of Chaetomium cochliodes, on the respiration of microsopores of Fusarium oxysporum. Can J Microbiol 17:83–86

Momesso LDS, Kawano CY, Ribeiro PH, Nomizo A, Goldman G, Pupo MT (2008) Chaetoglobosins produced by Chaetomium globosum, endophytic fungus found in association with Viguiera robusta Gardn (Asteraceae). Quim Nova 31:1680–1685

Müllbacher A, Waring P, Tiwari-Palni U, Eichner RD (1986) Structural relationship of epipolythiodioxopiperazines and their immunomodulating activity. Mol Immunol 23:231–235

Namgoong J, Yeon S-W, Paek N-S, Kim T-H, Kim Y-H, Kim C-J, Kim K-W (1998) Isolation and structural determination of anti-Helicobacter pylori compound, fungus 60686. Sanop Misaengmul Hakhoechi 26:137–142

Oikawa H, Murakami Y, Ichihara A (1991) New plausible precursors of chaetoglobosin a accumulated by treatment of chaetomium subaffine with cytochrome P-450 inhibitors. Tetrahedron Lett 32:4533–4536

Oikawa H, Murakami Y, Ichihara A (1992a) Biosynthetic study of chaetoglobosin A: origins of the oxygen and hydrogen atoms, and indirect evidence for biological Diels–Alder reaction. J Chem Soc Perkin Trans 1:2955–2959

Oikawa H, Murakami Y, Ichihara A (1992b) Useful approach to find the plausible biosynthetic precursors of secondary metabolites using P-450 inhibitors: postulated intermediates of chaetoglobosin A. J Chem Soc Perkin Trans 1:2949–2954

Oikawa H, Murakami Y, Ichihara A (1993) 20-ketoreductase activity of chaetoglobosin A and prochaetoglobosins in a cell-free system of Chaetomium subaffine and the isolation of new chaetoglobosins. Biosci Biotechnol Biochem 57:628–631

Probst A, Tamm C (1981a) 19-O-acetylchaetoglobosin B and 19-Oacetylchaetoglobosin D, two new metabolites of Chaetomium globosum. Helv Chim Acta 64:2056–2064

Probst A, Tamm C (1981b) Biosynthesis of the cytochalasans. Biosynthetic studies on chaetoglobosin A and 19-Oacetylchaetoglobosin A. Helv Chim Acta 64:2065–2077

Qin JC, Bai L, Li XM, Zhang YM, Gao JM, Laatsch H (2009a) Isolation and identification of the metabolites produced by an endophytic fungus Chaetomium globosum ZY-22 from Ginkgo biloba. Acta Botan Boreali-Occiden Sin 29:1264–1268

Qin JC, Zhang YM, Gao JM, Bai MS, Yang SX, Laatsch H, Zhang AL (2009b) Bioactive metabolites produced by Chaetomium globosum, an endophytic fungus isolated from Ginkgo biloba. Bioorg Med Chem Lett 19:1572–1574

Safe S, Taylor A (1972) Sporidesmins. Part XIII. Ovine 111-thrift in Nova Scotia. Part 11. The characterisation of chetomin, a toxic metabolite of Chaetomium cochliodes and Chaetomium globosum. J Chem Soc Perkin Trans 1:472–479

Saito T, Koyama K, Natori S, Iitaka Y (1985) Chetracin A, a new epipolythiodioxopiperazine having a tetrasulfide bridge from Chaetomium abuens and C. retardatum. Tetrahedron Lett 26:4731–4734

Saito T, Suzuki Y, Koyama K, Natori S, Iitaka Y, Kinoshita T (1988) Chetracin A and chaetocins B and C, three new epipolythiodioxopiperazines from Chaetomium spp. Chem Pharm Bull 36:1942–1956

Scherlach K, Boettger D, Remme N, Hertweck C (2010) The chemistry and biology of cytochalasans. Nat Prod Rep 27:869–886

Sekita S, Yoshihira K, Natori S, Kuwano H (1973) Structures of chaetoglobosin A and B, cytotoxic metabolites of Chaetomium globosum. Tetrahedron Lett 14:2109–2112

Sekita S, Yoshihira K, Natori S, Kuwano H (1976) Structures of chaetoglobosins C, D, E, and F, cytotoxic indol-3-yl[13]cytochalasans from Chaetomium globosum. Tetrahedron Lett 17:1351–1354

Sekita S, Yoshihira K, Natori S, Kuwano H (1977) Chaetoglobosins G and J, cytotoxic indol-3-yl[13]-cytochalasans from Chaetomium globosum. Tetrahedron Lett 18:2771–2774

Sekita S, Yoshihira K, Natori S, Udagawa S, Muroi T, Sugiyama Y, Kurata H, Umeda M (1981) Mycotoxin production by Chaetomium spp. and related fungi. Can J Microbiol 27:766–772

Sekita S, Yoshihira K, Natori S, Kuwano H (1982a) Chaetoglobosins, cytotoxic 10-(indo-3-yl)-[13] cytochalasans from Chaetomium spp. II. Structures of chaetoglobosins A, B, and D. Chem Pharm Bull 30:1618–1628

Sekita S, Yoshihira K, Natori S, Kuwano H (1982b) Chaetoglobosins, cytotoxic 10-(indol-3-yl)-[13] cytochalasans from Chaetomium spp. III. Structures of chaetoglobosins C, E, F, G, and J. Chem Pharm Bull 30:1629–1638

Sekita S, Yoshihira K, Natori S, Udagawa S, Sakabe F, Kurata H, Umeda M (1982c) Chaetoglobosins, cytotoxic 10-(indol-3-yl)[13] cytochalasans from Chaetomium spp. I. Production, isolation and some cytological effects of chaetoglobosins A-J. Chem Pharm Bull 30:1609–1617

Sekita S, Yoshihira K, Natori S (1983) Chaetoglobosins, cytotoxic 10(indol-3-yl)-[13]cytochalasans from Chaetomium spp. IV. Carbon13 nuclear magnetic resonance spectra and their application to a biosynthetic study. Chem Pharm Bull 31:490–498

Shinohara C, Chikanishi T, Nakashima S, Hashimoto A, Hamanaka A, Endo A, Hasumi K (2000) Enhancement of fibrinolytic activity of vascular endothelial cells by chaetoglobosin A, crinipellin B, geodin and triticone B. J Antibiot 53:262268

Silverton JV, Akiyama T, Kabuto C, Sekita S, Yoshihira K, Natori S (1976) X-ray analysis of chaetoglobosin A, an indol-3-yl[13]cytochalasan from Chaetomium globosum. Tetrahedron Lett 17:1349–1350

Staab A, Loeffler J, Said HM, Diehlmann D, Katzer A, Beyer M, Fleischer M, Schwab F, Baier K, Einsele H, Flentje M, Vordermark D (2007) Effects of HIF-1 inhibition by chetomin on hypoxia-related transcription and radiosensitivity in HT 1080 human fibrosarcoma cells. BMC Cancer 7:213219

Teng Y, Iuchi K, Iwasa E, Fujishiro S, Hamashima Y, Dodo K, Sodeoka M (2010) Unnatural enantiomer of chaetocin shows strong apoptosis-inducing activity through caspase8/caspase-3 activation. Bioorg Med Chem Lett 20:50855088

Thohinung S, Kanokmedhakul S, Kanokmedhakul K, Kukongviriyapan V, Tusskorn O, Soytong K (2010) Cytotoxic 10(indol-3-yl)-[13]cytochalasans from the fungus Chaetomium elatum ChE01. Arch Pharm Res 33:1135–1141

Tibodeau JD, Benson LM, Isham CR, Owen WG, Bible KC (2009) The anticancer agent chaetocin is a competitive substrate and inhibitor of thioredoxin reductase. Antioxid Redox Signal 11:1097–1106

Udagawa S, Muroi T, Kurata H, Sekita S, Yoshihira K, Natori S, Umeda M (1979) The production of chaetoglobosins, sterigmatocystin, O-methylsterigmatocystin, and chaetocin by Chaetomium spp. and related fungi. Can J Microbiol 25:170–177

Umeda M, Ohtsubo K, Saito M, Sekita S, Yoshihira K, Natori S, Udagawa S, Sakabe F, Kurata H (1975) Cytotoxicity of new cytochalasans from Chaetomium globosum. Experientia 31:435–438

von Arx JA, Guarro J, Figueras MJ (1986) The ascomycete genus *Chaetomium*. Beih Nova Hedwig 84:1–162

Von Wallbrunn C, Luftmann H, Bergander K, Meinhardt F (2001) Phytotoxic chaetoglobosins are produced by the plant pathogen Calonectria morganii (anamorph Cylindrocladium scoparium). J Gen Appl Microbiol 47:33–38

Walksman SA, Bugie E (1944) Chaetomin, a new antibiotic substance produced by Chaetomium cochliodes: I. Formation and properties. J Bacteriol 48:527–530

Wang S, Li XM, Teuscher F, Li DL, Diesel A, Ebel R, Proksch P, Wang BG (2006) Chaetopyranin, a benzaldehyde derivative, and other related metabolites from Chaetomium globosum, an endophytic fungus derived from the marine red alga Polysiphonia urceolata. J Nat Prod 69:1622–1625

Waring P, Eichner RD, Müllbacher A (1988) The chemistry and biology of the immunomodulating agent gliotoxin and related epipolythiodioxopiperazines. Med Res Rev 8:499–524

Weindling R (1941) Experimental consideration of the mold toxins of Gliocladium and Trichoderma. Phytopathology 31:991–1003

Weindling R, Emerson OH (1936) The isolation of a toxic substance from the culture filtrate of Trichoderma. Phytopathology 26:1068–1070

Wu ZJ, Li GY, Fang DM, Qi HY, Ren WJ, Zhang GL (2007) Electrospray tandem mass spectrometry of epipolythiodioxopiperazines. J Mass Spectrom 42:749–760

Xu G-B, He G, Bai H-H, Yang T, Zhang G-L, Wu L-W, Li G-Y (2015) Indole alkaloids from Chaetomium globosu. J Nat Prod 78(7):1479–1148. https://doi.org/10.1021/np5007235

Zhang J, Ge HM, Jiao RH, Li J, Peng H, Wang YR, Wu JH, Song YC, Tan RX (2010) Cytotoxic chaetoglobosins from the endophyte Chaetomium globosum. Planta Med 76:1910–1914

Zhang Q, Li H-Q, Zong S-C, Gao J-M, Zhang A-L (2012) Chemical and bioactive diversities of the genus Chaetomium secondary metabolites. Mini Rev Med Chem 12(2):127–148. https://doi.org/10.2174/138955712798995066

Chapter 8
Applications of *Chaetomium* Functional Metabolites with Special Reference to Antioxidants

Amira M. G. Darwish, Neveen M. Abdelmotilib, Ahmed M. Abdel-Azeem, Hebatallah H. Abo Nahas, and Marwa T. Mohesien

8.1 Introduction

Endophytic fungi are prominent components of plant microecosystem. Besides possessing a variety of potentials, these endophytic fungi have become an alternative source for antioxidant moieties. Antioxidants can be derived from natural sources as well as obtained synthetically. Among them, compound pestacin isolated from endophytes is considered as "keystone" in the research field, due to its extraordinarily higher antioxidant potential than trolax, a synthetic antioxidant. This has motivated the researcher to extract and study some more compounds from endophytes.

Although synthetic antioxidants such as butylated hydroxyanisole [BHA], butylated hydroxytoluene [BHT], and propyl gallate are being used in food and cosmetics industries, they do have their own drawbacks, such as probable carcinogenicity when administered in vivo, the concern of public safety, and economic values.

Therefore, compounds isolated from plants and other natural products are gaining momentum in cosmetics and food additives industry. Since endophytic fungi are an equally alternative source for such kind of products, exploring endophytic fungi

A. M. G. Darwish (✉) · N. M. Abdelmotilib
Food Technology Department, Arid Lands Cultivation Research Institute (ALCRI), City of Scientific Research and Technological Applications (SRTACity), New Borg El-Arab, Alexandria, Egypt

A. M. Abdel-Azeem
Botany Department, Faculty of Science, University of Suez Canal, Ismailia, Egypt
e-mail: ahmed_abdelazeem@science.suez.edu.eg

H. H. Abo Nahas
Zoology Department, Faculty of Science, University of Suez Canal, Ismailia, Egypt

M. T. Mohesien
Botany and Microbiology Department, Faculty of Science, University of Damietta, New Damietta, Egypt

A. M. Abdel-Azeem (ed.), *Recent Developments on Genus Chaetomium*, Fungal Biology, https://doi.org/10.1007/978-3-030-31612-9_8

for antioxidant compounds is becoming more important day by day. They are considered to have a strong protective mechanism against generation of free radicals which leads to several disorders such as aging, cancer, atherosclerosis, coronary heart ailment, and diabetes (Halliwell 1994), neurodegenerative disorders such as Alzheimer's and Parkinson's disease, and rheumatoid arthritis (Valko et al. 2007). Generation of free radicals and its deleterious effect in normal body function is the key element of oxidative stress. The brain is highly vulnerable to oxidative stress as it undergoes excessive physiological activities. The unsaturated lipids are targeted and oxidized.

This makes antioxidants an important component in treating brain disease. This chapter is focused on the antioxidant potential of *Chaetomium* endophytic taxa. A few important compounds that have been explored from endophytes have been discussed here.

8.2 Why Do Plants Synthesize Antioxidants?

Plants have been the major sources of natural products since the beginning of research, and it could also be stated that they are the natural craftsmen of molecules created in inexhaustible orders. Since plants are sedentary organisms, they undergo a number of changes in order to adapt to unfavorable conditions. These changes occur due to the formation of different important chemicals. These chemicals are accepted as a valuable resource of plants since they maintain their age and health. Endophytes and plants have a symbiotic relationship, where the endophytes acquire benefit in the form of nutrition and in return synthesize certain compounds that help the plant in metabolism and protect it from unfavorable conditions. These compounds produced by the endophytes present hidden repertoires of known and unknown medicinal significance.

Heinig et al. (2013) gave the concept of horizontal gene transfer which suggests that endophytes and the host plant contribute to the co-production of bioactive molecules. Some endophytes have been known to possess superior biosynthetic capabilities, owing to their presumable gene recombination with the host while residing and reproducing inside the healthy plant tissues (Li et al. 2005). Taxol, jesterone, ambuic acid, torreyanic acid, pestaloside, pestalotiopsins, and 2-a-hydroxydimeniol are few examples of such compounds (Strobel and Daisy 2003). These bioactive molecules synthesized by plants can be used for the treatment of several human diseases. Apart from plants, endophytes which are in symbiotic relationship with the plants are also considered to be a significant source of antioxidants (Huang et al. 2007).

8.3 Synthetic Antioxidants Versus Natural Antioxidants

Eating foods rich in antioxidants is the best way of acquiring antioxidants in a stressful environment. Studies suggest that antioxidants added in food materials have diverse positive effects (Willis et al. 2009). Synthetic phenolic antioxidants such as butylated hydroxyanisole [BHA], butylated hydroxytoluene [BHT], and propyl gallate are used as food additives to inhibit the oxidation of food materials. In addition to this, people consume synthetic antioxidants in the form of vitamins, colorants, flavoring agents (spice and herb), and preservatives. Long-term antioxidant stress may result in weak immunological response and the development of asthma, allergies, and obesity which have become serious concerns for public health. Moreover, the quest for safer supplements that are devoid of negative effects has motivated the researchers to search for consumer-friendly natural antioxidants derived from plants and other natural sources (Table 8.1).

After discussing natural antioxidants and their role in the body defense system, it can be suggested that nonenzymatic antioxidants are also added in the form of natural dietary components.

Table 8.1 Natural antioxidants

No	Source and function
1.	Enzymatic antioxidant (breaking down and removing free radicals): Superoxide dismutase (SOD) converts superoxide ion into dioxygen and hydrogen peroxide; this enzyme exists in three forms: cytosolic Cu, Zn-SOD, mitochondrial Mn-SOD, and extracellular SOD (Landis and Tower 2005). Catalase converts hydrogen peroxide to water and has the highest turnover rate (Mates et al. 1999). Glutathione peroxidase and glutathione reductase both involve themselves in secondary enzyme defense system. Glucose-6-phosphate dehydrogenase regenerate NADPH, thereby creating a reducing environment (Gamble and Burke 1984)
2.	Nonenzymatic antioxidants (interrupting free radical chain reactions): Minerals such as selenium, copper, iron, zinc, and manganese (acting as cofactor for antioxidants); vitamins such as vitamin A, vitamin C, and vitamin E; carotenoids such as β-carotene, lycopene, lutein, and zeaxanthin; and volatile oils found in nature such as eugenol, carvacrol, safrole, thymol, menthol, 1, 8-cineole, α-terpineol, p-cymene, cinnamaldehyde, myristicin, and piperine (Shan et al. 2005). Polyphenols: Phenolic acids (hydrocinnamide and hydrobenzoic acid). Flavonoids: They are scavengers of superoxide ions (Robak and Gryglewski 1988). Examples include flavones (luteolin, apigenin, tangeretin), flavanols (quercetin, kaemferol, myricetin, isorhamnetin, pachypodo), flavanones (hesteretin, naringenin, eriodictyol), isoflavanones (genistein, daidzein, glycitein), anthocyanin C (cyanidin, delphinidin, malvidin, pelargonidin, peonidin, petunidin), gingerol, and curcumin.
3.	Others: Transition metal-binding proteins such as albumin, ceruloplasmin, haptoglobin, and transferring nonprotein antioxidants include bilirubin, uric acid (Ames et al. 1981), and ubiquinol (coenzyme Q) (Stocker et al. 1987; Papas 1998)

8.4 Structure of Antioxidant Compounds and Their Role in Decreasing Oxidation

The chemical format of molecules suggests that compounds that possess any aromatic ring bonded to hydroxyl groups (–OH) as substituents act as effective antioxidants; these include phenols, flavonoids, etc. A phenolic compound develops resonance and delocalizes and converts itself into a phenoxide ion which further loses an electron and forms a corresponding free radical. Phenolic rings, which donate H to free radical during self-association, become a free radical. Later on, they are stabilized by internal delocalization of electrons within the aromatic ring (Brewer 2011).

Due to the presence of –OH group, they also act as an anti-inflammatory and contain antimicrobial properties. Hence, phenolic compounds have radical-scavenging and antioxidant activity (Waterman and Mole 1994). Among the diverse groups, flavonoids, tannins, and phenolic acids are the main phenolic compounds (Koes et al. 1994; Burns et al. 2001; Rababah et al. 2005). Antioxidants are not only eaters of free radicals but also suppress generation of superoxides and possess chemopreventive effects (Lippman et al. 1994).

Antioxidant compound uses different mechanisms to reduce the population of free radicals as it chelates metal ions, so they become unable to generate reactive species. They also aid in quenching O2, preventing formation of peroxides which break the auto-oxidative chain reactions, thus reducing localized O_2 concentrations (Nawar 1996).

8.5 Antioxidant Compounds Associated with Different Endophytic *Chaetomium*

Recently, the genus *Chaetomium* has received attention as a rich source producing more than 200 small secondary metabolite compounds with diverse bioactivities, such as antitumor, phytotoxicity on numerous plants, immunomodulatory, antifungal, nematicidal activities, antimalarial, enzyme inhibitory, antibiotic, and other activities which have significance for drug development (Hu et al. 2018; Soytong et al. 2001).

Chaetomium species can be antagonistic against various soil microorganisms and plant pathogens. Chaetoglobosins are well known for their robust cytotoxic bioactivity and potential pharmaceutical significance. Chaetoglobosins are grouped in the cytochalasin family of natural products and are actually polyketide derivatives found in fungi. They have unique biochemical property of binding eukaryotic actin proteins, disturbing the normal actin network in the cell. To date, more than eighty chaetoglobosins have been reported from different genera of filamentous fungi, including species of the genus *Chaetomium* (Wang et al. 2017).

Chaetomium globosum CDW7, an endophyte from *Ginkgo biloba*, exhibited strong inhibitory antifungal activity against phytopathogens such as *Fusarium graminearum*, *Rhizoctonia solani*, *Magnaporthe grisea*, *Pythium ultimum*, and *Sclerotinia sclerotiorum* both in vitro and in vivo.

8.5.1 Flavipin

Flavipin is a well-known natural product that is isolated from endophytes belonging to *Chaetomium* sp. associated with leaves of *Ginkgo biloba* (Ye et al. 2013). Yan et al. (2018) succeeded in isolating bioactive metabolites with antifungal activities from this fungus; the metabolites are flavipin, chaetoglobosins A and D, chaetoglobosins R (4) and T (5), new isocoumarin derivative prochaetoviridin A (1), new indole alkaloid, and chaetoindolin A (2) and chaetoviridin A (3) (Fig. 8.1). Flavipin is considered the major antioxidant component of CDW7's metabolites; it reacts by donating its electrons to the free radicals, leading to SOD and GSH-Px activity improvement and suppression of MDA content. This metabolite possesses three phenolic hydroxyl and two aldehyde groups, which are characteristic functional groups with antioxidant activity. When cultured under the optimal condition (25 °C, 100/250 mL flask, 12 discs/flask, 150 rpm, pH 6.5) for 14 days, *Chaetomium globosum* CDW7 was a highly yielded bio-source of antioxidant flavipin synthesizing a remarkable production of 315.5 mg/L (Ye et al. 2013).

Another endophytic fungus from *Ginkgo biloba*, *Chaetomium* sp. NJZTP21 (GenBank accession number: JN588553), isolated from the healthy leaf of the plant was able to produce flavipin, which significantly inhibited the growth of several

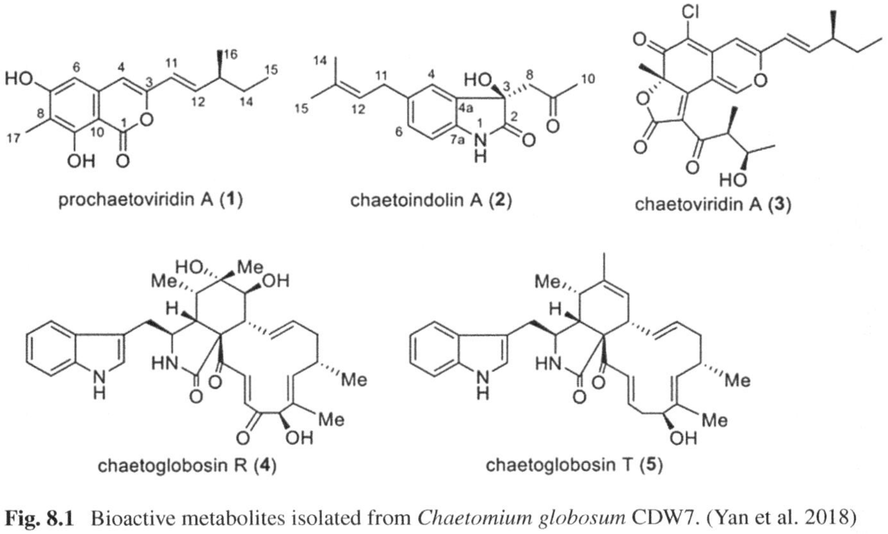

Fig. 8.1 Bioactive metabolites isolated from *Chaetomium globosum* CDW7. (Yan et al. 2018)

plant pathogenic fungi, especially *Fusarium graminearum*. But the extract from *C. globosum* CDW7, which had been deposited in the China General Microbiological Culture Collection Center (CGMCC) with an accession number 6658, has the strongest antioxidant activity among the studied endophytic fungi from *G. biloba* comparable to those of vitamin C and trolox, the well-known antioxidants (Ye et al. 2013).

Chaetomium globosum and *C. cochlioides* are antagonistic to species of *Fusarium* and *Helminthosporium*. They exhibited good control over many plant pathogens; seed coating treatments with viable spores of *Chaetomium globosum* were found to exert antagonistic effect controlled *Fusarium roseum* f. sp. *cerealis 'graminearum'* in corn; reduced disease incidence of apple scab caused by *Venturia inequalis*; suppressed damping-off of sugar beet caused by *Pythium ultimum*; had an antagonistic effect against *Macrophomina phaseolina*, *Pythium ultimum*, *Bipolaris sorokiniana*, *Rhizoctonia solani*, and *Alternaria brassicicola*; and reduced the quantity of sporulation of *Botrytis cinerea* on dead lily leaves exposed in the field (Biswas et al. 2012; Shternshis et al. 2005).

8.5.2 Chaetopyranin

The basic structure of chaetopyranin (I) is chromenol (I) (chromene carrying one or more hydroxyl substituents). It is chemically known as 3, 4-dihydro-2H-chromene substituted by a hydroxyl group at position 6, a 3-hydroxybut-1-en-1-yl at position 2, a formyl group at position 5, and a prenyl group at position 7 (Fig. 8.2) (Wang et al. 2006). These two compounds have been isolated from an endophytic fungus *Chaetomium globosum*, associated with *Polysiphonia urceolata*, and are found to possess antioxidant activity. The former compound also exhibits anticancer activity (Wang et al. 2006). Chaetopyranin also showed antioxidant activity.

8.5.3 Azaphilone

The most remarkable and valuable properties of azaphilones include their natural origin, yellow-red spectra, thermostability (in comparison with other natural red pigments), and water solubility. The azaphilone (Fig. 8.3) compounds produced

Fig. 8.2 Chaetopyranin (I). (Wang et al. 2006)

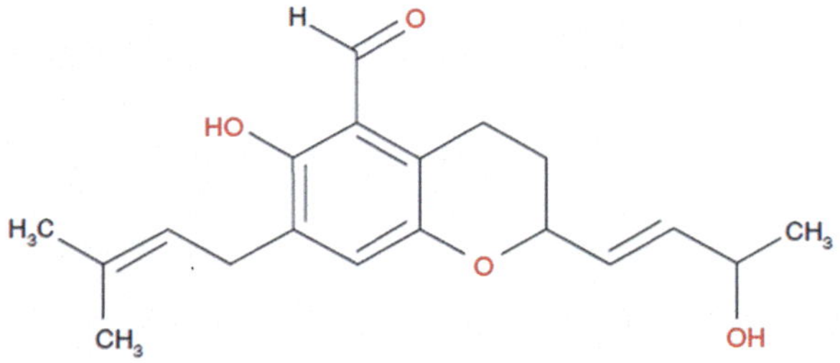

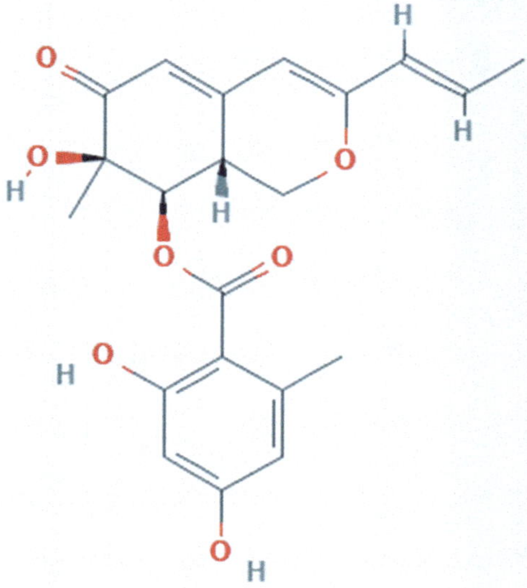

Fig. 8.3 Azaphilone

by different *Chaetomium* species display various biological activities such as antioxidant, nematicidal, antimicrobial, antifungal, anticancer, and inflammatory activities (Borges et al. 2011).

8.5.4 Hypericin and Emodin

Endophytic fungus *Chaetomium globosum* INFU/Hp/KF/34B isolated from *Hypericum perforatum* has been shown to produce hypericin (Fig. 8.4) and emodin of high medicinal value as antioxidants. This endophytic fungus has significant scientific and industrial potential to meet the pharmaceutical demands in a cost-effective, easily accessible, and reproducible way (Kusari et al. 2008; Zhao et al. 2011).

8.5.5 Mollicellins

Mollicellins O (1) isolated from the endophytic fungus *Chaetomium* sp., Eef-10, which was isolated from *Eucalyptus exserta* by Ouyang et al. (2018), showed antioxidant activity based on DPPH radical scavenging.

Fig. 8.4 Hypericin

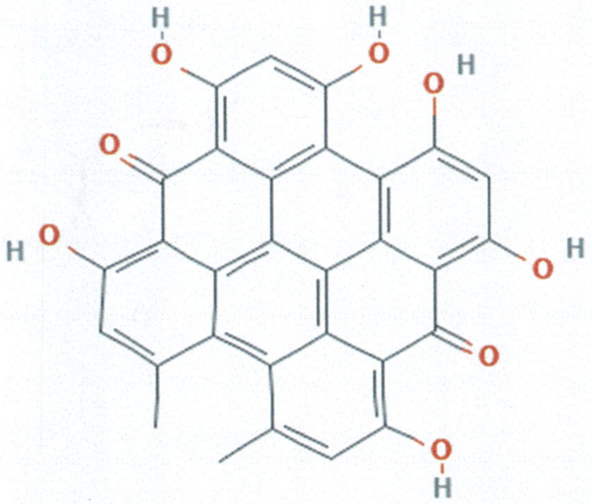

8.6 Other Metabolites

Chaetomium species are potential degraders of cellulosic and other organic material and can be antagonistic against various soil microbiota. *Chaetomium globosum* and *C. cochlioides* are antagonistic to species of *Fusarium* and *Helminthosporium* (Tveit and Moore 1954). It has been found that by using specific strains of *C. globosum*, it is possible to obtain good control over many plant pathogens. By coating seeds of corn with spores of *Chaetomium globosum*, it was possible to prevent seedling blight caused by *Fusarium roseum* f sp. *cerealis graminearum* (Chang and Kommedahl 1968; Kommedahl et al. 1981). Such seed coating treatments were also found to reduce disease incidence of apple scab caused by *Venturia inequalis* (Heye and Andrews 1983; Cullen and Andrews 1984; Cullen et al. 1984; Boudreau and Andrews 1987). It has also been reported that some isolates of *C. globosum* produce antibiotics that can suppress damping-off of sugar beet caused by *Pythium ultimum* (Di Pietro et al. 1991). A further isolate of *C. globosum* was found to be antagonist against *Rhizoctonia solani* (Wather and Gindrat 1988), *Alternaria brassicicola* (Vannacci and Harman 1987) and also reduced the quantity of sporulation of *Botrytis cinerea* on dead lily leaves exposed in the field (Kohl et al. 1995). One strain of *C. cupreum* has been reported to be antagonistic against *Phomopsis* and *Colletotrichum* spp. which are soybean pathogens (Manandhar et al. 1986).

Screening of *Chaetomium* species and strains isolated from soils for use as potential biological control agents commenced in Thailand in 1989. Reports indicated that strains of C. *cupreum* and *C. globosum* were able to suppress plant pathogens such as *Curvularia lunata, Pyricularia oryzae*, and *Rhizoctonia oryzae* in vitro (Soytong 1989; 1992a). Viable spores of *Chaetomium* spp. were able to reduce tomato wilt caused by *Fusarium oxysporum* f sp. *lycopersici* in greenhouse, field trials (Soytong 1990; 1992b) and prevent basal stem rot of corn caused by *Sclerotium rolfsii* (Soytong 1991).

Twenty-two strains of *C. cupreum* and *C. globosum* have been found to inhibit isolates of various plant pathogens (Soytong and Soytong 1997). These effective strains of *Chaetomium* have been formulated into biopellets and biopowders for the biological control of plant diseases and have been patented and registered under the name Ketomium®. Their efficacy against tomato wilt (*F. oxysporum* f. sp. *lycopersici*) and basal rot of corn (*S. rolfsii*) has been successfully evaluated in laboratory and greenhouse tests (Soytong 1992b).

Three research groups isolated at least 26 compounds from *Chaetomium globosum*, many of which were biologically active (Li et al. 2011; 2014; Xue et al. 2012). They represent a large number of structurally diverse metabolites, such as chaetoglobosins, azaphilones, xanthones, and steroids. These structures display a wide range of biological activities including anticancer, antimicrobial, immunosuppressive, and antioxidant. Different extracts of *C. cupreum* have significant scavenging activity in a dose-dependent manner. Thus, *C. cupreum* can be a new source of natural antioxidants useful for industrial applications (Wani and Tirumale 2018).

Twenty-two strains of *C. cupreum*, *C. globosum*, and *C. cochliodes* were reported to inhibit isolates of various plant pathogens such as *Curvularia lunata*, *Pyricularia oryzae*, *Rhizoctonia oryzae* in vitro and soybean pathogens, *Phomopsis* and *Colletotrichum* spp. (Soytong 2014).

Endophytic fungal species including *Chaetomium* were isolated from inner bark segments of ethno-pharmaceutically important tree species, especially *Azadirachta indica*, grown in different regions of Southern India that exhibited great antifungal, antioxidant, antibacterial, and antihypertensive activities (Tejesvi et al. 2009).

Chaetomium sp. was found to possess high tannin content (5.0 mg/g). The maximum reducing potential of *Chaetomium* sp. may be attributed to the potential of the extracts to act as reductants (Kumaresan et al. 2015).

Diketopiperazines are widely found as mold secondary metabolites for anticancer activity in several species of fungi belonging to the genus *Chaetomium* (Wang et al. 2017).

Chaetomium species also produce an interesting class of secondary metabolites called azaphilones. The most remarkable and valuable properties of azaphilones include their natural origin, yellow-red spectra, thermostability (in comparison with other natural red pigments), and water solubility. The azaphilone compounds are produced by different *Chaetomium* species which display various biological activities such as antioxidant, nematicidal, antimicrobial, antifungal, anticancer, and inflammatory activities (Qin et al. 2009).

Chaetomium globosum No. 04 was isolated from fresh bark and was the source for six cytochalasans, the chaetoglobosins A, C, D, E, G, and R. All of them had in vitro antifungal effects in agar diffusion assays against the phytopathogenic fungi *Rhizopus stolonifer* and *Coniothyrium diplodiella* at 20 μg/disc (Zhang et al. 2013).

Chaetomugilins D (active against brine shrimps) as well as chaetoglobosins A and C (active against *Mucor miehei*) were isolated from *Chaetomium globosum* species (Qin et al. 2009). From the same strain, 12 compounds were isolated (squalene, ergosterol peroxide, methylthiogliotoxin, chaetoglobosin G (Fig. 8.5), (22E,24R)-ergosta-7,22-diene-3β,5α,6β-triol, fumitremorgin C, chaetoglobosin C, gliotoxin (Fig. 8.5), 2,3,4-trimethyl-5,7-dihydroxy-2,3-dihydrobenzofuran, pseurotin A, 4-aminophenylacetic acid, 3,4-dihydroxyphenylacetic acid) and tested against different fungi (*Fusarium oxysporum* f. sp. *vasinfectum*, *Fusarium graminearum*, *Fusarium sulphureum*, *Cercospora sorghi*, *Botrytis cinerea*, and *Alternaria alternata*). Only gliotoxin was active against all of them (Li et al. 2011). Another cytochalasan chaetoglobosin V_b (Fig. 8.5) together with known derivatives chaetoglobosins G and V (Fig. 8.5) were isolated and characterized (Xue et al. 2012). The compound did not show any activity up to 100 μg/mL against bacteria and fungi (*E. coli, S. aureus, B. cereus, B. subtilis, Pseudomonas aeruginosa, Fusarium graminearum, Botrytis cinerea, Glomerella cingulata, Alternaria solani, Fusarium oxysporum* f. sp. *vasinfectum, Fusarium solani, Alternaria alternata, Fusarium oxysporium* f. sp. *niveum*), whereas the derivatives V and G had moderate antifungal and antibacterial effects. Using rice medium, further known metabolites (chaetoglobosins A, G, V, V_b, C, E, F, Fex, Fa, 20-dihydrochaetoglobosin A, fumigaclavine B, 5-(hydroxymethyl)-1H-pyrrole-2-carbaldehyde, 2′-O-methyladenosine, 5′-deoxy-5′-methylamino-adenosine) were isolated from this strain (He Li et al. 2014). The isolated chaetoglobosins were tested against human colorectal HCT116 cells and showed IC_{50} values of 3.15–65.6 μg/mL for chaetoglobosins A, G, V, V_b, F, Fa, Fex, 20-dihydrochaetoglobosin A but chaetoglobosin E had no activity up to 100 μg/mL. Cytochalasans like the chaetoglobosins possess diverse bioactivities (cytotoxicity, antimicrobial and antiparasitic activities) and are isolated from different fungi (Pirttila and Frank 2018).

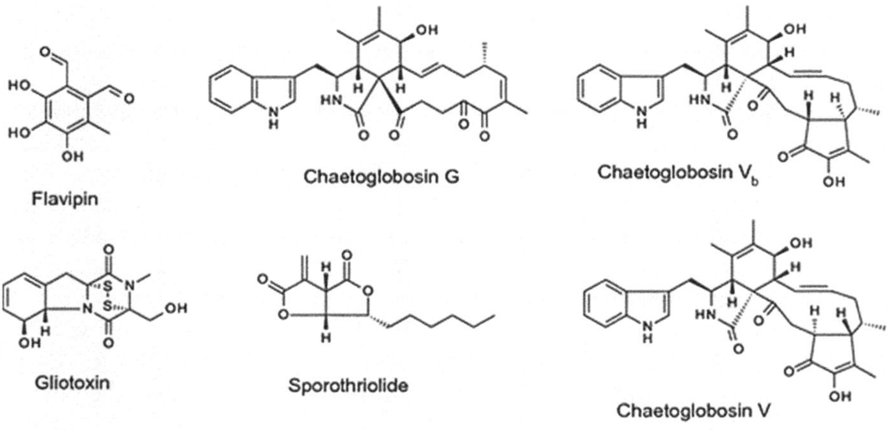

Fig. 8.5 Secondary metabolites isolated from *Ginkgo biloba* endophytes. (Pirttila and Frank 2018)

Furthermore, an endophytic fungus *Chaetomium fusiforme* was obtained from a liverwort, *Scapania verrucosa*. *C. fusiforme* has displayed a wider range of antimicrobial and antitumor activities, which were better than the host plant. These results could support the suggestion of endophytes as an alternative of the host for medicinal activity (Guo et al. 2008).

Since 2001, effective strains of *Chaetomium* have been formulated into biopellets and biopowders for the biological control of plant diseases and have been patented and registered under the name Ketomium®. Their efficacy against tomato wilt (*F. oxysporum* f. sp. *lycopersici*) and prevention against corn basal stem rot caused by *Sclerotium rolfsii* has been successfully evaluated in laboratory, greenhouse, and field trials (Soytong et al. 2001).

Nano-particles derived from *Chaetomium* sp. extracts exhibited significantly antifungal activity against *Magnaporthe* sp. Further investigation is necessary to formulate the nanoparticles from active compounds of *Chaetomium* sp. for plant immunity to be applied in the rice fields (Song and Soytong 2018).

In 2018, the Canadian Minister of the Environment and the Minister of Health have conducted a screening assessment of *Chaetomium globosum* strain ATCC 6205 which is found commonly on moldy building materials. This assessment considers the characteristics of *C. globosum* strain ATCC 6205 with respect to environmental and human health effects associated with consumer and commercial product use and in industrial processes. Based on the information available, it is concluded that *C. globosum* strain ATCC 6205 is not entering the environment in a quantity or concentration or under conditions that have or may have an immediate or long-term harmful effect on the environment or its biological diversity or that constitute or may constitute a danger to the environment on which life depends (Government of Canada 2018).

Upraised data merit further studies regarding extraction, identification, functionality, safety, precise site, and mechanism of action of *Chaetomium* functional metabolites for maximum benefit on safe basis.

8.7 Conclusions

The contribution of *Chaetomium* endophytes due to their outstanding bioactivities led to continued research of endophytically derived chemical moieties. Studies revealed that antioxidant bioactivities are prevalent in these microbial metabolites. Antioxidants are vital additives required in the food industry to maintain the healthy system of the human body, and they are used in the manufacture of processed food and natural food ingredients. Additionally, they are widely used in cosmetics and dermatology also. Apart from natural plant extract, *Chaetomium* endophytes are also the producer of many good antioxidants. Hence, the call for sustainable and eco-friendly antioxidant production is motivating cosmetics and pharmaceutical industries to continue their exploration.

References

Ames BN, Cathcart R, Schwiers E, Hochstein P (1981) Uric acid provides an antioxidant defense in humans against oxidant- and radical-caused aging and cancer: a hypothesis. Proc Nat AcadSci USA 73:6858–6862

Biswas SK, Aggarwal R, Srivastava KD, Gupta S, Dureja P (2012) Characterization of antifungal metabolites of Chaetomium globosum Kunze and their antagonism against fungal plant pathogens. J Biol Control 26(1):70–74

Borges WS, Mancilla G, Guimar-aes DO, Duran-Patron R, Collado IG, Pupo MT (2011) Azaphilones from the endophyte Chaetomium globosum. J Nat Prod 74:1182–1187

Boudreau MA, Andrews JH (1987) Factors influencing antagonism of Chaetomium globosum to Venturia inequalis: a case study in failed biocontrol. Phytopathology 77:1470–1475

Brewer MS (2011) Natural antioxidants: sources, compounds, mechanisms of action, and potential applications. Compr Rev Food Sci Food Saf 10:221–247

Burns J, Gardner PT, Matthews D, Duthie GG, Lean MEJ, Crozier A (2001) Extraction of phenolics and changes in antioxidant activity of red wines during vinification. J Agric Food Chem 49:5797–5808

Chang I, Kommedahl T (1968) Biological control of seedlings of corn by coating kernels with antagonistic microorganisms. Phytopathology 77:1470

Cullen D, Andrews JH (1984) Evidence for the role of antibiosis in the antagonism of Chaetomium globosum to the apple scab pathogen, Venturia inequalis. Can J Bot 62:1819–1823

Cullen D, Berbeem FM, Andrews JH (1984) Chaetomium globosum antagonizes the apple scab pathogen, Venturia inequalis, under field conditions. Can J Bot 62:1814–1818

Di Pietro A, Kung R, Gutrella M, Schwinn FJ (1991) Parameters influencing the efficacy of Chaetomium globosum in controlling Pythium ultimum damping-off of sugar-beet. J Plant Dis Protect 98:565–573

Gamble PE, Burke J (1984) Effect of water stress on the chloroplast antioxidant system. Plant Physiol 76:615–621

Government of Canada (2018) Final screening assessment of Chaetomium globosum strain ATCC 6205. Environment and Climate Change Canada Health Canada, Ottawa

Guo L, Wu J, Han T, Cao T, Rahman K, Qin L (2008) Chemical composition, antifungal and antitumor properties of ether extracts of Scapania verrucosa Heeg. and its endophytic fungus Chaetomium fusiforme. Molecules 13:2114–2125. https://doi.org/10.3390/molecules13092114

Halliwell B (1994) Free radicals, antioxidants, and human disease: curiosity, cause, or consequence? Lancet 344:721–724

Heinig S, Scholz S, Wein J (2013) Getting to the bottom of Taxol biosynthesis by fungi. Fungal Divers 60:161–170

Heye CC, Andrews IH (1983) Antagonism of Athelia bombacina and Chaetomium globosum to the apple scab pathogen, Venturia inequalis. Hytopathology 73:650–654

Hu Y, Hao X, Chen L, Akhberdi O, Yu X, Liu Y, Zhu X (2018) Gα-cAMP/PKA pathway positively regulates pigmentation, chaetoglobosin a biosynthesis and sexual development in Chaetomium globosum. PLoS One 13(4):1–18. https://doi.org/10.1371/journal.pone.0195553

Huang WY, Cai YZ, Xing J, Corke H, Sun MA (2007) Potential antioxidant resource: endophytic fungi from medicinal plants. Econ Bot 61:14–30

Koes RE, Quattrocchio F, Mol JNM (1994) The flavonoid biosynthetic pathway in plants: Function and Evolution. Bioessays 16:123

Kohl JW, Molhoek HL, van der Plas CH, Fokkema HJ (1995) Effect of U/ocladium atrum and other antagonists on sporulation of Botrytis cinerea on dead lily leaves exposed to field condition. Phytopathology 85:393–400

Kommedahl T, Windels CE, Sarbini G, Wiley HB (1981) Variability of performance of biological control and fungicidal seed treatments in corn, peas and soybean. Proc Ecol 3:5

Kumaresan S, Karthi V, Senthilkumar V, Balakumar BS, Stephen A (2015) Biochemical constituents and antioxidant potential of endophytic fungi isolated from the leaves of Azadirachta indica A. Juss (Neem) from Chennai, India. J Acad Ind Res 3(8):355–361
Kusari S, Lamshöft M, Zühlke S, Spiteller M (2008) An endophytic fungus from Hypericum perforatum that produces hypericin. J Nat Prod 71(2):26–29
Landis GN, Tower J (2005) Superoxide dismutase evolution and life span regulation. Mech Ageing Dev 126:365–379
Li Y, Song YC, Liu JY, Ma YM, Tan RX (2005) Anti-helicobacter pylori substances from endophytic fungal cultures. W J Microbiol Biotechnol 21:553–558
Li H, Li X, Wang Y, Zhang Q, Zhang A, Gao J (2011) Antifungal metabolites from Chaetomium globosum, an endophytic fungus in Ginkgo biloba. Biochem Syst Ecol 39(4–6):876–879. https://doi.org/10.1016/j.bse.2011.06.019
Li H, Xiao J, Gao Y, Zhang A, Gao J (2014) Chaetoglobosins from Chaetomium globosum, an endophytic fungus in Ginkgo biloba, and their phytotoxic and cytotoxic activities. J Agric Food Chem 62:3734–3741
Lippman SM, Benner SE, Hong WK (1994) Cancer chemoprevention. J Clin Oncol 12:851–873
Manandhar IB, Thapliyal N, Cavanaugh KJ, Sinclair IB (1986) Interaction between pathogenic and saprobic fungi isolated from soybean roots and seeds. Mycopathologia 98:69–75
Mates JM, Perez-Gomez C, De Castro IN (1999) Antioxidant enzymes and human diseases. Clin Biochem 32:595–603
Nawar WF (1996) Lipids. In: Fennema O (ed) Food Chem, 3rd edn. Marcel Dekker, New York, pp 225–320
Ouyang J, Mao Z, Guo H, Xie Y, Cui Z, Sun JWH, Wen X, Wang J, Shan T (2018) Mollicellins O–R, four new depsidones isolated from the endophytic fungus Chaetomium sp. Eef-10. Molecules 23:3218
Papas AM (ed) (1998) Antioxidant status, diet, nutrition and health. CRC Press, Boca Raton
Pirttila AM, Frank AC (2018) In: Jain SM, Haggman H (eds) Endophytes of Forest Trees: Biology and Application. Vol. 86, 2nd edn. Springer International Publishing AG, part of Springer Nature, Cham. https://doi.org/10.1007/978-94-007-1599-8
Qin JC, Zhang YM, Gao JM, Bai MS, Yang SX, Laatsch H, Zhang AL (2009) Bioactive metabolites produced by Chaetomium globosum, an endophytic fungus isolated from Ginkgo biloba. Bioorg Med Chem Lett 19(6):1572–1574. https://doi.org/10.1016/j.bmcl.2009.02.025
Rababah TM, Ereifej KI, Howard L (2005) Effect of ascorbic acid and dehydration on concentrations of total phenolics, antioxidant capacity, anthocyanins, and color in fruits. J Agric Food Chem 53:4444–4447
Robak J, Gryglewski RJ (1988) Flavonoids are scavengers of superoxide anions. Biochem Pharmacol 37:837–841
Shan B, Cai YZ, Sun M, Corke H (2005) Antioxidant capacity of 26 spice extracts and characterization of their phenolic constituents. J Agric Food Chem 53:7749–7759
Shternshis M, Tomilova O, Shpatova T, Soytong K (2005) Evaluation of Ketomium-mycofungicide on Siberian isolates of phytopathogenic fungi. J Agri Tech 1(2):247–254
Song J, Soytong K (2018) Nano-particles from Chaetomium against rice blast. BioRxiv – The Preprint Server for Biology. https://doi.org/10.1101/339283
Soytong K (1989) Antagonism of Chaetomium cupreum to Pyricularia oryzae: a case study to biocontrol of a rice blast disease. Thai Phytopathol 9:28–33
Soytong K (1990) A taxonomic study of Chaetomium spp. in Thailand. Abstracts, 4th Internatinal Mycological Congress, Regensburg, Germany
Soytong K (1991) Species of Chaetomium in Thailand and screening for their biocontrol properties against plant pathogens. Abstract. The 12th International Plant Protection Congress, Rio de Janeiro, Brazil
Soytong K (1992a) Antagonism of Chaetomium cupreum to Pyricularia oryzae. J Plant Prot Trop 9:17–24

Soytong K (1992b) Biological control of tomato wilt caused by Fusarium oxysporum f sp. lycopersici using Chaetomium cupreum. Kasetsart J (Nat Sci) 26:310–313
Soytong K (2014) Bio-formulation of Chaetomium cochliodes for controlling brown leaf spot of rice. Int J Agri Tech 10(2):321–337
Soytong K, Soytong, Kobboon (1997). Chaetomium as a new broad spectrum mycofungicide. Proceedings of the 1st Internatonal Symposium on Biopesticides. pp. 124–132
Soytong K, Kanokmedhakul S, Kukongviriyapa V, Isobe M (2001) Application of Chaetomium species (Ketomium) as a new broad spectrum biological fungicide for plant disease control: a review article. Fungal Divers 7:1–5
Stocker R, Yamamoto Y, McDonagh AF, Glazer A, Ames BN (1987) Bilirubin is an antioxidant of possible physiological importance. Science 235:1043–1046
Strobel GA, Daisy B (2003) Bioprospecting for microbial endophytes and their natural products. Microbiol Mol Biol Res 67:491–502
Tejesvi MV, Tamhankar SA, ,Kini KR, Rao VS, Prakash HS (2009) Phylogenetic analysis of endophytic Pestalotiopsis species from ethnopharmaceutically important medicinal trees. Fungal Divers 38: 167–183
Tveit M, Moore MB (1954) Isolates of Chaetomium that protect oats from Helminthosporium victoriae. Phytopathology 44:686–689
Valko M, Leibfritz D, Moncol J, Cronin MTD, Mazur M, Telser J (2007) Free radicals and antioxidants in normal physiological functions and human diseases. Int J Biochem Cell Biol 39:44–84
Vannacci G, Harman GE (1987) Biocontrol of seed-borne Alternaria raphani and A. brassicicola. Can J Microbiol 33:850–856
Wang S, Li XM, Teuscher F, Li DL, Diesel A, Ebel R, Proksch P, Wang BG (2006) Chaetopyranin, a benzaldehyde derivative, and other related metabolites from Chaetomium globosum, an endophytic fungus derived from the marine red alga Polysiphonia urceolata. J Nat Prod 69(11):1622–1625
Wang F, Jiang J, Hu S, Ma H, Zhu H, Tong Q, Zhang Y (2017) Secondary metabolites from endophytic fungus Chaetomium sp. induce colon cancer cell apoptotic death. Fitoterapia 121(2016):86–93. https://doi.org/10.1016/j.fitote.2017.06.021
Wani AN, Tirumale S (2018) Evaluation of antioxidant properties of different extracts of Chaetomium cupreum SS02. Bull Fac Pharm Cairo Univ 56(2):191–198. https://doi.org/10.1016/j.bfopcu.2018.08.001
Waterman PG, Mole S (1994) Analysis of phenolic plant metabolites. Alden Press Limited, Oxford, UK
Wather D, Gindrat D (1988) Biological control of damping-off of sugarbeet and cotton with Chaetomium globosum or a fluorescent Pseudomonas sp. Can J Microbiol 34:631–637
Willis LM, Shukitt-Hale B, Joseph JA (2009) Recent advances in berry supplementation and age-related cognitive decline. Curr Clin Nutr Metabol Care 12:91–94
Xue M, Zhang Q, Gao J-M, Li H, Tian J-M, Pescitelli G (2012) Chaetoglobosin V b from endophytic Chaetomium globosum : absolute configuration of chaetoglobosins. Chirality 24:668–674. https://doi.org/10.1002/chir
Yan W, Cao L-L, Zhang Y-Y, Zhao R, Zhao S-S, Khan B, Ye Y-H (2018) New metabolites from endophytic fungus Chaetomium globosum CDW7. Molecules 23(11):2873. https://doi.org/10.3390/molecules23112873
Ye Y, Xiao Y, Ma L, Li H, Xie Z, Wang M, Liu J (2013) Flavipin in Chaetomium globosum CDW7, an endophytic fungus from Ginkgo biloba, contributes to antioxidant activity. Appl Microbiol Biotechnol 97(16):7131–7139. https://doi.org/10.1007/s00253-013-5013-8
Zhang G, ZhangY QJ, Qu X, Liu J, Li X, Pan H (2013) Antifungal metabolites produced by Chaetomium globosum no.04, an endophytic fungus isolated from Ginkgo biloba. Indian J Microbiol 53(2):175–180. https://doi.org/10.1007/s12088-013-0362-7
Zhao J, Shan T, Mou Y, Zhou L (2011) Plant-derived bioactive compounds produced by endophytic fungi. Min Rev Med Chem 11:159–168

Chapter 9
Chaetomium Enzymes and Their Applications

Amira M. G. Darwish and Ahmed M. Abdel-Azeem

9.1 Introduction

The interest in *Chaetomium* enzymes almost has begun when one of the early researchers developed a method for testing the effectiveness of mildew proofing agents on cotton fabrics in which the fungus, *Chaetomium globosum*, is used as the test organism. This particular fungus was selected because it was found on nearly all outdoor fabrics as one of the most important organisms responsible for the loss of breaking strength of fabrics (Thom et al. 1934).

In the 1940s, researchers (Rogers et al. 1940) determined the action of *Chaetomium globosum* on bleached cotton duck by measuring the changes in breaking strength, weight per square yard, thickness, staple length, fluidity, methylene blue absorption, moisture content, ash content, and the rate of evolution of carbon dioxide as an indication of the rate of growth of the organisms on the fabric.

After three decades, the interest switched to studying of the cellulolytic system in more detail, such as the effect of temperature on growth and cellulase production in the thermophilic compost fungus *Chaetomium thermophile* var. *dissitum* (Eriksen and Goksoyr 1977) and the isolation, taxonomy, and growth rate of the genus *Chaetomium Kunze* ex Fr. as a wheat straw decomposer for mushroom growth (Chahal and Hawksworth 1976).

Later, stored useful books and important documents which were noticed to have moldy appearances were analyzed for isolation of cellulolytic fungi. Several fungi were isolated in pure form and identified, and the genus *Chaetomium* was the most dominant fungus. Thirteen different *Chaetomium* species were undertaken for

A. M. G. Darwish (✉)
Food Technology Department, Arid Lands Cultivation Research Institute (ALCRI), City of Scientific Research and Technological Applications (SRTACity), Alexandria, Egypt

A. M. Abdel-Azeem
Botany Department, Faculty of Science, University of Suez Canal, Ismailia, Egypt
e-mail: ahmed_abdelazeem@science.suez.edu.eg

A. M. Abdel-Azeem (ed.), *Recent Developments on Genus Chaetomium*,
Fungal Biology, https://doi.org/10.1007/978-3-030-31612-9_9

screening of their cellulase-producing capability by the filter paper degradation ability (Yadav and Bagool 2015).

Recent investigations were more specific in studying new *Chaetomium* cellulolytic fungal species, enzyme profiles and genotypes of *Chaetomium* isolates, and isolation and screening of other *Chaetomium* enzymes such as L-methioninase, laccase, polysaccharide monooxygenase (PMO), β-1,3-glucanase, dextranase, pectinolytic, lipolytic, amylolytic, proteolytic, and chitinolytic and new classes of cellulose-degrading enzymes and synergistic enzyme systems (Abdel-Azeem et al. 2016; Benhassine et al. 2016; Chen et al. 2018; Coronado-Ruiz et al. 2018; Hamed et al. 2016; Wanmolee et al. 2016).

Based on the abovementioned data that emphasize the enduring interest and importance of fungal enzymes for sustainability, this chapter is focused on presenting various enzymes produced by *Chaetomium* species and their miscellaneous applications.

9.2 *Chaetomium* Cellulase

Lignocellulosic materials consist mainly of three different types of biopolymers organized into a complex structure: (i) cellulose, a linear homopolymer of D-glucose organized into highly crystalline microfibers which are intimately associated with an intricate network of (ii) hemicellulose, an amorphous branched polymer comprising various pentoses, hexoses, and sugar acids, and (iii) lignin, a heteropolymer of phenolic alcohols which shields the polysaccharide microstructure from external physical, chemical, and biological attacks (Feldman 1985).

Cellulose is one of the most abundant renewable carbohydrates on earth. Enzymatic degradation of cellulose to glucose has great potential for economic biofuel production. Cellulases comprise three major groups of enzymes: (1) endoglucanases (EC 3.2.1.4), which attack regions of low crystallinity in cellulose fibers, creating free chain-ends; (2) exo-glucanases or cellobiohydrolases (EC 3.2.1.91) which further degrade the molecule by cleaving cellobiose from the free-chain ends; and (3) β-glucosidases (EC 3.2.1.21) which hydrolyze cellobiose to produce glucose. In addition to the three major groups of cellulases, there are a number of endo- and exo-acting enzymes that attack the heterogeneous hemicelluloses, such as endo-β-1,4-xylanase, β-xylosidase, galactomannanase, glucomannanase, and acetylesterase A (Zhang and Zhang 2013).

The applications of cellulases in various industries are exhibited in Table 9.1 (Kuhad et al. 2011).

Chaetomium is a *saprobic* fungus belonging to Ascomycota with high capability of degrading plant materials; it grows well and decomposes cellulose very rapidly, producing thermostable cellulases (Sajith et al. 2016). Studies have been conducted on various *Chaetomium* species such as *Chaetomium cellulolyticum, C. erraticum, C. fusisporale, C. globosum,* and *C. thermophile* to investigate their

Table 9.1 Applications of cellulases in various industries

Industry	Applications
Agriculture	Plant pathogen and disease control; generation of plant and fungal protoplasts; enhanced seed germination and improved root system; enhanced plant growth and flowering; improved soil quality; reduced dependence on mineral fertilizers
Bioconversion	Conversion of cellulosic materials to ethanol, other solvents, organic acids and single cell protein, and lipids; production of energy-rich animal feed; improved nutritional quality of animal feed; improved ruminant performance; improved feed digestion and absorption; preservation of high quality fodder
Detergents	Cellulase-based detergents; superior cleaning action without damaging fibers; improved color brightness and dirt removal; remove of rough protuberances in cotton fabrics; antiredeposition of ink particles
Fermentation	Improved malting and mashing; improved pressing and color extraction of grapes; improved aroma of wines; improved primary fermentation and quality of beer; improved viscosity and filterability of wort; improved clarification in wine production; improved filtration rate and wine stability
Food	Release of the antioxidants from fruit and vegetable pomace; improvement of yields in starch and protein extraction; improved maceration, pressing, and color extraction of fruits and vegetables; clarification of fruit juices; improved texture and quality of bakery products; improved viscosity of fruit purees; improved texture, flavor, aroma, and volatile properties of fruits and vegetables; controlled bitterness of citrus fruits
Pulp and paper	Co-additive in pulp bleaching; biomechanical pulping; improved draining; enzymatic deinking; reduced energy requirement; reduced chlorine requirement; improved fiber brightness, strength properties, and pulp freeness and cleanliness; improved drainage in paper mills; production of biodegradable cardboard, paper towels, and sanitary paper
Textile	Biostoning of jeans; biopolishing of textile fibers; improved fabrics quality; improved absorbance property of fibers; softening of garments; improved stability of cellulosic fabrics; removal of excess dye from fabrics; restoration of color brightness
Others	Improved carotenoids extraction; improved oxidation and cooler stability of carotenoids; improved olive oil extraction; improved malaxation of olive paste; improved quality of olive oil; reduced risk of biomass waste; production of hybrid molecules; production of designer cellulosomes

Source (Kuhad et al. 2011)

cellulolytic ability, localization, multiplicity, and characteristics of cellulase components.

Soni and Soni (2010) discussed a possible mechanism of regulation of cellulases and their existing polymorphism in *Chaetomium erraticum*. They concluded that *C. erraticum* appears to have the potential for the production of a complete cellulose enzyme complex and therefore can be exploited for hydrolysis of cellulosic waste. The regulation mechanism including catabolite repression and presence of multiple molecular forms of endoglunase and β-glucosidase suggests a level of complexity for the synthesis of cellulases in this fungus and are probably regulated by separate genes.

Alvarez and coauthors (2013) reported a novel cellulase belonging to GH5 family, named as CelE1, retrieved from a sugarcane soil metagenomic library, which is a promising biocatalyst in biofuel production. They provided details about the three-dimensional structure, catalytic properties, and stability of CelE1 that might encourage the use of sugarcane biomass as substrate. This enzyme was shown to be an endo-acting glucanase with high catalytic activity at a broad temperature range and under alkaline conditions that usually cause enzyme inactivation of classical acidic cellulases. Moreover, its crystal structure might confer higher conformational stability in comparison with its psychrophilic orthologs.

Five *Chaetomium* species were reported as potentially able to secrete high exoglucanase and endoglucanase cellulases, *Chaetomium dolichotrichum*, *C. funiculosum*, *C. globosum*, *C. anguistispirale*, and *Chaetomium* sp. which were found to perform as very good producers of total cellulase and endogluccanase (Yadav and Bagool 2015).

Wanmolee and coautors (2016) provided an approach for developing an active synergistic enzyme system of the soft-rot fungus *C. globosum* (BCC5776) applied for hydrolysis of alkaline-pretreated rice straw for lignocellulose saccharification and modification in feasible bio-industries. Moreover, the crude enzyme was characterized for its catalytic activities and its components identified by proteomics.

Recently, rampant biodeterioration by fungi was marked in the archive of the Universidad de Costa Rica which maintained a nineteenth-century French collection of drawings and lithographs especially in the nutritional conditions that encouraged their growth. Given the interest in the developing methods for protecting and preserving ancient documents from microbial degraders and the importance of obtaining microorganisms or enzymes with the capacity to degrade ligno-cellulosic wastes, Coronado-Ruiz and coauthors (2018) isolated and identified cellulose degrader fungi belonging to 19 fungal genera from this archive including *Chaetomium,* in addition to two new species, namely, *Periconia epilithographicola* sp. *nov*. and *Coniochaeta cipronana* sp. *nov*., that showed important cellulolytic activity.

9.3 Polysaccharide Monooxygenase (PMO)

Recently, a new class of cellulose-degrading enzymes, called Cu2 + −dependent lytic polysaccharide monooxygenases (PMOs), have been discovered and are attracting increasing interest because their oxidative degradation of cellulose dramatically boosts cellulase activity in cellulose hydrolysis. Based on their amino acid sequence similarities, PMO enzymes are classified into four families: auxiliary activity 9 (AA9), auxiliary activity 10 (AA10), auxiliary activity 11 (AA11), and auxiliary activity 13 (AA13). Additionally, genome analysis and transcriptomic studies revealed that cellobiose dehydrogenase (CDH) is a biological redox partner that is almost always co-expressed with PMOs in filamentous fungi (Corrêa et al. 2016; Vu and Ngo 2018).

Chaetomium thermophilum is a thermophilic fungus and can grow in temperatures up to 60 °C. Chen and coauthors isolated and characterized polysaccharide monooxygenase (PMO) from *C. thermophilum* isolated from China. *C. thermophilum* genome analysis reveals 19 genes encoding putative AA9 PMOs (http://www.fungalgenomics.cn). In addition they amplified a gene (KC441882) encoding a special putative AA9 PMO protein, CtPMO1. The CtPMO1 protein was predicted to be a secreted enzyme with a 17-amino acid potential signal peptide. The mature CtPMO1 protein is composed of 229 amino acids with a calculated molecular weight of 24.63 kDa (Chen et al. 2018).

9.4 *Chaetomium* *β*-1,3-Glucanase

β-1,3-glucanases belong to enzymes hydrolyzing the O-glycosidic linkages of 1,3-b-D glucan which are widely distributed in nature and isolated from many kinds of organisms. Increasing interest in enzymes from thermophilic fungi was expected to produce thermostable enzymes. Li and co-authors' success in purification and partial characterization of thermostable extracellular exo-*β*-1,3-glucanase from *Chaetomium thermophilum* may represent an important progression industrial-wise (Li et al. 2007).

In feed industry, thermostable *β*-1,3-glucanases are used for more complete utilization of feed components originating from plants reducing the antinutritional effect of non-starch polysaccharides NSP (Mourão et al. 2006).

The utilization of thermostable b-1,3-glucanases in malting and brewing processes replacing native barley b-glucanase is susceptible to irreversible inactivation at temperatures above 55 °C. During malting, b-glucans in barley are hydrolyzed by b-glucanase which leads to clarification of beer improvement. Genetic transformation of barley with thermostable b-glucanase genes has been attempted in order to ensure sufficient b-glucanase activity during mashing (Nuutila et al. 1999).

Thermostable b-1,3-glucanases can also be utilized in production processes of yeast extract and soluble 1,3-glucan, which is a potential immune-activator, in addition to its applications in biological control of plant pathogens and medical fields (Li et al. 2007).

9.5 *Chaetomium* Dextranase

Dextranase is an important industrial enzyme which has many applications, in the sugar industry in rendering the technical processing of alternated sugar beet, the preparation of clinical low molecular weight dextran, and the coupling of dextranase to tumor cell-specific antibodies followed by administration of cytotoxic dextran-conjugate (Erhardt and Jördening 2007). Dextranase is currently commercially produced from *Chaetomium erraticum* by Sigma-Aldrich®; its synonym is 1,6-α-D-glucan 6-glucanohydrolase, Dextranase Plus L.

For successful enzymatic synthesis of sucrose conversion, two enzymes must be present, dextransucrase for the conversion of sucrose to dextran and dextranase for dextran breakdown as the dextran hydrolysis generates about 20% of unwanted branched isomaltooligosaccharides (IMOs). Erhardt and Jördening (2007) applied dextranase from a submerged culture of *Chaetomium erraticum* strain in co-immobilization of dextran sucrase and dextranase. The immobilization of technical operation ought to be performed by entrapped dextranase immobilized in alginate, so the particle size of dextranase must fall far below the size of alginate beads in order to be entrapped within the alginate beads, which was achieved by *Chaetomium* dextranase.

9.6 *Chaetomium* Laccase

Laccase (benzenediol:oxygen oxidoreductase, EC 1.10.3.2) is a part of a broad group of enzymes called polyphenol oxidases containing copper atoms in the catalytic center; they are usually called multicopper oxidases, which are able to oxidize a variety of organic and inorganic compounds, including mono-, di-, and polyphenols, aromatic amines, carboxylic acids, and non-phenolic substrates (Khushal et al. 2010).

Fungal laccases play an important role in plant pathogenesis, pigment production, and degradation of lignocellulosic materials (Shraddha et al. 2011). Fungal laccases were isolated and characterized from *Chaetomium* species (Ref 051A) isolated from Chettaba Forest, Constantine, Algeria. Isolated laccase proved to be suitable for industrial production (Benhassine et al. 2016). In Egypt, Abdel-Azeem and Salem (2012) studied laccases from *Chaetomium globosum*.

9.7 *Chaetomium* L-Methioninase

L-Methioninase (E.C 4.4.1.11) is a pyridoxal phosphate-dependent enzyme; it fulfills several functions of an enzyme system because it stimulates the γ- and α, β-removal reactions of methionine and its derivatives. Physiologically, normal cells have the capability to grow on homocysteine, instead of methionine, due to their efficient methionine synthase. Unlike normal cells, tumor cells freed from efficient methionine synthase thus rely on external methionine supplementation from diet. Hence, there is reasonable scope of using L-methioninase as a therapeutic agent against different kinds of methionine-dependent tumors. Methionine reduction has a broad spectrum of antitumor activities (Sharma et al. 2014; Suganya et al. 2017).

A few studies were concentrated on the enzymes from eukaryotes especially fungi, due to the recurrent classification of L-methioninase as extracellular enzyme in the fungal extract; fungi can be considered as robust resources of this enzyme. Hamed and coauthors (2016) identified and produced L-methioninase from *Chaetomium globosum* isolated from Egyptian soil (gene bank under accession number KXO24450). The predicted specific activity of the produced L-methioninase was (≈2225 U/mg); it has been suggested as a good source for clinical therapeutic applications.

9.8 Other *Chaetomium* Enzymes

Abdel-Azeem and coauthors (2016) recovered ten *Chaetomium globosum* isolates, designated from TUCg1 to TUCg10 in the GenBank. All *Chaetomium globosum* isolates showed amylolytic, cellulolytic, and proteolytic activities; six isolates were chitinolytic and laccase producers, five were pectinolytic, and three isolates showed lipolytic activities with different arrays. Applied molecular techniques such as internal transcribed spacer (ITS) region sequencing and specific gene random primer polymerase chain reaction (SGRP-PCR) have shown high DNA polymorphism of *Chaetomium globosum*, so the authors suggested these techniques for identification of different fungal isolates.

9.9 Conclusion

Studies revealed that *Chaetomium* enzymes such as cellulases, L-methioninase, laccase, polysaccharide monooxygenase (PMO), β-1,3-glucanase, dextranase, pectinolytic, lipolytic, amylolytic, proteolytic, and chitinolytic represent a significant factor in sustainability in various fields. Future research in this era is crucial for a more thorough understanding of the mechanisms, structure, function, and substrate characteristics targeting efficient enzyme production to contribute in the development of a feasible bio-refinery industry.

References

Abdel-Azeem AM, Salem FM (2012) Biodiversity of laccase producing fungi in Egypt. Mycosphere 3(5):900–920

Abdel-Azeem AM, Gherbawy YA, Sabry AM (2016) Enzyme profiles and genotyping of Chaetomium globosum isolates from various substrates. Plant Biosyst 150(3):420–428. https://doi.org/10.1080/11263504.2014.984791

Alvarez TM, Paiva JH, Ruiz DM, Cairo JPLF, Pereira IO, Paixão DAA et al (2013) Structure and function of a novel cellulase 5 from sugarcane soil metagenome. PLoS One 8(12):1–9. https://doi.org/10.1371/journal.pone.0083635

Benhassine S, Kacem CN, Destain J (2016) Production of laccase without inducer by Chaetomium species isolated from Chettaba forest situated in the east of Algeria. Afr J Biotechnol 15(7):207–213. https://doi.org/10.5897/AJB2015.15001

Chahal DS, Hawksworth DL (1976) Chaetomium Cellulolyticum, a new thermotolerent and cellulolytic Chaetomium. I. Isolation, description and growth rate. Mycologia 68(3):600–610. https://doi.org/10.1103/PhysRevB.46.4681

Chen C, Chen J, Geng Z, Wang M, Liu N, Li D (2018) Regioselectivity of oxidation by a polysaccharide monooxygenase from Chaetomium thermophilum. Biotechnol Biofuels 11(1):1–16. https://doi.org/10.1186/s13068-018-1156-2

Coronado-Ruiz C, Avendaño R, Escudero-Leyva E, Conejo-Barboza G, Chaverri P, Chavarría M (2018) Two new cellulolytic fungal species isolated from a 19th-century art collection. Sci Rep 8(1):1–9. https://doi.org/10.1038/s41598-018-24934-7

Corrêa TLR, dos Santos LV, Pereira GAG (2016) AA9 and AA10: from enigmatic to essential enzymes. Appl Microbiol Biotechnol 100(1):9–16. https://doi.org/10.1007/s00253-015-7040-0

Erhardt FA, Jördening HJ (2007) Immobilization of dextranase from Chaetomium erraticum. J Biotechnol 131(4):440–447. https://doi.org/10.1016/j.jbiotec.2007.07.946

Eriksen J, Goksoyr J (1977) Cellulases from *Chaetomium thermophile* var. *dissitum*. Eur J Biochem 77:445–450

Feldman D (1985) Wood—chemistry, ultrastructure, reactions. J Polym Sci 23:601–602. https://doi.org/10.1038/icb.2010.85

Hamed SR, Abo Elsoud MM, Mahmoud MG, Asker MMS (2016) Isolation, screening and statistical optimizing of L-methioninase production by Chaetomium globosum. Afr J Microbiol Res 10(36):1513–1523. https://doi.org/10.5897/AJMR2016.8132

Khushal B, Rigdon A, Vadlani PV (2010) Fungal laccases: production, function, and applications in food processing. Enzyme Res 2010(September 2010):1. https://doi.org/10.4061/2010/149748

Kuhad RC, Gupta R, Singh A (2011) Microbial cellulases and their industrial applications. Enzyme Res 2011(1):1–10. https://doi.org/10.4061/2011/280696

Li H, Chen J, Li A, Li DC (2007) Purification and partial characterization of β-1,3-glucanase from Chaetomium thermophilum. World J Microbiol Biotechnol 23(9):1297–1303. https://doi.org/10.1007/s11274-007-9366-y

Mourão JL, Ponte PIP, Prates JAM, Centeno MSJ, Ferreira LMA, Soares MAC, Fontes CMGA (2006) Use of β-glucanases and β-1,4-xylanases to supplement diets containing alfalfa and rye for laying hens: effects on bird performance and egg quality. J Appl Poult Res 15(2):256–265. https://doi.org/10.1093/japr/15.2.256

Nuutila AM, Ritala A, Skadsen RW, Mannonen L, Kauppinen V (1999) Expression of fungal thermotolerant endo-1,4-β-glucanase in transgenic barley seeds during germination. Plant Mol Biol 41(6):777–783. https://doi.org/10.1023/A:1006318206471

Rogers RE, Wheeler HG, Humfeld H (1940) Physical and cherrlical changes produced in bleached cotrfon duck by Chaetomium globosum and Spirochaeta cytophaga. Technical Bulleten, United States Department of Agriculture, Washington, DC, No.726

Sajith S, Priji P, Sreedevi S, Benjamin S (2016) An overview on fungal cellulases with an industrial perspective. J Nutr Food Sci 06(01):1–13. https://doi.org/10.4172/2155-9600.1000461

Sharma B, Singh S, Kanwar SS (2014) L-Methionase: a therapeutic enzyme to treat malignancies. Biomed Res Int 2014:1–13. https://doi.org/10.1155/2014/506287

Shraddha SR, Sehgal S, Kamthania M, Kumar A (2011) Laccase: microbial sources, production, purification, and potential biotechnological applications. Enzyme Res 2011:1–11. https://doi.org/10.1111/1467-9809.12306

Soni SK, Soni R (2010) Regulation of cellulase synthesis in Chaetomium erraticum. Bioresources 5(1):81–98. https://doi.org/10.1128/JB.187.1.99-106.2005

Suganya K, Govindan K, Prabha P, Murugan M (2017) An extensive review on L-methioninase and its potential applications. Biocatal Agric Biotechnol 12(November 2018):104–115. https://doi.org/10.1016/j.bcab.2017.09.009

Thom C, Humfeld H, Holman HP (1934) Laboratory tests for mildew resistance of outdoor cotton fabrics. Amer Dyestuff Rptr Illus 23(581):586

Vu VV, Ngo ST (2018) Copper active site in polysaccharide monooxygenases. Coord Chem Rev 368:134–157. https://doi.org/10.1016/j.ccr.2018.04.005

Wanmolee W, Sornlake W, Rattanaphan N, Suwannarangsee S, Laosiripojana N, Champreda V (2016) Biochemical characterization and synergism of cellulolytic enzyme system from Chaetomium globosum on rice straw saccharification. BMC Biotechnol 16(1):1–12. https://doi.org/10.1186/s12896-016-0312-7

Yadav LS, Bagool RG (2015) Original research article isolation and screening of cellulolytic Chaetomium sp. from deteriorated paper samples. Int J Curr Microbiol Appl Sci 4(8):629–635

Zhang X-Z, Zhang YP (2013) Cellulases: characteristics, sources, production, and applications. In: Bioprocessing technologies in biorefinery for sustainable production of fuels, chemicals, and polymers. Wiley, pp 131–146. https://doi.org/10.1002/9781118642047

Chapter 10
The Use of *Chaetomium* Taxa as Biocontrol Agents

Adel K. Madbouly and Marwa T. A. Abdel-Wareth

10.1 Introduction

Crop plants suffer from many diseases caused by pathogenic fungi, bacteria and viruses both in the field and during storage. These diseases cause significant decrease in crops yields leading consequently to great economic losses. For a long period ago, farmers tend to use chemical control through synthetic pesticides to control these phytopathogens; however, these chemicals have proved to be harmful to human health and poultry, in addition to being non-ecofriendly. Recent trends nowadays focus on new control strategies including biological control using bioagents, and control through the use of nanopesticides. These methods are safe, easily applicable, economic, and ecofriendly. Biological control is an effecient alternative way to control plant diseases (Zhang et al. 2018).

The fungal bioagents exert several antagonistic potentials against phytopathogens directly through mycoparasitism, antibiosis, competition for nutrients and space, and indirectly through inducing plant resistance (Vujanovic and Goh 2011). In addition, they withstand unfavourable environmental conditions. According to Saldajeno and Hyakumachi (2011), bioagents were normally applied singly against most plant pathogens; however, recently several works have emphasized on using more than one bioagent simultaneously to be more effective and a more reliable way of biocontrol, as they will have multiple modes of action at different sites of the host plant.

Chaetomium is one of the largest genera in Chaetomiaceae family (Fatima et al. 2016) which includes about 100 species. *Chaetomium* spp. exist mainly in soil,

A. K. Madbouly (✉)
Microbiology Department, Faculty of Science, University of Ain Shams, Cairo, Egypt

M. T. A. Abdel-Wareth
Environmental Research and Medical Malacology Department,
Theodor Bilharz Research Institute, Giza, Egypt

A. M. Abdel-Azeem (ed.), *Recent Developments on Genus Chaetomium*,
Fungal Biology, https://doi.org/10.1007/978-3-030-31612-9_10

animals, and as endophytes in several plants through forming a symbiotic relationship without causing any harm to their hosts. Chaetomin mycotoxin was recorded for the first time from *Chaetomium* spp. in 1944 (Geiger et al. 1944); however, wide variety of bioactive compounds were reported later in *Chaetomium* spp. such as; orsellides (Schlörke et al. 2006), xanthenone (Pontius et al. 2008), chaetoglobins (Ge et al. 2008), polyhydroxylated steroids (Qin et al. 2009a), azaphilones (Yamada et al. 2011), and armochaetoglobins A–J (Chen et al. 2015). Some of these compounds presented significant biological activities including; antifungal, anti-inflammatory, cytotoxic, and enzyme inhibition (Li et al. 2016). Knowledge of the antagonistic potential and modes of action of some of the biocontrol agents such as *Chaetomium* spp. is highly recommended to be more effective against target pathogens in several ways.

Several strains of *Chaetomium* spp. showed different in vitro and in vivo antifungal potencies against many phytopathogens (Vitale et al. 2012) including; *Helminthosporium, Fusarium, Alternaria raphani, A. brassicicola, Phytophthora* and *Pythium ultimum* (Soytong et al. 2001). Later, Tomilova and Shternshis (2006) added that *Chaetomium* spp. have been used successfully to control damping off disease of sugar beet, root rot of citrus and black pepper. However, more than one factor should be considered before manipulation of *Chaetomium* spp. in the field such as the temperature of soil, moisture level and pH, because these factors affect their antagonistic potencies and effectiveness.

Chaetomium globosum is one of the most common species growing as a saprophyte in the rhizosphere, phyllosphere of plants and as a normal colonizer of the soil. This species has been applied successfully in vivo to control many epidemics of fruits such as; root rots in tangerine (*Citrus reticulata* Blanco) incited by *P. parasitica*, in strawberry (*Fragaria* spp.) caused by *P. cactorum*, in black pepper (*Piper nigram* L.) incited by *P. palmivora*, basal rot of corn caused by *Sclerotium rolfsii*, and tomato wilt (*Lycopersicon esculentum* L.) incited by *Fusarium oxysporum* f. sp. *lycopersici* (Zhang et al. 2010).

Recently, Dosen et al. (2017) pointed that *C. globosum* is able to produce several species-specific metabolites such as cochliodones and chaetoglobosins. Little is known about their human toxicity, although Udagawa et al. (1979) recorded the intravenous toxicity of chaetoglobosins A and C in animal tests.

10.2 *Chaetomium* spp. as Potent Producers of Antifungal Metabolites

Currently, more than 200 active metabolites are produced by *Chaetomium* spp., and some of these metabolites possess considerable biological activities including cytotoxicity, enzyme inhibition, and antibiosis (Yang et al. 2011a, b). As being an effective bioagent of several phytopathogens, *C. globosum* produces several types of bioactive compounds such as orsellides and globosumones A–C (Bashyal et al. 2005), steroids and chaetoviridins A and C (Qin et al. 2009b), cytoglobosins A–G

(Cui et al. 2010), chaetoglobosins (Zhang et al. 2010), in addition to pyrones and chaetoglocins A and B (Ge et al. 2011).

Previously, Park et al. (2005) succeeded in purifying two antifungal compounds from the broth culture of *C. globosum* strain F0142 which was isolated from barnyard grass, and showed potent disease control efficacy against the rice blast (*Magnaporthe grisea*) and the wheat leaf rust (*Puccinia recondita*). These antifungal metabolites were identified as chaetoviridins A and B (Fig. 10.1). Treatment with 62.5 μg/mL of chaetoviridin A suppressed the development of rice blast and wheat leaf rust by over 80%. Furthermore, moderate control of tomato late blight (50%) was observed following the application of 125 μg/mL of chaetoviridin A, while using 250 μg/mL caused higher control levels of about 87%.

In a previous work, Soytong et al. (2005) reported that several *Chaetomium* spp. such as *C. cupreum* and *C. globosum* produce resistance inducing substances which prevent many plant diseases including *Pythium aphanidermatum* disease in sugarcane, wilt symptoms in seedlings of grains, apple scab incited by *Venturia* spp., reduce tomato *Fusarium* wilt, and inhibit the growth of the pathogenic *R. solani* and *Botrytis* spp. In another study of Kaewchai and Soytong (2010), *C. cupreum* RY202 crude extracts (hexane, ethyl acetate and methanol) inhibited the growth of *Rigidoporus microporus*; the causal agent of white root disease of rubber with ED_{50} values of 170, 402 and 1220 μg/L, respectively. In addition, rotiorinol which is a bioactive compound produced by *C. cupreum*, inhibited the growth of the same pathogen with ED_{50} value of 26 μg/L.

In a later study, Sibounnavong et al. (2011) investigated the antifungal metabolites produced by *C. elatum* strain ChE01 and *C. lucknowense* strain CLT01 isolated from the soil in Thailand, and effectively controlled *F. oxysporum* f. sp. *lycopersici* NKSC02 the causal agent of wilt disease of tomato (*Lycopersicon esculentum* var *sida*). One of these antifungal substances called chaetoglobosin-C (Fig. 10.2) purified from both *Chaetomium* spp. exhibited pronounced antifungal potential against *F. oxysporum* f. sp. *lycopersici*, with ED_{50} of 5.98 μg/mL. In addition, inoculation of tomato seedlings with conidia of *F. oxysporum* f. sp. *lycopersici* mixed with a solution of chaetoglobosin-C did not exhibit wilt symptoms. Moreover, conidia of the

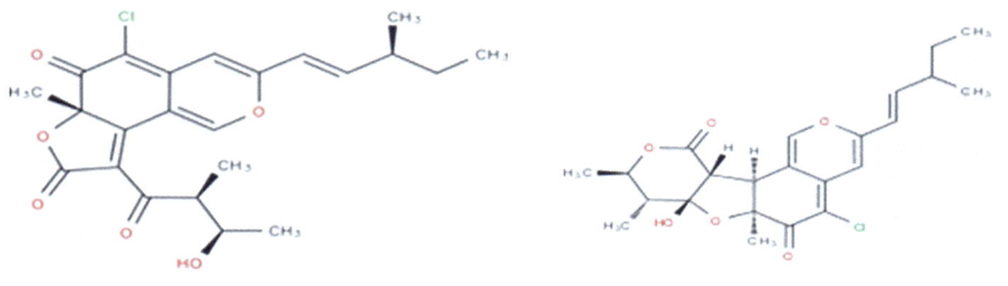

(A) (B)

Fig. 10.1 Chemical structures of chaetoviridins A and B

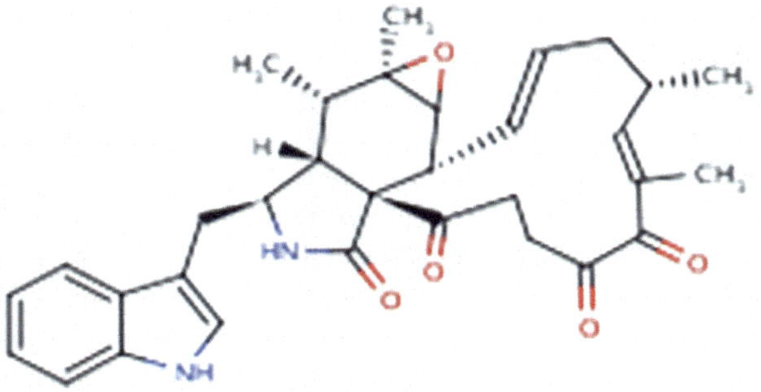

Fig. 10.2 Chemical structures of chaetoglobosin-C

same pathogen treated with this antifungal metabolite appeared abnormal and lost their pathogenicity.

Secondary metabolites from the culture filtrate of *C. globosum* Kunze were extracted by solvent extraction method using ethyl acetate, and then separated by Thin-layer chromatography (Biswas et al. 2012). GS-MS and NMR studies revealed that these metabolites were chetomin, BHT, mollicelin G, besides chaetoglobosin. In vitro bioassays with chaetoglobosin and chaetomin (Fig. 10.3) revealed their significant growth inhibitory activities against several phytopathogens including; *Bipolaris sorokiniana*, *Macrophomina phaseolina*, *R. solani,* and *Pythium ultimum*. The authors added that exposure of *F. graminearum* to 1000 ppm of chaetoglobosin decreased its colony diameter to 1.6 mm, compared with 33 mm of the non-treated control. Meanwhile, treatment of *Bipolaris sorokiniana* with 1000 ppm of chaetomin reduced its colony diameter to 13 mm compared with 69 mm of the control.

Generally, leaves carry more number of endophytic fungi compared with the stem, as a result of their larger surface area exposed to the outer environment, in addition to the presence of numerous stomata that facilitate entry of fungal hyphae. These fungi are considered as novel sources of bioactive compounds which have many medicinal and agricultural applications (Gond et al. 2012).

During production of the bioactive metabolites on a large scale in the fermentation process, the initial pH of the media is a critical factor because it affects several fungal physiological processes such as; growth, proliferation, enzyme production and stability, in addition to the efficacy of nutrient utilization (Taragano and Pilosof 1999). Recently, results of Pan et al. (2016) recorded that the initial pH of medium of about 6.9 is the most suitable for the production of antifungal compounds, and resulted in significant inhibition of pathogens growth by about 81.69%. This percentage increased significantly after 3 to 4 days of the fermentation time, and further increase from 4 to 8 days was non-significant. However, increasing the fermentation time more than 8 days was accompanied with a noticeable decrease in the inhibition percentage. This could be interpreted as production of the secondary metabolites such as the antifungal antibiotics begins near the onset of the fungal stationary growth phase where the growth rate is negligible, but further incubation

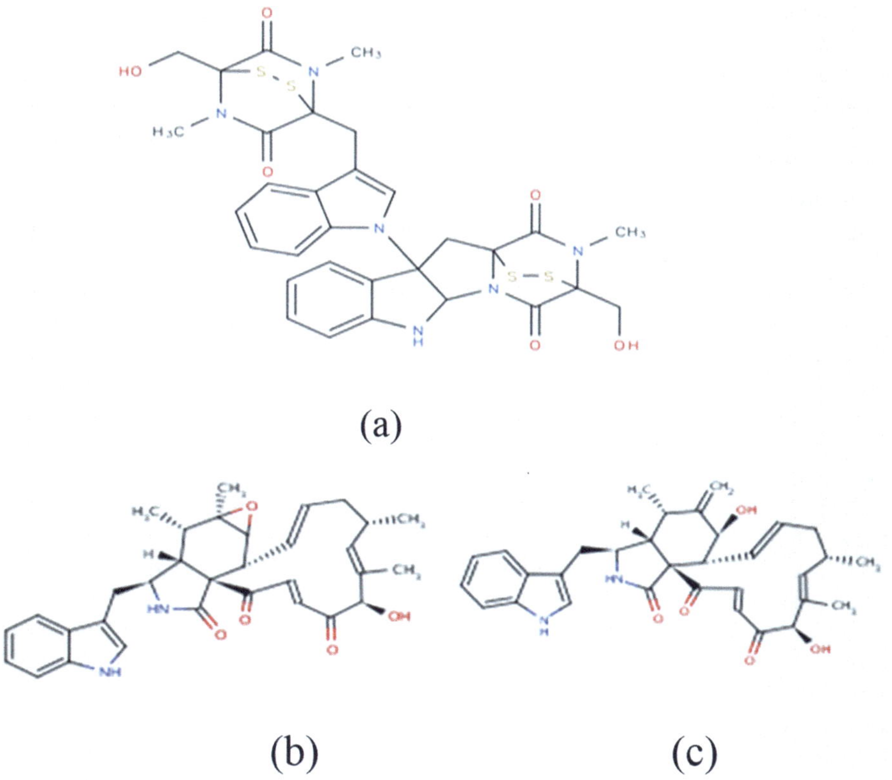

Fig. 10.3 Chemical structures of chaetomin (**a**), chaetoglobosin A (**b**) and chaetoglobosin D (**c**)

causes fungal cell burst or cell lysis. Temperature treatments during solvent extraction of the fungal metabolites are also critical, as several metabolites are temperature-sensitive and may have different levels of solubility at different extraction temperatures. Zhang et al. (2012) previously reported that the antimicrobial potential of the different fungal extracts treated at 80 °C or more may decline sharply as a result of the thermal degradation.

10.3 Endophytic *Chaetomium* spp. as Bioagents

Endophytes are microorganisms which live inside tissues of their host plants without causing them any harm (Petrini 1991). They play major roles in making plants healthy through providing them with resistance against different biotic and abiotic stresses (Gond et al. 2010), promoting their growth (Barka et al. 2002), protecting them from herbivores (Carroll 1988), and from many other pests (Vega et al. 2008). This is a type of mutualistic relationship as they inhabit plant tissues to get nutrients and protection from them, whereas in turn they benefit their hosts by producing several active metabolites that may enhance plants fitness and provide resistance against different stresses. The mechanisms of biocontrol of the different phytopathogens by the endophytic fungi mainly include; antibiosis (Kumar and Kaushik 2013), competition for nutrients and space (Narisawa et al. 2004), mycoparasitism (Gao et al. 2005), and induction of defence response in plants (Varma et al. 2012). Endophytes that showed effective biocontrol potency received much attention as alternatives to using the health hazard and non-ecofriendly synthetic pesticides during plant disease control (Porras-Alfaro and Bayman 2011).

C. globosum has been recorded as an endophyte of many plants and as a saprophyte in the soil (Lan et al. 2011). Earlier studies of Kay and Stewart (1994) and Pereira and Dhingra (1997) demonstrated that this species has the ability to suppress several soilborne pathogens including *Sclerotium cepivorum, Pythium ultimum,* and *Diapothe phaseolorum* f. sp. *meridionalis*. Gliotoxin produced by the endophytic *C. globosum* showed good antifungal potential against the wheat mycopathogen (*F. graminearum*) (Li et al. 2011). Moreover, a novel cytotoxin named chaetomugilin was obtained from *C. globosum* isolates originating from *Ginkgo biloba* (Qin et al. 2009b). Extract of the endophytic *C. globosum* EF18 that originated from *Withania somnifera* was recorded as a highly effective biocontrol agent against *Sclerotinia sclerotiorum*.

As a bioagent, the endophytic *C. aureum* HP047 significantly reduced the Pitch canker disease on *Pinus radiata* seedlings caused by *F. circinatum* (Wang et al. 2013). In addition, *C. aureum* also effectively controlled the pathogenic *Magnaporthe grisea* causing rice blast disease in vitro and in vivo, and the sheath blight disease incited by *R. solani*. Recently, more than 150 endophytic isolates were tested against *F. circinatum* in an in vitro antagonism assays. The six isolates that exhibited the most promising results were inoculated in the field together with the pathogen into seedlings of *P. radiata, P. sylvestris, P. pinaster, P. nigra* and *P. pinea*, to test their efficacy to reduce the damage caused by *F. circinatum*. Both of the endophytic *C. aureum* and *Alternaria* sp. reduced the area under disease progress curve (AUDPC) for the *P. radiata* seedlings, thus they are considered as suitable biocontrol agents of the Pitch canker disease (Martínez-Álvarez et al. 2016).

In the study of Thiep and Soytong (2015) carried out in Vietnam, three effective *Chaetomium* spp. namely, *C. cochliodes, C. bostrychodes,* and *C. gracile* were isolated. The authors have also isolated *F. roseum* (the pathogen causing coffee and tea wilts), and *Colletotrichum gloeosporioides* (the pathogen causing leaf anthracnose of coffee). In the dual culture assay, the three bioagents inhibited the colony growth and spore production of *F. roseum*. Moreover, the hexane, ethyl acetate and methanol extracts of *C. cochlioides* suppressed the spore production of *F. roseum* and *Colletotrichum gloeosporioides* by 60.87%, 78.16%; 74.57%, 74.76%; 76.50%, 67.63%, respectively. In another study, three *Chaetomium* spp. antagonistic to *F. oxysporum* (the causal agent of wilt and root-rot disease of tea in Vietnam) were isolated and identified as *C. cupreum* CC3003, *C. globosum* CG05, and *C. lucknowense* CL01 (Phong et al. 2016). These *Chaetomium* spp. expressed moderate in vitro inhibition of the pathogen mycelial growth by 31.69–34.03%, and reduced its conidial production by 67.25–75.92% after 30 days of incubation. The methanol extract of *C. cupreum* CC3003 was effective in inhibiting germination of conidia of the same pathogen with ED_{50} dose of 85.30 µg/mL. However, the hexane extract of *C. globosum* CG05, and the ethyl acetate extract of *C. lucknowense* CL01 had ED_{50} of 49.32 and 62.17 µg/mL, respectively.

The crude ethyl acetate (EtOAc) extract of the endophytic *C. globosum* (strain 28) isolated from *Houttuynia cordata* Thunb; a famous traditional Chinese medicine, was tested for its in vitro antifungal potential against *Botrytis cinerea* persoon and *B. cinerea* Pers. ex Fr. (Pan et al. 2016). Results demonstrated the high antifungal efficacy of this extract against both pathogens. However, optimal antifungal potency was recorded on using potato dextrose broth (PDB) as a basal culture medium, and on fermentation for 4–8 days (initial pH, 7.5), followed by extraction with EtOAc.

During the previous study of Fierro-Cruz et al. (2017), 355 different fungal endophytes were isolated from *Protium heptaphyllum* and *Trattinnickia rhoifolia* (Burseraceae) tree species, known of producing secondary metabolites of agronomic and industrial significance. The antifungal potential of these endophytes was tested using the in vitro dual culture assay against the pathogenic *F. oxysporum*. In addition, their ethyl acetate (EtOAc) extracts were also tested against conidial suspension of *F. oxysporum* by direct bioautography. Two of these extracts derived from *C. globosum* F211 UMNG and *Meyerozima* sp. F281 UMNG exhibited appreciable inhibitory activities against growth of *F. oxysporum*. The *C. globosum* F211 UMNG extract was subjected to chemical analysis by RP-HPLC-DAD-ESI-MS, and then the potent antifungal compounds such as cladosporin, chaetoatrosin A, and chaetoviridin A were identified. From the genetic point of view, Gophna et al. (2003) previously referred these results to the horizontal gene transfer between the plants and the endophytic fungi. Consequently, these fungi were able to synthesize several bioactive metabolites similar to those synthesized by their host plants.

10.4 *Chaetomium* spp. as Biocontrol Agents of Crop Diseases

10.4.1 Soilborne Diseases

The production of potato crop has been hindered by several diseases, the most destructive however is the late blight disease caused by *P. infestans* (Mont.) de Bary. Results of Shanthiyaa et al. (2013) showed that different isolates of *Chaetomium* spp. mainly *C. globosum* Cg-6 could reduce the incidence of *P. infestans*, through antagonizing its mycelial growth. This study also demonstrated that *Chaetomium* Cg-6 expressed high in vitro exo- and endo-glucanase potential. It is concluded earlier that the antagonistic *Chaetomium* spp. which are able to secrete higher levels of glucanase and cellulases enzymes would have greater biocontrol potency against *Phytophthora* spp., as their cell walls are composed mainly of glucan and cellulose constituents. Furthermore, *C. globosum* Cg-6 was then formulated as a liquid and applied in the fields of potato crop as tuber, foliar, and soil treatments, either individually or in combination with *P. infestans*. Among these different treatments, the combined application of *C. globosum* with the pathogen resulted in significant decrease in late blight infection to 72%, compared to 100% of the untreated control. Moreover, application of *C. globosum* resulted in higher tuber yield. These findings were in accordance with previous results of Soytong and Ratanacherdchai (2005) and Ahammed et al. (2008). *F. sporotrichioides* is a common soilborne pathogen causing dry rot of potato in North- east of China. The antagonistic *C. globosum* strain W7 significantly inhibited *F. sporotrichioides* in vitro without direct contact, suggesting that the antifungal substance produced was extracellular (Jiang et al. 2017). Metabolites of this bioagent were extracted with acetone and then separated with Thin-layer chromatography, where the active antifungal metabolite was identified as chaetoglobosin A. In vitro assay of the antifungal potency of this metabolite against *F. sporotrichioides* revealed that its minimum inhibitory concentration (MIC) was 9.45–10.50 g/mL, whereas its IC_{50} was 4.344 g/mL. Thus chaetoglobosin A expressed its excellent protective activity of the potato crop against the dry rot symptoms caused by this pathogen. Most trials to biocontrol *P. infestans* used a single method of application of the bioagents. This mainly accounts for the reported ineffectiveness of several biocontrol preparations, as a single method of application will not be effective during the entire growth period of the potato crop. Previous studies of Nandakumar et al. (2001) and Saravanakumar et al. (2007, 2009) emphasized on the application of more than one bioagent simultaneously, as they will be effective in more than one site of the host plant and would have multiple modes of action as well.

Maghazy et al. (2008) isolated *Pythium spinosum* var. *spinosum* the pathogen of clover, and soilborne mycoflora from the rhizosphere of many crop plants cultivated in several agricultural fields of Egypt. These mycoflora were tested for their in vitro antagonistic efficacy towards *Pythium spinosum*. Among these fungi, *Aspergillus carneus*, *A. cervinus*, *A. sulphureus*, *Penicillium funiculosum*, *P. islandicum*, *P. nigricans*, *C. globosum*, *Paecilomyces lilacinus,* and *Phoma pomorum* isolates were selected according to results of their in vitro antagonism, to examine their effectiveness for the in vivo biocontrol of this pathogen. Coating clover seeds with conidia and mycelia of these potent bioagents gave the growing seedlings significant protection from symptoms of pre-emergence damping-off caused by *Pythium spinosum*.

During the study of Zhang et al. (2013), *C. globosum* strain no.05 strongly inhibited the in vitro mycelial growth of several phytopathogenic fungi on PDA medium including; *Setosphaeria turcica*, *Verticillium mali*, *C. diplodiella*, *C. glocosporioides*, *C. fimbriata,* and *S. sclerotiorum*. In addition, culture filtrate of this strain inhibited the conidial germination of *S. turcica* causing northern corn leaf blight (NCLB) disease, suggesting the presence of antifungal metabolites in this filtrate. When the detached maize leaves were inoculated with conidia of *S. turcica* (5×10^4 cfu/mL) that were suspended in the PDB medium containing 10% culture filtrate of

C. globosum strain no.05, it caused complete suppression of the disease incidence of NCLB. The antifungal metabolites in culture filtrate of this strain were identified as chaetoglobosin A and chaetoglobosin C. Previous studies reported that chaetoglobosin A and C expressed high toxic activities against Hela cells (Sekita et al. 1982), mouse cells and chick embryos (Veselý et al. 1995). In accordance, studies of Jiao et al. (2004) and Scherlach et al. (2010) reported that chaetoglobosin A known also as cytochalasan alkaloid exhibited significant cytotoxic potential against different human cancer cells, and P388 murine leukemia cell lines. Furthermore, chaetoglobosin A and chaetoglobosin C also demonstrated high phytotoxic efficacy against alfalfa seedlings, and expressed pronounced antibacterial activities against *Staphylococcus aureus* (Jiao et al. 2004). The inhibitory efficacy of these antifungal substances of *C. globosum* strain no.05 could be enhanced by optimizing the fermentation conditions.

P. palmivora PHY02 is the causal agent of root rot of pomelo (*Citrus maxima*) in Thailand (Hung et al. 2015a). In the dual culture assay, *C. globosum* CG05, *C. cupreum* CC3003, and *C. lucknowense* CL01 bioagents caused noticeable inhibition of *P. palmivora* PHY02 mycelial growth, degraded its mycelia, and reduced 92–99% of its sporangial production. In addition, the methanol extract of *C. globosum* CG05 expressed significant inhibitory activity on the mycelial growth and formation of sporangia of this pathogen, with effective dose ED_{50} values of 26.5 μg/mL and 2.3 μg/mL, respectively. Meanwhile, in a similar study of Hung et al. (2015b) with another *Phytopthora* sp.; *P. nicotianae* a pathogenic oomycete which was isolated from rotted roots of pomelo in the same country. The In vitro and in vivo potencies of *C. globosum*, *C. lucknowense* and *C. cupreum* besides their crude extracts were tested as bioagents to control this pathogen. In the dual culture confrontation, these *Chaetomium* spp. suppressed the mycelial growth of this pathogen by 50~56%, and parasitized its hyphae resulting in degradation of the mycelia after 30 days of incubation (Fig. 10.4). In the greenhouse, application of the conidia and the methanol extracts of these *Chaetomium* spp. to pomelo seedlings infested with *P. nicotianae* reduced root rot incidence by 66~71%, and increased seedlings weight by 72~85%, compared with the non-treated control (Table 10.1).

In a recent study, Zhao et al. (2017) evaluated *C. globosum* strain CDW7 as an endophyte of *Ginkgo biloba* for its biocontrol potential against rape rot caused by *S. sclerotiorum*. The fermentation broth of CDW7 successfully inhibited disease development in *S. sclerotiorum*-infested rape with 57.8%, which was comparable with that of the synthetic carbendazim (59.8%) at 250 μg/mL. In addition, 10% of this broth enhanced the germination and growth of rape seedlings. The *C. globosum* CDW7 fermentation broth was actively stable when exposed to high temperature of 60 °C, UV illumination, and when stored at 4 °C. Seven known active metabolites were recovered from this broth. Among these metabolites, chaetoglobosin A and D expressed significant inhibitory efficacy against *S. sclerotiorum* with IC_{50} values of 0.35 and 0.62 μg/mL; respectively, compared with carbendazim (0.17 μg/mL).

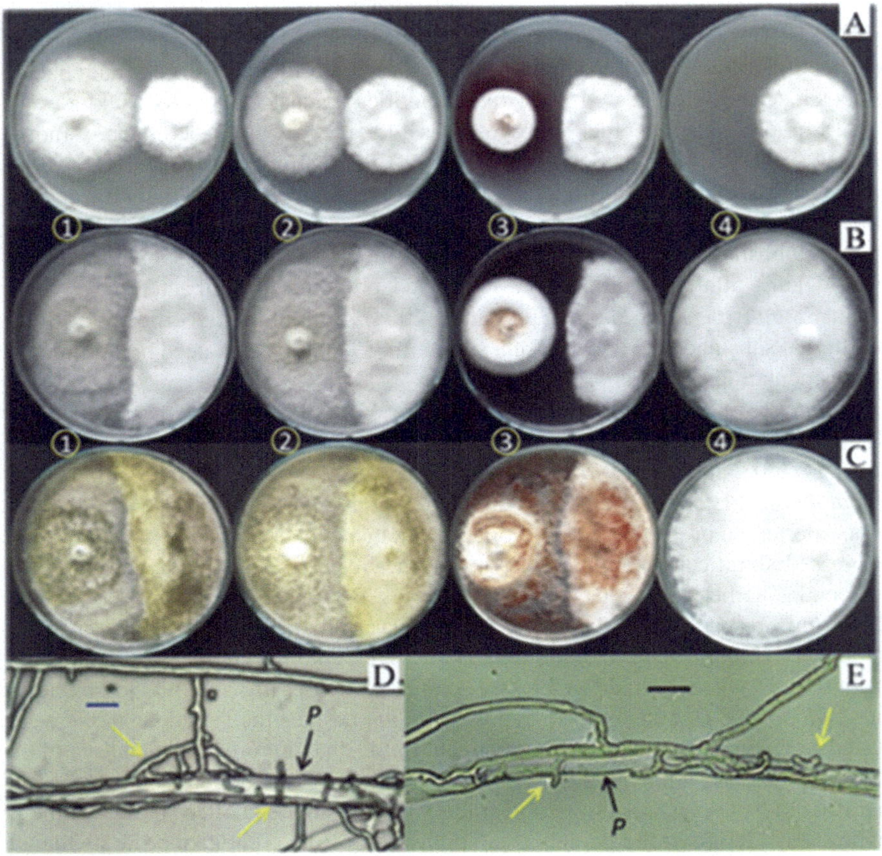

Fig. 10.4 Growth of *P. nicotianae* KA1 in bi-culture tests after 4 days (**a**), 10 days (**b**) and 30 days (**c**) after inoculation (1, 2 and 3, *P. nicotianae* KA1 (placed on the right side of the plates) inoculated with *Chaetomium* spp. strains CG05, CL01 and CC3003, respectively; 4, *P. nicotianae* KA1 alone); (**d**, **e**) *Chaetomium* hyphae (yellow arrows) coiled around and grown inside hypha of *P. nicotianae* KA1 (P) (scale bars, **d**, **e** = 10 μm). (From Hung et al. 2015b)

Crude extracts of *C. cupreum* CC3003 using hexane, ethyl acetate, and methanol solvents exhibited appreciable inhibition of the pathogenic fungus *Colletotrichum gloeosporioides* causing coffee anthracnose disease, with ED_{50} values of 13, 11 and 28 ppm, respectively. Bioactive substances extracted from this biocontrol agent caused abnormal appearance of the pathogen conidia. Moreover, application of the conidial suspension, nano-rotiorinol, nano-trichotoxin, and bio-formulation of the *C. cupreum* CC3003, reduced anthracnose incidence by 18.59%, 46.23%, 42.71% and 54.77%, respectively, compared with the non-treated control (Vilavong and Soytong 2017).

Table 10.1 Inhibition of *P. nicotianae* KA$_1$ growth at different concentrations of antagonistic crude extracts

Crude extract		Inhibition of *P. nicotianae* KA$_1$ growth (%) Concentration (μg/mL) 10	50	100	500	10,000	ED$_{50}$ (μg/mL)
C. globosum (CG05)	Hexane	21 ± 2.9^{a,f}	36 ± 3.6^{e}	66 ± 2.6^{d}	91 ± 2.3^{b}	97 ± 2.8^{a}	67.2
	EtOAC	71 ± 3.1^{c}	93 ± 2.4^{b}	98 ± 2.1^{a}	100 ± 0^{a}	100 ± 0^{a}	4.6
	MeOH	71 ± 2.1^{c}	93 ± 2.3^{b}	98 ± 2.0^{a}	100 ± 0^{a}	100 ± 0^{a}	4.6
C. lucknowense (CL01)	Hexane	6 ± 2.3^{g}	15 ± 4.3^{f}	38 ± 4.0^{e}	78 ± 2.6^{d}	94 ± 2.0^{b}	101.4
	EtOAC	81 ± 3.8^{c,d}	94 ± 2.1^{b}	99 ± 1.7^{a}	100 ± 0^{a}	100 ± 0^{a}	2.9
	MeOH	84 ± 2.6^{c}	95 ± 2.3^{b}	99 ± 1.0^{a}	100 ± 0^{a}	100 ± 0^{a}	2.6
C. cupreum (CC3003)	Hexane	34 ± 2.1^{e}	67 ± 2.1^{d}	86 ± 2.1^{c}	99 ± 1.1^{a}	100 ± 0^{a}	20.8
	EtOAC	26 ± 2.1^{f}	41 ± 2.1^{e}	70 ± 2.9^{d}	95 ± 2.3^{b}	100 ± 0^{a}	41.2
	MeOH	69 ± 4.1^{d}	91 ± 2.8^{c}	99 ± 1.6^{a}	100 ± 0^{a}	100 ± 0^{a}	4.5

(±) value represents the mean standard deviation of the four replicates of each concentration of the crude extracts. For the crude extracts of each antagonist, values indicated by the same superscript letter are not significantly different according to Duncan's multiple range test at p = 0.05. (EtOAc) is the ethyl acetate extract, and (MeOH) is the methanol extract

10.4.2 Foliar Diseases

Phylloplanes are colonized by different saprophytic microbes which have pronounced roles in protecting these plants' surfaces from different invading pathogens, by inducing defence responses in their hosts (Mitra et al. 2013). *C. globosum* is known to produce numerous antifungal metabolites. During the study of Cullen and Andrews (1984) and under growth chamber conditions, ascospore suspension of *C. globosum* strain NRRL 6296 controlled pathogenic *Venturia inaequalis* causing apple scab disease when applied as foliar treatment. When *C. globosum* was applied both singly and in association with cellulose formulation to the apple trees, dilution plating of the leaf washings and scanning electron microscopy (SEM) demonstrated that growth of *C. globosum* was significantly enhanced by this cellulose amendments. This has been attributed to the fact that the cellulose amendment supplied nutrients to this foliar bioagent (*C. globosum*), thus enhanced its survival and growth, leading finally to better biocontrol efficacy. Later, Boudreau and Andrews (1987) findings reported that *C. globosum* strain NRRL 6296 was unable to germinate when applied to non-amended apple leaves.

Dingle and McGee (2003) documented that *Chaetomium* spp. recovered from healthy wheat leaves reduced the number and development of rust pustules on these leaves caused by *Puccinia recondite* f. sp. *tritici*. Later, Istifadah et al. (2006) and Istifadah and McGee (2006) added that many *Chaetomium* spp. inhibited the growth and disease development caused by *Pyrenophora tritici-repentis* on wheat leaves in in vitro and in vivo assays.

A biocide was recovered from metabolites of *A. niger*, *T. viride*, *A. alternata*, and *C. globossum*, which are common phylloplane colonizers of tropical plants. This biocide reduced phyllosphere disease incidence in these crop plants through induction of alterations in their defence physiology (Mitra et al. 2013). Net blotch disease caused by *Drechslera teres*, and *Bipolaris* spot blotch incited by *Bipolaris sorokiniana* are foliar diseases of barley. Recently, two endophytes were isolated from barley seedlings identified as *Chaetomium* (C2 and C5), and their in vitro interactions with both pathogens were studied by Moya et al. (2016). In the dual culture assay, in vitro inhibition of *B. sorokiniana* and *D. teres* by the C2 and C5 strains accounted for 30–31.2% and 40–36%; respectively, compared with the control. Microscopic observation of the treated fungal pathogens demonstrated deformed conidia of *B. sorokiniana*, plasmolysis and coiling, whereas *D. teres* showed orange pigmentation. The mechanisms of action of both biocontrol agents were antibiosis, competition, and mycoparasitism.

10.5 Bioformulations of the Promising *Chaetomium* spp.

Results of Charoenporn et al. (2010) reported that 157 μg/mL of hexane extract from *C. globosum* N0802 which represents the ED_{50} value, greatly inhibited conidial production of *F. oxysporum* f. sp. *lycopersici* (the causal agent of tomato wilt). Similarly, crude hexane extract from *C. lucknowense* CLT expressed antifungal activity against the same pathogen with ED_{50} value of 188 μg/mL. The authors prepared oil formulations using conidia of both bioagents separately. Inocula of these *Chaetomium* spp. were adjusted to 2.5×10^6 conidia/mL and then added individually to sterilized palm oil. The in vivo assay results recorded that each of the N0802 and CLT formulations significantly reduced tomato wilt incidence by 44.68% and 36.28%; respectively, compared with the chemical fungicide prochoraz (21.95%). Moreover, both bioagents formulations noticeably increased the yield of tomato plant compared with the synthetic fungicide and the non-treated control.

References

Ahammed SK, Aggarwal R, Srivastava KD (2008) Production of extracellular proteins and cellulases by different isolates of *Chaetomium globosum* Kunze ex. Fr. Indian Phytopathol 614:437–440

Barka EA, Gognies S, Nowak J, Audran J-C, Belarbi A (2002) Inhibitory effect of endophyte bacteria on *Botrytis cinerea* and its influence to promote the grapevine growth. Biol Control 24:135–142

Bashyal BP, Wijeratne EM, Faeth SH, Gunatilaka AA (2005) Globosumones A–C, cytotoxic orsellinic acid esters from the sonoran desert endophytic fungus *Chaetomium globosum*. J Nat Prod 68:724–728

Biswas SK, Rashmi A, Srivastava KD, Sangeeta G, Prem D (2012) Characterization of antifungal metabolites of *Chaetomium globosum* Kunze and their antagonism against fungal plant pathogens. J Biol Control 26(1):70–74

Boudreau MA, Andrews JH (1987) Factors influencing antagonism of *Chaetomium globosum* to *Venturia inaequalis*: a case study in failed biocontrol. Phytopathology 77:1470–1475

Carroll G (1988) Fungal endophytes in stems and leaves from latent pathogen to mutualistic symbiont. Ecology 69:2–9

Charoenporn C, Kanokmedhakul S, Lin FC, Poeaim S, Soytong K (2010) Evaluation of bio-agent formulations to control *Fusarium* wilt of tomato. Afr J Biotechnol 9(36):5836–5844

Chen CM, Wang JP, Liu JJ, Zhu HC, Sun B, Wang J, Zhang JW, Luo ZW, Yao GM, Xue YB, Zhang YH (2015) Armochaetoglobins A–J: cytochalasan alkaloids from *Chaetomium globosum* TW1-1, a fungus derived from the terrestrial arthropod *Armadillidium vulgare*. J Nat Prod 78:1193–1201

Cui CM, Li XM, Li CS, Proksch P, Wang BG (2010) Cytoglobosins A–G, cytochalasans from a marine-derived endophytic fungus, *Chaetomium globosum* QEN-14. J Nat Prod 73:729–733

Cullen D, Andrews JH (1984) Evidence for the role of antibiosis in the antagonism of *Chaetomium globosum* to the apple scab pathogen, *Venturia inaequalis*. Can J Bot 62:1819–1823

Dingle J, McGee PA (2003) Some endophytic fungi reduce the density of *Puccinia recondita* f. sp. *tritici* in wheat. Mycol Res 107:310–316

Dosen I et al (2017) Potentially harmful secondary metabolites produced by indoor *Chaetomium* species on artificially and naturally contaminated building materials. Indoor Air 27(1):34–46. https://doi.org/10.1111/ina.12290

Fatima N, Muhammad SA, Khan I, Qazi MA et al (2016) *Chaetomium* endophytes: a repository of pharmacologically active metabolites. Acta Physiol Plant 38(6):136

Fierro-Cruz JE, Jiménez P, Coy-Barrera E (2017) Fungal endophytes isolated from *Protium heptaphyllum* and *Trattinnickia rhoifolia* as antagonists of *Fusarium oxysporum*. Rev Argent Microbiol 49(3):255–263

Gao K, Liu X, Kang Z, Mendgen K (2005) Mycoparasitism of *Rhizoctonia solani* by endophytic *Chaetomium spirale* ND35: ultrastructure and cytochemistry of the interaction. J Phytopathol 153:280–290

Ge HM, Zhang WY, Ding G, Saparpakorn P, Song YC, Hannongbua S, Tan RX (2008) Chaetoglobins A and B, two unusual alkaloids from endophytic *Chaetomium globosum* culture. Chem Commun 7(45):5978–5980

Ge HM, Zhang Q, Xu SH, Guo ZK, Song YC, Huang WY, Tan RX (2011) Chaetoglocins A–D, four new metabolites from the endophytic fungus *Chaetomium globosum*. Planta Med 77:277–280

Geiger WB, Conn JE, Waksman SA (1944) Chaetomin, a new antibiotic substance produced by *Chaetomium cochliodes*: I. Formation and properties. J Bacteriol 48(5):531–536

Gond SK, Verma VC, Mishra A, Kumar A, Kharwar RN (2010) Role of fungal endophytes in plant protection. In: Arya A, Perelló AE (eds) Management of fungal plant pathogens. CAB International, Wallingford, pp 183–197

Gond SK, Mishra A, Sharma VK, Verma SK, Kumar J, Ravindra N, Kharwar RN, Kumar A (2012) Diversity and antimicrobial activity of endophytic fungi isolated from *Nyctanthes arbor-tristis*, a well-known medicinal plant of India. Mycoscience 53(2):113–121

Gophna U, Ron EZ, Graur D (2003) Bacterial type III secretion systems are ancient and evolved by multiple horizontal-transfer events. Gene 312:151–163

Hung PM, Wattanachai P, Kasem S, Poaim S (2015a) Biological control of *Phytophthora palmivora* causing root rot of pomelo using *Chaetomium* spp. Mycobiology 43(1):63–70

Hung PM, Wattanachai P, Kasem S, Poeaim S (2015b) Efficacy of *Chaetomium* species as biological control agents against *Phytophthora nicotianae* root rot in citrus. Mycobiology 43(3):288–296

Istifadah N, McGee P (2006) Endophytic *Chaetomium globosum* reduces development of tan spot in wheat caused by *Pyrenophora tritici-repentis*. Australas Plant Pathol 35:411–418

Istifadah N, Saleeba J, McGee P (2006) Isolates of endophytic *Chaetomium* spp. inhibits the fungal pathogen *Pyrenophora tritici-repentis in vitro*. Can J Bot 84:1148–1155

Jiang C, Song J, Zhang J, Yang Q (2017) Identification and characterization of the major antifungal substance against *Fusarium sporotrichioides* from *Chaetomium globosum*. World J Microbiol Biotechnol 33(108):1–9

Jiao W, Feng Y, Blunt JW, Cole AL, Munro MH (2004) Chaetoglobosins Q, R, and T, three further new metabolites from *Chaetomium globosum*. J Nat Prod 67:1722–1725

Kaewchai S, Soytong K (2010) Application of biofungicides against *Rigidoporus microporus* causing white root disease of rubber trees. J Agric Technol 6(2):349–363

Kay SJ, Stewart A (1994) Evaluation of fungal antagonists for control of onion white rot in soil box trials. Plant Pathol 43:371–377

Kumar S, Kaushik N (2013) Endophytic fungi isolated from oil-seed crop *Jatropha curcas* produces oil and exhibit antifungal activity. PLoS One 8:56202

Lan N, Qi GF, Yu ZN, Zhao XY (2011) Isolation, identification and anti-fungal action of endophytic fungi of rape seed. J Huazhong Agric Univ 30:270–275

Li HQ, Li XJ, Wang YL et al (2011) Antifungal metabolites from *Chaetomium globosum*, an endophytic fungus in Ginkgo biloba. Biochem Syst Ecol 39:876–879

Li W, Yang XQ, Yang YB, Duang RT, Chen GY, Li XZ, Li QL, Qin SH, Li SQ, Zhao LX, Ding ZT (2016) Anti-phytopathogen, multi-target acetylcholinesterase inhibitory and antioxidant activities of metabolites from endophytic *Chaetomium globosum*. Nat Prod Res 30(22):2616–2619

Maghazy SMN, Abdelzaher HMA, Haridy MS, Moustafa SMN (2008) Biological control of damping-off disease of *Trifolium alexandrinum* L. caused by *Pythium spinosum Sawada* var. *spinosum* using some soil fungi. Arch Phytopathol Plant Protect 41(6):431–450

Martínez-Álvarez P, Fernández-González RA, Sanz-Ros AV, Pando V, Diez JJ (2016) Two fungal endophytes reduce the severity of pitch canker disease in *Pinus radiata* seedlings. Biol Control 94:1–10

Mitra J, Bhuvaneshwari V, Paul PK (2013) Broad spectrum management of plant diseases by phylloplane microfungal metabolites. Arch Phytopathol Plant Protect 46(16):1993–2001

Moya P, Pedemonte D, Amengual S, Franco MEE, Sisterna MN (2016) Antagonism and modes of action of *Chaetomium globosum* species group, potential biocontrol agent of barley foliar diseases. Bol Soc Argent Bot 51(4):569–578

Nandakumar R, Babu S, Viswanathan R, Sheela J, Raguchander T, Samiyappan R (2001) A new bio-formulation containing plant growth promoting rhizobacterial mixture for the management of sheath blight and enhanced grain yield in rice. BioControl 46:493–510

Narisawa K, Usuki F, Hashiba T (2004) Control of *Verticillium* yellows in Chinese cabbage by the dark septate endophytic fungus LtVB3. Phytopathology 94:412–418

Pan F, Liu Z-Q, Chen Q, Xu Y-W, Hou K, Wu W (2016) Endophytic fungus strain 28 isolated from *Houttuynia cordata* possesses wide-spectrum antifungal activity. Braz J Microbiol 47:480–488

Park JH, Choi GJ, Jang KS, Lim HK, Kim HT, Cho KY, Kim JC (2005) Antifungal activity against plant pathogenic fungi of chaetoviridins isolated from *Chaetomium globosum*. FEMS Microbiol Lett 252:309–313

Pereira J, Dhingra OD (1997) Suppression of *Diapothe phaseolorum* f. sp. *meridionalis* in soybean stems by *Chaetomium globosum*. Plant Pathol 46:216–223

Petrini O (1991) Fungal endophytes of tree leave. In: Andrews JA, Hirano SS (eds) Microbial ecology of leaves. Springer, New York, pp 179–197

Phong NH, Pongnak W, Soytong K (2016) Antifungal activities of *Chaetomium* spp. against *Fusarium* wilt of tea. Plant Prot Sci 52(1):10–17

Pontius A, Krick A, SKehraus S, Brun R, König GM (2008) Antiprotozoal activities of heterocyclic-substituted xanthones from the marine-derived fungus *Chaetomium* sp. J Nat Prod 71(9):1579–1584

Porras-Alfaro A, Bayman P (2011) Hidden fungi, emergent properties: endophytes and microbiomes. Annu Rev Phytopathol 49:291–315

Qin JC, Gao JM, Zhang YM, Yang SX, Bai MS, Ma YT, Laatsch H (2009a) Polyhydroxylated steroids from an endophytic fungus, *Chaetomium globosum* ZY-22 isolated from Ginkgo biloba. Steroids 74(9):786–790

Qin JC, Zhang YM, Gao JM, Bai MS, Yang SX, Laatsch H, Zhang AL (2009b) Bioactive metabolites produced by *Chaetomium globosum*, an endophytic fungus isolated from Ginkgo biloba. Bioorg Med Chem Lett 19:1572–1574

Saldajeno MGB, Hyakumachi M (2011) The plant growth-promoting fungus *Fusarium equiseti* and the arbuscular mycorrhizal fungus *Glomus mosseae* stimulate plant growth and reduce severity of anthracnose and damping-off diseases in cucumber (*Cucumis sativus*) seedlings. Ann Appl Biol 159:28–40

Saravanakumar D, Vijayakumar C, Kumar N, Samiyappan R (2007) PGPR induced defense responses in tea plants against blister blight disease. Crop Prot 26:556–565

Saravanakumar D, Lavanya N, Muthumeena K, Raguchander T, Samiyappan R (2009) Mixture of fluorescent *Pseudomonads* mediated disease resistance in rice plants against sheath rot (*Sarocladium oryzae*) pathogen. BioControl 54:273–286

Scherlach K, Boettger D, Remme N, Hertweck C (2010) The chemistry and biology of cytochalasans. Nat Prod Rep 27:869–886

Schlörke O, Zeeck A, Orsellides A-E (2006) An example for 6-deoxyhexose derivatives produced by fungi. Eur J Org Chem 2006:1043–1049

Sekita S, Yoshihira K, Natori S, Kuwano H (1982) Chaetoglobosins, cytotoxic 10-(indol-3-yl)-[13] cytochalasans from *Chaetomium* spp. III. Structures of chaetoglobosins C, E, F, G, and J. J Chem Pharm Bull 30:1629–1638

Shanthiyaa V, Saravanakumar D, Rajendran L, Karthikeyan G, Prabakar K, Raguchander T (2013) Use of *Chaetomium globosum* for biocontrol of potato late blight disease. Crop Prot 52:33–38

Sibounnavong P, Charoenporn C, Kanokmedhakul S, Soytong K (2011) Antifungal metabolites from antagonistic fungi used to control tomato wilt fungus *Fusarium oxysporum* f. sp. *lycopersici*. Afr J Biotechnol 10(85):19714–19722

Soytong K, Ratanacherdchai K (2005) Application of mycofungicide to control late blight of potato. J Agric Technol 1:19–32

Soytong K, Kanokmedhakul S, Kukongviriyapa V, Isobe M (2001) Application of *Chaetomium* species (Ketomium R) as a new broad spectrum biological fungicide for plant disease control. Fungal Divers 7:1–15

Soytong K, Srinon W, Rattanacherdchai K, Kanokmedhakul S, Kanokmedhakul K (2005) Application of antagonistic fungi to control anthracnose disease of grape. J Agric Sci Technol 1:33–41

Taragano VM, Pilosof AMR (1999) Application of Doehlert designs for water activity, pH, and fermentation time optimization for *Aspergillus niger* pectinolytic activities production in solid-state and submerged fermentation. Enzym Microb Technol 25:411–419

Thiep NV, Soytong K (2015) *Chaetomium* spp. as biocontrol potential to control tea and coffee pathogens in Vietnam. J Agric Technol 11(6):1381–1392

Tomilova OG, Shternshis MV (2006) The effect of a preparation from *Chaetomium* fungi on the growth of phytopathogenic fungi. Appl Biochem Microbiol 42:76–80

Udagawa S et al (1979) The production of chaetoglobosins, sterigmatocystin, O-methyl sterigmatocystin, and chaetocin by *Chaetomium* spp. and related fungi. Can J Microbiol 25:170–177

Varma A, Bakshi M, Lou BG, Hartmann A, Oelmueller R (2012) *Piriformospora indica*: a novel plant growth-promoting mycorrhizal fungus. Agric Res 1:117–131

Vega FE, Posada F, Aime MC, Pava-Ripoll M, Infante F, Rehner SA (2008) Entomopathogenic fungal endophytes. Biol Control 46:72–82

Veselý D, Veselá D, Jelínek R (1995) *Penicillium aurantiogriseum* dierckx produces chaetoglobosin A toxic to embryonic chickens. Mycopathologia 132:31–33

Vilavong S, Soytong K (2017) Application of a new bio-formulation of *Chaetomium cupreum* for biocontrol of *Colletotrichum gloeosporioides* causing coffee anthracnose on Arabica variety in Laos. AGRIVITA J Agric Sci 39(3):303–310

Vitale A, Cirvilleri G, Castello I, Aiello D, Polizzi G (2012) Evaluation of *Trichoderma harzianum* strain T22 as biological control agent of *Calonectria pauciramosa*. BioControl 57(5):687–696. https://doi.org/10.1007/s10526-011-9423-1

Vujanovic V, Goh YK (2011) *Sphaerodes mycoparasitica* biotrophic mycoparasite of 3-acetyldeoxynivalenol and 15-acetyldeoxynivalenol-producing toxigenic *Fusarium graminearum* chemotypes. FEMS Microbiol Lett 316:136–143

Wang YL, Liu SY, Mao XQ, Zhang Z, Jiang H, Chai RY, Qiu HP, Wang JY, Du XF, Li B, Sun GC (2013) Identification and characterization of rhizosphere fungal strain MF-91 antagonistic to rice blast and sheath blight pathogens. J Appl Microbiol 114:1480–1490

Yamada T, Muroga Y, Jinno M, Kajimoto T, Usami Y, Numata A, Tanaka R (2011) New class azaphilone produced by a marine fish-derived *Chaetomium globosum*. The stereochemistry and biological activities. Bioorg Med Chem 19(13):4106–4113

Yang SX, Gao JM, Zhang Q, Laatsch H (2011a) Toxic polyketides produced by *Fusarium* sp., an endophytic fungus isolated from *Melia azedarach*. Bioorg Med Chem Lett 21:1887–1889

Yang SZ, Peng LT, Su XJ, Chen F, Cheng YJ, Fan G, Pan SY (2011b) Bioassay guided isolation and identification of antifungal components from propolis against *Penicillium italicum*. Food Chem 127:210–215

Zhang J, Ge HM, Jiao RH, Li J, Peng H, Wang YR, Wu JH, Song YC, Tan RX (2010) Cytotoxic chaetoglobosins from the endophyte *Chaetomium globosum*. Planta Med 76:1910–1914

Zhang Q, Li HQ, Zong SC, Gao JM, Zhang AL (2012) Chemical and bioactive diversities of the genus *Chaetomium* secondary metabolites. Mini Rev Med Chem 12:127–148

Zhang G, Wang F, Qin J, Wang D, Zhang J, Zhang Y, Zhang S, Pan H (2013) Efficacy assessment of antifungal metabolites from *Chaetomium globosum* no.05, a new biocontrol agent, against *Setosphaeria turcica*. Biol Control 64:90–98

Zhang S, Xu B, Zhang J, Gan Y (2018) Identification of the antifungal activity of *Trichoderma longibrachiatum* T6 and assessment of bioactive substances in controlling phytopathgens. Pestic Biochem Physiol 147:59–66

Zhao S-S, Zhang Y-Y, Yan W, Cao L-L, Xiao Y, Ye Y-H (2017) *Chaetomium globosum* CDW7, a potential biological control strain and its antifungal metabolites. FEMS Microbiol Lett 364:1–6

Chapter 11
Nanoparticle-Mediated *Chaetomium*, Unique Multifunctional Bullets: What Do We Need for Real Applications in Agriculture?

Mohamed A. Mohamed and Nivien A. Nafady

11.1 Introduction

Nanotechnology has established itself as one of the fastest and biggest research and development fields in recorded history. Although previous technological revolutions as space exploration, semiconductors, and biotechnology have made it big, they were, and still are, severely confined to their respective domains and find fewer interest from other disciplines. Nanotechnology, on the other hand, though starting off like the aforementioned fields of study and research, limited mostly to the electronics industry, has expanded its horizons of application by the amalgamation of nearly all subjects of science, viz. biology, chemistry, and physics. The rapid strides of nanotechnology in electronics and manufacturing have recently been paralleled by advances in medical nanotechnology, with an ever-increasing list of scientific publications, patents, and products being commercially released. The profuse amount of nanoparticles in personal care, industrial, food, and agricultural products, as well as in soil and water remediation technologies, has raised concerns about contamination of ecosystems and food supply (Gardea-Torresdey et al. 2014). Different methods have been used in the synthesis process of metallic nanoparticles (Bankar et al. 2010; Vaseeharan et al. 2010; Nagajyothi and Lee 2011; Karthiga 2017). Microwave, vapour deposition, spray pyrolysis, hydrothermal, solvothermal, seed-mediated, wet chemical, and biological methods are among the most efficient methods exploited to produce nanoparticles with numerous diverse characters regarding their size, morphology, and shape. The biological synthesis of metallic nanoparticles has been explored in recent years. Plants, fungi, and bacteria are

M. A. Mohamed (✉)
Nanotechnology & Advanced Nano-Materials Laboratory (NANML), Plant Pathology Research Institute, Agricultural Research Center, Giza, Egypt

N. A. Nafady
Botany and Microbiology Department, Faculty of Science, Assiut University, Assiut, Egypt

A. M. Abdel-Azeem (ed.), *Recent Developments on Genus Chaetomium*, Fungal Biology, https://doi.org/10.1007/978-3-030-31612-9_11

among the commonly used biological resources that have been employed in the synthesis process of different nanoparticles (Elavazhagan and Elavazhagan 2011; Mehmood et al. 2017). Compared to other biological resources, fungi are considered the best efficient options to produce different nanomaterials, due to the vast repertoire of proteins, enzymes, and other bioactive secondary metabolites they produce and the redox activity they possess. Furthermore, they also contribute to the high productivity and stability of the nanoparticles (Thakkar et al. 2010; Ramanathan et al. 2013; Metuku et al. 2014; Schröfel et al. 2014). In this regard, different scientists suggested that the crude extracts derived from most organisms including fungi contain unique bioactive secondary metabolites including phenolic acid, flavonoids, alkaloids, and terpenoids which are constantly involved in the redox reaction and responsible for the reduction of bulk material in its ionic state to synthesize eco-friendly nanosized metallic particles (Huang et al. 2007; Jyoti et al. 2016). On the other hand and in comparison to plants and other microorganisms, fungal mycelia can withstand severe environments in bioreactors or chambers. Lastly, fungi can be easily handled with and fabricated in downstream processing. Consequently, they are approved as an effective biological alternative to conventional chemical synthesis (Fig. 11.1). To date, different fungi have been found to synthesize diverse metallic nanoparticles, and *Fusarium, Alternaria, Aspergillus,* and *Penicillium* species are broadly used in this issue (Duan et al. 2015; Salem et al. 2015; Maruyama et al. 2016). However, very few studies are being published till now on exploiting other important fungi like *Chaetomium* species in the field of nanotechnology (Duan et al. 2015). *Chaetomium* was reported as an efficient biological fungicide to control several plant pathogens. The metabolites from *Chaetomium* spp. were published to control different plant pathogens. Consequently, we envision that it could be a good step for exploiting *Chaetomium* species to produce efficient and safe eco-friendly nanomaterials that can be loaded with efficient biomolecules and functionalized in many biological fields in particular agricultural applications. On the other hand, more effort and more research knowledge are still in need to accomplish those goals in agriculture science particularly as fertilizers and pesticide alternatives to fight plant pathogens, not only for the development and synthesis of nanomaterials but also for studying their interactions with plants and the environment. Indeed, it is necessary to answer the question "what do we need for real application in agriculture?" We need to understand broadly how those nano-bullets are bio-produced, how the fungal system is the better producer and what factors govern this process, how the plants absorb and uptake nanoparticles, how they move and track inside the plant cells and the factors that govern this critical process and how they interact with the plant cell. In this chapter, we finally discuss on the positive as well as negative impacts of those nanoparticles on the plant system, which briefly answer the above-mentioned questions.

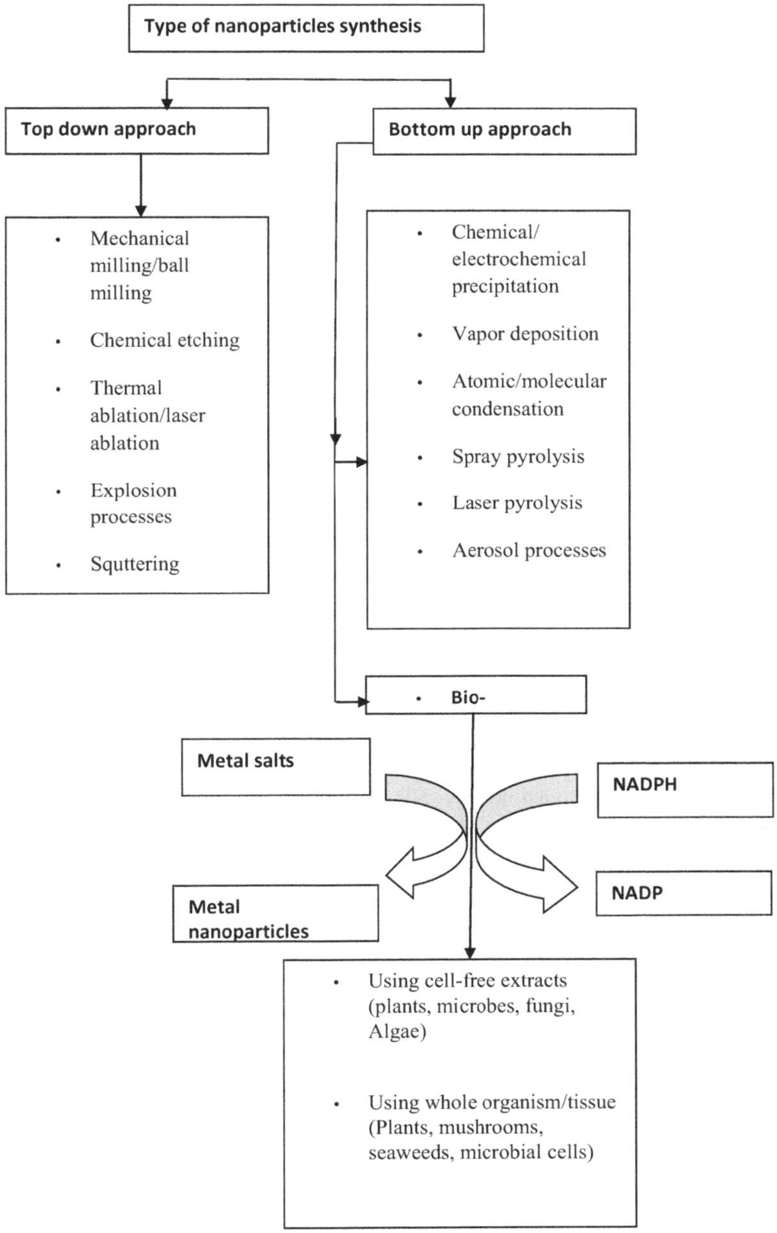

Fig. 11.1 Major approaches in nanoparticle synthesis

11.2 Biosynthesis of Nanoparticles Using Fungi

The biological synthesis of nanoparticles has been explored in recent years. A wide array of microorganisms such as bacteria, fungi, yeast, algae, and actinomycetes are majorly employed as biological agents for the synthesis process (Kumar and Yadav 2009; Mallikarjuna et al. 2014). This has led to the development of "green nanotechnology", which combines biological principles with physical and chemical procedures, to generate eco-friendly, nanosized particles with specific functions. The synthesis of metal nanoparticles employs both intracellular and extracellular methods (Mallikarjuna et al. 2014; Ghorbani 2017). Some examples of these microbial agents include bacteria (Shahverdi et al. 2007), fungi (Mohanpuria et al. 2007; Parikh et al. 2008), actinomycetes (Golinska et al. 2014; Manimaran and Kannabiran 2017), lichens (Mie et al. 2012), and algae (Parial et al. 2012). These diverse groups of biological agents have many advantages over physical and chemical methods such as easy and simple scale-up, easy downstream processing, simpler biomass handling and recovery, and economic viability (Thakkar et al. 2010; Rai et al. 2012). Compared to other naturally available biological resources, fungi are more efficient, and thus they are more suitable for the synthesis of nanoscale metal particles. For example, fungi produce large amounts of protein that contribute to the high productivity and stability of the particles. Moreover, compared with plant materials and other microorganisms, fungal mycelia can withstand severe environments in bioreactors or chambers. Lastly, they are easy to handle and fabricate in downstream processing. Consequently, as an effective alternative to chemical synthesis, the fungal system for nanomaterial biosynthesis has gained considerable popularity in recent years. To date, different fungi have been found to synthesize nanoparticles. For example, silver nanoparticles (AgNPs) could be synthesized by *Fusarium oxysporum* (Husseiny et al. 2015), *Fusarium solani* (Birla et al. 2013), *Aspergillus niger* (Sagar and Ashok 2012), *Phoma glomerata* (Birla et al. 2009), *Cladosporium cladosporioides* (Moazeni et al. 2012), *Penicillium fellutanum* (Kathiresan et al. 2009), *Phoma* sp.3.2883 (Chen et al. 2003), *Aspergillus flavus* (Vigneshwaran et al. 2007), *Epicoccum nigrum* (Qian et al. 2013), and *Phoma macrostoma* (Sheikhloo and Salouti 2012).

11.3 Biosynthesis of Nanoparticles Using *Chaetomium* spp.

Chaetomium species are strictly saprobic antagonists and have been shown to be efficient biocontrol agent against several plant pathogens, e.g. *Botrytis cinerea* (Paul et al. 1997), *Fusarium oxysporum f. sp. Lycopersici* (Tongon and Soytong 2015), *Phytophthora palmivora* (Hung et al. 2015), and *Phytophthora parasitica* (Meng et al. 2014). Screening *Chaetomium* spp. as biological control agents has been carried out firstly in Thailand since 1989, resulting in the development of a biological formulation from *Chaetomium cupreum* CC1–10 and *Chaetomium globosum*

CG1–12. The product has now been developed into pellets and powder formulations and registered for Patent Right No. 6266, Intl. cl. 5 AO 1 N 25/12 in 1994 (Servin et al. 2015). Although *Chaetomium* is a large genus and considered one of the most rich fungi producing large amounts of proteins, enzymes, and effective secondary metabolites having antimicrobial and redox activity and can be easily exploited in biosynthesis of different nanoparticles (Abdel-Raouf et al. 2017), however, very few studies on *Chaetomium* spp. were used in the field of nanotechnology (Servin et al. 2015). In 2014, Joselito and his co-workers constructed and characterized a copolymer nanoparticle that carried bioactive compounds from the crude extract of *Chaetomium globosum* and *Chaetomium cupreum* species. Other studies exploited *Chaetomium globosum* isolated from a healthy bean plant to produce a huge amount of xylanase, pectinase, and chitinase in order to functionalize those enzymes in the biosynthesis of zinc nanoparticles (Król et al. 2017). Their results were promising particularly after investigation when the antifungal activity of the produced nanoparticles displayed the highest inhibition against *Fusarium solani*, *Rhizoctonia solani*, and *Sclerotium rolfsii* as shown by the zone of inhibition (13.0, 15.0, and 17.0 mm, respectively). More importantly, it was observed that the antifungal activities of zinc nanoparticles were most promising on the pathogenic fungi used at pH 8.0. The authors also observed modifications in the fungal hyphae with morphological alterations and conidial and reduction mortality. Consequently, we consider that *Chaetomium* species are a very promising biological precursor for the biosynthesis of different multifunctional nanoparticles which can be efficiently exploited in diverse applications.

11.4 Mechanisms of Nanoparticle Biosynthesis in Liquid Phase

Generally speaking, the synthesis process of metallic nanoparticles is categorized under two main approaches: (1) "top-down" approach and (2) "bottom-up" approach. In top-down synthesis, the nanoparticles are biologically synthesized following the reduction route for the involved bulk material's size using different physical, chemical, or mechanical processes (Fig. 11.2). However, this method mostly introduces distortions on the surface of the produced nanomaterials, causing a major biological and electrical limitation as the physical, optical, and surface chemistry of most nanoparticles is highly dependent on the nanomaterial surface structure. In the case of the bottom-up approach, the formed nanoparticles are built from very small entities by joining different small atoms together, where the process often requires starting from molecules, atoms, or ions to obtain the target nanoparticles (Ulbrich and Lamprecht 2010). In the past few years, most scientists tend to follow the bottom-up approach in the synthesis of nanoparticles particularly in liquid-phase media (Mühlig et al. 2013; Yang et al. 2016; Liu et al. 2017), and nucleation theories and mechanisms have been extensively described by Finney and Finke (2008) and Liu et al. (2017).

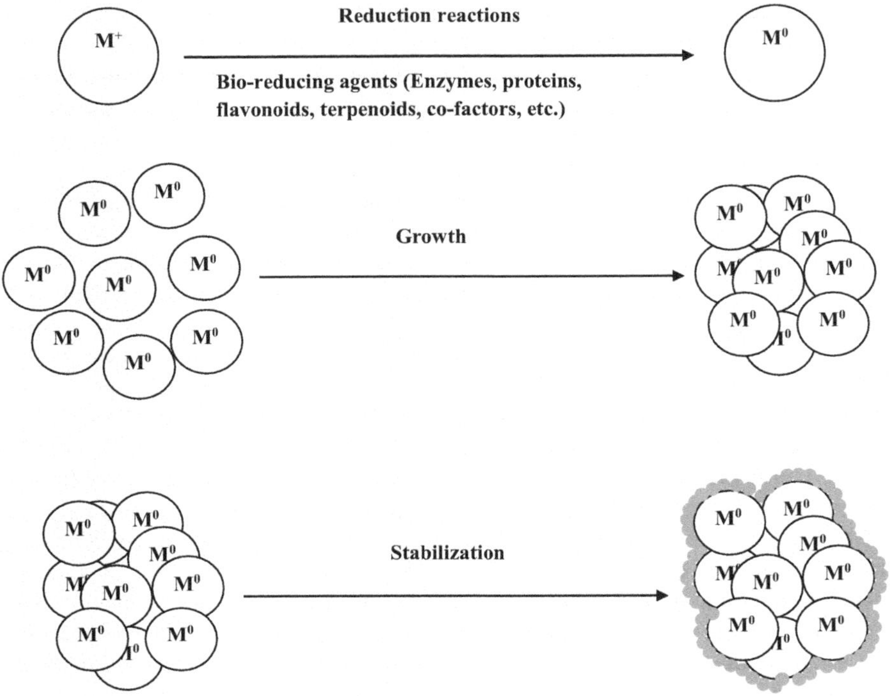

Fig. 11.2 Stages of nanoparticle synthesis and cofactor-dependent bio-reduction reactions

11.5 Factors Affecting the Mycosynthesis Process

In recent years, the development of reliable methods for the synthesis of different nanoparticles with a desired shape and size has been the subject of significant interest (Tran et al. 2013; Liu et al. 2017). Optimization of the bio-reduction conditions in the mixture solution considering both the substrate and biocatalyst concentration, electron donor and its concentration, pH, time of exposure, temperature, buffer strength, and mixing speed is among the most important issues playing very core roles in obtaining the desired nanoparticle size with different shapes such as spherical, triangular, hexagonal, and rectangular with large-scale production and good monodispersity in the aqueous phase.

In this context, different types of conventional chemical methods with reducing agents like citrate (Kumar et al. 2006), $NaBH_4$ (Lee and Meisel 1982), gallic acid (Consoli et al. 2016), polyol (Kim et al. 2006), organic solvents, and photochemical (Kasthuri et al. 2009), electrochemical, and sono-chemical methods (Wongpisutpaisan et al. 2011) have been utilized for this goal. However, very few were efficient to achieve the accurate nanoparticle size and the desired shape in need safely. *Consequently*, it is necessary to settle some reaction conditions that facilitate reproducible synthesis of the target nanoparticles with uniform size distributions. In this regard,

(i) the reducing and stabilizing agents, their type, and the concentration used (Dadosh et al. 2009); (ii) the addition of complexing agents (i.e. NH_3) for removing precursor agents and decreasing particle size (Yang et al. 2010); and (iii) the addition of alkaline co-reducers using strong and/or weak reducing agents (Agnihotri et al. 2014) are the most crucial conditions that can be exploited to tune the desired nanoparticle size and shape in the synthesis process. The stabilizing agent of nanomaterials is also a necessary factor for their compatibility across different ranges of their applications (Haider and Kang 2015) and will impact the interaction in the environment. In general terms, the stabilization processes decrease the NP surface energy making the colloidal system thermodynamically stable (Kraynov and Müller 2011). The stabilizing ligands, which may be alkaloids, terpenoids and antioxidants, are derived from the fungal cell coat and bound to the nanoparticles (NPs) surface not only to manage their growth during the synthesis process but also to prevent their aggregation and conversion again to their ionic state. The main mechanisms of interaction between these molecules and ligands with the surface of the NPs are mostly through chemisorption processes, electrostatic attractions, or hydrophobic interactions (Calderón-Jiménez et al. 2017). This can be easily demonstrated through the noticeable strong surface interactions occurring between some metallic nanomaterials like silver nanoparticles and sulphur, amide, carboxylic, and ketonic (-SH, -NH, -COOH, -C=O) groups that allow for functionalization and further stabilization (Sperling and Parak 2010). Different reports also indicated that the polymeric synthetic ligands used in nanoparticle stabilization tend to generate more contact points with the NP surface, creating better adsorption or ligand/surface interactions (Toshima and Yonezawa 1998).

On the other hand, hydrophilic polymer chain interactions generate external loops which can interact with the solvent and sterically stabilize NPs (see Figs. 11.3 and 11.4). On the other hand, storage temperatures are also a critical factor to the stability of these materials. In this regard, it was revealed that incubation of the nanoparticles in aqueous solution under different temperatures can produce extra oxidation processes promoting unwanted shapes like nanorods and nanoprisms and may also lead to aggregation and/or agglomeration. Also, light may play an important modifying role with some sensitive nanoparticles like AgNPs and modify them through photochemical reactions. This also was indicated by (Gorham et al. 2012) and revealed that AgNPs coated with citrate can be destabilized with UV radiation exposure. Other indirect factors like pH of the medium should also be taken into account with regard to the destabilization of the formed nanomaterials, where it was shown that the stability of silver nanoparticles coated with a particular amino acid increased under acidic conditions ($pH \approx 3$), resulting in noticeable elimination of agglomerate formation due to suppression of intermolecular interactions between solvent and ligand (Bayram and Bayram 2016). Using high concentrations from the reducing agents in the synthesis liquid phase irregularly accelerates the probability of nanoparticle transformation into undesired new shapes like nanorods, nanocubes, and nano-triangles (Dadosh et al. 2009; Jana et al. 2001). Therefore, extensive

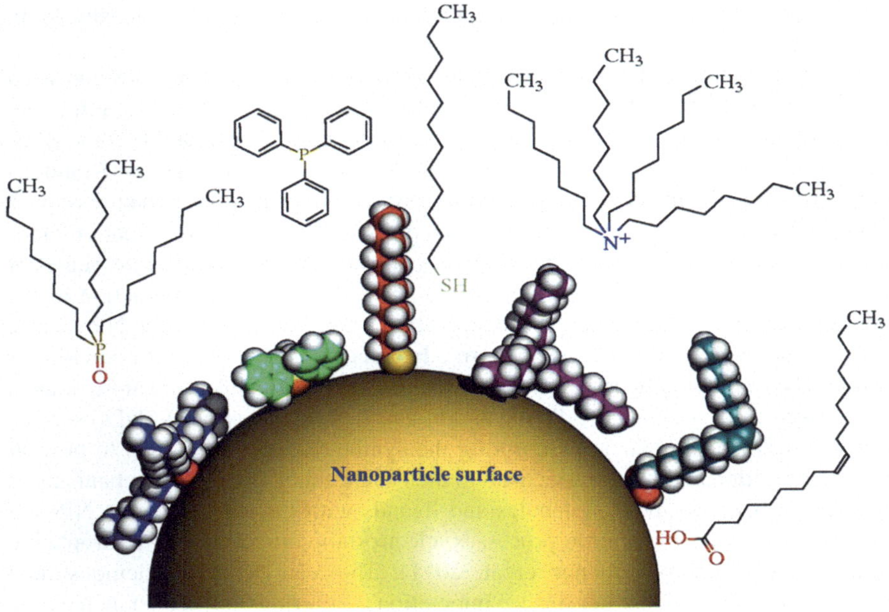

Fig. 11.3 Schematic illustration demonstrating the cross-linking and conjugation ways of different chemical ligands on the nanoparticle surfaces

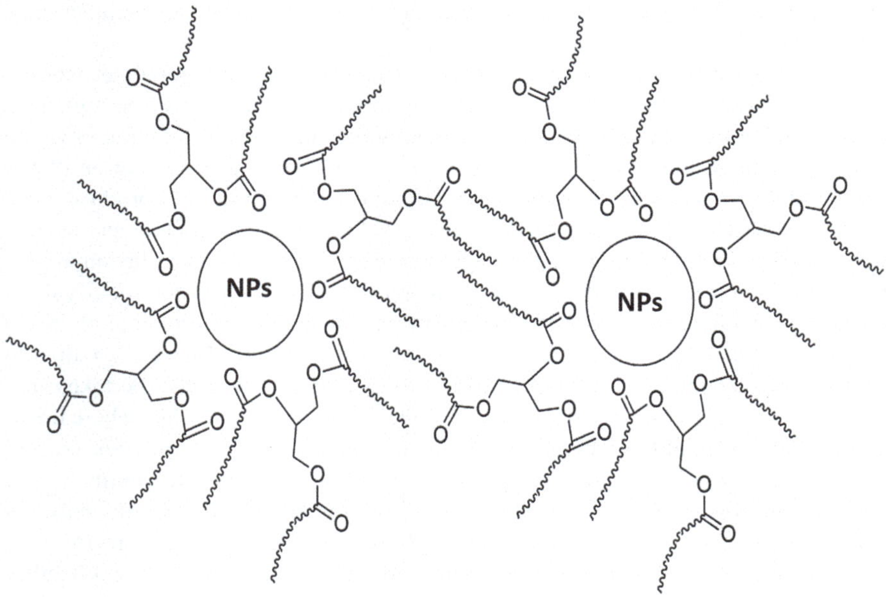

Fig. 11.4 Steric stabilization of metallic nanoparticles

ultimate work is still in need for further large-scale nanomaterial production research following biological routes in order to define well and optimize those parameters for this purpose (Iravani 2011).

11.6 Multiple Applications of Metallic Nanomaterials in Agriculture: Current and Future Trends

More than 50 years later, nanotechnology is starting to reshape our lives positively and straightforwardly towards our future. Excellent reviews have been published continually demonstrating the broad diversity of biological agents involved in the synthesis of many types of nanomaterials demonstrating their potential use in diverse application including material sciences, medicine, electronics, and pharmaceutical sciences (Mandal et al. 2006; Bhattacharya and Mukherjee 2008; Chen and Schluesener 2008; Rai et al. 2009; Sharma et al. 2009; Durán et al. 2014; Abdel-Hafez et al. 2016; Mohamed et al. 2016). Some nanomaterials like silver, gold, titanium, zinc, and copper may be of particular relevance to new and emerging technologies. The use of silver nanoparticles as an antimicrobial coating agent in solar absorption systems is the best example. Other applications in such areas also include the use of gold nanoparticles as efficient carrier agent to deliver drugs in cancer treatments and precursors in electronic applications (Chen et al. 2008; Sugunan et al. 2005) and platinum nanoparticles in the production of fuel cells (Riddin et al. 2006). But what about agriculture? Will this tiny revolution help to solve the critical agriculture problems regarding low food production, emerging phytopathogen resistance, water treatment, environmental challenges within a climate change scenario, etc.? That is something we want to believe in, but science is about facts and we need to work hard in order to prove it more clearly. Indeed, the unique properties of those materials at nanoscale make them suitable candidates for the design and development of novel tools in support of a sustainable agriculture.

11.6.1 Antimicrobial Activity

Different types of organic and inorganic nanomaterials have an immense potential as antimicrobial agents in fighting many phytopathogens and plant disease management, providing an unusual economic cost-effective tool in crop protection with ignorable limitations regarding environment risk profile. Till date, extensive research is under way on the application of nanomaterials as new antimicrobial agents to treat microbial species with resistance against traditional antibiotics and pesticides in medicine and agriculture, respectively (Singh Duhan et al. 2017; Sun et al. 2018).

The unique small size together with the high surface area of those nanoparticles permits them to have broad-spectrum microbial activity (V. L. Das et al. 2014b).

Silver nanoparticles are the best example and recently were considered one of the most common antimicrobial agents than other metals having diverse industrial applications in many consumer products including clothes, sprays, cell phones, detergents, shoes, laptop keyboards, contraceptives, cosmetics, dietary supplements and pesticides (Danwanichakul et al. 2016). In 2013, Kanmani and Lim suggested that the unique antimicrobial activity of silver nanoparticles may be supported by different scenarios (Fig. 11.5): (1) their potential to produce reactive oxygen species (ROS); (2) denaturation of DNA, proteins, and enzymes by combining with sulphur and phosphorus; (3) making pits and pores in the microbial cell walls resulting in complete destruction; and (4) promotion of intracellular ion efflux, resulting in complete cell death (Kanmani and Lim 2013).

Moreover, different types of biogenic metallic nanoparticles such as zinc oxide (35–45 nm), silver (20–80 nm), and titanium dioxide (85–100 nm) were also investigated as antifungal agents against the phytopathogenic fungus *Macrophomina phaseolina*, a major soil-borne pathogen of pulse and oilseed crops. The results indicated that silver nanoparticles have the higher antifungal potential at lower concentrations and zinc oxide comes in second followed by titanium dioxide nanoparticles (Shi et al. 2013). Similarly, another study investigated the antifungal activity of silica nanoparticles (20–40 nm size) against *Fusarium oxysporum* and *Aspergillus niger*, as compared with that of bulk silica. Interestingly, it was noticed that

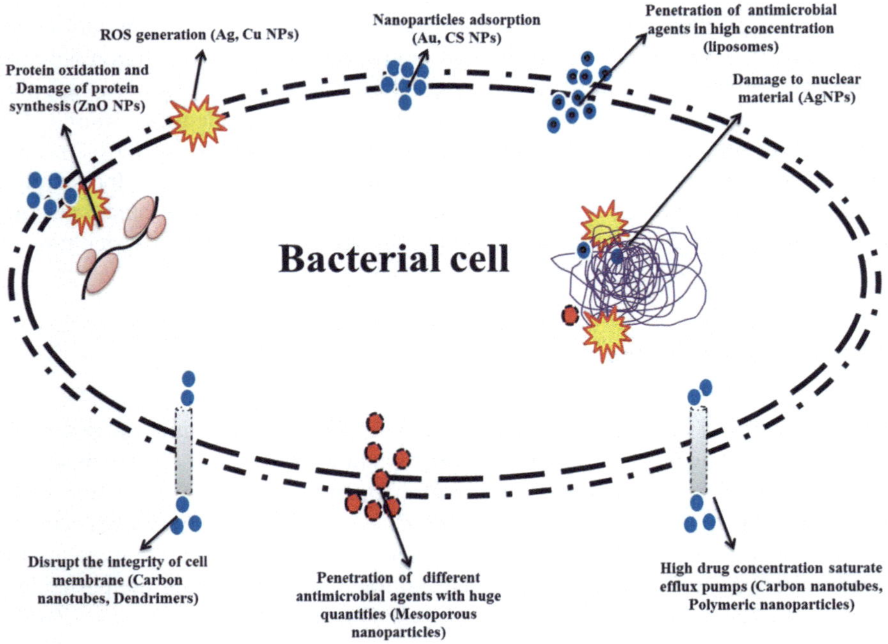

Fig. 11.5 A summary of the mechanisms associated with the antimicrobial behaviour of metal nanoparticles (Mohamed and Abd–Elsalam 2018)

silica-nanoparticle-treated plants showed a higher expression of phenolic compounds (2056 and 743 mg/mL, respectively) in collected leaf extracts and a low expression of stress-responsive enzymes against these fungi. Those findings significantly clarified a higher resistance in maize treated with silica nanoparticles than its bulk material in terms of disease index and expression of total phenols, phenylalanine ammonia lyase, peroxidase, and polyphenol oxidase, at 10 and 15 kg/ha. Consequently, silica nanoparticles can be also used as unique alternative as potent antifungal agent against phytopathogens (Suriyaprabha et al. 2014). This is because silver (Ag^+) ions have the potential to interact with the thiol functional groups in the microbial cell wall, resulting in significant disturbance in the electron transport chain and metabolism activities of fungal cells. Also, in 2017 Tripathi and his co-workers demonstrated another mechanism in the controlling activity of those nanoparticles based on their potential to deform the nucleic acid causing mutations and dissociation of the cell enzyme complexes, disturbed cell permeability and ion efflux, and finally cell lysis (Abdel-Hafez et al. 2016; Tripathi et al. 2017).

Consequently, it was observed that the antimicrobial effect of silver nanoparticles is dependent on particle size and shape and decreases with increasing particle size. In more details, Pal and his co-workers stated that the nanoparticles with truncated triangular shape show greater "cidal" effect compared with those having rod and spherical shapes (Pal et al. 2007). Furthermore, the antimicrobial potential of nanoparticles is also dependent in size, whereas the nanoparticle size decreases as their surface area increases, which permits better chance of interaction with the bacterial cell, producing electronic effects (Raimondi et al. 2005; Shahverdi et al. 2007). This was indicated by different scientists in different reports. (Morones et al. 2005) as an example indicated that the nanoparticle effects vary with their different shapes (truncated triangular =1 μg, spherical =12. 5 μg, and rod = 50–100 μg) (Morones et al. 2005), which have different effects on bacterial cell inhibition. This work was extended by Kim and co-workers and revealed that a well-formed silver nanoparticle (AgNP) solution can act as an excellent nano-fungicide against different microbial pathogens and attributed that to the good adhesion of AgNPs on the microbial cell surface (Kim et al. 2009).

11.6.2 Antiviral Activity

Different types of engineered metal nanoparticles have been studied for their antimicrobial potential and have proven to be lead antibacterial (Bankier et al. 2018; Kim et al. 2006; Morones et al. 2005; Rai et al. 2012; Shahverdi et al. 2007; Singh et al. 2018; Sondi and Salopek-Sondi 2004) and antifungal agents (Kim et al. 2009). However, despite the growing interest in these nanoparticles, there are very few reports on the effectiveness of nano-based materials against plant viruses (Cordero et al. 2017; El-Say and El-Sawy 2017; Elbeshehy et al. 2015). Theoretically, any metal could be analysed for antiviral activity; however, little effort has been done as mentioned above to determine the interactions of metal nanoparticles with viruses in all biological fields. Mainly, only three types of metallic nanoparticles are commonly used – silver, gold, and zinc nanoparticles – in testing the antiviral activity of nanomaterials. In medicine, few studies recently emerged to study the antiviral activity of some gold and silver nanoparticles against different human viruses like monkeypox virus (Rogers et al. 2008), herpes simplex virus (Baram-Pinto et al. 2009), HIV (Elechiguerra et al. 2005; Lara et al. 2010; Sun et al. 2016), hepatitis B virus (Lu et al. 2008), respiratory syncytial virus (Zheng et al. 2008), influenza virus (Papp et al. 2010), and Tacaribe virus (Speshock et al. 2010). Similarly, very few reports were published on the effectiveness of nanomaterials, particularly silver and zinc nanoparticles, against plant viruses like bean yellow mosaic virus (BYMV) (Elbeshehy et al. 2015), potato virus Y (PVY) (Cordero et al. 2017), and cucumber mosaic virus (CMV) (El-Say and El-Sawy 2017) and their use as antiviral agents in agricultural systems (Cordero et al. 2017; El-Say and El-Sawy 2017) (Fig. 11.6).

It is interesting to note that while in case of report published by Elbeshehy and his co-workers in controlling CMV, the authors did not detect any beneficial effect of silver nanoparticles in pre-infection treatment and only remarkable positive results were observed in postinfection treatment (Elbeshehy et al. 2015), in contrast to the reported study published by Cordero and her co-workers that clearly observed a significant beneficial effect of silver nanoparticles in treatments 48 h before and ahead of PVY viral challenging as well. On the other hand, although there are

Fig. 11.6 Effect of mycosynthesized silver nanoparticles on potato Y virus (PVY-Ros1) infectivity: Pictures of representative leaves of tobacco plants mock-treated with different concentrations of silver nanoparticles at 100 and 1000 ppm, as indicated. Two days after treatment, leaves were mechanically inoculated with PVY-Ros1. Pictures were taken at 6 dpi with PVY-Ros1 (Mohamed and Abd–Elsalam 2018)

noticeable antiviral properties of those nanomaterials, the mechanism of AgNP action as an antiviral and virucidal is still an unexplored field (Cordero et al. 2017; El-Say and El-Sawy 2017; Elbeshehy et al. 2015).

In 2015, Elbeshehy and his co-workers studied the effect of silver nanoparticles on bean yellow mosaic virus and suggested their action may come from their ability to bind with a viral envelope glycoprotein and inhibit the virus by binding to the disulphide bond regions of the CD4 binding domain within the yellow mosaic viral envelope glycoprotein gp120 (Elbeshehy et al. 2015). Besides the immediate interaction with glycoprotein of the virus surface, AgNPs may also enter the cell and fulfil their antiviral activity through interactions with the viral nucleic acids. Continually, this strategy concluding in fusion inhibition was later elegantly demonstrated by Lara and her co-workers in their report (Lara et al. 2010). Their obtained results introduced an evidence for the high binding affinity of the used nanoparticles for virus DNA and extracellular virions with different sizes (10 and 50 nm). Moreover, the authors also revealed that the silver nanoparticles also inhibit the production of the virus RNA and extracellular virions in vitro, which was determined using a UV-vis absorption titration assay.

In another study, Sun and his co-workers indicated that AgNPs were superior to gold nanoparticles for cytoprotective activities towards viruses. It is generally understood that Ag, in various forms, inactivates viruses by denaturing enzymes via reactions with sulfhydryl, amino, carboxyl, phosphate, and imidazole groups (Borkow and Gabbay 2004; Rai et al. 2012; Ruparelia et al. 2008). Also, silver nanoparticles were indicated for their ability to interfere with the fusion of the viral membrane, inhibiting viral penetration into the host cell (Mehrbod et al. 2012). It has also been suggested that the antiviral activity of AgNPs depends on the particle size, as well as on the distribution of interacting ligand/receptor molecules (Elbeshehy et al. 2015). However, it is necessary to design studies in vivo to increase their application benefits and minimize adverse effects.

11.6.2.1 Factors Involved in the Control Activity

Different reports indicated that the antimicrobial activity of any nanomaterial is mainly governed by their chemical and physical properties regarding their chemical constitution, size, shape, stability, and surface area and its charge (Kreibig and Vollmer 1995). The unique large surface area of the smaller-sized nanoparticles permits contact with the microbial cell's surface, consequently covering a large area leading to more efficient antimicrobial activity in contrast to the larger ones (Morones et al. 2005). In this regard, different reports confirmed this concept, for example, it was observed that silver nanoparticles (AgNPs) with size less than 10 nm have more antibacterial affinity compared to larger ones, proving that antibacterial activity is size-dependent (Morones et al. 2005; Raimondi et al. 2005).

On the other hand, the nanoparticle's shape (spherical, rod-shaped, nanoshells, nanocages, nanowires, triangular, dimensional, etc.) also plays a critical role in their antimicrobial activity. In 2007, Pan and his co-workers indicated that the nanoparticle's capacity in holding the metallic content varies based on their shape (spherical =12.5 μg, triangular =1 μg, rod = 50–100 μg); consequently, the authors recorded different effects on the growth of the bacterial cell inhibition based on the AgNP shape. Those findings were also confirmed again after studying the cytotoxic effect of gold nanoparticles with different sizes (1.2 nm, 1.8 nm and 15 nm) on four cell lines (HeLa, Sk-Mel-28, L929, J774A1). The MTT assay results revealed that gold nanoparticles with 15 nm sizes have less cytotoxic activity in contrast to those with 1.2 and 1.8 nm size (Pan et al. 2007), which can irreversibly bind to the nucleic acid (DNA) and other cytoplasmic contents as well (Pan et al. 2007). Also, Dasgupta and his co-workers studied the correlation between the nanoparticles size (60, 80 nm) of AgNPs and their antimicrobial efficiency prepared by thermal co-reduction approach. Although Dasgupta et al.'s results indicated that both sizes of AgNPs have broad-spectrum antimicrobial and anticancer activity, the antimicrobial and cytotoxic behaviour of AgNPs with 60-nm size was higher than that with 85 nm (Dasgupta et al. 2016). Furthermore, the surface plasmon resonance of the metallic nanoparticles also plays a major functional role in the determination of the optical density of those nanoparticles, where the nanoparticle size is directly proportional to the wavelength in nm, where the large nanoparticle surface area leads to increase in the intensity of interaction than larger particle sizes in micrometres or more.

11.6.3 Nanoparticles as Delivery System

Smart nano-delivery systems functionalized as carriers to deliver organic/inorganic agrochemicals, nucleic acid, and genes into the living cells have received great attention recently (Zarei et al. 2018). Different types of nanomaterials with potential advantages are used nowadays as reliable vehicle systems to deliver therapeutic and agrochemical agents (i.e. mesoporous silica nanoparticles, magnetic nanoparticles, liposomes, polymersomes, microspheres, nanoclay sheets and polymer conjugates) (Zarei et al. 2018).

The strategy of nano-delivery systems is mainly based on functionalization of the active sites on the nano-delivery's surface, cross-linked with the biomolecules needed to be carried through electrostatic attraction without any modification in its nature. This permits scientists to exploit those nanomaterials that penetrate through cell walls and membranes of plant cells, carrying the biomolecules that need to be delivered accurately, thus solving major critical problems in agriculture sciences. Magnetic nanoparticles like Fe_2O_3 are injected into the drug molecule which is to be delivered; these particles are then guided towards the chosen site under a localized magnetic field. These magnetic carriers can carry large doses of drugs (Bava et al. 2013; Martínez-Banderas et al. 2016). Similarly, silica-coated nanoparticles are also used in drug delivery due to their high stability, surface properties, and compatibility. Silica nanoparticles are also used in biological applications such as artificial implants (Martínez-Pérez et al. 2018).

Furthermore, carbon nanotube is also a best example showing safe interaction with bio-macromolecules and a remarkable potential nano-vector to transfect plant cells with genes of interest (Wang et al. 2014). Liu and his co-workers revealed the efficient ability of carbon-based nanomaterials, particularly carbon nanotubes, to deliver nucleic acids (DNA or RNA) and other small molecules into healthy tobacco plants. Different reports suggested that the multiwalled CNTs (MWCNTs) have a more fascinating ability to influence seed germination and plant growth and work as a delivery system of nucleic acids and other biomolecules and proteins into plant tissue cells (Lahiani et al. 2017). On the other hand, Serag and his co-workers observed that single-stranded DNA molecules wrapped around CNTs were able to target the cytoplasm of walled plant cells. The authors suggested that this unique property can be exploited in introducing RNA pieces into the nucleus to activate or silence the genes of interest. Similarly, protoplast could be a target for delivering larger DNA molecules such as the delivery of plasmids into the plant cell genome. Thus, it is important to take into consideration all possible effects of carbon-based materials used as the DNA-delivery machine on plant genome and proteome (Serag et al. 2011).

11.6.4 Nano-nutrient Supply

Although different reports established the crucial functional role of micronutrients like copper, iron, zinc, and manganese in plant growth and development, a substantial increase of crop yields with the green revolution and new farming practices have progressively decreased the soil micronutrients (Alloway 2008). In this regard, nanotechnology has been used to make micronutrients available to plants. Different types of nano-formulations were produced with efficient potential to deliver reliable amounts of micronutrients either by spray or by supply to soil for root uptake in order to improve soil health and vigour. Different types of nanomaterials have been evaluated for their potential to deliver the desired level of micronutrients into the plants. Chitosan nanoparticle is a best example which is exploited as a natural carrier system to deliver different required plant growth hormones such as 1-naphthylacetic acid at different pH and temperature. The experiment introduced good potential to deliver the hormone in a slow-release manner (Tao et al. 2016). Also, one of the most critical and widespread problems in plant growth comes from iron deficiency in plants growing in high pH and calcareous soils. For this reason, in 2015, Bakhtiari and his co-workers used iron oxide nanoparticles and investigated their effect on wheat growth, yield, and quality at different concentrations (0, 0.01%, 0.02%, 0.03%, and 0.04%) under foliar application. Similarly, Delfani and his co-workers used about 500 mg L^{-1} iron nanoparticles to control black-eyed peas disease and their investigation clearly indicated a significant increase in the number of pods per plant (by 47%), weight of 1000 seeds (by 7%), iron content in leaves (by 34%) and chlorophyll content (by 10%) over those of the controls. More interestingly, their study revealed that the foliar application of iron nanoparticles significantly improved the crop performance better than that by application of a regular iron salt and showed that only 1–5 mg L^{-1} iron in soil solution is required for optimal growth of most plants (Delfani et al. 2014). On the other hand, manganese and zinc nanoparticles have been reported to improve the growth and photosynthesis of mung bean (*Vigna radiata*) and chickpea (*Cicer arietinum*) seedlings (Mahajan et al. 2011); however, it was also observed that the plant growth rates of roots and shoots started to decrease beyond optimal concentrations (Mahajan et al. 2011).

11.6.5 Nanopesticides

The significant broad-spectrum antimicrobial activity of many nanomaterials such as of carbon, silver, silica and alumina silicate nanoparticles encourages scientists to design different in vivo trials in the past recent years to produce nanopesticides in order to fight microbial phytopathogens with more efficient antimicrobial alternatives. This route aims to control or reduce the growing pesticide microbial resistance and avoid the unwise use of synthetic pesticides in large quantities in our environment (Patel et al. 2014). Besides zinc, copper, silver, and titanium metallic nanoparticles as the most common nanomaterials that show efficient antimicrobial agents, different promising material forms from carbon nanomaterials like hollow spheres, ellipsoids (fullerenes) or cylindrical [nanotubes such as single-walled carbon nanotube (SWCNT) and multiwalled carbon nanotubes (MWCNTs)] have been used for the development of nano-antifungal pesticides (Pankaj et al. 2015). In 2014, Gorczyca and his co-workers indicated the significant effect of commercial multiwalled carbon nanotubes (MWCNTs) on the conidial production of the entomopathogenic fungus *Paecilomyces fumosoroseus* (Gorczyca et al. 2014). Their promising results were also confirmed by (Sarlak et al. 2014), who also used the multiwalled carbon nanotube coated with polycitric acid as a carrier for the water-soluble dithiocarbanate fungicides. The encapsulation of pesticide was optimized in the pH range of 6–8 with a stirring time of 30–80 min. The pesticide-loaded nanocarriers exhibited superior fungicidal activity against the leaf spot fungi *Alternaria alternata*, when compared to the bulk form of the fungicides. Similarly, the antifungal activities of MWCNTs with different surface groups against the fungal pathogen *Fusarium graminearum* were explored by (Wang and Cheng 2017). The authors' findings clearly indicate that fungal spore germination was remarkably inhibited by surface-modified MWCNTs, with germination rate being 18%, threefold lower than for pristine MWCNTs.

11.6.6 Nanofertilizers

Different nano-based strategies are already used by agronomists aiming to reduce the excessive losses of plant nutrients as a result of conventional agricultural practices (leaching, evaporation or foliar spraying). Carbon-based nanomaterial is an example showing a promising significant solution regarding this issue, because they can provide a controlled-release technique for fertilizer delivery precisely functionalized to adapt the nutrient supply to the current demand of the target plant, extending the time of function and inhibiting losses by leaching with no risk of facing overdose effects. Most of these fertilizers are based on amendments of mineral and organic fertilizers with nano-carbon, which in most cases acts as a fertilizer synergist with the aim of improving plant nutrient availability, reducing nutrient losses and stimulating plant growth. Several carbon-based nanomaterials have found applications in patents on nanofertilizer formulations (Khodakovskaya et al. 2012; Zhang et al. 2012; Lemraski et al. 2013; LIU et al. 2015; Lahiani et al. 2016). More interestingly, using graphene oxide as a cheap carbon nanomaterial aids the process of commercial large-scale encapsulated fertilizer production with controlled-release technique to be possible with more economic cost (Zhang et al. 2010). Upon successful delivery into the cells, the absorption and utilization of N by the plant can be increased leading to enhanced plant growth. Carbon nanotubes could also be used as a nutrient carrier for macro- and microelements and also as slow-release fertilizers that may reduce their higher concentrations which are usually used (Yatim et al. 2015). Moreover, the nano-carbon material can also be used as a capping or coating agent for slow-release to the target fertilizer. Additionally, incorporation of nano-carbon into slow-release fertilizer introduces a significant merit for reducing water pollution (Wu et al. 2014). In 2012, the scientist Fan and his-co-workers studied the effect of combined application of nitrogen fertilizer and nano-carbon on nitrogen use of soil and rice yield through six treatments. Their results showed a clear increase in nitrogen fertilizer utilization rate after combined application of nano-carbon, which can save the N fertilizer in production practice. Consequently, the authors advised that using combined treatments is so suitable for application and dissemination in soda saline-alkali soil in agriculture. Although significant findings were obtained using those nanomaterials, much work and more knowledge about nanofertilizers in agriculture are still in need. Also, much work should be done by scientists to define the optimized concentrations of different types of nano-carbons in different plant species as organic fertilizer to reach positive impacts that promote plant growth and increase crop yields (Prasad et al. 2017).

11.6.7 Nanoparticles in Water Treatment

Remediation of agricultural pollutants is considered one of the major global challenges up to date. Different conventional and classical protocols have been used in a way to improve the pollutant remediation strategies for environmental sustainability. In the past few decades, activated carbon (AC) has been used as an effective organic sorbent in wastewater treatments due to its unique ability to adsorb a broad spectrum of organic and inorganic contaminants. However, the effectiveness of activated carbon against microbes is relatively low plus its adsorption kinetics is also slow. Furthermore, it is not effective at all in the treatment of heavy wastewater contaminated with oils, greases and hard solids. Furthermore, activated carbon itself is frequently removed together with the adsorbed pollutants and therefore needs to be replaced in regular intervals. In this regard, more interesting types of nanomaterials can be exploited in solving this issue. Carbon-based nanomaterials and gold and magnetic nanoparticles offer an attractive and promising alternative in improving wastewater filtration systems with numerous examples in the available literature (R. Das et al. 2014a; Smith and Rodrigues 2015). Magnetic nanoparticles can be harnessed as adsorbents for separating and removing contaminants in water by applying an external magnetic field (Goyal et al. 2018). Also, the nano-gold bioconjugate was firstly described in water hygiene management by Das and his co-workers. The authors described the single-step removal of some model organo-phosphorus pesticide from contaminated water along with some microorganisms (Das and Kaya 2009). Carbon nanotubes (CNTs) are best example of carbon nanomaterials that have received wide attention due to their unique properties, like large specific surface area with high absorption potential, high thermal stability and high chemical stability. Different reports revealed the unique efficient adsorption capacity of CNTs towards diverse biological and non-biological contaminants like microcystins (cyanobacterial toxins), lead and copper (III), which are even more stronger than that of activated carbon. Furthermore, they can be used for sorption of herbicides or nutrients like nitrogen (N) and phosphorus (Ph) elements in wastewater. On the other hand, fullerenes as well as CNTs exhibit a mobilization potential for various organic pollutants (Misra et al. 2013), such as lindane (agricultural insecticide) (Singh et al. 2011) and persistent polychlorinated biphenyls. More interestingly, Husen and Siddiqi stated that carbon nanotubes and fullerenes have the potential to increase the water-retaining capacity, biomass and fruit yield in plants up to 118% (Husen and Siddiqi 2014).

11.7 Strategies of Nanoparticles Uptake by Plants

Many researchers stated that the process of nanoparticle plant uptake is governed by a number of crucial factors related to the nanoparticles itself (type and morphology, chemical composition, size, and the coating agent). On the other hand, the plant type, its age and the target plant organ are also other important factors that have a predetermined role in the nanoparticle uptake success (Raliya et al. 2016; Schwab et al. 2015). Regarding the nanoparticle size, some scientists could not report the exact nanoparticle size reliable to penetrate and enter the plant cell easily as this depends on the plant type and its age stage. However, other studies reported about the maximum nanoparticle dimensions that should not be more than 40–50 nm as a size exclusion limit that is able to penetrate the plant cell walls and move and accumulate inside the cells (Corredor et al. 2009; Sabo-Attwood et al. 2012; Taylor et al. 2014). Additionally, the coating agent of the used nanoparticles and its functionalization can greatly change and alter cell's properties for its absorption and accumulation by the plant (Blum et al. 2015; Pérez-de-Luque 2017).

On the other hand, the physiological plant variation between two species even in the same botanical family can play a significant role in the nanoparticle uptake as reported by Z. Cifuentes et al. (2010b). Their findings clearly indicated how crops species belonging to different botanical families, and exposed to different types of nanomaterials like silver, gold, titanium, zinc, or magnetic carbon-coated nanoparticles, presented diverse absorption and accumulation patterns inside the plants. Also, the protocol of nanoparticle application plays a crucial role in determining how effectively a plant will internalize the nanomaterials: the plant root cells are particularly specialized in water and nutrient absorption, whereas plant leaves are functionalized for gas exchange and present a cuticle which hampers penetration of substances (Schwab et al. 2015). Most importantly, the possibility of the applied nanoparticles to interact firstly with other environmental components before their absorption by the target plant can affect their properties and their traits for being assimilated by plants. As an example, organic substances found in the soil like humic acids positively improve the nanoparticle function and increase their stability, in contrast to soil salt ions which might induce their precipitation. Even more, the presence of symbiotic microorganisms in the applied soil like mycorrhizal fungi may also help in the improvement of plant nanoparticle uptake (Navarro et al. 2008; Wang et al. 2016).

11.8 How Do Nanoparticles Move Inside Plant Cells?

After penetration of the nanoparticles inside the plant cells, they move through two pathways inside the plant cell tissues: (**1**) the apoplast and (**2**) the symplast pathways. The apoplastic transport pathway occurs outside the plasma cell membrane through the extracellular spaces, cell walls of adjacent cells and xylem vessels (Schwab et al. 2015). The apoplastic pathway allows nanomaterials to reach the vascular tissues and the root central cylinder, for further movement upwards the upper parts of the target plant (González-Melendi et al. 2008; T. Sun et al. 2014a). Once the nanomaterials enter the root central cylinder, they can easily transport towards the upper plant part though the xylem, following the transpiration stream (Z. Cifuentes et al. 2010b; T. Sun et al. 2014a). Most importantly, we should consider

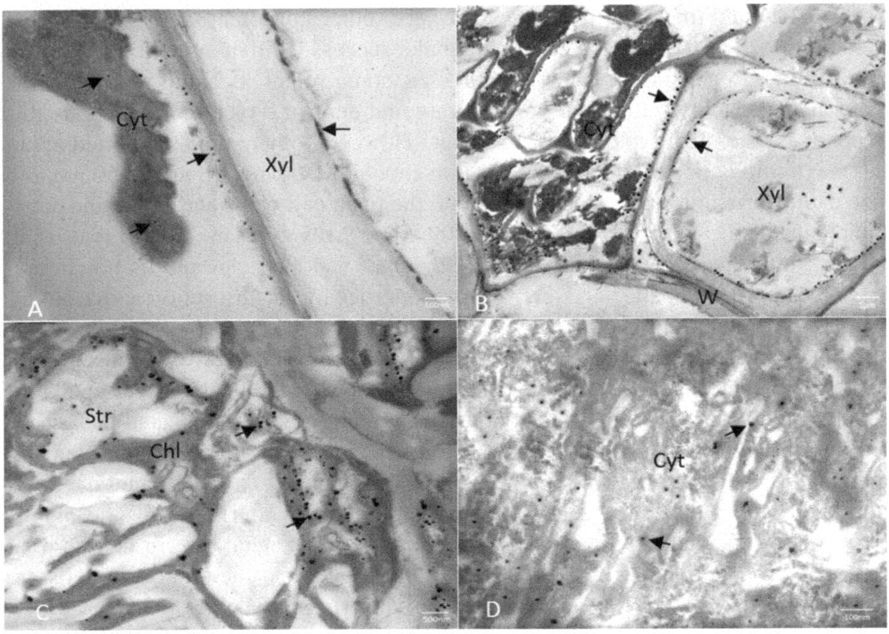

Fig. 11.7 HR-TEM images for ultrathin sections in different parts of tomato plants showing the localization of AgNPs inside their tissues. *Abbreviations*: cytoplasm (*Cyt*), xylem (*Xyl*), cell wall (*W*), stroma (*Str*), and chloroplast (*Chl*) (A, B, stem; C, D, leaves)

that reaching the nanoparticles to the xylem vessels through the root implies crossing a barrier to the apoplastic pathway (Fig. 11.7), the Casparian strip, which must be done following a symplastic way (Pérez-de-Luque 2017) via endodermal cells. Consequently, some nanomaterials may accumulate and stop moving at the Casparian strip (Y. Sun et al. 2014b). In contrast the symplastic transport pathway occurs inside the cytoplasm and involves movement of water and nutrients between the cytoplasm of adjacent cells through specialized structures called plasmodesmata (Schwab et al. 2015). The symplastic transport pathway usually uses the sieve tube elements in the phloem, allowing distribution towards non-photosynthetic tissues and organs (Raliya et al. 2016). This kind of transport is reliable for foliar application; however, the applied nanoparticles must firstly cross the barrier the cuticle presents, following the lipophilic or the hydrophilic pathway. The lipophilic one involves diffusion through cuticle waxes, whereas the hydrophilic pathway is accomplished through polar aqueous pores presented in the cuticle and/or stomata (Eichert et al. 2008). This is because the cuticular pores have a diameter within 2 nm as suggested by Eichert et al. (2008); furthermore, it seems that the stomatal transport pathway for nanoparticle penetration and movement appears as the most likely route, with a size exclusion limit above 10 nm (Eichert et al. 2008). We suggest that determination and understanding the way of nanoparticle penetration and their movement inside the host plant is a really crucial factor for the success of the nanoparticle's job and explaining their results, because it gives indications about the real pathway inside living tissues and what parts of the plant they can reach and where they might end and accumulate. For example, if a kind of nanoparticle is transported mainly through the xylem and not the phloem, they will likely move mainly from root to shoot and leaves and not downwards, so they should be applied to the roots in order to get a good distribution in the plant. On the contrary, if the nanoparticles show a good translocation through the phloem, application should be done via foliar spraying. In addition, nanomaterials moving along the phloem will likely accumulate in plant organs acting as sink, such as fruits and grains, so it is another important consideration when trying to avoid further human or animal ingestion of nanomaterials. However, translocation is not necessarily restricted to a specific cell type, and lateral movement of nanomaterials between xylem and phloem is possible (Gagliardi et al. 2017).

On the other hand, different studies reported that the translocation of nanoparticles inside a plant varies according to the type of the nanoparticles and their surface characteristics and the plant species (Z. Cifuentes et al. 2010b; Zhu et al. 2012). Varna and his co-workers attributed this phenomenon to the negative charges that are found in the plant cell walls, attracting the nanoparticles with positive surface charges and permitting it to be accumulated in the plant tissues and hampering their movement through the plant (Pérez-de-Luque 2017). But our question is, where do the applied nanomaterials go after being translocated to other parts? Typically, nanoparticles will accumulate in certain tissues and organs, in the same way it happens with animals (Pérez-de-Luque 2017). This view was also supported by others and indicated that the plant organs act as strong sinks for fluids, salts and nutrients; hence, the applied nanoparticles are proposed to accumulate while traveling through the plant vascular system, mainly in fruits, grains (Zhao et al. 2015), flowers and young leaves (D. Cifuentes et al. 2010a). This issue is very important for use particularly in evaluating the efficiency of the nanomaterials, and we evaluate and also track their residue if we have no desire to persist for human or animal consumption after the treatment. Most importantly, some of those nanoparticles could be degraded by the plant after some time (Bolaños-Villegas et al. 2017; Wang et al. 2012), or simply being stored in tissues that will not be used after harvesting. In this last case, they are possibly recovered and recycled for further use (Liu et al. 2017).

11.9 Advantages and Disadvantages of Nanoparticles in Agriculture Ecosystems

No one can ignore the great opportunities and unique merits offered by nanotechnology in all aspects of the agricultural fields and food production because of their unique physicochemical properties. Although those innovative studies allow us to constitute an important step forward in elucidating the mechanisms of interaction between plant cells and NPs and thus in designing strategies for using NPs for targeted delivery of many substances and there are many exciting potential applications of NPs, considerable challenges and issues remain to be resolved. For example, nanomaterial remains a major problem if not used wisely, and it is hard to precisely control the number of functional molecules on the surface of those magic bullets. For this reason, researchers are really in need to develop better strategies for producing NPs that have precise composition, uniform surface modification, and reproducible functionalization. For applications, the purity, dispersity, and stability of the NPs in a physiological environment are highly important. Therefore, it is necessary to further study and explore physical and chemical properties for creating successful nano-biotechnology. On the other hand, although the broad-spectrum antimicrobial activity of nanoparticles against different microbial plant pathogens is a well-known and accepted fact worldwide, their impact on soil biota is still less documented. As an example, the negative effect of nanoparticles is found to be more pronounced on denitrifying bacteria, disrupting the process of denitrification in soil (VandeVoort and Arai 2012). As a result, nanoparticles in soil have been used as a model system to evaluate the dose-dependent effects of the metal itself (Mishra et al. 2017). In this viewpoint, Yang et al. (2013) studied the interaction of carbon-coated AgNPs with 35 nm in size and Ag + (provided as $AgNO_3$) with *Pseudomonas stutzeri* (denitrifier), *Azotobacter vinelandii* (nitrogen fixer) and *Nitrosomonas europaea* (nitrifier). They concluded lower toxicity of AgNPs towards these bacteria compared to 20–48 times higher toxicity exerted by the Ag + ions. Conversely, low and sublethal concentrations of Ag + and AgNPs (20–25 μg/L) yielded no significant impact on the expression pattern of denitrifying genes and nitrogen-fixing genes but showed 2.1- to 3.3-fold upregulation in nitrifying genes, indicating the sensitivity of the nitrification process towards silver. Another study representing the impact of nano-silver on aerobic denitrification process by (Mishra et al. 2017) advocated that a low dose of AgNPs had no adverse effect on nitrate reductase activity of *Rhizobium* and *Azotobacter*. Interestingly, 0.2 ppm of AgNP treatment enhanced the process of nitrate reduction in *Azotobacter*. In another study, when nanosized silica-silver particles were applied under in vivo condition to control the fungal disease cucurbit powdery mildew, 100% control was achieved after 25 days. Consequently, many gaps need to be filled in our knowledge regarding the toxicity of those nanomaterials on the environment and the ecological systems. Therefore, when talking about the application of nanomaterials in agro-ecosystems, their major interaction with residing soil biota cannot be ignored. Similarly, more studies are still in need to obtain more explanation and precisely clarify the NP's mode of action, their interaction and location in plant biomass. On the other hand, many researchers discussed safety concerns associated with nanomaterial giving emphasis on the possibility of nanoparticle migration from the packaging material into the food and their impact on consumer's health (Singh et al. 2017). Although a material is being considered as GRAS (generally regarded as safe) substance, additional studies must be acquired to examine the risk of its nano counterparts because the physiochemical properties in nanostates are completely different from that in the macrostate. Moreover, the small size of these nanomaterials may increase the risk for bioaccumulation within body organs and tissues (Savolainen et al. 2010). For example, silica nanoparticles which are used as anti-caking agents can be cytotoxic in human lung cells when subjected to exposure (Periasamy et al. 2015). There are a lot of factors that affect dissolution including surface morphology of the particles, concentration, surface energy and aggregation.

Altogether, those findings clearly indicate a dose-dependent effect of the used nanomaterials on the microbial process of the nitrogen cycle, giving us a clue that entering an optimum concentration of applied nanomaterials into the environment could be favourable for microbial processes with no hindrance in beneficial plant-microbe interactions in agro-ecosystems. On the other hand, size-dependent toxicity of the nanoparticles has also been evidenced by Choi et al. (2008), where they found that AgNPs of size less than 5 nm were more toxic to nitrification bacteria. So, besides the need to understand the possible benefits of applying nanotechnology to agriculture, there is also an urgent need to feel secure about nanomaterial phytotoxicity when applied on crop plants. In consequence, the first step should be to analyse penetration of those particles and the pathway of their transport inside plant cells. Formulation stability is also an important aspect of the biosafety of nanomaterials. Liu et al. (2008) successfully formulated a stable nanopesticide (bifenthrin) using a polymer stabilizer such as poly(acrylic acid)-b-poly(butyl acrylate) (PAA-b-PBA), polyvinylpyrrolidone (PVP) and polyvinyl alcohol (PVOH). While using such techniques commercially, the stability profile of the polymers over an extended time period needs to be firstly considered.

11.10 Conclusion and Future Road Map

Nano-science is in its beginning phase, providing enormous possibility to transform the way of agriculture and food production and lure most scientists to contribute in improving the food safety with innovative green chemistry approaches. Recently, we clearly see how nanotechnology can facilitate additional advantage in food processing, plant disease management and plant growth functional food. In this chapter, we try to offer clean, non-toxic, relatively cost-effective, and environmentally acceptable green chemistry procedures using fungi that have potential to secrete a huge amount of fungal protein and enzymes (like *Chaetomium* spp.) in a large scale and are used as reducing/capping agents in the biosynthesis process of different nanomaterials. The application of green synthesized NPs to control phytopathogens and the formulation of nanopesticides and their smart delivery systems to enhance effectiveness have also been addressed. Importantly, we are still in need of different analytical techniques that permit us real-time, in situ monitoring to optimize production processes, thus reducing waste and controlling energy costs as well as providing mechanistic information.

Despite the progress described within this chapter, there are still considerable research challenges within this newly born field that remain to be addressed. Firstly, we do not yet know which aspects of nanomaterials should be measured, e.g. number, surface area, charges or mass concentration, a combination of these or something else entirely. Secondly, have we learned anything new demonstrating the biological toxicity mechanism more accurately? The answer, with several exceptions, is not really. Consequently, once the biomedical community embraces those nanomaterials as new tools that can be applied safely to any in vivo applications, over longer course of time, we speculate that new knowledge will be obtained on how the living cells work and how they interact with those nanomaterials, both internally and externally.

Acknowledgments MAM gratefully acknowledges the financial support under the EU-WELCOME program of the European Commission. We also thank the Instituto de Biología Molecular y Celular de Plantas (Consejo Superior de Investigaciones Científicas – Universidad Politécnica de Valencia), Valencia, Spain, for the valuable help in recovering relevant articles.

References

Abdel-Hafez SII, Nafady NA, Abdel-Rahim IR, Shaltout AM, Daròs J-A, Mohamed MA (2016) Assessment of protein silver nanoparticles toxicity against pathogenic *Alternaria solani*. 3 Biotech 6:199. https://doi.org/10.1007/s13205-016-0515-6

Abdel-Raouf N, Al-Enazi NM, Ibraheem IBM (2017) Green biosynthesis of gold nanoparticles using *Galaxaura elongata* and characterization of their antibacterial activity. Arab J Chem 10:S3029–S3039. https://doi.org/10.1016/J.ARABJC.2013.11.044

Agnihotri S, Mukherji S, Mukherji S (2014) Size-controlled silver nanoparticles synthesized over the range 5–100 nm using the same protocol and their antibacterial efficacy. RSC Adv 4:3974–3983. https://doi.org/10.1039/C3RA44507K

Alloway BJ (2008) Micronutrient deficiencies in global crop production. Springer, Dordrecht

Bankar A, Joshi B, Kumar AR, Zinjarde S (2010) Banana peel extract mediated novel route for the synthesis of silver nanoparticles. Colloids Surfaces A Physicochem Eng Asp 368:58–63. https://doi.org/10.1016/J.COLSURFA.2010.07.024

Bankier C, Cheong Y, Mahalingam S, Edirisinghe M, Ren G, Cloutman-Green E, Ciric L (2018) A comparison of methods to assess the antimicrobial activity of nanoparticle combinations on bacterial cells. PLoS One 13:e0192093. https://doi.org/10.1371/journal.pone.0192093

Baram-Pinto D, Shukla S, Perkas N, Gedanken A, Sarid R (2009) Inhibition of herpes simplex virus type 1 infection by silver nanoparticles capped with mercaptoethane sulfonate. Bioconjug Chem 20:1497–1502. https://doi.org/10.1021/bc900215b

Bava A, Cappellini F, Pedretti E, Rossi F, Caruso E, Vismara E, Chiriva-Internati M, Bernardini G, Gornati R (2013) Heparin and carboxymethylchitosan metal nanoparticles: an evaluation of their cytotoxicity. Biomed Res Int 2013:314091. https://doi.org/10.1155/2013/314091

Bayram E, Bayram H (2016) Fracture resistance of immature teeth filled with mineral trioxide aggregate, bioaggregate, and biodentine. Eur J Dent 10:220. https://doi.org/10.4103/1305-7456.178310

Bhattacharya R, Mukherjee P (2008) Biological properties of "naked" metal nanoparticles☆. Adv Drug Deliv Rev 60:1289–1306. https://doi.org/10.1016/j.addr.2008.03.013

Birla SS, Tiwari VV, Gade AK, Ingle AP, Yadav AP, Rai MK (2009) Fabrication of silver nanoparticles by *Phoma glomerata* and its combined effect against *Escherichia coli*, *Pseudomonas aeruginosa* and *Staphylococcus aureus*. Lett Appl Microbiol 48:173–179. https://doi.org/10.1111/j.1472-765X.2008.02510.x

Birla SS, Gaikwad SC, Gade AK, Rai MK (2013) Rapid synthesis of silver nanoparticles from *Fusarium oxysporum* by optimizing physicocultural conditions. Sci World J 2013:796018. https://doi.org/10.1155/2013/796018

Blum JL, Edwards JR, Prozialeck WC, Xiong JQ, Zelikoff JT (2015) Effects of maternal exposure to cadmium oxide nanoparticles during pregnancy on maternal and offspring kidney injury markers using a murine model. J Toxicol Environ Health A 78:711–724. https://doi.org/10.1080/15287394.2015.1026622

Bolaños-Villegas P, De K, Pradillo M, Liu D, Makaroff CA (2017) In favor of establishment: regulation of chromatid cohesion in plants. Front Plant Sci 8:846. https://doi.org/10.3389/fpls.2017.00846

Borkow G, Gabbay J (2004) Putting copper into action: copper-impregnated products with potent biocidal activities. FASEB J 18:1728–1730. https://doi.org/10.1096/fj.04-2029fje

Calderón-Jiménez B, Johnson ME, Montoro Bustos AR, Murphy KE, Winchester MR, Vega Baudrit JR (2017) Silver nanoparticles: technological advances, societal impacts, and metrological challenges. Front Chem 5:6. https://doi.org/10.3389/fchem.2017.00006

Chen X, Schluesener HJ (2008) Nanosilver: a nanoproduct in medical application. Toxicol Lett 176:1–12. https://doi.org/10.1016/j.toxlet.2007.10.004

Chen JC, Lin ZH, Ma XX (2003) Evidence of the production of silver nanoparticles via pretreatment of Phoma sp.3.2883 with silver nitrate. Lett Appl Microbiol 37:105–108. https://doi.org/10.1046/j.1472-765X.2003.01348.x

Chen PC, Mwakwari SC, Oyelere AK (2008) Gold nanoparticles: From nanomedicine to nanosensing. Nanotechnol Sci Appl 1:45–65

Choi O, Deng KK, Kim N-J, Ross L, Surampalli RY, Hu Z (2008) The inhibitory effects of silver nanoparticles, silver ions, and silver chloride colloids on microbial growth. Water Res 42:3066–3074. https://doi.org/10.1016/j.watres.2008.02.021

Cifuentes D, Xue H, Taylor DW, Patnode H, Mishima Y, Cheloufi S, Ma E, Mane S, Hannon GJ, Lawson ND, Wolfe SA, Giraldez AJ (2010a) A novel miRNA processing pathway independent of Dicer requires Argonaute2 catalytic activity. Science 328:1694–1698. https://doi.org/10.1126/science.1190809

Cifuentes Z, Custardoy L, de la Fuente JM, Marquina C, Ibarra MR, Rubiales D, Pérez-de-Luque A (2010b) Absorption and translocation to the aerial part of magnetic carbon-coated nanoparticles through the root of different crop plants. J Nanobiotechnol 8:26. https://doi.org/10.1186/1477-3155-8-26

Consoli L, Grimaldi R, Sartori T, Menegalli FC, Hubinger MD (2016) Gallic acid microparticles produced by spray chilling technique: production and characterization. LWT- Food Sci Technol 65:79–87. https://doi.org/10.1016/J.LWT.2015.07.052

Cordero T, Mohamed MA, López-Moya J-J, Daròs J-A (2017) A recombinant potato virus Y infectious clone tagged with the rosea1 visual marker (PVY-Ros1) facilitates the analysis of viral infectivity and allows the production of large amounts of anthocyanins in plants. Front Microbiol 8:611. https://doi.org/10.3389/fmicb.2017.00611

Corredor E, Testillano PS, Coronado M-J, González-Melendi P, Fernández-Pacheco R, Marquina C, Ibarra MR, de la Fuente JM, Rubiales D, Pérez-de-Luque A, Risueño M-C (2009) Nanoparticle penetration and transport in living pumpkin plants: in situ subcellular identification. BMC Plant Biol 9:45. https://doi.org/10.1186/1471-2229-9-45

Dadosh T, Sperling J, Bryant GW, Breslow R, Shegai T, Dyshel M, Haran G, Bar-Joseph I (2009) Plasmonic control of the shape of the raman spectrum of a single molecule in a silver nanoparticle dimer. ACS Nano 3:1988–1994. https://doi.org/10.1021/nn900422w

Danwanichakul P, Suwatthanarak T, Suwanvisith C, Danwanichakul D (2016) The role of ammonia in synthesis of silver nanoparticles in skim natural rubber latex. J Nanosci 2016:1–6. https://doi.org/10.1155/2016/7258313

Das YK, Kaya S (2009) Organophosphorus insecticide residues in honey produced in Turkey. Bull Environ Contam Toxicol 83:378–383. https://doi.org/10.1007/s00128-009-9778-5

Das R, Bee S, Hamid A, Ali E, Fauzi Ismail A, Annuar M, Ramakrishna S (2014a) Multifunctional carbon nanotubes in water treatment: the present, past and future. DES 354:160–179. https://doi.org/10.1016/j.desal.2014.09.032

Das VL, Thomas R, Varghese RT, Soniya EV, Mathew J, Radhakrishnan EK (2014b) Extracellular synthesis of silver nanoparticles by the Bacillus strain CS 11 isolated from industrialized area. 3 Biotech 4:121–126. https://doi.org/10.1007/s13205-013-0130-8

Dasgupta N, Ranjan S, Rajendran B, Manickam V, Ramalingam C, Avadhani GS, Kumar A (2016) Thermal co-reduction approach to vary size of silver nanoparticle: its microbial and cellular toxicology. Environ Sci Pollut Res 23:4149–4163. https://doi.org/10.1007/s11356-015-4570-z

Delfani M, Baradarn Firouzabadi M, Farrokhi N, Makarian H (2014) Some physiological responses of black-eyed pea to iron and magnesium nanofertilizers. Commun Soil Sci Plant Anal 45:530–540. https://doi.org/10.1080/00103624.2013.863911

Duan H, Wang D, Li Y (2015) Green chemistry for nanoparticle synthesis. Chem Soc Rev 44:5778–5792. https://doi.org/10.1039/c4cs00363b

Durán N, Cuevas R, Cordi L, Rubilar O, Diez MC (2014) Biogenic silver nanoparticles associated with silver chloride nanoparticles (Ag@AgCl) produced by laccase from Trametes versicolor. Springerplus 3:645. https://doi.org/10.1186/2193-1801-3-645

Eichert T, Kurtz A, Steiner U, Goldbach HE (2008) Size exclusion limits and lateral heterogeneity of the stomatal foliar uptake pathway for aqueous solutes and water-suspended nanoparticles. Physiol Plant 134:151–160. https://doi.org/10.1111/j.1399-3054.2008.01135.x

Elavazhagan T, Elavazhagan T (2011) Memecylon edule leaf extract mediated green synthesis of silver and gold nanoparticles. Int J Nanomedicine 6:1265. https://doi.org/10.2147/IJN.S18347

Elbeshehy EKF, Elazzazy AM, Aggelis G (2015) Silver nanoparticles synthesis mediated by new isolates of Bacillus spp., nanoparticle characterization and their activity against Bean Yellow Mosaic Virus and human pathogens. Front Microbiol 6:453. https://doi.org/10.3389/fmicb.2015.00453

Elechiguerra J, Burt JL, Morones JR, Camacho-Bragado A, Gao X, Lara HH, Yacaman M (2005) Interaction of silver nanoparticles with HIV-1. J Nanobiotechnology 3:6. https://doi.org/10.1186/1477-3155-3-6

El-Say KM, El-Sawy HS (2017) Polymeric nanoparticles: promising platform for drug delivery. Int J Pharm 528:675–691. https://doi.org/10.1016/j.ijpharm.2017.06.052

Finney EE, Finke RG (2008) Nanocluster nucleation and growth kinetic and mechanistic studies: a review emphasizing transition-metal nanoclusters. J Colloid Interface Sci 317:351–374. https://doi.org/10.1016/j.jcis.2007.05.092

Gagliardi M, Bertero A, Bifone A (2017) Molecularly imprinted biodegradable nanoparticles. Sci Rep 7:40046. https://doi.org/10.1038/srep40046

Gardea-Torresdey JL, Rico CM, White JC (2014) Trophic transfer, transformation, and impact of engineered nanomaterials in terrestrial environments. Environ Sci Technol 48:2526–2540. https://doi.org/10.1021/es4050665

Ghorbani HR (2017) Biosynthesis of nanosilver particles using extract of Salmonella typhirium. Arab J Chem 10:S1699–S1702. https://doi.org/10.1016/J.ARABJC.2013.06.017

Golinska P, Wypij M, Ingle AP, Gupta I, Dahm H, Rai M (2014) Biogenic synthesis of metal nanoparticles from actinomycetes: biomedical applications and cytotoxicity. Appl Microbiol Biotechnol 98:8083–8097. https://doi.org/10.1007/s00253-014-5953-7

González-Melendi P, Fernández-Pacheco R, Coronado MJ, Corredor E, Testillano PS, Risueño MC, Marquina C, Ibarra MR, Rubiales D, Pérez-de-Luque A (2008) Nanoparticles as smart treatment-delivery systems in plants: assessment of different techniques of microscopy for their visualization in plant tissues. Ann Bot 101:187–195. https://doi.org/10.1093/aob/mcm283

Gorczyca DA, Younger S, Meltzer S, Kim SE, Cheng L, Song W, Lee HY, Jan LY, Jan YN (2014) Identification of Ppk26, a DEG/ENaC channel functioning with Ppk1 in a mutually dependent manner to guide locomotion behavior in drosophila. Cell Rep 9:1446–1458. https://doi.org/10.1016/J.CELREP.2014.10.034

Gorham JM, MacCuspie RI, Klein KL, Fairbrother DH, Holbrook RD (2012) UV-induced photochemical transformations of citrate-capped silver nanoparticle suspensions. JNanoparticle Res 14:1139. https://doi.org/10.1007/s11051-012-1139-3

Goyal P, Chakraborty S, Misra SK (2018) Multifunctional Fe_3O_4-ZnO nanocomposites for environmental remediation applications. Environ Nanotechnol Monit Manag 10:28–35. https://doi.org/10.1016/J.ENMM.2018.03.003

Haider A, Kang I-K (2015) Preparation of silver nanoparticles and their industrial and biomedical applications: a comprehensive review. Adv Mater Sci Eng 2015:1–16. https://doi.org/10.1155/2015/165257

Nanotechnology and plant sciences: nanoparticles and their impact on plants – Google Books [WWW Document] (n.d.) URL https://books.google.com.eg/books?id=wd9sBgAAQBAJ&pg=PA261&lpg=PA261&dq=Renugadevi+and+Aswini,+2012,+nanoparticles&source=bl&ots=QGWwN2orf_&sig=NXNjBZmCrj1xhvnC_O1gLsGU7Vk&hl=en&sa=X&ved=0ahUKEwidhcKgyc7aAhXMGewKHZJOAKcQ6AEIMjAB#v=onepage&q=Renugadevi a (Accessed 4.22.18)

Huang J, Li Q, Sun D, Lu Y, Su Y, Yang X, Wang H, Wang Y, Shao W, He N, Hong J, Chen C (2007) Biosynthesis of silver and gold nanoparticles by novel sundried *Cinnamomum camphora* leaf. Nanotechnology 18:105104. https://doi.org/10.1088/0957-4484/18/10/105104

Hung P, Wattanachai P, Kasem S, Poaim S (2015) Biological control of phytophthora palmivora causing root rot of pomelo using *Chaetomium* spp. Mycobiology 43:63–70. https://doi.org/10.5941/MYCO.2015.43.1.63

Husen A, Siddiqi KS (2014) Phytosynthesis of nanoparticles: concept, controversy and application. Nanoscale Res Lett 9:229. https://doi.org/10.1186/1556-276X-9-229

Husseiny SM, Salah TA, Anter HA (2015) Biosynthesis of size controlled silver nanoparticles by *Fusarium oxysporum*, their antibacterial and antitumor activities. Beni-Suef Univ J Basic Appl Sci 4:225–231. https://doi.org/10.1016/J.BJBAS.2015.07.004

Iravani S (2011) Green synthesis of metal nanoparticles using plants. Green Chem 13:2638. https://doi.org/10.1039/c1gc15386b

Jana NR, Gearheart L, Murphy CJ (2001) Seeding growth for size control of 5–40 nm diameter gold nanoparticles. Langmuir 17:6782–6786. https://doi.org/10.1021/LA0104323

Jyoti K, Baunthiyal M, Singh A (2016) Characterization of silver nanoparticles synthesized using Urtica dioica Linn. leaves and their synergistic effects with antibiotics. J Radiat Res Appl Sci 9:217–227. https://doi.org/10.1016/J.JRRAS.2015.10.002

Kanmani P, Lim ST (2013) Synthesis and characterization of pullulan-mediated silver nanoparticles and its antimicrobial activities. Carbohydr Polym 97:421–428. https://doi.org/10.1016/j.carbpol.2013.04.048

Karthiga P (2017) Preparation of silver nanoparticles by *Garcinia mangostana* stem extract and investigation of the antimicrobial properties. Biotechnol Res Innov. https://doi.org/10.1016/J.BIORI.2017.11.001

Kasthuri J, Veerapandian S, Rajendiran N (2009) Biological synthesis of silver and gold nanoparticles using apiin as reducing agent. Colloids Surf B Biointerfaces 68:55–60. https://doi.org/10.1016/j.colsurfb.2008.09.021

Kathiresan K, Manivannan S, Nabeel MA, Dhivya B (2009) Studies on silver nanoparticles synthesized by a marine fungus, *Penicillium fellutanum* isolated from coastal mangrove sediment. Colloids Surf B Biointerfaces 71:133–137. https://doi.org/10.1016/j.colsurfb.2009.01.016

Khodakovskaya MV, de Silva K, Biris AS, Dervishi E, Villagarcia H (2012) Carbon nanotubes induce growth enhancement of tobacco cells. ACS Nano 6:2128–2135. https://doi.org/10.1021/nn204643g

Kim JS, Yoon T-J, Yu KN, Kim BG, Park SJ, Kim HW, Lee KH, Park SB, Lee J-K, Cho MH (2006) Toxicity and tissue distribution of magnetic nanoparticles in mice. Toxicol Sci 89:338–347. https://doi.org/10.1093/toxsci/kfj027

Kim S, Choi JE, Choi J, Chung K-H, Park K, Yi J, Ryu D-Y (2009) Oxidative stress-dependent toxicity of silver nanoparticles in human hepatoma cells. Toxicol In Vitro 23:1076–1084. https://doi.org/10.1016/j.tiv.2009.06.001

Kraynov A, Müller T (2011) Concepts for the stabilization of metal nanoparticles in ionic liquids. In: Applications of ionic liquids in science and technology. InTech, Rijeka. https://doi.org/10.5772/22111
Kreibig U, Vollmer M (1995) Optical properties of metal clusters. Springer, Berlin
Król A, Pomastowski P, Rafińska K, Railean-Plugaru V, Buszewski B (2017) Zinc oxide nanoparticles: synthesis, antiseptic activity and toxicity mechanism. Adv Colloid Interf Sci 249:37–52. https://doi.org/10.1016/j.cis.2017.07.033
Kumar V, Yadav SK (2009) Plant-mediated synthesis of silver and gold nanoparticles and their applications. JChem Technol Biotechnol 84:151–157. https://doi.org/10.1002/jctb.2023
Kumar S, Gandhi KS, Kumar R (2006) Modeling of formation of gold nanoparticles by citrate method. https://doi.org/10.1021/IE060672J
Lahiani MH, Dervishi E, Ivanov I, Chen J, Khodakovskaya M (2016) Comparative study of plant responses to carbon-based nanomaterials with different morphologies. Nanotechnology 27:265102. https://doi.org/10.1088/0957-4484/27/26/265102
Lahiani MH, Eassa S, Parnell C, Nima Z, Ghosh A, Biris AS, Khodakovskaya MV (2017) Carbon nanotubes as carriers of *Panax ginseng* metabolites and enhancers of ginsenosides Rb1 and Rg1 anti-cancer activity. Nanotechnology 28:15101. https://doi.org/10.1088/0957-4484/28/1/015101
Lara HH, Ayala-Nuñez NV, Ixtepan-Turrent L, Rodriguez-Padilla C (2010) Mode of antiviral action of silver nanoparticles against HIV-1. J Nanobiotechnol 8:1. https://doi.org/10.1186/1477-3155-8-1
Lee PC, Meisel D (1982) Adsorption and surface-enhanced Raman of dyes on silver and gold sols. J Phys Chem 86:3391–3395. https://doi.org/10.1021/j100214a025
Lemraski MG, Normohamadi G, Madani H, Heidari H, Abad S, Mobasser HR (2013) Two Iranian rice cultivars' response to nitrogen and nano-fertilizer. Open J Ecol 7:591–603. https://doi.org/10.4236/oje.2017
Liu H, Zhao M, Zhang C, Ma Y, Liu W (2008) Enantioselective cytotoxicity of the insecticide bifenthrin on a human amnion epithelial (FL) cell line. Toxicology 253:89–96. https://doi.org/10.1016/j.tox.2008.08.015
Liu Y, Li Y, Peng Z, Wang Y, Ma S, Guo L, Lin E, Han X (2015) Effects of different nitrogen fertilizer management practices on wheat yields and N_2O emissions from wheat fields in North China. J Integr Agric 14:1184–1191. https://doi.org/10.1016/S2095-3119(14)60867-4
Liu H, Zhang H, Wang J, Wei J (2017) Effect of temperature on the size of biosynthesized silver nanoparticle: deep insight into microscopic kinetics analysis. Arab J Chem. https://doi.org/10.1016/J.ARABJC.2017.09.004
Lu L, Sun RW-Y, Chen R, Hui C-K, Ho C-M, Luk JM, Lau GKK, Che C-M (2008) Silver nanoparticles inhibit hepatitis B virus replication. Antivir Ther 13:253–262
Mahajan P, Dhoke SK, Khanna AS (2011) Effect of nano-ZnO particle suspension on growth of mung (*Vigna radiata*) and gram (*Cicer arietinum*) seedlings using plant agar method. J Nanotechnol 2011:1–7. https://doi.org/10.1155/2011/696535
Mallikarjuna K, John Sushma N, Narasimha G, Manoj L, Deva Prasad Raju B (2014) Phytochemical fabrication and characterization of silver nanoparticles by using Pepper leaf broth. Arab J Chem 7:1099–1103. https://doi.org/10.1016/J.ARABJC.2012.04.001
Mandal D, Bolander ME, Mukhopadhyay D, Sarkar G, Mukherjee P (2006) The use of microorganisms for the formation of metal nanoparticles and their application. Appl Microbiol Biotechnol 69:485–492. https://doi.org/10.1007/s00253-005-0179-3
Manimaran M, Kannabiran K (2017) Actinomycetes-mediated biogenic synthesis of metal and metal oxide nanoparticles: progress and challenges. Lett Appl Microbiol 64:401–408. https://doi.org/10.1111/lam.12730
Martínez-Banderas AI, Aires A, Teran FJ, Perez JE, Cadenas JF, Alsharif N, Ravasi T, Cortajarena AL, Kosel J (2016) Functionalized magnetic nanowires for chemical and magneto-mechanical induction of cancer cell death. Sci Rep 6:35786. https://doi.org/10.1038/srep35786

Martínez-Pérez B, Quintanar-Guerrero D, Tapia-Tapia M, Cisneros-Tamayo R, Zambrano-Zaragoza ML, Alcalá-Alcalá S, Mendoza-Muñoz N, Piñón-Segundo E (2018) Controlled-release biodegradable nanoparticles: from preparation to vaginal applications. Eur J Pharm Sci 115:185–195. https://doi.org/10.1016/j.ejps.2017.11.029

Maruyama CR, Guilger M, Pascoli M, Bileshy-José N, Abhilash PC, Fraceto LF, de Lima R (2016) Nanoparticles based on chitosan as carriers for the combined herbicides imazapic and imazapyr. Sci Rep 6:19768. https://doi.org/10.1038/srep19768

Mehmood A, Murtaza G, Bhatti TM, Kausar R (2017) Phyto-mediated synthesis of silver nanoparticles from *Melia azedarach* L. leaf extract: characterization and antibacterial activity. Arab J Chem 10:S3048–S3053. https://doi.org/10.1016/J.ARABJC.2013.11.046

Mehrbod P, Ideris A, Omar AR, Hair-Bejo M, Tan SW, Kheiri MT, Tabatabaian M (2012) Attenuation of influenza virus infectivity with herbal-marine compound (HESA-A): an in vitro study in MDCK cells. Virol J 9:44. https://doi.org/10.1186/1743-422X-9-44

Meng Y, Zhang Q, Ding W, Shan W (2014) *Phytophthora parasitica*: a model oomycete plant pathogen. Mycology 5:43–51. https://doi.org/10.1080/21501203.2014.917734

Metuku RP, Pabba S, Burra S, Hima Bindu NSV, Gudikandula K, Singara Charya MA (2014) Biosynthesis of silver nanoparticles from *Schizophyllum radiatum* HE 863742.1: their characterization and antimicrobial activity. 3 Biotech 4:227–234. https://doi.org/10.1007/s13205-013-0138-0

Mie R, Samsudin MW, Din LB, Ahmad A (2012) Green synthesis of silver nanoparticles using two lichens species: *Parmotrema praesorediosum* and *Ramalina dumeticola*. Appl Mech Mater 229–231:256–259. https://doi.org/10.4028/www.scientific.net/AMM.229-231.256

Mishra S, Keswani C, Abhilash PC, Fraceto LF, Singh HB (2017) Integrated approach of agri-nanotechnology: challenges and future trends. Front Plant Sci 8:471. https://doi.org/10.3389/fpls.2017.00471

Misra R, Gautam P, Mobin SM (2013) Aryl-substituted unsymmetrical benzothiadiazoles: synthesis, structure, and properties. J Org Chem 78:12440–12452. https://doi.org/10.1021/jo402111q

Moazeni M, Rashidi N, Shahverdi AR, Noorbakhsh F, Rezaie S (2012) Extracellular production of silver nanoparticles by using three common species of dermatophytes: *Trichophyton rubrum*, *Trichophyton mentagrophytes* and *Microsporum canis*. Iran Biomed J 16:52–58. https://doi.org/10.6091/IBJ.1001.2012

Mohamed MA, Abd–Elsalam KA (2018) Nanoantimicrobials for plant pathogens control: potential applications and mechanistic aspects. Springer, Cham, pp 87–109. https://doi.org/10.1007/978-3-319-91161-8_4

Mohamed AE-MA, Mohamed MA, Vega V, Hernando B, Ahmed AM (2016) Tuning magnetoresistive and magnetocaloric properties via grain boundaries engineering in granular manganites. RSC Adv 6:77284–77290. https://doi.org/10.1039/C6RA15874A

Mohanpuria P, Rana NK, Sudesh A, Yadav K (2007) Biosynthesis of nanoparticles: technological concepts and future applications. https://doi.org/10.1007/s11051-007-9275-x

Morones JR, Elechiguerra JL, Camacho A, Holt K, Kouri JB, Ramírez JT, Yacaman MJ (2005) The bactericidal effect of silver nanoparticles. Nanotechnology 16:2346–2353. https://doi.org/10.1088/0957-4484/16/10/059

Mühlig S, Cunningham A, Dintinger J, Farhat M, Hasan SB, Scharf T, Bürgi T, Lederer F, Rockstuhl C (2013) A self-assembled three-dimensional cloak in the visible. Sci Rep 3:2328. https://doi.org/10.1038/srep02328

Nagajyothi PC, Lee KD (2011) Synthesis of plant-mediated silver nanoparticles using *Dioscorea batatas* rhizome extract and evaluation of their antimicrobial activities. J Nanomater 2011:1–7. https://doi.org/10.1155/2011/573429

Navarro E, Baun A, Behra R, Hartmann NB, Filser J, Miao A-J, Quigg A, Santschi PH, Sigg L (2008) Environmental behavior and ecotoxicity of engineered nanoparticles to algae, plants, and fungi. Ecotoxicology 17:372–386. https://doi.org/10.1007/s10646-008-0214-0

Pal S, Tak YK, Song JM (2007) Does the antibacterial activity of silver nanoparticles depend on the shape of the nanoparticle? A study of the Gram-negative bacterium *Escherichia coli*. Appl Environ Microbiol 73:1712–1720. https://doi.org/10.1128/AEM.02218-06

Pan Y, Neuss S, Leifert A, Fischler M, Wen F, Simon U, Schmid G, Brandau W, Jahnen-Dechent W (2007) Size-dependent cytotoxicity of gold nanoparticles. Small 3:1941–1949. https://doi.org/10.1002/smll.200700378

Pankaj SG, Anita Sharma PK, Khati P, Sharma A (2015) Mycoremediation of imidaclopridin the presence of different soil amendments using *Trichoderma longibrachiatum* and aspergillusoryzae isolated from pesticide contaminated agricultural fields of Uttarakhand. J Bioremed Biodegr 6:1–5. https://doi.org/10.4172/2155-6199.1000310

Papp I, Sieben C, Ludwig K, Roskamp M, Böttcher C, Schlecht S, Herrmann A, Haag R (2010) Inhibition of influenza virus infection by multivalent sialic-acid-functionalized gold nanoparticles. Small 6:2900–2906. https://doi.org/10.1002/smll.201001349

Parial D, Patra HK, Dasgupta AK, Pal R (2012) Screening of different algae for green synthesis of gold nanoparticles. Eur J Phycol 47:22–29. https://doi.org/10.1080/09670262.2011.653406

Parikh RY, Singh S, Prasad BLV, Patole MS, Sastry M, Shouche YS (2008) Extracellular synthesis of crystalline silver nanoparticles and molecular evidence of silver resistance from *Morganella* sp.: towards understanding biochemical synthesis mechanism. Chembiochem 9:1415–1422. https://doi.org/10.1002/cbic.200700592

Patel AP, Tirosh I, Trombetta JJ, Shalek AK, Gillespie SM, Wakimoto H, Cahill DP, Nahed BV, Curry WT, Martuza RL, Louis DN, Rozenblatt-Rosen O, Suva ML, Regev A, Bernstein BE (2014) Single-cell RNA-seq highlights intratumoral heterogeneity in primary glioblastoma. Science 344:1396–1401. https://doi.org/10.1126/science.1254257

Paul B, Girard I, Bhatnagar T, Bouchet P (1997) Supression of Botrytis cinerea causing grey mould disease of grape vine (Vitis vinifera) and its pectinolytic activities by a soil bacterium. Microbiol Res 152:413–420. https://doi.org/10.1016/S0944-5013(97)80060-2

Pérez-de-Luque A (2017) Interaction of nanomaterials with plants: what do we need for real applications in agriculture? Front Environ Sci 5:12. https://doi.org/10.3389/fenvs.2017.00012

Periasamy VS, Athinarayanan J, Al-Hadi AM, Al Juhaimi F, Alshatwi AA (2015) Effects of titanium dioxide nanoparticles isolated from confectionery products on the metabolic stress pathway in human lung fibroblast cells. Arch Environ Contam Toxicol 68:521–533. https://doi.org/10.1007/s00244-014-0109-4

Prasad R, Bhattacharyya A, Nguyen QD (2017) Nanotechnology in sustainable agriculture: recent developments, challenges, and perspectives. Front Microbiol 8:1014. https://doi.org/10.3389/fmicb.2017.01014

Qian Y, Yu H, He D, Yang H, Wang W, Wan X, Wang L (2013) Biosynthesis of silver nanoparticles by the endophytic fungus *Epicoccum nigrum* and their activity against pathogenic fungi. Bioprocess Biosyst Eng 36:1613–1619. https://doi.org/10.1007/s00449-013-0937-z

Rai M, Yadav A, Gade A (2009) Silver nanoparticles as a new generation of antimicrobials. Biotechnol Adv 27:76–83. https://doi.org/10.1016/j.biotechadv.2008.09.002

Rai MK, Deshmukh SD, Ingle AP, Gade AK (2012) Silver nanoparticles: the powerful nanoweapon against multidrug-resistant bacteria. J Appl Microbiol 112:841–852. https://doi.org/10.1111/j.1365-2672.2012.05253.x

Raimondi F, Scherer GG, Kötz R, Wokaun A (2005) Nanoparticles in energy technology: examples from electrochemistry and catalysis. Angew Chemie Int Ed 44:2190–2209. https://doi.org/10.1002/anie.200460466

Raliya R, Franke C, Chavalmane S, Nair R, Reed N, Biswas P (2016) Quantitative understanding of nanoparticle uptake in watermelon plants. Front Plant Sci 7:1288. https://doi.org/10.3389/fpls.2016.01288

Ramanathan R, Field MR, O'Mullane AP, Smooker PM, Bhargava SK, Bansal V (2013) Aqueous phase synthesis of copper nanoparticles: a link between heavy metal resistance and nanoparticle synthesis ability in bacterial systems. Nanoscale 5:2300–2306. https://doi.org/10.1039/c2nr32887a

Riddin TL, Gericke M, Whiteley CG (2006) Analysis of the inter- and extracellular formation of platinum nanoparticles by *Fusarium oxysporum* f. sp. *lycopersici* using response surface methodology. Nanotechnology 17:3482–3489. https://doi.org/10.1088/0957-4484/17/14/021

Rogers JV, Parkinson CV, Choi YW, Speshock JL, Hussain SM (2008) A preliminary assessment of silver nanoparticle inhibition of monkeypox virus plaque formation. Nanoscale Res Lett 3:129–133. https://doi.org/10.1007/s11671-008-9128-2

Ruparelia JP, Chatterjee AK, Duttagupta SP, Mukherji S (2008) Strain specificity in antimicrobial activity of silver and copper nanoparticles. Acta Biomater 4:707–716. https://doi.org/10.1016/j.actbio.2007.11.006

Sabo-Attwood T, Unrine JM, Stone JW, Murphy CJ, Ghoshroy S, Blom D, Bertsch PM, Newman LA (2012) Uptake, distribution and toxicity of gold nanoparticles in tobacco (*Nicotiana xanthi*) seedlings. Nanotoxicology 6:353–360. https://doi.org/10.3109/17435390.2011.579631

Sagar G, Ashok B (2012) Green synthesis of silver nanoparticles using *Aspergillus niger* and its efficacy against human pathogens. Eur J Exp Biol 2:1654–1658

Salem W, Leitner DR, Zingl FG, Schratter G, Prassl R, Goessler W, Reidl J, Schild S (2015) Antibacterial activity of silver and zinc nanoparticles against Vibrio cholerae and enterotoxic *Escherichia coli*. Int J Med Microbiol 305:85–95. https://doi.org/10.1016/j.ijmm.2014.11.005

Sarlak N, Taherifar A, Salehi F (2014) Synthesis of nanopesticides by encapsulating pesticide nanoparticles using functionalized carbon nanotubes and application of new nanocomposite for plant disease treatment. J Agric Food Chem 62:4833–4838. https://doi.org/10.1021/jf404720d

Savolainen K, Alenius H, Norppa H, Pylkkänen L, Tuomi T, Kasper G (2010) Risk assessment of engineered nanomaterials and nanotechnologies—a review. Toxicology 269:92–104. https://doi.org/10.1016/j.tox.2010.01.013

Schröfel A, Kratošová G, Šafařík I, Šafaříková M, Raška I, Shor LM (2014) Applications of biosynthesized metallic nanoparticles – a review. Acta Biomater 10:4023–4042. https://doi.org/10.1016/J.ACTBIO.2014.05.022

Schwab F, Zhai G, Kern M, Turner A, Schnoor JL, Wiesner MR (2015) Barriers, pathways and processes for uptake, translocation and accumulation of nanomaterials in plants – critical review. Nanotoxicology:1–22. https://doi.org/10.3109/17435390.2015.1048326

Serag MF, Kaji N, Gaillard C, Okamoto Y, Terasaka K, Jabasini M, Tokeshi M, Mizukami H, Bianco A, Baba Y (2011) Trafficking and subcellular localization of multiwalled carbon nanotubes in plant cells. ACS Nano 5:493–499. https://doi.org/10.1021/nn102344t

Servin A, Elmer W, Mukherjee A, De la Torre-Roche R, Hamdi H, White JC, Bindraban P, Dimkpa C (2015) A review of the use of engineered nanomaterials to suppress plant disease and enhance crop yield. J Nanopart Res 17:1–21. https://doi.org/10.1007/s11051-015-2907-7

Shahverdi AR, Fakhimi A, Shahverdi HR, Minaian S (2007) Synthesis and effect of silver nanoparticles on the antibacterial activity of different antibiotics against *Staphylococcus aureus* and *Escherichia coli*. Nanomed Nanotechnol Biol Med 3:168–171. https://doi.org/10.1016/j.nano.2007.02.001

Sharma J, Chhabra R, Cheng A, Brownell J, Liu Y, Yan H (2009) Control of self-assembly of DNA tubules through integration of gold nanoparticles. Science 323:112–116. https://doi.org/10.1126/science.1165831

Sheikhloo Z, Salouti M (2012) Intracellular biosynthesis of gold nanoparticles by fungus *Phoma macrostoma*. Synth React Inorganic Met Nano-Metal Chem 42:65–67. https://doi.org/10.1080/15533174.2011.609230

Shi H, Magaye R, Castranova V, Zhao J (2013) Titanium dioxide nanoparticles: a review of current toxicological data. Part Fibre Toxicol 10:15. https://doi.org/10.1186/1743-8977-10-15

Singh Duhan J, Kumar R, Kumar N, Kaur P, Nehra K, Duhan S (2017) Nanotechnology: the new perspective in precision agriculture. Biotechnol Rep 15:11–23. https://doi.org/10.1016/j.btre.2017.03.002

Singh R, Singh A, Misra V, Singh RP (2011) Degradation of lindane contaminated soil using zero-valent iron nanoparticles. J Biomed Nanotechnol 7:175–176

Singh T, Shukla S, Kumar P, Wahla V, Bajpai VK, Rather IA (2017) Application of nanotechnology in food science: perception and overview. Front Microbiol 8:1501. https://doi.org/10.3389/fmicb.2017.01501

Singh R, Vora J, Nadhe SB, Wadhwani SA, Shedbalkar UU, Chopade BA (2018) Antibacterial activities of bacteriagenic silver nanoparticles against nosocomial *Acinetobacter baumannii*. J Nanosci Nanotechnol 18:3806–3815. https://doi.org/10.1166/jnn.2018.15013

Smith SC, Rodrigues DF (2015) Carbon : an international journal sponsored by the American carbon committee. Pergamon Press, Carbon

Sondi I, Salopek-Sondi B (2004) Silver nanoparticles as antimicrobial agent: a case study on E. coli as a model for Gram-negative bacteria. J Colloid Interface Sci 275:177–182. https://doi.org/10.1016/j.jcis.2004.02.012

Sperling RA, Parak WJ (2010) Surface modification, functionalization and bioconjugation of colloidal inorganic nanoparticles. Philos Trans R Soc A Math Phys Eng Sci 368:1333–1383. https://doi.org/10.1098/rsta.2009.0273

Speshock JL, Murdock RC, Braydich-Stolle LK, Schrand AM, Hussain SM (2010) Interaction of silver nanoparticles with Tacaribe virus. J Nanobiotechnol 8:19. https://doi.org/10.1186/1477-3155-8-19

Sugunan A, Thanachayanont C, Dutta J, Hilborn JG (2005) Heavy-metal ion sensors using chitosan-capped gold nanoparticles. Sci Technol Adv Mater 6:335–340. https://doi.org/10.1016/j.stam.2005.03.007

Sun T, Zhang YS, Pang B, Hyun DC, Yang M, Xia Y (2014a) Engineered nanoparticles for drug delivery in cancer therapy. Angew Chem Int Ed Engl 53:12320–12364. https://doi.org/10.1002/anie.201403036

Sun Y, Wang L, Li J, Zhao C, Zhao J, Liu M, Wang S, Lu C, Shang G, Jia Y, Wen A (2014b) Synergistic efficacy of meropenem and rifampicin in a murine model of sepsis caused by multidrug-resistant *Acinetobacter baumannii*. Eur J Pharmacol 729:116–122. https://doi.org/10.1016/j.ejphar.2014.02.015

Sun X, Shi Y, Akahoshi T, Fujiwara M, Gatanaga H, Schönbach C, Kuse N, Appay V, Gao GF, Oka S, Takiguchi M (2016) Effects of a single escape mutation on T Cell and HIV-1 Co-adaptation. Cell Rep 15:2279–2291. https://doi.org/10.1016/J.CELREP.2016.05.017

Sun C, Li Y, Li Z, Su Q, Wang Y, Liu X (2018) Durable and washable antibacterial copper nanoparticles bridged by surface grafting polymer brushes on cotton and polymeric materials. J Nanomater 2018:1–7. https://doi.org/10.1155/2018/6546193

Suriyaprabha R, Karunakaran G, Yuvakkumar R, Rajendran V, Kannan N (2014) Foliar application of silica nanoparticles on the phytochemical responses of maize (*Zea mays* L.) and its toxicological behavior. Synth React Inorganic Met Nano-Metal Chem 44:1128–1131. https://doi.org/10.1080/15533174.2013.799197

Tao R, Krishnan S, Bhosale PR, Javle MM, Aloia TA, Shroff RT, Kaseb AO, Bishop AJ, Swanick CW, Koay EJ, Thames HD, Hong TS, Das P, Crane CH (2016) Ablative radiotherapy doses lead to a substantial prolongation of survival in patients with inoperable intrahepatic cholangiocarcinoma: a retrospective dose response analysis. J Clin Oncol 34:219–226. https://doi.org/10.1200/JCO.2015.61.3778

Taylor AF, Rylott EL, Anderson CWN, Bruce NC (2014) Investigating the toxicity, uptake, nanoparticle formation and genetic response of plants to gold. PLoS One 9:e93793. https://doi.org/10.1371/journal.pone.0093793

Thakkar KN, Mhatre SS, Parikh RY (2010) Biological synthesis of metallic nanoparticles. Nanomed Nanotechnol Biol Med 6:257–262. https://doi.org/10.1016/j.nano.2009.07.002

Tongon R, Soytong K (2015) Application of nano-particles from *Chaetomium globosum* to control leaf spot of rice. J Agric Technol 11:1919–1926

Toshima N, Yonezawa T (1998) Bimetallic nanoparticles—novel materials for chemical and physical applications. New J Chem 22:1179–1201. https://doi.org/10.1039/a805753b

Tran TH, Poudel BK, Marasini N, Chi S-C, Choi H-G, Yong CS, Kim JO (2013) Preparation and evaluation of raloxifene-loaded solid dispersion nanoparticle by spray-drying technique without an organic solvent. Int J Pharm 443:50–57. https://doi.org/10.1016/j.ijpharm.2013.01.013

Tripathi DK, Tripathi A, Shweta S, Singh Y, Vishwakarma K, Yadav G, Sharma S, Singh VK, Mishra RK, Upadhyay RG, Dubey NK, Lee Y, Chauhan DK (2017) Uptake, accumulation and toxicity of silver nanoparticle in autotrophic plants, and heterotrophic microbes: a concentric review. Front Microbiol 8:7. https://doi.org/10.3389/fmicb.2017.00007

Ulbrich W, Lamprecht A (2010) Targeted drug-delivery approaches by nanoparticulate carriers in the therapy of inflammatory diseases. J R Soc Interface 7(Suppl 1):S55–S66. https://doi.org/10.1098/rsif.2009.0285.focus

VandeVoort AR, Arai Y (2012) Environmental chemistry of silver in soils. pp. 59–90. https://doi.org/10.1016/B978-0-12-394275-3.00005-5

Vaseeharan B, Ramasamy P, Chen JC (2010) Antibacterial activity of silver nanoparticles (AgNps) synthesized by tea leaf extracts against pathogenic *Vibrio harveyi* and its protective efficacy on juvenile *Feneropenaeus indicus*. Lett Appl Microbiol 50:352–356. https://doi.org/10.1111/j.1472-765X.2010.02799.x

Vigneshwaran N, Ashtaputre NM, Varadarajan PV, Nachane RP, Paralikar KM, Balasubramanya RH (2007) Biological synthesis of silver nanoparticles using the fungus *Aspergillus flavus*. Mater Lett 61:1413–1418. https://doi.org/10.1016/J.MATLET.2006.07.042

Wang C-L, Cheng Y-H (2017) Identification and trichothecene genotypes of *Fusarium graminearum* species complex from wheat in Taiwan. Bot Stud 58:4. https://doi.org/10.1186/s40529-016-0156-4

Wang C-J, Yang W, Wang C, Gu C, Niu D-D, Liu H-X, Wang Y-P, Guo J-H (2012) Induction of drought tolerance in cucumber plants by a consortium of three plant growth-promoting *Rhizobacterium* strains. PLoS One 7:e52565. https://doi.org/10.1371/journal.pone.0052565

Wang JN, Luo XG, Wu T, Chen Y (2014) High-strength carbon nanotube fibre-like ribbon with high ductility and high electrical conductivity. Nat Commun 5:3848. https://doi.org/10.1038/ncomms4848

Wang C, Lu J, Zhou L, Li J, Xu J, Li W, Zhang L, Zhong X, Wang T (2016) Effects of long-term exposure to zinc oxide nanoparticles on development, zinc metabolism and biodistribution of minerals (Zn, Fe, Cu, Mn) in mice. PLoS One 11:e0164434. https://doi.org/10.1371/journal.pone.0164434

Wongpisutpaisan N, Charoonsuk P, Vittayakorn N, Pecharapa W (2011) Sonochemical synthesis and characterization of copper oxide nanoparticles. Energy Procedia 9:404–409. https://doi.org/10.1016/j.egypro.2011.09.044

Wu M, Zhang Z, Chiu Y (2014) Life-cycle water quantity and water quality implications of biofuels. Curr Sustain Energy Rep 1:3–10. https://doi.org/10.1007/s40518-013-0001-2

Yang Z, Liu ZW, Allaker RP, Reip P, Oxford J, Ahmad Z, Ren G (2010) A review of nanoparticle functionality and toxicity on the central nervous system. J R Soc Interface 7(Suppl 4): S411–S422. https://doi.org/10.1098/rsif.2010.0158.focus

Yang H, Wang Y, Huang H, Gell L, Lehtovaara L, Malola S, Häkkinen H, Zheng N (2013) All-thiol-stabilized Ag44 and Au12Ag32 nanoparticles with single-crystal structures. Nat Commun 4(2422). https://doi.org/10.1038/ncomms3422

Yang H, Ren Y, Wang T, Wang C (2016) Preparation and antibacterial activities of $Ag/Ag^{+}/Ag^{3+}$ nanoparticle composites made by pomegranate (*Punica granatum*) rind extract. Results Phys 6:299–304. https://doi.org/10.1016/J.RINP.2016.05.012

Yatim NM, Shaaban A, Dimin MF, Yusof F (2015) ScienceDirect statistical evaluation of the production of urea fertilizer- multiwalled carbon nanotubes using plackett burman experimental design. procedia -social. Behav Sci 195:315–323. https://doi.org/10.1016/j.sbspro.2015.06.358

Zarei H, Kazemi Oskuee R, Hanafi-Bojd MY, Gholami L, Ansari L, Malaekeh-Nikouei B (2018) Enhanced gene delivery by polyethyleneimine coated mesoporous silica nanoparticles. Pharm Dev Technol:1–6. https://doi.org/10.1080/10837450.2018.1431930

Zhang L, Xia J, Zhao Q, Liu L, Zhang Z (2010) Functional graphene oxide as a nanocarrier for controlled loading and targeted delivery of mixed anticancer drugs. Small 6:537–544. https://doi.org/10.1002/smll.200901680

Zhang Z, Zhang J, Chen N, Qu L (2012) Graphene quantum dots: an emerging material for energy-related applications and beyond. Energy Environ Sci 5:8869. https://doi.org/10.1039/c2ee22982j

Zhao L, Tan L, Zhu Z, Xiao L, Xie D, Sun C (2015) PAY1 improves plant architecture and enhances grain yield in rice. Plant J 83:528–536. https://doi.org/10.1111/tpj.12905

Zheng Y, Cloutier P, Hunting DJ, Sanche L (2008) Radiosensitization by gold nanoparticles: comparison of DNA damage induced by low and high-energy electrons. J Biomed Nanotechnol 4:469–473. https://doi.org/10.1166/jbn.2008.012

Zhu X, Tian S, Cai Z (2012) Toxicity assessment of Iron oxide nanoparticles in Zebrafish (*Danio rerio*) early life stages. PLoS One 7:e46286. https://doi.org/10.1371/journal.pone.0046286

Chapter 12
Biodegradation of Agricultural Wastes by *Chaetomium* Species

Ahmed M. Abdel-Azeem, Gihan A. Hasan, and Marwa T. Mohesien

12.1 Introduction

Cellulose is the most abundant and renewable component of plant biomass (Srivastava et al. 2015a, b, c; Srivastava and Jaiswal 2016). It is considered as the main product of photosynthesis, and ~100 billion dry tons/year of cellulosic biomass is produced in the biosphere (Wang et al. 2016). Besides plants, some animals and bacteria also produce cellulose. However, it is the major component of the plant cell walls and abundantly available in the environment. In the lignocellulosic biomass structure, cellulose is mainly associated with hemicelluloses and lignin and sometimes with silica (e.g., rice straw, rice husk). Further, it accounts for ~35–50% of plant dry weight, whereas hemicelluloses as well as lignin cover ~20–35 and ~5–30% of plant dry weight, respectively (Zabeda et al. 2016). Cellulose is a linear polysaccharide, made up of combined units of glucose monomers bound by β-1,4-glycosidic linkage (Fig. 12.1). To release these monomeric molecules, cellulase enzymes are required for carrying out enzymatic hydrolysis. Cellulases are the combination of enzymes capable of degrading the insoluble cellulose polymer present in the lignocellulosic biomass into fermentable sugars, predominantly small chain of cellobiose and glucose molecules (Taherzadeh and Karimi 2007). Endoglucanases, cellobiohydrolases, and β-glucosidases are the main subcomponents which make the cellulase enzyme system (Thota et al. 2017). Nevertheless,

A. M. Abdel-Azeem (✉)
Department of Botany, Faculty of Science, Suez Canal University, Ismailia, Egypt
e-mail: ahmed_abdelazeem@science.suez.edu.eg

G. A. Hasan
Department of Zoology, Faculty of Science, Suez Canal University, Ismailia, Egypt

M. T. Mohesien
Department of Botany and Microbiology, Faculty of Science, Damietta University, New Damietta, Egypt

A. M. Abdel-Azeem (ed.), *Recent Developments on Genus Chaetomium*, Fungal Biology, https://doi.org/10.1007/978-3-030-31612-9_12

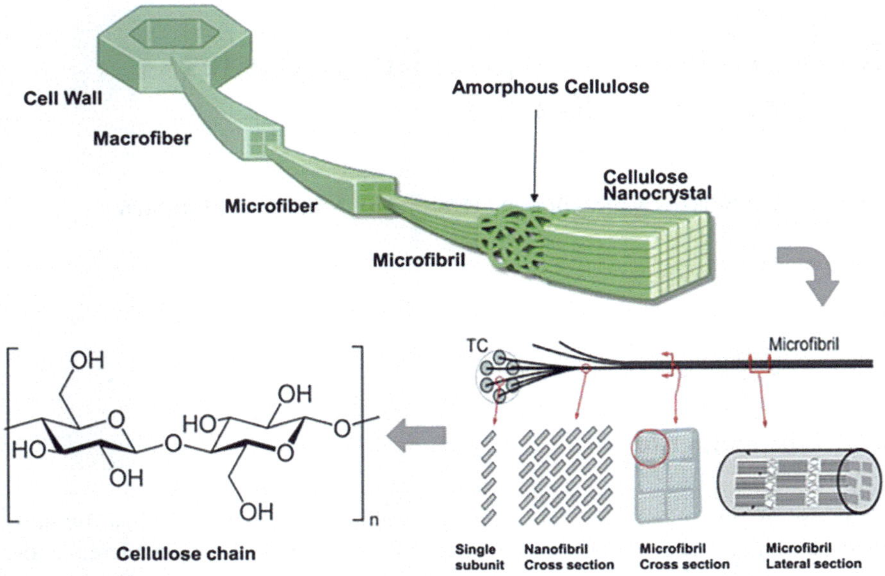

Fig. 12.1 Hierarchical structure of cellulose from wood. (Kim et al. 2015)

intensive researches are going on to achieve novel cellulase producing systems, not only to increase the yield and economic feasibility but also to expand the use of such systems by progressing toward more and more industrially flexible bacteria or fungi production systems. Additionally, many enzymatic hydrolysis researches have also been aimed at the production of enzymes with high enzymatic activity for improvement in the current technology to produce cellulase enzyme (Chandel et al. 2012; Garvey et al. 2013). Besides biofuel production, other major industrial applications of cellulases are bio-polishing; bio-stoning; bio-finishing in textile industry; starch processing; grain alcohol fermentation; malting and brewing in beer and wine industry; extraction and processing of fruit and vegetable juices in food, pulp, and paper industry (Cavacopaulo 1998; Gao et al. 2008a, b; Ibrahim et al. 2015); controlling plant pathogen and disease in agriculture as well as in household laundry detergents for improving the fabrics' softness and brightness; etc. (Wilson 2009).

This chapter deals with the production of fungal cellulases, their type, and efficiency to degrade lignocellulosic biomass. Additionally, recent advancements in cellulase production and future prospects have also been discussed.

12.2 Agricultural Lignocellulosic Wastes, Environment, and Human

Many of the agricultural lignocellulosic wastes which are produced every day in the world cause serious environmental pollution effects if they are allowed to accumulate in the agro-ecosystems or, much worse, burned for uncontrolled domestic purposes. Every year, large amounts of lignocellulosic wastes are generated through forestry and agricultural practices, in timber industries and many agroindustries, generating environmental pollution problems by their burning on the soil surface or their incorporation into the soil matrix.

We have two different types of agro-industrial wastes, i.e., agriculture residues and industrial residues. Agriculture residues can be further divided into field residues and process residues. Field residues are residues that present in the field after the process of crop harvesting. These field residues consist of leaves, stalks, seed pods, and stems, whereas the process residues are residues present even after the crop is processed into alternate valuable resource (Table 12.1). These residues consist of molasses, husks, bagasse, seeds, leaves, stem, straw, stalk, shell, pulp, stubble, peel, roots, etc. and used for animal feed, soil improvement, fertilizers, manufacturing, and various other processes. Huge amount of field residues are generated and most of them are underutilized. Controlled use of field remains can enhance the proficiency of irrigation and control of erosion. In Middle East region, wheat and barley are the major crops. In addition to this, various other crops like rice, lentils, maize, chickpeas, fruits, and vegetables are also produced all over the world. Agricultural residues are differentiated on the basis of their availability as well as characteristics that can be different from other solid fuels like charcoal, wood, and char briquette (Zafar 2014).

The greenhouse effect is our planet's ability to pass the Sun's incoming radiation and to reflect the reradiated long wavelengths from the Earth's surface, so increasing our planet's surface and atmosphere lower-layer temperature. As a result, the melting of glaciers and snow cover intensifies, and the water level of seas and oceans increases, flooding islands and coastlines. Increase in temperature intensifies water evaporation and increases the possibility of downpour and cyclone formation in

Table 12.1 Composition of agro-industrial wastes

	Chemical Composition (%)						
Agro-industrial wastes	C	HC	L	A (%)	TS (%)	M (%)	References
Sugarcane bagasse	30.2	56.7	13.4	1.9	91.66	4.8	El-Tayeb et al. (2012) and Nigam et al. (2009)
Rice straw	39.2	23.5	36.1	12.4	98.62	6.58	El-Tayeb et al. (2012)
Corn stalks	61.2	19.3	6.9	10.8	97.78	6.40	El-Tayeb et al. (2012)
Sawdust	45.1	28.1	24.2	1.2	98.54	1.12	El-Tayeb et al. (2012) and Martin et al. (2012)
Sugar beet waste	26.3	18.5	2.5	4.8	_	12.4	El-Tayeb et al. (2012)
Barley straw	33.8	21.9	13.8	11	_	_	Nigam et al. (2009)
Cotton stalks	58.5	14.4	21.5	9.98	_	7.45	Nigam et al. (2009)
Oat straw	39.4	27.1	17.5	8	_	_	Martin et al. (2012)
Soya stalks	34.5	24.8	19.8	10.39	_	11.84	Motte et al. (2013)
Sunflower stalks	42.1	29.7	13.4	11.17	_	_	Motte et al. (2013)
Wheat straw	32.9	24.0	8.9	6.7	95.6	7	Nigam et al. (2009) and Martin et al. (2012)

**C* cellulose, *HC* hemicellulose, *L* lignin, *TS* total solids, *M* moisture seaside regions, while on continents—droughts, heat waves, and forest fires. So lowering of the greenhouse effect is one of the main global problems today (Kazragis 2005).

The main greenhouse effect agents are carbon dioxide (CO_2) (53%), Freon, ozone, methane (CH_4), and other substances. CO_2 is formed during breathing of live organisms, activity of microbes in the soil, and also the combustion of various organic substances. CH_4 and CO_2 are formed when cellulosic matter (wood, peat moss, agricultural production waste, vegetation) decay at treatment places and landfills. From all the mentioned processes, mankind controls only the cellulose matter combustion and decay processes which should decrease as much as possible to weaken the greenhouse effect. However, at present it is unrealistic to lower the amount of CO2 formed by burning fossil fuel in industrial enterprises and heating systems.

However, the excess lignocellulosic waste is often disposed of by biomass burning, causing an environmental pollution problem through smoke and development of CO2 which is not restricted to developing countries, but is considered a global phenomenon (Levine 1996). Burning also causes almost complete losses of soil N, P, K, and S (Dobermann and Fairhurst 2002).

On the global scale, biomass burning (BB) is the main source of primary organic carbon (OC) (Bond et al. 2004; Huang et al. 2015), black carbon (BC) (Bond et al. 2013; Cheng and Yang 2016), and brown carbon (BrC) (Laskin et al. 2015). It is also the second largest source of non-methane organic gases (NMOGs) in the atmosphere (Yokelson et al. 2008; Stockwell et al. 2014). In addition, atmospheric aging of biomass burning plumes produces substantial secondary pollutants.

The increase in tropospheric ozone (O_3) in aged biomass burning plumes could last for days and even months (Thompson et al. 2001; Duncan et al. 2003; Real et al. 2007) with complex atmospheric chemistry (Arnold et al. 2015; Müller et al. 2016). Moreover, biomass and biofuel burning could contribute up to 70% of the global secondary organic aerosol (SOA) burden (Srivastava et al. 2015a, b, c) and hence influence the seasonal variation of global SOA (Tsigaridis et al. 2014). Since it produces large amounts of primary and secondary pollutants, it is essential to characterize primary emissions and photochemical evolution of biomass burning in order to better understand its impacts on air quality (Huang et al. 2014), human health (Alves et al. 2015) and climate change (Andreae et al. 2004; Koren et al. 2004; Laskin et al. 2015; Huang et al. 2016).

Open burning of agricultural residues, a convenient and inexpensive way to prepare for the next crop planting, could induce severe regional haze events (Cheng et al. 2013; Tariq et al. 2016). Among all the biomass burning types, agricultural residue burning in the field is estimated to contribute ~10% of the total mass burned globally (Andreae and Merlet 2001), and its relative contribution is even larger in Asia (~34%), and especially in China (>60%) (Streets et al. 2003), where >600 million people live in the countryside (NBSPRC 2015). Agricultural residues burned in China were estimated to be up to 160 million ton in 2012, accounting for ~40% of the global agricultural residues burned (Li et al. 2016). As estimated by Tian et al. (2011), agricultural residue burning contributed 70–80% of non-methane hydrocarbons (NMHCs) and particulate matter (PM) emitted by biomass burning in China during 2000–2007. A better understanding of the role agricultural residual burning plays in air pollution in China and elsewhere requires better characterization of primary emission and atmospheric aging of emitted trace gases and particles for different types of agricultural residues under different burning conditions. In the past two decades, there have been increasing numbers of characterizations of biomass burning emissions. Andreae and Merlet (2001) summarized emission factors (EFs) for both gaseous and particulate compounds from seven types of biomass burning. Akagi et al. (2011) updated the emission data for 14 types of biomass burning, and newly identified species were included. Since biomass types and combustion conditions may differ in different studies, reported emission factors are highly variable, especially for agricultural residue burning (Li et al. 2007, 2017; Cao et al. 2008; Zhang et al. 2008; Yokelson et al. 2011; Brassard et al. 2014; Sanchis et al. 2014; Wang et al. 2014; Ni et al. 2015; Kim Oanh et al. 2015; Stockwell et al. 2016; Bruns et al. 2017; Tkacik et al. 2017). Moreover, previous studies on agricultural residue burning were mostly carried out near fire spots or in chambers with low dilution ratios. Since biomass burning organic aerosols (BBOAs) are typically semi-volatile (Grieshop et al. 2009b; May et al. 2013), it is expected that measured BBOA emission factors would be affected by dilution processes (Lipsky and Robinson 2006), and BBOA emission factors under ambient dilution conditions are still unclear. Furthermore, knowledge of NMOGs emitted from agricultural residue burning is very limited. As reported by Stockwell et al. (2015), ~21% (in weight) of NMOGs in biomass burning plumes have not been identified yet. Therefore, comprehensive measurement and characterization of gaseous and particulate species emitted by agricultural residue burning under ambient dilution conditions are urgently needed.

Great attention has been drawn to SOA formation and transformation in biomass burning plumes recently, since a significant increase in mass and apparent change in physicochemical characteristics of aerosols have been observed during atmospheric aging of biomass burning plumes in both field and laboratory studies (Grieshop et al. 2009a, b; Hennigan et al. 2011; Heringa et al. 2011; Lambe et al. 2011; Jolleys et al. 2012; Giordano et al. 2013; Martin et al. 2013; Ortega et al. 2013; Ding et al. 2016a, b, 2017). For agricultural residue burning, evolution processes have not been well characterized yet. To our knowledge, up to now, there has only been a chamber study (Li et al. 2015) which has investigated the evolution of aerosol particles emitted by wheat straw burning under dark conditions. Although field studies (Adler et al. 2011; Liu et al. 2016) witnessed the evolution in mass concentrations, size distribution, oxidation state, and optical properties of aerosol particles emitted by agricultural residue burning, these changes could be also influenced by other emission sources and meteorological conditions as well. Since NMOGs emitted by agricultural residue burning are not fully quantified, it is still challenging to predict the concentration and physicochemical properties of SOA that resulted from biomass burning (Spracklen et al. 2011; Jathar et al. 2014; Srivastava et al. 2015a, b, c; Hatch et al. 2017). Bruns et al. (2016) suggested that the 22 major NMOGs identified in residential wood combustion could explain the majority of observed SOA, but it remains unclear whether identified NMOGs emitted by agricultural residue burning could fully (or at least largely) explain the SOA formed. In addition, aerosol mass spectrometry (AMS) has been widely used to characterize sources and evolution of ambient OA (Zhang et al. 2011). Although agricultural residue burning is an important type of biomass burning in Asia and especially in China, the lack of AMS spectra for primary and aged OA from agricultural residue burning significantly limits the further application of AMS in BBOA research.

In 2017, indoor chamber experiments were conducted to investigate primary emissions from open burning of rice, corn, and wheat straws and their photochemical aging as well by Fang et al. Emission factors of NOx, NH3, SO2, 67 non-methane hydrocarbons (NMHCs), particulate matter (PM), organic aerosol (OA), and black carbon (BC) under ambient dilution conditions were determined. Olefins accounted for >50% of the total speciated NMHC emission (2.47–5.04 g kg^{-1}), indicating high ozone formation potential of straw burning emissions. Emission factors of PM (3.73–6.36 g kg^{-1}) and primary organic carbon (POC, 2.05–4.11 gC kg^{-1}), measured at dilution ratios of 1300–4000, were lower than those reported in previous studies at low dilution ratios, probably due to the evaporation of semi-volatile organic compounds under high dilution conditions. After photochemical aging with an OH exposure range of (1.97–4.97) × 10^{10} molecule cm^{-3} s in the chamber, large amounts of secondary organic aerosol (SOA) were produced with OA mass enhancement ratios (the mass ratio of total OA to primary OA) of 2.4–7.6. The 20 known precursors could only explain 5.0–27.3% of the observed SOA mass, suggesting that the major precursors of SOA formed from open straw burning remain unidentified. Aerosol mass spectrometry (AMS) signaled that the aged OA contained less hydrocarbons but more oxygen- and nitrogen-containing compounds than primary OA, and carbon oxidation state (OS_c) calculated with AMS resolved O/C and H/C ratios increased linearly ($p < 0.001$) with OH exposure with quite similar slopes.

In Egypt and elsewhere, large quantities of organic matter from agricultural wastes (AWs) and the removal of exotic plants (EPs) are burned without treatment of the combusts directly on agricultural fields. This practice causes important environmental issues that have been identified as major health risks for the local population by reducing local air quality and by contributing to the black-cloud phenomenon in the region (El-Askary and Kafatos 2008).

A research carried by Awasthi et al. (2010) showed that the smoke produced by crop burning could have a lasting effect on children's lung function. Professor Ravinder Agarwal, head of the University Science Instrumentation Centre at Thapar University in Patiala, India, and colleagues used portable spirometers to regularly test the lung function of children aged 10–13 and adults aged 20–35 over the course of a year. The 40 participants were healthy nonsmokers living in a village surrounded by farmland, with little traffic and no industry within 10 km (Awasthi et al. 2010). Children's force vital capacity (FVC) dropped from a mean 98% in August 2008 to 92% in July 2009. Mean FVC dipped as low as 88% in October and November, when farmers burned their rice crop residue, and in April and May, when they burned wheat stubble. The children's mean lung function remained significantly lower throughout the test period. The mean lung function of the adult study participants declined during the burn seasons as well, but largely returned to original levels by the end of the study (Awasthi et al. 2010). Decreases in lung function correlated with increases in the concentration of particulate matter, which exceeded India's national air quality standards during the burn season (Awasthi et al. 2010). Small particles (PM2.5 and PM10)—which make up the majority of the smoke produced by crop burning—were more closely associated with decreases in lung function than suspended particulate matter (SPM), which can contain particles 100 μm or larger (EPA 1999). The findings linking seasonal burning with health issues "coincide with the anecdotal evidence that have been recorded in the Canadian prairies," notes Kate Letkemann, environmental issues coordinator of The Lung Association, Manitoba, and a member of the provincial Crop Residue Burning Advisory Committee. Argawal's work "builds a relationship between pulmonary function tests and the concentration of SPM, PM10, and PM2.5," notes Shijian Yang of the School of Environmental Science and Engineering at China's Shanghai Jiao Tong University. But he would like to see further research that looks closely at the dose–effect relationship between lung function and crop residue burning. Yang's work has shown that the peak concentration of PM10 and its duration may be more important than average concentrations for estimating the health effects of burning crops (Yang et al. 2008).

Cellulose-containing waste can be reprocessed without emitting CO_2 or CH_4 by manufacturing building materials, thermal as well as acoustic insulating composites. Wooden waste has been used for these purposes for a long time. At present it is recommended to use other cellulose raw materials—straw, reeds, boon, peat moss, and other materials. The utilization of straws given composites, containing straws, Portland cement (in some cases—construction gypsum and sand and polymeric additives—vinyl acetate (e.g., polyvinyl acetate PVA) or cellulosic materials (e.g., carboxymethylcellulose (CMC) dispersions. The utilization of reeds gives composites containing reeds, Portland cement, and PVA or CMC. The utilization of boon or chaff gives composites containing boons or chaffs, anhydrite or aluminate cements, and PVA or CMC. Optimal composition composites distinguish themselves by good physico-mechanical as well as thermal and acoustic properties and can find applications as building materials as well as thermal and acoustic insulating materials.

12.3 Lignocellulose Biodegradation

12.3.1 Cellulolytic Enzymes

Most known cellulolytic enzymes are produced and excreted by filamentous fungi, among which *Trichoderma reesei* has received special attention for its hyperenzymatic production capabilities (Vinzant et al. 2001; Sun et al. 2008a, b). It has been subsequently used as a model organism for investigations of enzymatic hydrolysis mechanisms. Proteomic analysis of these enzymes with high-resolution two-dimensional gel electrophoresis revealed many glycoside hydrolases with this species (Vinzant et al. 2001; Sun et al. 2008a, b), which can be divided into two main groups: cellulases and hemicellulases. Cellulases are enzymes that hydrolyze cellulose, a linear polysaccharide molecule composed of repeated β(1 → 4) linked D-glucopyranosyl (Glc) units. Multiple types of cellulases have been discovered, including at least two exo-β-glucanases, or cellobiohydrolases (CBHs; EC 3.2.1.91) (CBH I and CBH II), four endo-β-glucanases (EGs; EC 3.2.1.4) (EG I, EG II, EG III, EG V), and one β-glucosidase (β G; EC 3.2.1.21) (Lynd et al. 2002).

Hemicellulases are enzymes that degrade hemicelluloses, a group of polysaccharides and one of major plant cell wall components. Unlike cellulose, which is composed entirely of glucosyl moieties linked by only β-1,4-glycosidic bonds, various types of sugar moieties linked by different bonds, intramolecular architecture, and intermolecular interactions can be found within hemicelluloses. Hemicelluloses also differ from the major plant cell wall structural component cellulose in their much smaller polysaccharide chains. Hemicelluloses can serve to cross-link cellulose microfibrils by interconnecting them as well as linking cellulose molecules to other cell wall components (Kumar et al. 2008). Common hemicelluloses include β-glucan (different from cellulose), xylan, xyloglucan, arabinoxylan, mannan, galactomannan, arabinan, galactan, and polygalacturonan. Corresponding to these hemicelluloses, various types of hemicellulases exist, each specific to a kind of hemicellulose (Polizeli et al. 2005; Collins et al. 2005; Kumar et al. 2008). These enzymes can be clustered into two groups: hemicellulases that attack the polysaccharide backbone and those that attack the side chains.

Xylan, whose structure differs from plant to plant, is the second most abundant polysaccharide in herbs and hardwoods, demanding the collaboration of a group of enzymes during their degradation. Multiple enzymes including endo-β-xylanase (EC 3.2.1.8), β-xylosidase (EC 3.2.1.37), α-glucuronidase (EC 3.2.1.139), α-L-arabinofuranosidase (EC 3.2.1.55), and acetylxylan esterase (EC 3.1.1.6) act synergistically in this process (Fig. 12.2). Endo-1,4-xylanases cleave internal β-1,4-xylosidic bonds on the xylan polysaccharide backbone. Unlike EGs, whose cleavage sites are random, endo-1,4-xylanases recognize specific bonds for cleavage on the basis of polysaccharides properties such as chain length and branching levels (Polizeli et al. 2005). Endoxylanases were initially classified into two groups by their ability to hydrolyze the 1,3-α-L-arabinofuranosyl branching points of arabinoxylans: hydrolyzing and nonhydrolyzing endoxylanases, which have different pI values and molecular weights (Wong et al. 1988). However, these patterns were shown to account for only 70% of all endoxylanases, and a classification system of all glycoside hydrolases (glycoside hydrolase families) is better recognized now (Collins et al. 2005). Products from xylan degradation by endoxylanases are a mixture of β-D-xylopyranosyl oligomers of various lengths, which serve as substrates for β-xylosidases that subsequently hydrolyze them to xylose from the nonreducing end of these oligomers (Polizeli et al. 2005). Multiple other enzymes are also involved in xylan degradation, primarily due to the complex nature of these polysaccharides: α-L-arabinofuranosidase cleaves the α-glycosidic

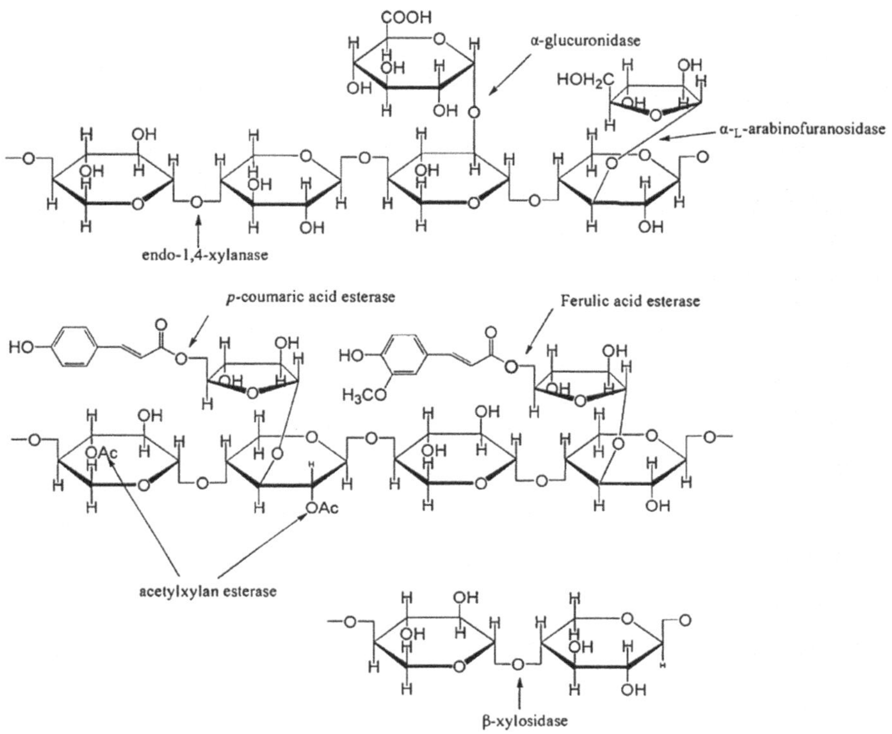

Fig. 12.2 Chemical structure and degradation of hemicellulose

bonds between arabinose and xylose moieties in xylan; α-glucuronidases cleave the α(1 → 2) bonds linking the (methyl) GlcU units in xylan (Kumar et al. 2008); acetylxylan esterase removes the O-acetyl groups at the 2- and 3-positions of β-D-xylopyranosyl residues; ferulic acid esterase (EC 3.1.1.73) cleaves the ester bond between the arabinose and ferulic acid side chains; and p-coumaric acid esterase (EC 3.1.1.73) cleaves the ester bond between the arabinose and ρ-coumaric acid (Polizeli et al. 2005).

12.3.2 Fungal Cellulases

Cellulase enzyme plays a key role in the hydrolysis of cellulosic substrate and converts it into monomeric sugars. For the effective hydrolysis of cellulosic substrate, three types of synergistically acting subcomponent enzymes are essential: endoglucanases (EG), exoglucanases (CBH), and beta-glucosidase (BGL). Cellobiohydrolases or exoglucanases which attack the crystalline ends of cellulosic substrate produce cellobiose, while endoglucanases divide glycosidic

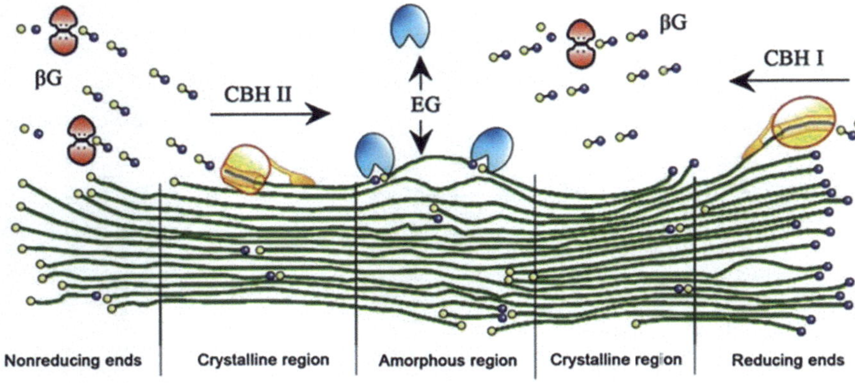

Fig. 12.3 Mechanism of cellulose biodegradation. CBH cellobiohydrolase (or exo-β-glucanase), EG endo-β-glucanase, β-G β-glucosidase

bonds within the amorphous part of the cellulosic substrate (Zhang and Lynd 2004; Yoon et al. 2014). Further, the liberated cellobiose is sliced by β-glucosidases (BGL) and releases glucose molecules (Liu et al. 2012; Wang et al. 2013). Figure 12.3 shows all three components and functions of cellulase responsible for enzymatic hydrolysis of agriculture waste. Cellulase enzymes are widely distributed in nature, and fungi are known as the potential producer of cellulase. Additionally, cellulolytic fungi have the major advantages to utilize secretory pathways as well as the production of high yields of protein. Intense research on other fungi like *Penicillium*, *Acremonium*, and *Chrysosporium* are underway for the potential production of cellulase. Table 12.2 summarizes cellulase production from different fungal species using variety of wastes via solid-state fermentation (SSF) and submerged fermentation (SmF).

12.3.3 Production of Fungal Cellulases

Fungal cellulase production can be carried out by two methods: (i) solid-state fermentation (SSF) and (ii) submerged fermentation (SmF) (Pandey 2003; Singhania et al. 2009; Bansal et al. 2012a, b). In case of SSF, solid substrates are used such as agriculture waste of rice straw, wheat bran, sugarcane bagasse, etc. for the production of cellulases (Xia and Cen 1994; Subramaniyam and Vimala 2012; Cherian et al. 2016). SSF involves the fermentation process carried out in absence or nearly in absence of free water using solid substrate. On the other hand, SmF involves fermentation in presence of water. Moreover, SmF uses primarily free molecules soluble in water as liquid substrates, like molasses in broth (Subramaniyam and Vimala 2012). The main advantage of SSF technique is easy recycling of cheap waste material and less cost, whereas SmF offers ease of purification and product recovery (Pandey et al. 2000; Couto and Sanromán 2006). Further, SSF is mainly

Table 12.2 Ratio of different used biomass for cellulase production

Fungi	Type of agricultural waste	Cellulase productivity	Fermentation process (SmF/SSF)	References
Aspergillus terreus	Rice straw	FPase: 10.96 IU/g	SSF	Narra et al. (2012)
Aspergillus fumigatus ABK9	Wheat bran: rice straw(1.1:1)[a]	CMCase: 826.2 IU/g FPase: 102.5 U/g β-glucosidase: 255.16 IU/g	SSF	Das et al. (2013)
Aspergillus protuberus	Rice husk	β-glucosidase: 26.06 IU/g	SSF	Yadav et al. (2016)
Trichoderma asperellum SR1–7	Wheat bran	FPase: 2.2 IU/gds CMCase: 13.2 IU/gds β-glucosidase 9.2 IU/gds	SSF	Raghuwanshi et al. (2014)
Aspergillus niger N402	Wheat straw	FPase: 24 IU/g	SSF	Pensupa et al. (2013)
Aspergillus fumigatus NITDGPKA3	Rice straw	CMCase: 64.18 IU/gds FPase: 3.1 IU/gds	SSF	Sarkar and Aikat (2012)
Myceliophthora thermophila JCP 1–4	Sugar cane bagasse: wheat bran (1: 1)[a]	CMCase: 357.51 IU/g β-glucosidase: 45.42 IU/g	SSF	Pereira et al. (2015)
Aspergillus nige KK2	Rice straw	FPase: 19.5 IU/g CMCase: 129 IU/g β-glucosidase: 100 IU/g	SSF	Kang et al. (2004)
Aspergillus niger NCIM 548	Wheat bran: corn bran, kinnow peel (2:1:2)[a]	FPase: 5.54 IU/g	SmF	Kumar et al. (2011)
Aspergillus fumigatus P40M2	Soybean bran	CMCase: 160.1 IU/g	SSF	Delabona et al. (2012)
Aspergillus ellipticus	Wheat straw	FPase: 117.25 IU/g CMCase: 725.11 IU/g β-glucosidase: 29.65 IU/g	SSF	Agrawal and Matkar (2016)

(continued)

Table 12.2 (continued)

Fungi	Type of agricultural waste	Cellulase productivity	Fermentation process (SmF/SSF)	References
Penicillium echinulatum 9A02S1	Sugar cane Bagasse	FPase: 12.5 IU/g	SmF	Camassola and Dillon (2014)
Trichoderma viride VKF3	Sugar cane Bagasse	CMCase: 33 IU/g FPase: 10.09 IU/g	SSF	Nathan et al. (2014)
Rhizopus oryzae CCT 7560	Rice husk and rice bran	CMCase: 5.1 U/g FPase: 2.3 IU/g	SSF	Kupski et al. (2014)

[a]Ratio of different used biomass for cellulase production

favorable for microorganisms which require less moisture content, while due to high water activity, SmF is suited to bacteria for cellulase production (Babu and Satyanarayana 1996). Additionally, SSF has been known for fermentation in Asian and Western countries since the ancient time (Ryu and Mandels 1980; Zhuang et al. 2007; Swain and Ray 2007). However, SSF has gained importance in Western countries after the discovery of penicillin via SmF technique in the 1940s. Though, in the last two decades, SSF has gained attention because of many biotechnological advantages like high fermentation ability, more stable end product, subordinate catabolic repression, as well as cost-effective technology (Sukumaran et al. 2009; Kasana et al. 2008; Liang et al. 2010; Dhillon et al. 2013). In the past 10 years, interest in SSF has been renewed because microorganisms including genetically modified organisms (GMO) may produce cellulase effectively through the SSF (Chahal 1983). Additionally, cellulase production via SSF is preferred over SmF because of two to three times higher enzyme production, high protein rate, and direct accessibility of dried fermentable solids as source of enzyme which easily eliminate the cost involved in downstream processing (Sánchez 2009; Hendriks and Zeeman 2009; Sadhu and Maiti 2013; El-Bakry et al. 2015).

12.3.4 Source of Fungal Cellulase Production

Degradation of cellulose via fungal cellulase is well documented, and several cellulose degrading fungi like *Aspergillus niger*, *Cladosporium cladosporioides*, *Cladosporium sphaerospermum*, *Penicillium chrysogenum*, *Scopulariopsis brevicaulis*, *Stachybotrys chartarum*, *Verticillium cycolsporum*, and *Chaetomium hamadae* have been investigated for cellulase production (El-Morsy 2000; Luo et al. 2005). Further, various fungi have also been screened for the cellulase production based on their habitat (El-Morsy 2000; Luo et al. 2005). In one of the study by Maria et al. (2005), twenty-nine different fungal isolates were reported for cellulase production. Moreover, most of these fungal isolates produce endoglucanases where most of them belong to *Ascomycetes*. Kathiresan and Manivannan (2006) isolated seven fungal species such as *Acremonium* sp., *Alternaria*, chlamydospore, *Aspergillus* sp., *Fusarium* sp., as well as *Pestalotiopsis* sp. from the southwest coast of India and used for cellulase production. Among all these fungi, *Aspergillus* sp. was found to be efficient cellulase producers.

Besides *Aspergillus* sp., *T. reesei* are also known as prominent cellulase producers that possess a complete cellulase system (Rasmussen et al. 2010). Though, action of fungi is different in terms of degradation of lignocellulosic biomass in decaying pattern and the structural changes found in the degraded substrates, category belonging to white-rot fungi (WRF) degrades all the components of biomass, namely, lignin, cellulose, and hemicellulose by colonizing themselves on the lignocellulosic substrate (Kuhad and Singh 2007). On the other hand, fungi belonging to category brown-rot fungi (BRF) favorably degrade the cellulosic component and hemicellulosic part with the modified lignin part (Schwarze et al. 2000; Schmidt 2006). Apart from the fungi type, production of cellulases might also be dependent on the initial amount of cellulose, hemicellulose, and lignin present in biomass (Liu et al. 2014). Figure 12.4 shows different subcomponents of lignocellulosic biomass and their obtained product as sugars after enzymatic hydrolysis. A separate group of fungi are responsible for degrading these biomass via bioconversion reaction of cellulases.

12.3.5 Chaetomium' *Cellulases*

Eriksen and Goksøyr (1977) cultivated *Chaetomium thermophile* var. *dissitum* in a liquid medium with cellulose. They recorded that taxon produced extracellular cellulolytic enzymes. By concentration of the culture filtrate, followed by ion-exchange chromatography on DEAE-Sephadex A-50 and gel filtration on Biogel P-100, three electrophoretically pure components were obtained. Of these, one was a typical C, enzyme (endoglucanase) causing rapid decrease of the viscosity of carboxymethylcel-

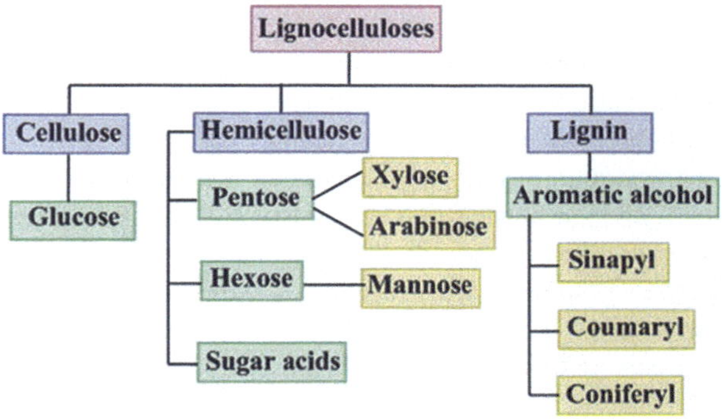

Fig. 12.4 Schematic representation showing the contents present in the lignocelluloses. (Sajith et al. 2016)

lulose solutions, while showing low effect on native cellulose. The other was active toward native cellulose but had little effect on the viscosity of carboxymethylcellulose. It is concluded that this enzyme is an exoglucanase (CI enzyme), possibly a cellobiohydrolase. The third component showed only cellobiase (P-glucosidase) activity and had no effect on cotton or carboxymethylcellulose. The three components, when mixed, showed synergistic effects on highly ordered cellulose. The endo- and exoglucanases were characterized with regard to molecular size and isoelectric point (pl). Both cellulases had pl near 4.55, but their molecular weights were different: 67000 (exoglucanase) and 41,000 (endoglucanase). The effect of temperature on the activity of the cellulases was examined with both cotton and carboxymethylcellulose as substrate. Arrhenius activation energies, Q_{10}, and temperature optima for the different reactions were determined.

The cellulolytic properties of a *Chaetomium crispatum* strain were investigated by Geeraerts and Vandamme (2008). They found that the cellulolytic enzyme complex, i.e., exo-1,4-β-glucosidase (EC 3.2.1.74), endo-l,4-β-glucanase (EC 3.2.1.4.), and 8-glucosidase or cellobiase (EC 3.2.1.21), displayed optimal activity at pH 5.0 and 25 °C. Although carboxymethyl-celluloses are the usual pseudo-substrates for this enzyme complex, those with a high degree of substitution gave rise to poor growth and low cellulase activity. Insoluble crude cellulosics such as newsprint, recycled paper, rice, and flax straw were substantially solubilized at 28 °C within 3–5 days of fermentation. A study of the cellulase-complex formation during the growth cycle revealed that β-glucosidase was produced mainly intracellularly in the early exponential phase, while the overall exo-l,4-β-glucosidase and endo-1,4-β-glucanase formation gradually increased during the total fermentation cycle. The mycelial protein of *Chaetomium crispatum* grown on crude cellulosics displayed a favorable amino acid pattern, indicating its potential value as a source of single-cell protein (SCP).

Soni et al. in 1999 studied the distribution pattern of cellulases in the extracellular and cell-associated fractions of *Chaetomium erraticum* varied depending upon the cultural conditions. The extracellular fractions revealed two forms each of endoglucanase (EG I, EG II) and β-glucosidase (β-Glu I, β-Glu II) under static and shake conditions. However, the appearance of an additional form of endoglucanase, EG III, and β-glucosidase, β-Glu III, in static intracellular and cell debris fractions showed a relation of cellulase production with the perithecia development in this fungus. The maximal production of enzymes was observed at 37 °C, pH range 5.0–10.0 in the presence of 1–2% carboxy methyl cellulose (CMC) in static cultures (for exoglucanase and β-glucosidase) and shake cultures (for endoglucanase).

Chaetomium globosum has been a well-known potential antagonist of several seed- and soilborne fungi. Eight isolates of *C. globosum* were obtained from different sources and were identified by morphological characters. *C. globosum* isolates were examined for the presence of extracellular proteins, cellulases, and antifungal metabolites in culture filtrate by sodium dodecyl sulfate-polyacrylamide gel electrophoresis (SDSPAGE), thin-layer chromatography, and high-performance liquid chromatography by Shanthiyaa et al. (2014). Variation in the mycelial protein of *C. globosum* isolates was noted in the SDSPAGE analysis. Different *C. globosum* isolates that showed more number of bands in protein profile were further screened for the production of cellulases in culture filtrate. Cellulase activity of *C. globosum* isolates revealed that maximum activity was observed in the isolate Cg-6 after 11 days of incubation, while Cg-2 had least activity.

Yadav and Bagool (2015) used books and important documents in storage having moldy appearance to analyze and isolate cellulolytic fungi. They found that most dominated genus was *Chaetomium* with 13 different species and were screened for their cellulase producing capability by the filter paper degradation ability. Eight *Chaetomium* isolates were chosen for exoglucanase and endoglucanase enzyme activity assay on the basis of percentage loss of filter paper. Five *Chaetomium* species were selected as potentially able to secret high exoglucanase and endoglucanase cellulases. *Chaetomium dolichotrichum*, *C. funiculosum*, *C. globosum*, *C. anguistispirale*, and *Chaetomium* sp. were found very good producer of total cellulase and endogluccanase during study.

Recently Cuomo et al. (2015) sequenced a strain of *Chaetomium globosum* (CBS 148.51) isolated from stored cotton in Washington, DC. This strain is commonly used in testing paper and polymers for fungal resistance (Gu and Gu 2005). Three size-selected libraries were constructed from genomic DNA. These included a 4-kb plasmid, a 10-kb plasmid, and a 40-kb fosmid library; each library was paired-end sequenced using Sanger technology. The resulting 568,566 reads were assembled using Arachne version 3.0, with an average of 8.9 sequence depth in the final assembly. Based on the assembly, the genome size was estimated to be 34.3 Mb with a GC content of 55.6%. The assembly was organized in 1245 contigs, which are linked by paired-end reads into 37 scaffolds. The average base is found in a scaffold of N50 size 4.72 Mb and a contig of N50 size 50.76 kb. The assembly is highly contiguous; the largest 8 scaffolds account for 98% of the assembly bases. This genome sequence of *C. globosum* will serve as an important reference for further studies of the basis of its cellulose specificity, for genes that enable human infection and for further comparative studies with other fungi.

Flannigan and Sellars (1972) evaluated 30 thermophilous fungi for their ability to produce CMCase and found that 16 were able to degrade cellulose. Similarly, Rosenberg (1978) tested 21 species of thermophilic and thermotolerant fungi for their cellulolytic activity. It was observed by the author that *Chaetomium thermophile* var. *coprophile*, *Chaetomium thermophile* var. *dissitum*, *Humicola grisea*, *Humicola insolens*, *Myriococcum albomyces*, *Sportotrichum thermophile*, *Malbranchea pulchella*, *Allescheria terrestris*, *Allescheria fumigatus*, *Talaromyces thermophilus*, *Torula thermophila*, *Thielavia thermophile*, and *Chrysosporium pruinosum* showed positive cellulolytic activity. On the contrary, Jain et al. (1979) observed that *Humicola lanuginosa*, *Mucor meihei*, *Malbranchea pulchella* var. *sulfurea*, and *Talaromyces dupontii* did not degrade cellulose and filter paper when used as a sole source of carbon, but showed filter paper-degrading activity when grown on wheat straw. Based on these observations, they suggested that these fungi have some specific requirements for cellulose production that were fulfilled by growing on wheat straw. *Humicola lanuginosa* (syn. *Thermomyces lanuginosus*) was unable to synthesize cellulose, as observed by Chang and Hudson (1967), Fergus (1969), and Deacon (1985). However, some recent reports (Lee et al. 2014) indicate that *Thermomyces lanuginosus* is able to produce cellulase.

Of the 15 thermophilc fungi tested by Srivastava et al. (1981), 6 were found to be celluloytic. They observed variations in the decomposition of cellulose by these fungi. Some of the fungal isolates were high decomposers, while others were weak decomposers. Thermophilic fungi such as *Mucor miehei*, *Mucor pusillus*, and *Rhizopus rhizopodiformis*, which were considered secondary sugar fungi, are confirmed to be moderately cellulolytic (Johri and Pandey 1982). Tong and Cole (1982) observed *Thermoascus aurantiacus* as the most active cellulose producer among the several thermophilic fungi evaluated by them. Sandhu et al. (1985) screened 98 strains of thermophilous fungi for their ability to produce cellulases. Most of these fungi belonged to the genera *Acremonium*, *Aspergillus*, *Chaetomium*, *Penicillium*, *Thermoascus*, and *Thielavia*. They also revealed that cellulases produced by true thermophiles were more stable than those produced by thermotolerant and microthermophiles.

Chaetomium thermophilum var. *coprophilum* produced large quantities of extracellular as well as intracellular b-glucosidase when grown on cellulose or cellobiose (Venturi et al. 2002). The purification and characterization of cellulases from *Humicola grisea* and *Aspergillus fumigatus* have been studied by Takashima et al. (1996) and Ximenes et al. (1996).

van Noort et al. (2013) investigated the genome of *Chaetomium thermophilum* and observed that there are fewer genes that encode complex carbohydrate-degrading enzymes, in particular, in thermophilic mold genomes than their mesophilic counterpart. The genome of *Chaetomium thermophilum* encodes three CDHs, while those of *Chaetomium globosum* and *Neurospora crassa* encode only two, which depicts a higher cellulolytic ability of the former.

12.4 Mechanisms of Cellulase Synthesis

Apart from the development of economically feasible systems for cellulose degradation, there has also been continuing interest in understanding the mechanisms of cellulase synthesis and production to identify feasible approaches for increasing cellulase production (Li et al. 2010).

Sun et al. (2008a, b) investigated the proteome profiling map of the cellulases secreted by Trichoderma reesei Rut C-30 using two-dimensional gel electrophoresis. CBH I and CBH II were found to represent about 37% of the total extracellular proteins, and the CBH II concentration produced with nonpretreated rice straw powder was about threefold higher than that with alkali-treated straw. This interesting result suggests that the synthesis of CBH II is controlled by other factors aside from cellulose. Sun et al. (2008a, b) reported the differences in the composition and expression levels of *P. decumbens* cellulases under induced and basal conditions. The basal cellulase in *P. decumbens* was demonstrated to be composed of CBH I, CBH II, EG I, EG II, and bG, whereas two EGs were expressed only under induction conditions. Furthermore, the basal and induced EGs from *Penicillium decumbens* were encoded by different genes.

To understand the mechanism of the two transcription repressors Cre1 and ACE I in T. reesei, Su et al. (2009) developed a new strategy wherein a plasmid that encodes a chimeric transcription activator containing the DNA binding domains from Cre1 and ACE I and the effector domain from the activator ACE II was constructed and transformed into T. reesei. The recombinant strain had higher cellulase activity than its parent strain and had a different colony appearance. The results also provide an overview of the set of genes that might be regulated by Cre1 or ACE I. These results contribute to further understanding the regulatory roles of these two repressors in cellular pathways and provided a new method for strain improvement through genetic manipulation.

Liu et al. (2008) studied the differences in gene sequences of CBH I gene (cbh1) from wild-type and mutant *P. decumbens* strains and found that the mutant strain JU-A10 is a multiple mutant of the wild-type strain in the sequences upstream of the gene. The enhanced CBH activity of the mutant may be due to a single-base mutation of the upstream sequence of cbh1, which affects the transcription regulation of the mutant instead of the protein-coding sequences. This discovery suggests the critical role of the promoter region of cellulose-encoding genes, which is helpful in constructing hyperproducing strains of *P. decumbens*. An apparent abolishment of glucose repression was also identified in strain JU-A10, with enhanced observed cellulase and hemicellulase production in glucose-containing media. Genomic analysis of this strain revealed a single nucleotide deletion at the +1205 position in the creA gene, which encodes a carbon catabolite repressor protein. This frameshift mutation changes the amino acid sequence downstream from the site of the mutation (unpublished data). Numerous other mutations have also been identified from this mutant strain through genomic analysis, in addition to the changes in the creA gene.

Another focus for Chinese scientists is identifying inducers for cellulolytic enzyme production, which could potentially benefit the cellulase industry. Wang et al. (1995) observed that the concentrations of ATP and cyclic AMP (cAMP) influence cellulase production. Cellulase synthesis is repressed by high concentrations of intracellular ATP, whereas exogenous cAMP increases cellulase synthesis. The effects of wheat bran on the hydrolysis of extracellular biomass were investigated in *P. decumbens* by Sun et al. (2008a, b). The soluble cello-oligosaccharide composition of wheat bran was shown to be one of the most significant factors in cellulase production. This significant discovery may be critical in the cellulase industry because wheat bran, as an inducer in cellulase and xylanase production, is inexpensive.

12.5 Molecular Weight of Fungal Cellulases

The MW of cellulase produced by different fungal species may vary from 12 kDa to 126 kDa (Parry et al. 1983; Bai et al. 2013). SDS-polyacrylamide gel electrophoresis (SDS-PAGE) is the most commonly used method for judging the apparent MW of enzymes (Joo et al. 2010; Lee et al. 2010; Ramani et al. 2012). Fungal cellulase may be of monomeric (Naika et al. 2007) or dimeric (Chaabouni et al. 2005) in nature. Cellulase produced by T. viride was purified to homogeneity using DEAE-sepharose column and the MW was estimated as 87 kDa by SDS-PAGE (Yasmin et al. 2013). P. pinophilum MS 20 produced a monomeric cellulase with MW of 42 kDa, which appeared as a single band on SDS-PAGE gel (Pol et al. 2012). It was reported that A. awamori VTCC-F099 and Fomitopsis pinicola produced monomeric thermo active cellulase with a MW of 32 kDa (Yoon et al. 2008; Van Tuan Nguyen 2010). The cellulase produced by *A. niger* revealed a MW of 60 kDa on SDS-PAGE gel (Baraldo et al. 2014). In all the aforesaid studies, the purified celluase appeared as single band upon SDS-PAGE, indicating that the cellulase produced by these fungi are active in solution as monomers or homodimers, consequently migrates through the SDS-PAGE gel according to their MW so as to segregate as single band. In contrast, some studies reported the identification of hetero-dimeric cellulases or isoforms which appeared as separate bands upon SDS-PAGE. For instance, *A. niger* Z10 produced two cellulase bands on SDS-PAGE gel with MWs 50 and 83 kDa (Coral et al. 2002a, b). Similarly, another strain of *A. niger* is also reported as producing dimeric cellulase with MWs of 23 and 36 kDa, whereas *A. fumigatus* produced dimeric cellulase with MWs of 21 and 32 kDa. Kaur et al. (2007) reported 40 and 50 kDa isoforms of cellulase produced by Melanocarpus sp. MTCC 3922 with MW as judged by SDS-PAGE.

12.6 Factor Affecting Cellulase Production

Several fermentation conditions play fundamental roles on cellulases production, among which fermentation method, carbon source, nitrogen source, pH, temperature, salt/metal ions effect, incubation time, aerations, and fungal species (Norouzian 2008; Okoye et al. 2013; Saini et al. 2017).

12.6.1 Fermentation Method

Fungal cellulases have been produced through solid-state fermentation (SSF) and submerged fermentation (SmF). In SSF, the fungal species is grown on one or more solid substrate such as rice straw, wheat bran, corn husk, cassava cake, or sugar cane bagasse without or very low water content. The grown microorganism utilized the solid substrate steadily and slowly thus under SSF condition, the microorganism can be grown for long period of time, for instance, for several days (Ahmed et al. 2017a, b). The high productivity, cheap substrate utilization, and low energy requirement are the advantages of SSF. Moreover, under SSF conditions, there is minimal water output as well as lack of foam up which makes it economically feasible (Faisal and Benjamin 2016). SSF shortcomings are limited to heat generation and lack of knowledge on automation (Ahmed et al. 2017a, b, Shweta 2015; Soccol et al. 2017). SSF has been utilized for cellulase production from several fungal species such as *lichtheimia romosa* (Garcia et al. 2015), *Phaffomycetaceae* (Cerda et al. 2017), Dipodascaceae (Cerda et al. 2017), *Trichoderma citrinoviride* AUKAR04 (Periyasamy et al. 2017), *Humicola insolens* MTCC 1433 (Singla and Taggar 2017), among many others. On the other hand, in SMF, free-flowing liquid like molasses and/or broths supplemented with different nutrients is used to cultivate microorganisms. The enzymes including cellulase and metabolic byproducts are secreted into fermentation medium, and medium supplements or nutrients are rapidly utilized and a continuous supply is needed. SMF has several advantages such as simplicity of sterilization, heat and mass transfer, process monitoring (pH, temperature, and soluble molecules) and automation, and extraction and recovery of enzymes and bioactives (Ahmed et al. 2017a, b). Several cellulase enzymes have been produced by SMF from different fungal species including *Aspergillus flavus* (Gomathi et al. 2012), *Aspergillus niger* FC-1 (Jiang 2013), and *Aspergillus niger* (Reddy et al. 2015), among many others.

12.6.2 Carbon Source

Carbon source is the major factor affecting the cellulases production, attributing to the fact that cellulases are inducible enzymes that are expressed by cells in response to different carbon source present in the fermentation medium (Saini et al. 2017; Zhang et al. 2017). For instance, optimal cellulase production from *Hypocrea jecorina* QM6a, QM9414, and RUTC-30 was attained in medium containing microcrystalline celluloses as the sole carbon source (Dashtban et al. 2011). *Penicillium* sp. produced the highest cellulase activity on lactose-containing media among different carbon sources tested such as sarbose, maltose, sucrose, lactose, dextrose, galactose, cellobiose, and CMC (Prasanna et al. 2016). Expression of different cellulase isoforms in response to carbon source has also been reported (Amore et al. 2013). For example, *Aspergillus terreus* expressed four endoglucanase (EG) isoforms in the presence of rice straw as solid substrate or corn cobs as liquid substrate. Similarly, supplementation of fructose and cellobiose to corn cobs medium upregulates at least one of EG, while adding mannitol, ethanol, and glycerol selectively suppressed the expression of three EG isoforms. Similarly four isoforms of β-glucosidase (βG) were expressed in presence of corn cob containing medium, and addition of glucose, cellobiose, mannitol, fructose, sucrose, or glycerol repressed one or more βG isoforms (Nazir et al. 2010). *Aspergillus fumigatus* Z5 grown on culture media containing glucose, avicel, and rice straw secreted 61, 125, and 152 proteins, respectively. Proteomic analysis suggested that glycoside hydrolases including cellulases and hemicellulases were overexpressed on rice straw and avicel-containing media compared to glucose used as carbon sources (Liu et al. 2013). The molecular mechanisms by which the expression of these different isoforms are regulated and different carbon sources influence the quantity and isoforms expression are not well established which hampered genetic engineering of these fungi for industrial purpose (Coradetti et al. 2012). Thus understanding the mechanisms of cellulase expression hold a critical significance for enhancement of cellulase enzymes production and have been investigated in *Aspergillus* and *Trichoderma* (Gautam et al. 2011). These fungi produce extracellular cellulase enzymes when they are grown on media containing plant polymers, or short oligosaccharides as an energy source, and when cultivated on media containing easily metabolizable sugar such as glucose, the expression of these enzymes is repressed. Carbon catabolite repression is considered the most acceptable mechanism to repress cellulase production when grown on easily metabolizable sugars (Amore et al. 2013).

Recently, Zhang et al. published a study demonstrating that *Rhizopus stolonifera* host a gene which encodes for cellobiose synthetase (CBS) to synthesize cellobiose from uridine diphosphate glucose (UDPG). CBS was found to play a fundamental role in expression of cellulase gene through the induction of the cellobiose-responsive regulators CLR1 and CLR2 and thus inducing the transcription of cellulase genes. The authors suggested that minimal constitutive expression of cellulase may be driven by cellobiose synthesized by CBS from carbohydrate metabolites (Zhang et al. 2017).

12.6.3 Nitrogen Source

Another important factor that affects protein secretion in fungi is nitrogen. Different nitrogen source can be included in fermentation medium for cellulase production. Among organic nitrogen sources that can be used are peptone, yeast or beef extract, and tryptone or soybean meal. Inorganic nitrogen sources like ammonium sulfate, ammonium chloride, and ammonium hydrogen phosphate can also be used as a nitrogen source (Ahmed et al. 2017b; Kachlishvili et al. 2006). Optimum cellulase activity was achieved from *Penicillium* sp. when cultivated on yeast extract containing medium (Prasanna et al. 2016). *Trichoderma reesei* showed optimum production of cellulase when cultivated on *Parthenium* biomass containing ammonium molybdate, peptone, or yeast extracts as nitrogen source (Saini et al. 2017).

12.6.4 pH and Temperature

Optimization of parameters such as pH and temperature is also of crucial significance for enzyme production since these physicochemical parameters affect the growth of microorganism hence the bioactive production. The optimal cellulase production from *Penicillium* sp. was attained on Czapek-Dox medium at pH 5.0 and

30 °C (Prasanna et al. 2016). Similarly, optimum production of celllulase by *Aspergillus tubingensis* KY615746 was achieved at pH 4 and temperature of 30 °C (El-Nahrawy et al. 2017).

12.6.5 Incubation Time

Myceliophthora heterothallica produced the highest endoglucanase on SSF containing wheat bran or sugarcane occurred at 192 hours and on SmF containing cardboard at 168 hours (da Teixeira et al. 2016). Optimal production of carboxymethylcellulase (CMCase) from *Aspergillus hortai* under SMF was achieved after 96 hours (El-Hadi et al. 2014).

12.7 Statistical Approach for Optimization of Cellulase Production

Optimization of cellulase production is a critical process for efficient and cost-effective cellulase production. Traditionally optimization of cellulase production is carried out by employing One Variable at A time (OVAT) approach. OVAT involves varying one parameter at a time keeping other factors constant. OVAT is regarded as a laborious technique and time-consuming and misleading approach because these parameters are independent and OVAT tends to ignore the interactions between them, in addition to extensive time needed to perform a large number of experiments.

Statistical approaches such as surface response methodology and Plackett-Burman design are efficient approaches employed for optimization of fermentation parameters (Shajahan et al.; Singh et al. 2014). Several studies have employed statistical methods for optimization of cellulase production. Cellulase production from *Trichoderma reesei* was optimized using Plackett-Burman design of 9 nutrients for their influence on cellulase secretion using Response Surface Methodology (RSM). The study demonstrated that the optimal concentration of avicel, soybean cake flour, KH2PO4, and CoCl2·6H2O for cellulase production were 25.30 g/l, 23.53 g/l, 4.90 g/L, and 0.95 g/l, respectively (Saravanan et al. 2012). In another study, using statistical Full Factorial Design (FFD), optimal cellulase production from *Penicillium funiculosum* ATCC11797 was achieved on culture media containing avicel (10 g/l) as carbon source, urea (1.2 g/l), yeast extract (1.0 g/l), KH_2PO_4 (6.0 g/l), and $MgSO_4$. $7H_2O$ (1.2 g/l) with an agitation speed of 220 rpm and aeration rate of 0.6 vvm. These conditions resulted in activities of 508 U/l for FPase, 9,204 U/l for endoglucanase, and 2,395 U/l for β-glucosidase which are 3.6–9.5 times higher than production using nonoptimized conditions (de Albuquerque de Carvalho et al. 2014). Cellulase production from *Trichoderma reesei* RUT C-30 was optimized employing a two-stage statistical design, namely, fractional factorial design and response surface Box-Behnken design, on wheat bran and cellulose under SSF. This approach resulted in a 3.2-fold increase in CMCase production to

959.53 IU/gDS (Idris et al. 2017). The statistical approaches for optimization of fermentation conditions are considered efficient because the interactions of multiple variables are taken into consideration and the number of experiments needed to be performed is reduced to minimum (Ahmed et al. 2017a, b; Shajahan et al. 2017; Singh et al. 2014).

12.8 Application of Cellulases

Cellulases, over many decades, are used in various industrial applications, securing the third rank among enzymes annual sale and expected to exceed the protease in the near future (Menendez et al. 2015). Cellulase enzymes have got tremendous applications in different industries including biofuel production, paper and pulp industry, detergent industries, animal feeds among others.

12.8.1 Biomass Hydrolysis and Biofuel Production

Cellulase along with other enzymes is used in the hydrolysis of biomass into sugar and other chemicals. Hexoses or pentoses are then fermented to bioethanol or other fuel (Sun and Cheng 2002).

With the rapid increase in world population accompanied by increased demand of energy, depletion of fossil fuel, and enhanced greenhouse effect from traditional fuel, there is crucial need to develop or search for cheap, renewable, and sustainable sources of energy. Thus cellulases are involved in biofuel productions and minimization of energy crisis and environmental pollution (Horn et al. 2012; Sharada et al. 2014). However, the bioconversion of pretreated cellulose-based materials at the industrial level into fermentable sugars employs a mixture of enzymes for complete hydrolysis, the cost of which is very high, making biorefining processes economically unfeasible. Thus the search of biocatalysts such as cellulases with novel properties exemplified by high thermostabilty, acidophilicity, and high solvent tolerance could help to overcome the cost hurdles. Cellulases application in biomass hydrolysis and biofuel productions is currently the subject of numerous studies supported by different agencies across the world (Budihal et al. 2016; Srivastava et al. 2015a, b, c).

12.8.2 Paper and Pulp Industry

Cellulases are used in the paper and pulp industry which has expanded significantly in the last decades from 320 to 395 million tons (Przybysz Buzała et al. 2016). Pulping process can be achieved either through mechanical or biomechanical manners.

Mechanical pulping such as refining and grinding of the woody raw material results in pulps containing a high content of fines, bulk, and stiffness. On the other hand, biomechanical pulping employing enzymes such as cellulases results in around 20–40% energy savings during refining making the process economically feasible and significantly improved hand-sheet strength properties (Demuner et al. 2011; Sharada et al. 2014). It has also been reported that the addition of cellulases enhanced the bleachability of softwood kraft pulp and improved the final brightness score comparable to that of xylanase treatment (Kuhad et al. 2011).

12.8.3 Waste Management

Cellulase can be used in waste management. For instance, cellulases are used in the conversion of cellulosic municipal solid wastes to desirable chemicals and energy. Cellulases benefits in minimizing the effect of cellulose waste on our environment and driving the conversion of the pollutants to an alternative source of energy and chemicals thus displacing our growing dependence on fossil fuels (Bayer et al. 2007; Gautam et al. 2011; Kuhad et al. 2011).

12.8.4 Animal Feed Industry

Cellulase has great potential to be used in the animal feeds industry. Cellulase can be used in the pretreatment of agricultural silage and grain feed to enhance nutritional value and performance of animals (Kuhad et al. 2011). Similarly, addition of cellulase, along with other enzymes, can eliminate anti-nutritional factors present in the feed grains such as arabinoxylans, cellulose, dextrins, inulin, lignin, pectins, β-glucan, and oligosaccharides by degrading them. This in turn enhances the nutritional value and improves animal's health and performance (Asmare 2014; Murad and Azzaz 2010; Sharada et al. 2014).

12.8.5 Laundry and Detergent Industry

Cellulases are also used in the laundry and detergent industry which is one of the most popular markets for enzymes sale accounting for 20–30%, with lipase and proteases as major enzymatic components. An innovative approach recently adopted in this industry is the use of alkaline cellulases, protease, and lipase results in a crucial improvement of color brightness and dirt removal from the cotton blend garments (Juturu and Wu 2014b; Olsen and Falholt 1998).

12.8.6 Textile Industry

The most successful and popular application of cellulases is the textile industry. Cellulases are used in textile wet processing such as finishing of cellulose-based textiles, biostoning of jeans, and biopolishing of cotton and other cellulosic fabrics in order to improve hand and appearance (Arja 2007; Duran and Duran 2000; Juturu and Wu 2014b).

12.8.7 Wine and Beverage Industry

Cellulase enzymes along with glucanase can be used to improve both quality and yields of the fermented products such as wine and beverages. For example, during wine production, cellulase, pectinases, glucanases, and hemicellulases are used to improve color extraction, skin maceration, must clarification, filtration, and finally the wine quality and stability. Addition of β-glucosidases can increase the aroma of wines by hydrolyzing glycosylated precursors into their aglycones and glucose (Araujo et al. 2008; Kuhad et al. 2011).

12.8.8 Other Applications

Cellulases have also been applied in agriculture where they are used to hydrolyze the cell wall of plant pathogens thus controlling the plant infection and diseases. Many cellulolytic fungi including *Trichoderma* sp., *Geocladium* sp., *Chaetomium* sp., and *Penicillium* sp. are known to play a key role in agriculture by enhancing the seed germination, rapid plant growth and flowering, improving root system, and increasing crop yields (Behera et al. 2016; Kuhad et al. 2011). Cellulases have also been used for the improvement of the soil quality (Phitsuwan et al. 2013). In addition, cellulases are used in food processing during fruit and vegetable juices manufacturing to improve extraction (Sharada et al. 2014; Zhang and Zhang 2013).

Furthermore, applications of cellulases along with macerating enzymes have been found to increase extraction of olive oil under cold processing conditions and to improve its antioxidants and vitamin E contents (Aliakbarian et al. 2011; Sharma et al. 2015). Moreover, humans are known to poorly digest cellulose fiber, and taking a digestive enzyme product containing cellulases like Digestin helps to relieve digestive problems such as malabsorption (Gurung et al. 2013; Sharada et al. 2014). Finally, an interest in applying cellulases enzymes in chemical analysis such as diagnostic and food analysis has been considered (Li et al. 2012).

12.9 Biodegradation of Agricultural Biomass

Bioconversion of lignocellulosic biomass via cellulase enzyme is also known as enzymatic hydrolysis or saccharification. For effective conversion of cellulose into monomers, a complete cellulase system is required (endoglucanase, exoglucanase, and β-glucosidase) to act synergistically. Biomass, which undergoes pretreatment before the enzymatic hydrolysis can be easily converted into sugars, effectively (Megan et al. 2013; Jose and Arnold 2014). Moreover, released sugars can undergo fermentation to produce biofuel. Although pretreatment process removes lignin effectively, the hemicellulose and cellulose part is converted by the synergistic action of hemicellulases and cellulases (Dashtban et al. 2009). Enzymatic hydrolysis of lignocellulosic biomass is divided into primary hydrolysis and secondary hydrolysis. Primary hydrolysis generally takes place on the surface of substrate, and cellobiose is released due to the catalytic action of exo- and endoglucanases. The cellobiose is further converted into glucose via β-glucosidase in secondary hydrolysis (Zhang et al. 2006).

12.10 Biofuel Production from Biomass Waste Degradation

Over the past decade, several research groups have focused on biofuel production using biomass-based process as cost-effective technology. Cellulosic biomass to sugar conversion is the key step for biofuel production, and efforts have been made to explore the technology related to biomass conversion via cellulase such as improving the efficiency of cellulase-producing microorganisms and cellulase efficiency (Garvey et al. 2013). Figure 12.5 explains the efficient destruction of biomass via cellulase to release sugar and produce biofuels. Additionally, non-efficient cellulase is incapable to release sugars from this biomass and hydrolysis becomes incomplete (Fig. 12.5). Although commercial cellulases are manufactured from native microorganism, economic viability seems far away. Currently, for efficient enzymatic hydrolysis of biomass, different cellulases (EG, CBH, and BGL) have been achieved from different microbial sources, which makes the process economically sustainable. Therefore, intense research is underway to find out novel cellulase-producing systems in order to increase the flexibility of available microbial strain, yield, and processing. The current enzymatic hydrolysis research is aimed to improve the production of cellulase via screening of effective microorganisms that would provide new opportunities to face this challenges (Peterson and Nevalainen 2012; Chandel et al. 2012).

The fermentable sugars obtained from the effective hydrolysis via cellulase enzymes are used for the production of renewable energy (Fig. 12.6). In addition, fermentable sugars can be converted into bioethanol, biohydrogen, and methane through different processes with the help of specific microorganisms. Effective production and yield of biofuels are directly dependent on the amount of fermentable

Fig. 12.5 Cartoon depiction of enzymatic cellulose deconstruction. Processive reducing (*orange*) and nonreducing end (*green*) cellulases move along the cellulose fibers liberating cellobiose units. The cellobiose released is then converted to individual glucose sugars by β-glucosidases (*pink*). Endo-acting cellulases (*red*) introduce chain breaks that the exo-acting enzymes can act upon. All of the aforementioned enzymes use hydrolytic mechanisms, while the new players in this process, LPMOs (*blue*), further potentiate their action using an oxidative mechanism to introduce further chain breaks on which the processive cellulases can initiate further degradation (Hemsworth et al. 2015)

sugars present in reaction medium, and the fair sugar amount is directed by viable cellulase system. Apart from cellulase system, the targeted process can be improved by removing the structural weakness of biomass via effective pretreatment strategies for biofuel production.

12.11 Conclusion

Continuous and significant studies have been made to improve the production and efficiency of cellulase enzyme for low-cost biofuel production. Although cellulase enzymes have versatile industrial applications, improvement in efficiency, reduction in cost, and energy consumptions are always selective parameters for the bioconversion of cellulosic biomass into biofuels. Isolation, screening, and cultivation of thermophiles/thermotolerant fungi for obtaining thermophilic/thermostable cellulase systems at commercial scale to produce biofuels are still a roll-back factor. Low cell yield and submerged production processes of enzyme are also cost-intensive for biomass to biofuel conversion processes. Limited knowledge of protein engineering and its implementation are also adding an extra obstacle in the overall bioconversion process apart from the purification of enzyme and end products separation. In

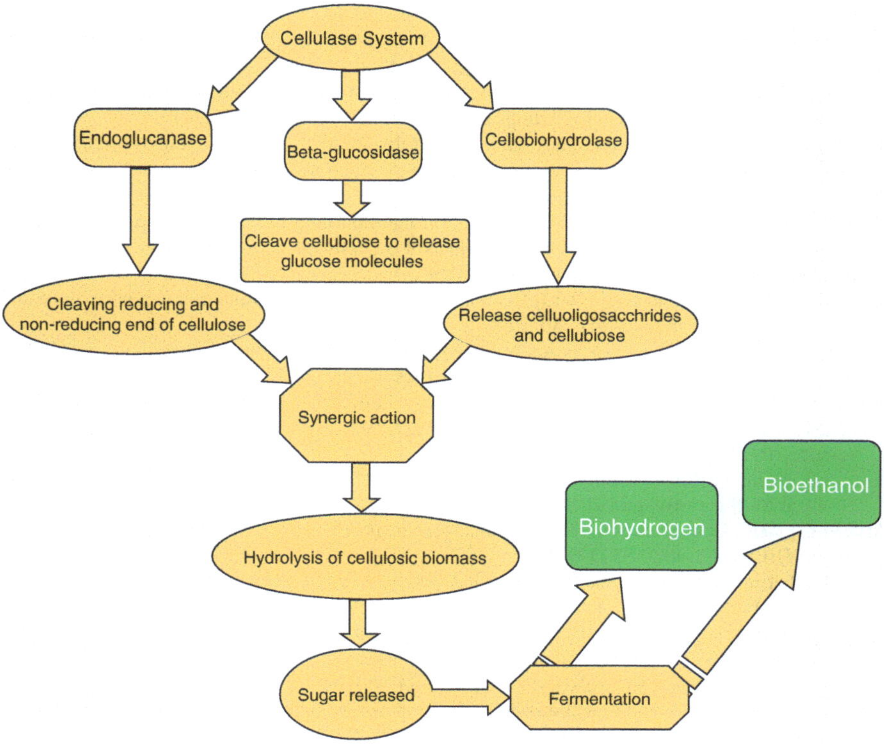

Fig. 12.6 Role of cellulases in complete biofuel production processes

view of the above issues, there is need of more and vast research for the production and optimization of cost-effective cellulases from potential fungal strain using SSF and biomass as a low-cost process. Additionally, research on more thermostable cellulase enzymes, their stability, and protein engineering may also support to achieve novel and more economical processes for biofuel production compared to the existing process technologies.

References

Adler G, Flores JM, Abo Riziq A, Borrmann S, Rudich Y (2011) Chemical, physical, and optical evolution of biomass burning aerosols: a case study. Atmos Chem Phys 11:1491–1503. https://doi.org/10.5194/acp-11-1491-2011

Agrawal D, Matkar K (2016) Enhancement of cellulase production by substrate manipulation in three Aspergillus sp. Indian J Appl Res 6(3). ISSN – 2249-555X

Ahmed A, Aslam M, Ashraf M, ul-Hassan Nasim F, Batool K, Bibi A (2017a) Microbial ß- glucosidases: screening, characterization, cloning and applications. J Appl Environ Microbiol 5:57–73. https://doi.org/10.12691/jaem-5-2-2

Ahmed A, Nasim F-H, Batool K, Bibi A (2017b) Microbial β-Glucosidase: sources, production and applications. J Appl Environ Microbiol 5:31–46. https://doi.org/10.12691/jaem-5-1-4

Akagi SK, Yokelson RJ, Wiedinmyer C, Alvarado MJ, Reid JS, Karl T, Crounse JD, Wennberg PO (2011) Emission factors for open and domestic biomass burning for use in atmospheric models. Atmos Chem Phys 11:4039–4072. https://doi.org/10.5194/acp-11-4039-2011

Aliakbarian B, De Faveri D, Casazza A, Oliveira R, Oliveira M, Converti A, Perego P (2011) Bio-extraction of olive oil: improvement of quality and extraction outputs: Citeseer

Alves N d O, Brito J, Caumo S, Arana A, Hacon S d S, Artaxo P, Hillamo R, Teinila K, Batistuzzo de Medeiros SR, Vasconcellos P d C (2015) Biomass burning in the Amazon region: Aerosol source apportionment and associated health risk assessment. Atmos Environ 120:277–285. https://doi.org/10.1016/j.atmosenv.2015.08.059

Amore A, Giacobbe S, Faraco V (2013) Regulation of cellulase and hemicellulase gene expression in fungi. Curr Genomics 14(4):230–249

Andreae MO, Merlet P (2001) Emission of trace gases and aerosols from biomass burning. Global Biogeochem Cy 15:955–966. https://doi.org/10.1029/2000GB001382

Andreae MO, Rosenfeld D, Artaxo P, Costa AA, Frank GP, Longo KM, Silva-Dias MAF (2004) Smoking rain clouds over the Amazon. Science 303:1337. https://doi.org/10.1126/science.1092779

Araujo R, Casal M, CavacoPaulo A (2008) Application of enzymes for textile fibres processing. Biocatal Biotransformation 26:332–349. https://doi.org/10.1080/10242420802390457

Arja M-O (2007) Cellulases in the textile industry. Industrial enzymes. Springer, pp 51–63. https://doi.org/10.1007/1-4020-5377-0_4

Arnold SR, Emmons LK, Monks SA, Law KS, Ridley DA, Turquety S, Tilmes S, Thomas JL, Bouarar I, Flemming J, Huijnen V, Mao J, Duncan BN, Steenrod S, Yoshida Y, Langner J, Long Y (2015) Biomass burning influence on high-latitude tropospheric ozone and reactive nitrogen in summer 2008: a multi-model analysis based on POLMIP simulations. Atmos Chem Phys 15:6047–6068. https://doi.org/10.5194/acp-15-6047-2015

Asmare B (2014) Biotechnological advances for animal nutrition and feed improvement. World J Agric Res 2:115–118. https://doi.org/10.12691/wjar-2-3-5

Awasthi A, Singh N, Mittal S, Gupta PK, Agarwal R (2010) Effects of agriculture crop residue burning on children and young on PFTs in North West India. Sci Total Environ 408(20):4440–4445

Babu KR, Satyanarayana T (1996) Production of bacterial enzymes by solid state fermentation. J Sci Ind Res 55:464–467. ISSN 0022-4456

Bai H, Wang H, Sun J, Irfan M, Han M et al (2013) Production, purification and characterization of novel beta glucosidase from newly isolated *Penicillium simplicissimum* H-11 in submerged fermentation. EXCLI J 12:528–540

Bansal N, Tewari R, Soni R, Soni SK (2012a) Production of cellulases from *Aspergillus niger* NS-2 in solid state fermentation on agricultural and kitchen waste residues. Waste Manag 32:1341–1346. https://doi.org/10.1016/j.wasman.2012.03.006

Bansal N, Tewari R, Soni R, Soni SK (2012b) Production of cellulases from *Aspergillus niger* NS-2 in solid state fermentation on agricultural and kitchen waste residues. Waste Manag 32:1341–1346. https://doi.org/10.1016/j.wasman.2012.03.006

Baraldo A Jr, Borges DG, Tardioli PW, Farinas CS (2014) Characterization of ß-Glucosidase produced by *Aspergillus niger* under solid-state fermentation and partially purified using MANAE-Agarose. Biotechnol Res Int 2014:8

Bayer EA, Lamed R, Himmel ME (2007) The potential of cellulases and cellulosomes for cellulosic waste management. Curr Opin Biotechnol 18:237–245. https://doi.org/10.1016/j.copbio.2007.04.004

Behera BC, Sethi BK, Mishra RR, Dutta SK, Thatoi HN (2016) Microbial cellulases diversity & biotechnology with reference to mangrove environment: a review. J Genet Eng Biotechnol 15:197–210. https://doi.org/10.1016/j.jgeb.2016.12.001

Bond TC, Streets DG, Yarber KF, Nelson SM, Woo JH, Klimont Z (2004) A technology-based global inventory of black and organic carbon emissions from combustion. J Geophys Res Atmos 109:D14203. https://doi.org/10.1029/2003JD003697

Bond TC, Doherty SJ, Fahey DW, Forster PM, Berntsen T, DeAngelo BJ, Flanner MG, Ghan S, Kärcher B, Koch D et al (2013) Bounding the role of black carbon in the climate system: A scientific assessment. J Geophys Res Atmos 118(11):5380–5552. https://doi.org/10.1002/jgrd.50171

Brassard P, Palacios JH, Godbout S, Bussieres D, Lagace R, Larouche JP, Pelletier F (2014) Comparison of the gaseous and particulate matter emissions from the combustion of agricultural and forest biomasses. Bioresour Technol 155:300–306. https://doi.org/10.1016/j.biortech.2013.12.027

Bruns EA, El Haddad I, Slowik JG, Kilic D, Klein F, Baltensperger U, Prevot ASH (2016) Identification of significant precursor gases of secondary organic aerosols from residential wood combustion. Sci Rep 6:27881. https://doi.org/10.1038/srep27881

Bruns EA, Slowik JG, El Haddad I, Kilic D, Klein F, Dommen J, Temime-Roussel B, Marchand N, Baltensperger U, Prévôt ASH (2017) Characterization of gas-phase organics using proton transfer reaction time-of-flight mass spectrometry: fresh and aged residential wood combustion emissions. Atmos Chem Phys 17:705–720. https://doi.org/10.5194/acp-17-705-2017

Budihal SR, Agsar D, Patil SR (2016) Enhanced production and application of acidothermophilic Streptomyces cellulase. Bioresour Technol 200:706–712. https://doi.org/10.1016/j.biortech.2015.10.098

Camassola M, Dillon AJP (2014) Effect of different pretreatment of sugar cane bagasse on cellulase and xylanases production by the mutant *Penicillium echinulatum* 9A02S1 grown in submerged culture. Biomed Res Int 2014:9. https://doi.org/10.1155/2014/720740

Cao G, Zhang X, Gong S, Zheng F (2008) Investigation on emission factors of particulate matter and gaseous pollutants from crop residue burning. J Environ Sci 20:50–55. https://doi.org/10.1016/S1001-0742(08)60007-8

Cavacopaulo A (1998) Mechanism of cellulase action in textile processes. Carbohydr Polym 37:273–277. https://doi.org/10.1016/S0144-8617(98)00070-8

Cerda A, Gea T, Vargas-García MC, Sánchez A (2017) Towards a competitive solid state fermentation: Cellulases production from coffee husk by sequential batch operation and role of microbial diversity. Sci Total Environ 589:56–65. https://doi.org/10.1016/j.scitotenv.2017.02.184

Chaabouni SE, Mechichi T, Limam F, Marzouki N (2005) Purification and characterization of two low molecular weight endoglucanases produced by *Penicillium occitanis* mutant pol 6. Appl Biochem Biotechnol 125:99–112

Chahal OS (1983) Growlh characteristics of microorganisms in solid state fermentation for upgrading of protein value of lignocelluloses and cellulase production. In: Blanch HW, Poputsakis ET, Stephanopoulos G (eds) Foundation of biochemical engineering kinetics and thermody- namics in biological systems, ACS Symposium. Vol. 207, Chapter 20. American Chemical Society, Washington, DC, pp 421–442. https://doi.org/10.1021/bk-1983-0207.ch020

Chandel AK, Chandrasekhar G, Silva MB, Silva SSD (2012) The realm of cellulases in biorefinery development. Crit Rev Biotechnol 32:187–202. https://doi.org/10.3109/07388551.2011.595385

Chang Y, Hudson HJ (1967) The fungi of wheat straw compost. II. Biochemical and physiological studies. Trans Br Mycol Soc 50:649–666

Cheng YH, Yang LS (2016) Characteristics of ambient black carbon mass and size-resolved particle number concentrations during corn straw open-field burning episode observations at a rural site in Southern Taiwan. Int J Environ Res Public Health 13(7):688. https://doi.org/10.3390/ijerph13070688

Cheng Y, Engling G, He K-B, Duan F-K, Ma Y-L, Du Z-Y, Liu J-M, Zheng M, Weber RJ (2013) Biomass burning contribution to Beijing aerosol. Atmos Chem Phys 13:7765–7781. https://doi.org/10.5194/acp-13-7765-2013

Cherian E, Kumar MD, Baskar G (2016) Production and optimization of cellulase from agricultural waste and its application in bioethanol production by simultaneous saccharification and fermentation. Manag Environ Qual Int J 27:22–35. https://doi.org/10.1108/MEQ-07-2015-0128

Collins T, Gerday C, Feller G (2005) Xylanases, xylanase families and extremophilic xylanases. FEMS Microbiol Rev 29:3–23

Coradetti ST, Craig JP, Xiong Y, Shock T, Tian C, Glass NL (2012) Conserved and essential transcription factors for cellulase gene expression in ascomycete fungi. Proc Natl Acad Sci 109:7397–7402. https://doi.org/10.1073/pnas.1200785109

Coral G, Arikan B, Ünaldi MN, Guvenmez H (2002a) Some properties of crude carboxymethyl cellulase of Aspergillus niger Z10 wild-type strain. TJB 26:209–213

Coral G, Arikan B, Ünlandi MN, Guvenmez H (2002b) Some properties of crude carboxymethyl cellulase of Aspergillus niger Z10 wild-type strain. TJB 26:209–213

Couto SRG, Sanromán MA (2006) Application of solid-state fermentation to food industry—A review. J Food Eng 76:291–302. https://doi.org/10.1016/j.jfoodeng.2005.05.022

Cuomo CA, Untereiner WA, Ma LJ, Grabherr M, Birren BW (2015) Draft genome sequence of the cellulolytic fungus *Chaetomium globosum*. Genome Announc 3(1):e00021–e00015. https://doi.org/10.1128/genomeA.00021-15. PMID: 25720678; PMCID: PMC4342419

da Teixeira S, de Cássia V, de Souza Coto AL, de Carvalho Souza R, Neves MBS, Gomes E, Bonilla-Rodriguez GO (2016) Effect of pH, temperature, and chemicals on the Endoglucanases and β-Glucosidases from the Thermophilic fungus *Myceliophthora heterothallica* F.2.1.4. Obtained by solid-state and submerged cultivation. Biotechnol Res Int 2016:9. https://doi.org/10.1155/2016/9781216

Das A, Paul T, Halder SK, Jana A, Maity C, Mohapatra PKD, Pati BR, Mondal KC (2013) Production of cellulolytic enzymes by *Aspergillus fumigatus* ABK9 in wheat bran-rice straw mixed substrate and use of cocktail enzymes for deinking of waste office paper pulp. Bio Resour Technol 128:290–296. https://doi.org/10.1016/j.biortech.2012.10.080

Dashtban M, Schraft H, Qin W (2009) Fungal bioconversion of lignocellulosic residues; opportunities & perspectives. Int J Biol Sci 5(6):578–595. https://doi.org/10.7150/ijbs.5.578

Dashtban M, Buchkowski R, Qin W (2011) Effect of different carbon sources on cellulase production by *Hypocrea jecorina* (*Trichoderma reesei*) strains. Int J Biochem Mol Biol 2:274–286

De Albuquerque de Carvalho ML, Carvalho DF, de Barros Gomes E, Nobuyuki Maeda R, Melo Santa Anna LM, AMd C, Pereira N (2014) Optimisation of cellulase production by *Penicillium funiculosum* in a stirred tank bioreactor using multivariate response surface analysis. Enzyme Research 2014:8. https://doi.org/10.1155/2014/703291

Deacon JW (1985) Decomposition of filter paper cellulose by thermophilic fungi acting singly, in combination, and in sequence. Trans Br Mycol Soc 85:663–669

Delabona PDS, Rosangela DP, Pirota B, Codima CA, Tremacoldi CR, Rodrigues A, Farinas CS (2012) Using Amazon forest fungi and agricultural residues as a strategy to produce cellulolytic enzymes. Biomass Bioenergy 37:243–250. https://doi.org/10.1016/j.biombioe.2011.12.006

Demuner BJ, Pereira Junior N, Antunes AMS (2011) Technology prospecting on enzymes for the pulp and paper industry. J Technol Manag Innov 6:148–158. https://doi.org/10.4067/S0718-27242011000300011

Dhillon GS, Kaur S, Brar SK (2013) Perspective of apple processing wastes as low-cost substrates for bioproduction of high value products: a review. Renew Sust Energ Rev 27:789–805. https://doi.org/10.1016/j.rser.2013.06.046

Ding X, He Q-F, Shen R-Q, Yu Q-Q, Zhang Y-Q, Xin J-Y, Wen T-X, Wang X-M (2016a) Spatial and seasonal variations of isoprene secondary organic aerosol in China: Significant impact of biomass burning during winter. Sci Rep 6:20411. https://doi.org/10.1038/srep20411

Ding X, Zhang Y-Q, He Q-F, Yu Q-Q, Shen R-Q, Zhang Y, Zhang Z, Lyu S-J, Hu Q-H, Wang Y-S, Li L-F, Song W, Wang X-M (2016b) Spatial and seasonal variations of secondary organic aerosol from terpenoids over China. J Geophys Res Atmos 121:14661–14678. https://doi.org/10.1002/2016JD025467

Ding X, Zhang Y-Q, He Q-F, Yu Q-Q, Wang J-Q, Shen R-Q, Song W, Wang Y-S, Wang X-M (2017) Significant increase of aromatics-derived secondary organic aerosol during fall to winter in China. Environ Sci Technol 51:7432–7441. https://doi.org/10.1021/acs.est.6b06408

Dobermann A, Fairhurst TH (2002) Rice straw management. Better Crop Int 16:1–11. Special Supplement

Duncan BN, Bey I, Chin M, Mickley LJ, Fairlie TD, Martin RV, Matsueda H (2003) Indonesian wildfires of 1997: impact on tropospheric chemistry. J Geophys Res Atmos 108:4458. https://doi.org/10.1029/2002jd003195

Duran N, Duran M (2000) Enzyme applications in the textile industry. Color Technol 30:41–44. https://doi.org/10.1111/j.1478-4408.2000.tb03779.x

El-Askary H, Kafatos M (2008) Dust storm and black cloud influence on aerosol optical properties over cairo and the greater delta region, Egypt. Int J Remote Sens 29(24):7199–7211. ISSN 0143-1161

El-Bakry M, Abraham J, Cerda A, Barrena R, Ponsa S, Gea T, Sanchez A (2015) From wastes to high value added products: novel aspects of SSF in the production of enzymes. Crit Rev Environ Sci Technol 45:1999–2042. https://doi.org/10.1080/10643389.2015.1010423

El-Hadi AA, El-Nour SA, Hammad A, Kamel Z, Anwar M (2014) Optimization of cultural and nutritional conditions for carboxymethylcellulase production by *Aspergillus hortai*. J Radiat Res Appl Sci 7:23–28. https://doi.org/10.1016/j.jrras.2013.11.003

El-Morsy E (2000) Fungi isolated from the endorhizosphere of halophytic plants from the red sea coast of Egypt. Fung Divers 5:43–54

El-Nahrawy S, Metwally M, Abd El-Kodoos RY, Belal ESB, Shabana SA, El-Refai IM (2017) Optimization of culture conditions for production of cellulase by *Aspergillus tubingensis* KY615746 using rice straw waste. Environ Biodivers Soil Secur 1:177–189. https://doi.org/10.21608/jenvbs.2017.1525.1007

El-Tayeb TS, Abdelhafez AA, Ali SH, Ramadan EM (2012) Effect of acid hydrolysis and fungal biotreatment on agro-industrial wastes for obtainment of free sugars for bioethanol production. Braz J Microbiol 43(4):1523–1535

Eriksen J, Goksøyr J (1977) Cellulases from *Chaetomium thermophile* var. *dissitum*. Eur J Biochem 77:445–450

Faisal PA, Benjamin S (2016) Optimization of parameters for the production of cellulase from *Achromobacter xylosoxidans* BSS4 by solid-state fermentation. Electron J Biol 12:4

Fergus CL (1969) The production of amylase by some thermophilic fungi. Mycologia 61:1171–1175

Flannigan B, Sellars PN (1972) Activities of thermophilous fungi from barley kernels against arabinoxylan and carboxymethyl cellulose. Trans Br Mycol Soc 58:338–341

EPA (1999) Overview. Cincinnati, OH: Center for Environmental Research Information, Office of Research and Development, U.S. Environmental Protection Agency; Jun, 1999. Compendium of Methods for the Determination of Inorganic Compounds in Ambient Air. Chapter IO-2. Integrated Sampling of Suspended Particulate Matter (SPM) in Ambient Air. EPA/625/R-96/010a

Gao JH, Zhu WD, Yuan M, Guan F, Xi Y (2008a) Production and characterization of cellulolytic enzymes from the thermos acidophilic fungal *Aspergillus terreus* M11 under solid-state cultivation of corn stover. Bioresour Technol 99:7623–7629. https://doi.org/10.1016/j.biortech.2008.02.005

Gao J, Weng H, Zhu D, Yuan M, Guan F, Xi Y (2008b) Production and characterization of cellulolytic enzymes from the thermoacidophilic fungal *Aspergillus terreus* M11 under solid-state cultivation of corn stover. Bioresour Technol 99:7623–7629. https://doi.org/10.1016/j.biortech.2008.02.005

Garcia NFL, da Silva Santos FR, Gonçalves FA, da Paz MF, Fonseca GG, Leite RSR (2015) Production of β-glucosidase on solid-state fermentation by *Lichtheimia ramosa* in agroindustrial residues: Characterization and catalytic properties of the enzymatic extract. Electron J Biotechnol 18(4):314–319

Garvey M, Klose H, Fischer R, Lambertz C, Commandeur U (2013) Cellulases for biomass degradation: comparing recombinant cellulase expression platforms. Trends Biotechnol 31(10):581–593. https://doi.org/10.1016/j.tibtech.2013.06.006

Gautam SP, Bundela PS, Pandey AK, Khan J, Awasthi MK, Sarsaiya S (2011) Optimization for the production of cellulase enzyme from municipal solid waste residue by two novel cellulolytic fungi. Biotechnol Res Int 2011:8. https://doi.org/10.4061/2011/810425

Geeraerts HAM, Vandamme EJ (2008) Cellulolytic properties of *Chaetomium crispatum*. J Chem Technol Biotechnol 33(2):107–113

Giordano MR, Short DZ, Hosseini S, Lichtenberg W, Asa-Awuku AA (2013) Changes in droplet surface tension affect the observed hygroscopicity of photochemically aged biomass burning aerosol. Environ Sci Technol 47:10980–10986. https://doi.org/10.1021/es401867j

Gomathi D, Muthulakshmi C, Kumar DG, Ravikumar G, Kalaiselvi M, Uma C (2012) Submerged fermentation of wheat bran by *Aspergillus flavus* for production and characterization of carboxy methyl cellulase. Asian Pac J Trop Biomed 2:S67–S73. https://doi.org/10.1016/S2221-1691(12)60132-4
Grieshop AP, Donahue NM, Robinson AL (2009a) Laboratory investigation of photochemical oxidation of organic aerosol from wood fires 2: analysis of aerosol mass spectrometer data. Atmos Chem Phys 9:2227–2240. https://doi.org/10.5194/acp-9-2227-2009
Grieshop AP, Logue JM, Donahue NM, Robinson AL (2009b) Laboratory investigation of photochemical oxidation of organic aerosol from wood fires 1: measurement and simulation of organic aerosol evolution. Atmos Chem Phys 9:1263–1277. https://doi.org/10.5194/acp-9-1263-2009
Gu J-G, Gu J-D (2005) Methods currently used in testing microbiological degradation and deterioration of a wide range of polymeric materials with various degree of degradability: a review. J Polym Environ 13:65–74. https://doi.org/10.1007/s10924-004-1230-7
Gurung N, Ray S, Bose S, Rai V (2013) A broader view: microbial enzymes and their relevance in industries, medicine, and beyond. Biomed Res Int 2013:18. https://doi.org/10.1155/2013/329121
Hatch LE, Yokelson RJ, Stockwell CE, Veres PR, Simpson IJ, Blake DR, Orlando JJ, Barsanti KC (2017) Multi-instrument comparison and compilation of non-methane organic gas emissions from biomass burning and implications for smoke-derived secondary organic aerosol precursors. Atmos Chem Phys 17:1471–1489. https://doi.org/10.5194/acp17-1471-2017
Hemsworth GR, Johnston EM, Davies GJ, Walton PH (2015) Lytic polysaccharide monooxygenases in biomass conversion. Trends Biotechnol 33:747–761. https://doi.org/10.1016/j.tibtech.2015.09.006
Hendriks ATWM, Zeeman G (2009) Pretreatments to enhance the digestibility of lignocellulosic biomass. Bioresour Technol 100(1):10–18. https://doi.org/10.1016/j.biortech.2008.05.027
Hennigan CJ, Miracolo MA, Engelhart GJ, May AA, Presto AA, Lee T, Sullivan AP, McMeeking GR, Coe H, Wold CE, Hao W-M, Gilman JB, Kuster WC, de Gouw J, Schichtel BA, Collett JL Jr, Kreidenweis SM, Robinson AL (2011) Chemical and physical transformations of organic aerosol from the photo-oxidation of open biomass burning emissions in an environmental chamber. Atmos Chem Phys 11:7669–7686. https://doi.org/10.5194/acp11-7669-2011
Heringa MF, DeCarlo PF, Chirico R, Tritscher T, Dommen J, Weingartner E, Richter R, Wehrle G, Prévôt ASH, Baltensperger U (2011) Investigations of primary and secondary particulate matter of different wood combustion appliances with a high-resolution time-of-flight aerosol mass spectrometer. Atmos Chem Phys 11:5945–5957. https://doi.org/10.5194/acp11-5945-2011
Horn SJ, Vaaje-Kolstad G, Westereng B, Eijsink V (2012) Novel enzymes for the degradation of cellulose. Biotechnol Biofuels 5:45. https://doi.org/10.1186/1754-6834-5-45
Huang RJ, Zhang YL, Bozzetti C, Ho KF, Cao JJ, Han YM, Daellenbach KR, Slowik JG, Platt SM, Canonaco F, Zotter P, Wolf R, Pieber SM, Bruns EA, Crippa M, Ciarelli G, Piazzalunga A, Schwikowski M, Abbaszade G, Schnelle-Kreis J, Zimmermann R, An ZS, Szidat S, Baltensperger U, El Haddad I, Prevot ASH (2014) High secondary aerosol contribution to particulate pollution during haze events in China. Nature 514:218–222. https://doi.org/10.1038/nature13774
Huang Y, Shen HZ, Chen YL, Zhong QR, Chen H, Wang R, Shen GF, Liu JF, Li BG, Tao S (2015) Global organic carbon emissions from primary sources from 1960 to 2009. Atmos Environ 122:505–512. https://doi.org/10.1016/j.atmosenv.2015.10.017
Huang X, Ding A, Liu L, Liu Q, Ding K, Niu X, Nie W, Xu Z, Chi X, Wang M, Sun J, Guo W, Fu C (2016) Effects of aerosol-radiation interaction on precipitation during biomassburning season in East China. Atmos Chem Phys 16:10063–10082. https://doi.org/10.5194/acp-16-10063-2016
Ibrahim MF, Abd-Aziz S, Yusoff MEM, Phang LY, Hassan MA (2015) Simultaneous enzymatic saccharification and ABE fermentation using pretreated oil palm empty fruit bunch as substrate to produce butanol and hydrogen as biofuel. Renew Energy 77:447–455. https://doi.org/10.1016/j.renene.2014.12.047

Idris ASO, Pandey A, Rao SS, Sukumaran RK (2017) Cellulase production through solid-state tray fermentation, and its use for bioethanol from sorghum stover. Bioresour Technol 242:265–271. https://doi.org/10.1016/j.biortech.2017.03.092

Jain MK, Kapoor KK, Mishra MM (1979) Cellulase activity, degradation of cellulose and lignin, and humus formation by thermophilic fungi. Trans Br Mycol Soc 73:85–89

Jathar SH, Gordon TD, Hennigan CJ, Pye HOT, Pouliot G, Adams PJ, Donahue NM, Robinson AL (2014) Unspeciated organic emissions from combustion sources and their influence on the secondary organic aerosol budget in the United States. Proc Natl Acad Sci U S A 111:10473–10478. https://doi.org/10.1073/pnas.1323740111

Jiang SY (2013) Cellulase production by submerged fermentation using biological materials of corncob residue with aspergillus niger FC-1. Adv Mater Res 648:116–119. https://doi.org/10.4028/www.scientific.net/AMR.648.116

Johri BN, Pandey AR (1982) Cellulose breakdown by *Rhizopus rhizopodiformis* (Cohn) Zopf., a thermophilic mould. Indian J Microbiol 22:76–78

Jolleys MD, Coe H, McFiggans G, Capes G, Allan JD, Crosier J, Williams PI, Allen G, Bower KN, Jimenez JL, Russell LM, Grutter M, Baumgardner D (2012) Characterizing the aging of biomass burning organic aerosol by use of mixing ratios: a meta-analysis of four regions. Environ Sci Technol 46:13093–13102. https://doi.org/10.1021/es302386v

Joo A-R, Jeya M, Lee K-M, Lee K-M, Moon H-J et al (2010) Production and characterization of ß-1, 4-glucosidase from a strain of Penicillium pinophilum. Process Biochem 45:851–858

Jose LA, Arnold LD (2014) Microbial enzymes: tools for biotechnological processes. Biomol Ther 4:117–139. https://doi.org/10.3390/biom4010117

Juturu V, Wu JC (2014) Microbial cellulases: engineering, production and applications. Renew Sust Energ Rev 33:188–203. https://doi.org/10.1016/j.rser.2014.01.077

Kachlishvili E, Penninckx MJ, Tsiklauri N, Elisashvili V (2006) Effect of nitrogen source on lignocellulolytic enzyme production by white-rot basidiomycetes under solid-state cultivation. World J Microbiol Biotechnol 22:391–397. https://doi.org/10.1007/s11274-005-9046-8

Kang SW, Park YS, Lee JS, Hong SI, Kim SW (2004) Production of cellulases and hemicellulases by *Aspergillus niger* KK2 from lignocellulosic biomass. Bioresour Technol 91:153–156. https://doi.org/10.1016/S0960-8524(03)00172-X

Kasana RC, Salawan R, Dhar H, Dutt S, Gulati A (2008) A rapid and easy method for the detection of microbial celluloses on agar plates using gram's iodine. Curr Microbiol 57(5):53–507. https://doi.org/10.1007/s00284-008-9276-8

Kathiresan K, Manivannan S (2006) α-Amylase production by Penicillium fellutanum isolated from mangrove rhizosphere soil. Afr J Biotechnol 5:829–832

Kazragis A (2005) Minimization of atmosphere pollution by utilizing cellulose waste. J Environ Eng Landsc Manag 13(2):81–90

Kim JH, Shim BS, Kim HS et al (2015) Cellulose nanocrystals, nanofibers, and their composites as renewable smart materials. Int J PR Eng Man GT 2:197. https://doi.org/10.1007/s40684-015-0024-9

Kim Oanh NT, Tipayarom A, Bich TL, Tipayarom D, Simpson CD, Hardie D, Sally Liu LJ (2015) Characterization of gaseous and semi-volatile organic compounds emitted from field burning of rice straw. Atmos Environ 119:182–191. https://doi.org/10.1016/j.atmosenv.2015.08.005

Kaur J, Chadha BS, Kumar BA, Saini HS (2007) Purification and characterization of two endoglucanases from Melanocarpus sp. MTCC 3922. Bioresour Technol 98:74–81

Koren I, Kaufman YK, Remer LA, Martins JV (2004) Measurement of the effect of Amazon smoke on inhibition of cloud formation. Science 303(5662): 1342–1345

Kuhad R, Singh A (2007) Lignocellulose biotechnology: future prospects. IK International Publishing House Pvt Ltd, New Delhi

Kuhad RC, Gupta R, Singh A (2011) Microbial cellulases and their industrial applications. Enzym Res 2011:10. https://doi.org/10.4061/2011/280696

Kumar J, Singh S, Singh OV (2008) Bioconversion of lignocellulose biomass: biochemical and molecular perspectives. J Ind Microbiol Biotechnol 35:377–391

Kumar S, Sharma HK, Sarkar BC (2011) Effect of substrate and fermentation conditions on pectinase and cellulase production by aspergillus niger NCIM 548 in submerged (SmF) and solid state fermentation (SSF). Food Sci Biotechnol 20(5):1289–1298. https://doi.org/10.1007/s10068-011-0178-3

Kupski L, Pagnussatt FA, Buffon JG, Furlong EB (2014) Endoglucanase and total cellulase from newly isolated *Rhizopus oryzae* and *Trichoderma reesei*: production, characterization, and thermal stability. Appl Biochem Biotechnol 172:458–468. https://doi.org/10.1007/s12010-013-0518-2

Lambe AT, Ahern AT, Williams LR, Slowik JG, Wong JPS, Abbatt JPD, Brune WH, Ng NL, Wright JP, Croasdale DR, Worsnop DR, Davidovits P, Onasch TB (2011) Characterization of aerosol photooxidation flow reactors: heterogeneous oxidation, secondary organic aerosol formation and cloud condensation nuclei activity measurements. Atmos Meas Tech 4:445–461. https://doi.org/10.5194/amt-4-445-2011

Laskin A, Laskin J, Nizkorodov SA (2015) Chemistry of atmospheric brown carbon. Chem Rev 115:4335–4382. https://doi.org/10.1021/cr5006167

Lee K-M, Jeya M, Joo A-R, Singh R, Kim I-W et al (2010) Purification and characterization of a thermostable endo-ß-1, 4-glucanase from a novel strain of *Penicillium purpurogenum*. Enzym Microb Technol 46:206–211

Lee H, Lee YM, Jang Y, Lee S, Lee H, Ahn BJ, Kim GH, Kim JJ (2014) Isolation and analysis of the enzymatic properties of thermophilic fungi from compost. Mycobiology 42:181–184

Levine JS (1996) Biomass burning and global change. In: Levine JS (ed) Remote sensing and inventory development and biomass burning in Africa. The MIT Press, Cambridge, p 35

Li XG, Wang SX, Duan L, Hao J, Li C, Chen YS, Yang L (2007) Particulate and trace gas emissions from open burning of wheat straw and corn stover in China. Environ Sci Technol 41:6052–6058. https://doi.org/10.1021/es0705137

Li Z, Du C, Zhong Y et al (2010) Development of a highly efficient gene targeting system allowing rapid genetic manipulations in *Penicillium decumbens*. Appl Microbiol Biotechnol 87(3):1065–1076

Li S, Yang X, Yang S, Zhu M, Wang X (2012) Technology prospecting on enzymes: application, marketing and engineering. Comput Struct Biotechnol J 2:e201209017. https://doi.org/10.5936/csbj.201209017

Li C, Ma Z, Chen J, Wang X, Ye X, Wang L, Yang X, Kan H, Donaldson DJ, Mellouki A (2015) Evolution of biomass burning smoke particles in the dark. Atmos Environ 120:244–252. https://doi.org/10.1016/j.atmosenv.2015.09.003

Li Q, Li XH, Jiang JK, Duan L, Ge S, Zhang Q, Deng JG, Wang SX, Hao JM (2016) Semi-coke briquettes: towards reducing emissions of primary PM2.5, particulate carbon, and carbon monoxide from household coal combustion in China. Sci Rep 6:19306. https://doi.org/10.1038/srep19306

Li C, Hu Y, Zhang F, Chen J, Ma Z, Ye X, Yang X, Wang L, Tang X, Zhang R, Mu M, Wang G, Kan H, Wang X, Mellouki A (2017) Multi-pollutant emissions from the burning of major agricultural residues in China and the related healtheconomic effects. Atmos Chem Phys 17:4957–4988. https://doi.org/10.5194/acp-17-4957-2017

Liang Y, Feng Z, Yesuf J, Blackburn JW (2010) Optimization of growth medium and enzyme assay conditions for crude cellulases produced by a novel thermophilic and cellulolytic bacterium, *Anoxybacillus* sp. 527. Appl Biochem Biotechnol 160:1841–1852. https://doi.org/10.1007/s12010-009-8677-x

Lipsky EM, Robinson AL (2006) Effects of dilution on fine particle mass and partitioning of semi-volatile organics in diesel exhaust and wood smoke. Environ Sci Technol 40:155–162. https://doi.org/10.1021/es050319p

Liu Z, Sun X, Qu Y (2008) Cloning cellobiohydrolase I from *Penicillium decumbens* 114-2 with TAIL-PCR and comparing with its derepressed mutant JU-A10. Acta Microbiol Sin 48(5):667–671

Liu HQ, Feng Y, Zhao DQ, Jiang JX (2012) Evaluation of cellulases produced from four fungi cultured on furfural residues and microcrystalline cellulose. Biodegradation 23:465–472. https://doi.org/10.1007/s10532-011-9525-6

Liu D, Li J, Zhao S, Zhang R, Wang M, Miao Y, Shen Y, Shen Q (2013) Secretome diversity and quantitative analysis of cellulolytic *Aspergillus fumigatus* Z5 in the presence of different carbon sources. Biotechnol Biofuels 6:149. https://doi.org/10.1186/1754-6834-6-149

Liu J, Ju M, Wu W, Liu B, Zhan L, Wu M (2014) Lignocellulolytic enzyme production in solid-state fermentation of corn stalk with ammoniation pretreatment by *Lentinusedodes* L-8. Bioresources 9(1):1430–1444

Liu Z, Ho S-H, Sasaki K, Haan RD, Inokuma K, Ogino C, Van Zyl WH, Hasunuma T, Kondo A (2016) Engineering of a novel cellulose-adherent cellulolytic *Saccharomyces cerevisiae* for Cellulosic biofuel production. Sci Rep 6:24550. https://doi.org/10.1038/srep24550. http://www.novozymes.com

Luo W, Vrijmoed LLP, Jones EBG (2005) Screening of marine fungi for lignocellulose-degrading enzyme activities. Bot Mar 48(5–6):379–386. https://doi.org/10.1515/bot.2005.051

Lynd LR, Weimer PJ, van Zyl WH, Pretorius IS (2002) Microbial cellulose utilization: fundamentals and biotechnology. Microbiol Mol Biol Rev 66:506–577. https://doi.org/10.1128/MMBR.66.3.506-577.2002

Maria GL, Sridhar KR, Raviraja NS (2005) Antimicrobial and enzyme activity of mangrove endophytic fungi of southwest coast of India. J Agric Technol 1:67–80

Martin JGP, Porto E, Correa CB, Alencar SM, Gloria EM, Cabral ISR, Aquino LM (2012) Antimicrobial potential and chemical composition of agroindustrial wastes. J Nat Prod 5:27–36

Martin M, Tritscher T, Juranyi Z, Heringa MF, Sierau B, Weingartner E, Chirico R, Gysel M, Prevot ASH, Baltensperger U, Lohmann U (2013) Hygroscopic properties of fresh and aged wood burning particles. J Aerosol Sci 56:15–29. https://doi.org/10.1016/j.jaerosci.2012.08.006

May AA, Levin EJT, Hennigan CJ, Riipinen I, Lee T, Collett JL Jr, Jimenez JL, Kreidenweis SM, Robinson AL (2013) Gas-particle partitioning of primary organic aerosol emissions: 3. Biomass burning. J Geophys Res Atmos 118:11327–11338. https://doi.org/10.1002/jgrd.50828

Megan G, Holger K, Rainer F, Camilla L, Ulrich C (2013) Cellulases for biomass degradation: comparing recombinant cellulase expression platforms. Trends Biotechnol 31(10):581–593. https://doi.org/10.1016/j.tibtech.2013.06.006

Menendez E, Garcia-Fraile P, Rivas R (2015) Biotechnological applications of bacterial cellulases. AIMS Bioeng 2:163–182. https://doi.org/10.3934/bioeng.2015.3.163

Motte JC, Trably E, Escudié R, Hamelin J, Steyer JP, Bernet N, Delgenes JP, Dumas C (2013) Total solids content: a key parameter of metabolic pathways in dry anaerobic digestion. Biotechnol Biofuels 6:164

Müller M, Anderson BE, Beyersdorf AJ, Crawford JH, Diskin GS, Eichler P, Fried A, Keutsch FN, Mikoviny T, Thornhill KL, Walega JG, Weinheimer AJ, Yang M, Yokelson RJ, Wisthaler A (2016) In situ measurements and modeling of reactive trace gases in a small biomass burning plume. Atmos Chem Phys 16:3813–3824. https://doi.org/10.5194/acp-16-3813-2016

Murad H, Azzaz H (2010) Cellulase and dairy animal feeding. Biotechnology 9:238–256. https://doi.org/10.3923/biotech.2010.238.256

Naika GS, Kaul P, Prakash V (2007) Purification and characterization of a new endoglucanase from *Aspergillus aculeatus*. J Agric Food Chem 55:7566–7572

Narra M, Dixit G, Divecha J, Madamwar D, Shah AR (2012) Production of cellulases by solid state fermentation with Aspergillus terreus and enzymatic hydrolysis of mild alkali-treated rice straw. Bioresour Technol 121:55–361. https://doi.org/10.1016/j.biortech.2012.05.140

Nathan VK, Rani ME, Rathinasamy G, Dhiraviam KN, Jayavel S (2014) Process optimization and production kinetics for cellulase production by *Trichoderma viride* VKF3. Springer Plus 3:92. https://doi.org/10.1186/2193-1801-3-92

Nazir A, Soni R, Saini HS, Kaur A, Chadha BS (2010) Profiling differential expression of cellulases and metabolite footprints in *Aspergillus terreus*. Appl Biochem Biotechnol 162:538–547. https://doi.org/10.1007/s12010-009-8775-9

NBSPRC (National bureau of statistics of people's republic of China) (2015). The statistical communiqu on national economy and social development in 2015. Available at: http://www.stats.gov.cn/tjsj/zxfb/201602/t20160229_1323991.html (last access: 27 Sep 2017)

Ni HY, Han YM, Cao JJ, Chen LWA, Tian J, Wang XL, Chow JC, Watson JG, Wang QY, Wang P, Li H, Huang RJ (2015) Emission characteristics of carbonaceous particles and trace gases from open burning of crop residues in China. Atmos Environ 123:399–406. https://doi.org/10.1016/j.atmosenv.2015.05.007

Nigam PS, Gupta N, Anthwal A (2009) Pre-treatment of agro-industrial residues. In: Nigam PS, Pandey A (eds) Biotechnology for agro-industrial residues utilization. Springer, Heidelberg, pp 13–33

Norouzian D (2008) Effect of different factors on fermentative production of enzymes by fungi. Dyn Biochem Process Biotechnol Mol Biol 2:14–18

Okoye I, Ezugwu A, Udenwobele D, Eze S, Anyawu C, Chilaka F (2013) Production and partial characterization of cellulases from *Apergillus fumigatus* using two distinct parts of corn cob as carbon sources. Niger J Biotechnol 26:50–59

Olsen HS, Falholt P (1998) The role of enzymes in modern detergency. J Surfactant Deterg 1:555–567. https://doi.org/10.1007/s11743-998-0058-7

Ortega AM, Day DA, Cubison MJ, Brune WH, Bon D, de Gouw JA, Jimenez JL (2013) Secondary organic aerosol formation and primary organic aerosol oxidation from biomassburning smoke in a flow reactor during FLAME-3. Atmos Chem Phys 13:11551–11571. https://doi.org/10.5194/acp-13-11551-2013

Pandey A (2003) Solid-state fermentation. Biochem Eng J 13:81–84. https://doi.org/10.1016/S1369-703X(02)00121-3

Pandey A, Soccol CR, Mitchell D (2000) New developments in solid state fermentation: i-bioprocesses and products. Process Biochem 35:1153–1169. https://doi.org/10.1016/S0032-9592(00)00152-7

Parry JB, Stewart JC, Heptinstall J (1983) Purification of the major endoglucanase from *Aspergillus fumigatus* Fresenius. Biochem J 213:437–444

Pensupa N, Jin M, Kokolski M, Archer DB, Du C (2013) A solid state fungal fermentation-based strategy for the hydrolysis of wheat straw. Bioresour Technol 149:261–267. https://doi.org/10.1016/j.biortech.2013.09.061

Pereira JDC, Marques NP, Rodrigues A, Oliveira TBD, Boscolo M, Silva RD, Gomes E, Martins DAB (2015) Thermophilic fungi as new sources for production of cellulases and xylanases with potential use in sugarcane bagasse saccharification. J Appl Microbiol 118:928–939. https://doi.org/10.1111/jam.12757

Periyasamy K, Santhalembi L, Mortha G, Aurousseau M, Guillet A, Dallerac D, Sivanesan S (2017) Production, partial purification and characterization of enzyme cocktail from *Trichoderma citrinoviride* AUKAR04 through solid-state fermentation. Arab J Sci Eng 42:53–63. https://doi.org/10.1007/s13369-016-2110-x

Peterson R, Nevalainen H (2012) *Trichoderma reesei* RUT-C30 – thirty years of strain improvement. Microbiology 158:58–68. https://doi.org/10.1099/mic.0.054031-0

Phitsuwan P, Laohakunjit N, Kerdchoechuen O, Kyu KL, Ratanakhanokchai K (2013) Present and potential applications of cellulases in agriculture, biotechnology, and bioenergy. Folia Microbiol 58:163–176. https://doi.org/10.1007/s12223-012-0184-8

Pol D, Laxman RS, Rao M (2012) Purification and biochemical characterization of endoglucanase from *Penicillium pinophilum* MS 20. Indian J Biochem Biophys 49:189–194

Polizeli MLTM, Rizzatti ACS, Monti R, Terenzi HF, Jorge JA, Amorim DS (2005) Xylanases from fungi: properties and industrial applications. Appl Microbiol Biotechnol 67:577–591

Prasanna HN, Ramanjaneyulu G, Rajasekhar Reddy B (2016) Optimization of cellulase production by *Penicillium* sp. 3 Biotech 6:162. https://doi.org/10.1007/s13205-016-0483-x

Przybysz Buzała K, Przybysz P, Kalinowska H, Derkowska M (2016) Effect of Cellulases and Xylanases on refining process and Kraft pulp properties. PLoS One 11:e0161575. https://doi.org/10.1371/journal.pone.0161575

Raghuwanshi S, Deswal D, Karp M, Kuhad RC (2014) Bioprocessing of enhanced cellulase production from a mutant of *Trichoderma asperellum* RCK2011 and its application in hydrolysis of cellulose. Fuel 124:183–189. https://doi.org/10.1016/j.fuel.2014.01.107

Ramani G, Meera B, Vanitha C, Rao M, Gunasekaran P (2012) Production, purification, and characterization of a β-glucosidase of *Penicillium funiculosum* NCL1. Appl Biochem Biotechnol 167:959–972

Rasmussen ML, Shrestha P, Khanal SK, Pom-etto III, Van AL, Leeuwen J (2010) Sequential sacchar- ification of corn fiber and ethanol production by the brown rot fungus *Gloeophyllum trabeum*. Bioresour Technol 101(10):3526–3533. https://doi.org/10.1016/j.biortech.2009.12.115

Real E, Law KS, Weinzierl B, Fiebig M, Petzold A, Wild O, Methven J, Arnold S, Stohl A, Huntrieser H, Roiger A, Schlager H, Stewart D, Avery M, Sachse G, Browell E, Ferrare R, Blake D (2007) Processes influencing ozone levels in Alaskan forest fire plumes during long-range transport over the North Atlantic. J Geophys Res Atmos 112:D10S41. https://doi.org/10.1029/2006jd007576

Reddy GPK, Narasimha G, Kumar KD, Ramanjaneyulu G, Ramya A, Kumari BS, Reddy BR (2015) Cellulase production by *Aspergillus niger* on different natural lignocellulosic substrates. Int J Curr Microbiol App Sci 4:835–845

Rosenberg SL (1978) Cellulose and lignocellulose degradation by thermophilic and thermotolerant fungi. Mycologia 70:1–13

Ryu DDY, Mandels M (1980) Cellulases biosynthesis and applications. Enzym Microb Technol 2:91–102. https://doi.org/10.1016/0141-0229(80)90063-0

Sadhu S, Maiti TK (2013) Cellulase production by bacteria: a review. Br Microbiol Res J 3(3):235–258. http://imsear.hellis.org/handle/123456789/162913

Saini A, Aggarwal NK, Yadav A (2017) Cost-effective cellulase production using Parthenium hysterophorus biomass as an unconventional lignocellulosic substrate. 3 Biotech 7:12. https://doi.org/10.1007/s13205-017-0604-1

Sajith S, Priji P, Sreedevi S, Benjamin S (2016) An overview on fungal cellulases with an industrial perspective. J Nutr Food Sci 6:661. https://doi.org/10.4172/2155-9600.1000461

Sánchez C (2009) Lignocellulosic residues: biodegradation and bioconversion by fungi. Biotechnol Adv 27:185–194. https://doi.org/10.1016/j.biotechadv.2008.11.001

Sanchis E, Ferrer M, Calvet S, Coscolla C, Yusa V, Cambra-Lopez M (2014) Gaseous and particulate emission profiles during controlled rice straw burning. Atmos Environ 98:25–31. https://doi.org/10.1016/j.atmosenv.2014.07.062

Sandhu DK, Bagga PS, Singh S (1985) Cellulolytic activity of thermophilous fungi isolated from soils. Kavaka 13:21–31

Saravanan P, Muthuvelayudham R, Viruthagiri T (2012) Application of statistical design for the production of cellulase by *Trichoderma reesei* using mango peel. Enzyme Research 2012:7. https://doi.org/10.1155/2012/157643

Sarkar N, Aikat K (2012) Cellulase and xylanase production from rice straw by a locally isolated fungus *Aspergillus fumigatus* NITDGPKA3 under solid state fermentation – statistical optimization by response surface methodology. J Technol Innov Renew Energy 1:54–62. ISSN: 1929-6002

Schmidt O (2006) Wood and tree fungi: biology, damage, protection, and use. Springer-Verlag, Berlin/Heidelberg

Schwarze F, Engels J, Mattheck C (2000) Fungal strategies of wood decay in trees. Springer-Verlag, Berlin/Heidelberg

Shajahan S, Moorthy IG, Sivakumar N, Selvakumar G (2017) Statistical modeling and optimization of cellulase production by *Bacillus licheniformis* NCIM 5556 isolated from the hot spring, Maharashtra, India. J King Saud Univ Sci 29:302–310. https://doi.org/10.1016/j.jksus.2016.08.001

Shanthiyaa V, Karthikeyan G, Raguchander T (2014) Production of extracellular proteins, cellulases and antifungal metabolites by *Chaetomium globosum* Kunze ex. Fr. Arch Phytopathol Plant Protect 47(5):517–528. https://doi.org/10.1080/03235408.2013.813124

Sharada R, Venkateswarlu G, Venkateswar S, AnandRao M (2014) Applications of cellulases: review. Int J Pharm Chem Biol Sci 4:424–437

Sharma R, Sharma PC, Rana JC, Joshi VK (2015) Improving the olive oil yield and quality through enzyme-assisted mechanical extraction, antioxidants and packaging. J Food Process Preserv 39:157–166. https://doi.org/10.1111/jfpp.12216

Shweta A (2015) Solid state fermentation for cellulase production. Biotechnol Res 1:108–112
Singh K, Richa K, Bose H, Karthik L, Kumar G, Bhaskara Rao KV (2014) Statistical media optimization and cellulase production from marine Bacillus VITRKHB. 3 Biotech 4:591–598. https://doi.org/10.1007/s13205-013-0173-x
Singhania RR, Patel AK, Soccol CR, Pandey A (2009) Recent advances in solid-state fermentation. Biochem Eng J 44:13–18. https://doi.org/10.1016/j.bej.2008.10.019
Singla D, Taggar MS (2017) Production of cellulases by solid state fermentation of different agricultural residues using *Humicola insolens* MTCC 1433. Int J Curr Microbiol App Sci 6:1409–1418. https://doi.org/10.20546/ijcmas.2017.611.168
Soccol CR, ESFd C, Letti LAJ, Karp SG, Woiciechowski AL, Vandenberghe LPS (2017) Recent developments and innovations in solid state fermentation. Biotechnol Res Innov 1:52–71. https://doi.org/10.1016/j.biori.2017.01.002
Soni R, Sandhu DK, Sonic SK (1999) Localisation and optimisation of cellulase production in *Chaetomium erraticum*. J Biotechnol 73:43–51
Spracklen DV, Jimenez JL, Carslaw KS, Worsnop DR, Evans MJ, Mann GW, Zhang Q, Canagaratna MR, Allan J, Coe H, McFiggans G, Rap A, Forster P (2011) Aerosol mass spectrometer constraint on the global secondary organic aerosol budget. Atmos Chem Phys 11:12109–12136. https://doi.org/10.5194/acp-11-12109-2011
Srivastava N, Jaiswal P (2016) Production of cellulases using agriculture waste: application in biofuels production. Environmental biotechnology a new approach. Daya Publishing House® A Division of Astral International Pvt. Ltd. New Delhi–110 002.233–425 244
Srivastava RB, Narain R, Mehrotra BS (1981) Comparative cellulolytic ability of mesophilic and thermophilic fungi isolated in India from manure and plant refuse. Indian J Mycol Plant Pathol 11:66–72
Srivastava N, Singh J, Srivastava M, Ramteke PW, Mishra PK (2015a) Improved production of reducing sugars from rice straw using crude cellulase activated with Fe_3O_4/alginate nanocomposite. Bioresour Technol 183:262–266. https://doi.org/10.1016/j.biortech.2015.02.059
Srivastava N, Rawat R, Oberoi HS, Ramteke PW (2015b) A review on fuel ethanol production from lignocellulosic biomass. Int J Green En 12:949–960. https://doi.org/10.1080/15435075.2014.890104
Srivastava N, Srivastava M, Mishra PK, Singh P, Ramteke PW (2015c) Application of cellulases in biofuels industries: an overview. J Biofuels Bioenerg 1(1):55–63. https://doi.org/10.5958/2454-8618.2015.00007.3
Stockwell CE, Yokelson RJ, Kreidenweis SM, Robinson AL, DeMott PJ, Sullivan RC, Reardon J, Ryan KC, Griffith DWT, Stevens L (2014) Trace gas emissions from combustion of peat, crop residue, domestic biofuels, grasses, and other fuels: configuration and Fourier transform infrared (FTIR) component of the fourth Fire Lab at Missoula Experiment (FLAME-4). Atmos Chem Phys 14:9727–9754. https://doi.org/10.5194/acp-14-9727-2014
Stockwell CE, Veres PR, Williams J, Yokelson RJ (2015) Characterization of biomass burning emissions from cooking fires, peat, crop residue, and other fuels with high-resolution proton-transfer-reaction time-of-flight mass spectrometry. Atmos Chem Phys 15:845–865. https://doi.org/10.5194/acp-15-845-2015
Stockwell CE, Christian TJ, Goetz JD, Jayarathne T, Bhave PV, Praveen PS, Adhikari S, Maharjan R, DeCarlo PF, Stone EA, Saikawa E, Blake DR, Simpson IJ, Yokelson RJ, Panday AK (2016) Nepal Ambient Monitoring and Source Testing Experiment (NAMaSTE): emissions of trace gases and light-absorbing carbon from wood and dung cooking fires, garbage and crop residue burning, brick kilns, and other sources. Atmos Chem Phys 16:11043–11081. https://doi.org/10.5194/acp-16-11043-2016
Streets DG, Yarber KF, Woo JH, Carmichael GR (2003) Biomass burning in Asia: annual and seasonal estimates and atmospheric emissions. Global Biogeochem Cy 17:1099. https://doi.org/10.1029/2003gb002040
Su X, Chu X, Dong Z (2009) Identification of elevated transcripts in a *Trichoderma reesei* strain expressing a chimeric transcription activator using suppression subtractive hybridization. World J Microbiol Biotechnol 25(6):1075–1084

Subramaniyam R, Vimala R (2012) Solid state and submerged fermentation for the production of bioactive substances: a comparative study. Int J Sci Nat 3(3):480–486

Sukumaran RK, Singhania RR, Mathew GM, Pandey A (2009) Cellulase production using biomass feed stock and its application in lignocellulose saccharification for bio-ethanol production. Renew Energy 34:421–424. https://doi.org/10.1016/j.renene.2008.05.008

Sun Y, Cheng J (2002) Hydrolysis of lignocellulosic materials for ethanol production: a review. Bioresour Technol 83:1–11. https://doi.org/10.1016/S0960-8524(01)00212-7

Sun X, Liu Z, Qu Y et al (2008a) The effects of wheat bran composition on the production of biomass-hydrolyzing enzymes by *Penicillium decumbens*. Appl Biochem Biotechnol 146(1–3):119–128

Sun X, Liu Z, Zheng K et al (2008b) The composition of basal and induced cellulase systems in *Penicillium decumbens* under induction or repression conditions. Enzym Microb Technol 42(7):560–567

Swain MR, Ray RC (2007) Alpha-amylase production by Bacillus subtilis CM3 in solid state fermentation using cassava fibrous residue. J Basic Microbiol 47:417–425. https://doi.org/10.1002/jobm.200710132

Taherzadeh MJ, Karimi K (2007) Enzyme-based hydrolysis processes for ethanol from lignocellulosic materials: a review. Bioresources 2:707–738. ISSN 1930-2126

Takashima S, Nakamura A, Masaki H, Uozumi T (1996) Purification and characterization of cellulases from *Humicola grisea*. Biosci Biotechnol Biochem 60:77–82

Tariq S, ul-Haq Z, Ali M (2016) Satellite and ground-based remote sensing of aerosols during intense haze event of October 2013 over lahore, Pakistan. Asia-Pac J Atmos Sci 52:25–33. https://doi.org/10.1007/s13143-015-0084-3

Tian H, Zhao D, Wang Y (2011) Emission inventories of atmospheric pollutants discharged from biomass burning in China. Acta Sci Circumst 31:349–357

Thompson AM, Witte JC, Hudson RD, Guo H, Herman JR, Fujiwara M (2001) Tropical tropospheric ozone and biomass burning. Science 291:2128–2132. https://doi.org/10.1126/science.291.5511.2128

Thota SP, Badiya PK, Yerram S, Vadlani PV, Pandey M, Golakoti NR, Belliraj SK, Dandamudi RB, Ramamurthy SS (2017) Macro-micro fungal cultures synergy for innovative cellulase enzymes production and biomass structural analyses. Renew Energy 103:766–773. https://doi.org/10.1016/j.renene.2016.11.010

Tkacik DS, Robinson ES, Ahern A, Saleh R, Stockwell C, Veres P, Simpson IJ, Meinardi S, Blake DR, Yokelson RJ, Presto AA, Sullivan RC, Donahue NM, Robinson AL (2017) A dual-chamber method for quantifying the effects of atmospheric perturbations on secondary organic aerosol formation from biomass burning emissions. J Geophys Res Atmos 122:6043–6058. https://doi.org/10.1002/2016JD025784

Tong CC, Cole AL (1982) Cellulase production by the thermophilic fungus, *Thermoascus aurantiacus*. Pertanika 5:255–262

Tsigaridis K, Daskalakis N, Kanakidou M, Adams PJ, Artaxo P, Bahadur R, Balkanski Y, Bauer SE, Bellouin N, Benedetti A, Bergman T, Berntsen TK, Beukes JP, Bian H, Carslaw KS, Chin M, Curci G, Diehl T, Easter RC, Ghan SJ, Gong SL, Hodzic A, Hoyle CR, Iversen T, Jathar S, Jimenez JL, Kaiser JW, Kirkevåg A, Koch D, Kokkola H, Lee YH, Lin G, Liu X, Luo G, Ma X, Mann GW, Mihalopoulos N, Morcrette J-J, Müller J-F, Myhre G, Myriokefalitakis S, Ng NL, O'Donnell D, Penner JE, Pozzoli L, Pringle KJ, Russell LM, Schulz M, Sciare J, Seland Ø, Shindell DT, Sillman S, Skeie RB, Spracklen D, Stavrakou T, Steenrod SD, Takemura T, Tiitta P, Tilmes S, Tost H, van Noije T, van Zyl PG, von Salzen K, Yu F, Wang Z, Wang Z, Zaveri RA, Zhang H, Zhang K, Zhang Q, Zhang X (2014) The AeroCom evaluation and intercomparison of organic aerosol in global models. Atmos Chem Phys 14:10845–10895. https://doi.org/10.5194/acp-14-10845-2014

van Noort V, Bradatsch B, Arumugam M, Amlacher S, Bange G, Creevey C, Falk S, Mende DR, Sinning I, Hurt E, Bork P (2013) Consistent mutational paths predict eukaryotic thermostability. BMC Evol Biol 13:7. https://doi.org/10.1186/1471-2148-13-7

Van Tuan Nguyen DTQ (2010) Purification and properties of a novel thermoactive endoglucanase from *Aspergillus awamori* VTCC-F099. Aust J Basic Appl Sci 4:6211–6216

Venturi LL, Polizeli LM, Terenzi HF, Furriel R, Jorge JA (2002) Extracellular beta-D-glucosidase from *Chaetomium thermophilum* var. *coprophilum*: production, purification and some biochemical properties. J Basic Microbiol 42:55–66

Vinzant TB, Adney WS, Decker SR, Baker JO, Kinter MT, Sherman NE, Fox JW, Himmel ME (2001) Fingerprinting *Trichoderma reesei* hydrolases in a commercial cellulase preparation. Appl Biochem Biotechnol 91–93:99–107

Wang D, Qu Y, Gao P (1995) Regulation of cellulase synthesis in mycelial fungi. Biotechnol Lett 17(6):593–598

Wang M, Liu K, Dai L, Zhang J, Fang X (2013) The structural and biochemical basis for cellulose biodegradation. J Chem Technol Biotechnol 88:491–500. https://doi.org/10.1002/jctb.3987

Wang H, Lou S, Huang C, Qiao L, Tang X, Chen C, Zeng L, Wang Q, Zhou M, Lu S, Yu X (2014) Source profiles of volatile organic compounds from biomass burning in Yangtze River Delta, China. Aerosol Air Qual Res 14:818–828. https://doi.org/10.4209/aaqr.2013.05.0174

Wang Y, Fan C, Hu H, Li Y, Suna D, Wang Y, Peng L (2016) Genetic modification of plant cell walls to enhance biomass yield and biofuel production in bioenergy crops. Biotechnol Adv 34:997–1017. https://doi.org/10.1016/j.biotechadv.2016.06.001

Wilson DB (2009) Cellulases and biofuels. Curr Opin Biotechnol 20:295–299. https://doi.org/10.1016/j.copbio.2009.05.007

Wong KK, Tan LU, Saddler JN (1988) Multiplicity of β-1,4-xylanase in microorganisms: functions and applications. Microbiol Rev 52:305–317

Xia L, Cen P (1994) Cellulase production by solid state fermentation on lignocellulosic waste from the xylose industry. Process Biochem 34:909–912. https://doi.org/10.1016/S0032-9592(99)00015-1

Ximenes EA, Felix CR, Ulhoa CJ (1996) Production of cellulases by *Aspergillus fumigatus* and characterization of one b-glucosidase. Curr Microbiol 32:119–123

Yadav LS, Bagool RG (2015) Isolation and screening of cellulolytic *Chaetomium* sp. from deteriorated paper samples. Int J Curr Microbiol Appl Sci 4(8):629–635

Yadav PS, Shruthi K, Prasad BVS, Chandra MS (2016) Enhanced production of β-glucosidase by new strain *Aspergillus protuberus* on solid state fermentation in rice husk. Int J Curr Microbiol App Sci 5(12):551–564. https://doi.org/10.20546/ijcmas.2016.512.061

Yang SJ, He HP, Lu SL, Chen D, Zhu JX (2008) Quantification of crop residue burning in the field and its influence on ambient air quality in Suqian. China Atmos Environ 42(9):1961–1969

Yasmin S, Mattoo R, Nehvi F (2013) Isolation, characterization and molecular weight determination of cellulase from *Trichoderma viride*. Afr J Biotechnol 12:4503–4511

Yokelson RJ, Christian TJ, Karl TG, Guenther A (2008) The tropical forest and fire emissions experiment: laboratory fire measurements and synthesis of campaign data. Atmos Chem Phys 8:3509–3527. https://doi.org/10.5194/acp-8-3509-2008

Yokelson RJ, Burling IR, Urbanski SP, Atlas EL, Adachi K, Buseck PR, Wiedinmyer C, Akagi SK, Toohey DW, Wold CE (2011) Trace gas and particle emissions from open biomass burning in Mexico. Atmos Chem Phys 11:6787–6808. https://doi.org/10.5194/acp-11-6787-2011

Yoon JJ, Cha CJ, Kim YS, Kim W (2008) Degradation of cellulose by the major endoglucanase produced from the brown-rot fungus *Fomitopsis pinicola*. Biotechnol Lett 30:1373–1378

Yoon LW, Ang TN, Ngoh GC, Chua ASM (2014) Fungal solid-state fermentation and various methods of enhancement in cellulase production. Biomass Bioenergy 67:319–338. https://doi.org/10.1016/j.biombioe.2014.05.013

Zabeda H, Sahu JN, Boyce AN, Faruq G (2016) Fuel ethanol production from lignocellulosic biomass: an overview on feed-stocks and technological approaches. Renew Sust Energ Rev 66:751–774. https://doi.org/10.1016/j.rser.2016.08.038

Zafar S (2014) Waste management, waste-to-energy. https://www.bioenergyconsult.com/tag/waste-to-energy

Zhang YHP, Lynd LR (2004) Toward an aggregated understanding of enzymatic hydrolysis of cellulose: non-complexed cellulase systems. Biotechnol Bioeng 88:797–824. https://doi.org/10.1002/bit.20282

Zhang XZ, Zhang YHP (2013) Cellulases: characteristics, sources, production, and applications. In: Bioprocessing technologies in biorefinery for sustainable production of fuels, chemicals, and polymers, vol 1, pp 131–146. https://doi.org/10.1002/9781118642047.ch8

Zhang YHP, Himmel ME, Mielenz JR (2006) Outlook for cellulase improvement: screening and selection strategies. Bio technol Adv 24(5):452–481. https://doi.org/10.1016/j.biotechadv.2006.03.003

Zhang H, Ye X, Cheng T, Chen J, Yang X, Wang L, Zhang R (2008) A laboratory study of agricultural crop residue combustion in China: Emission factors and emission inventory. Atmos Environ 42:8432–8441. https://doi.org/10.1016/j.atmosenv.2008.08.015

Zhang H, Hu D, Chen J, Ye X, Wang SX, Hao JM, Wang L, Zhang R, An Z (2011) Particle size distribution and polycyclic aromatic hydrocarbons emissions from agricultural crop residue burning. Environ Sci Technol 45:5477–5482. https://doi.org/10.1021/es1037904

Zhang Y, Tang B, Du G (2017) Self-induction system for cellulase production by cellobiose produced from glucose in *Rhizopus stolonifer*. Sci Rep 7:10161. https://doi.org/10.1038/s41598-017-10964-0

Zhuang J, Marchant MA, Nokes SE, Strobel HJ (2007) Economic analysis of cellulase production methods for bio-ethanol. Appl Eng Agric 23(5):679–687. https://doi.org/10.13031/2013.23659

Chapter 13
Bioconversion of Lignocellulosic Residues into Single-Cell Protein (SCP) by *Chaetomium*

Ahmed M. Abdel-Azeem and Donia H. Sheir

13.1 Introduction

The term single-cell protein (SCP) refers to dead, dry cells of microorganisms such as yeast, bacteria, fungi and algae which serve as food and/or feed supplements. SCP will be an alternative to conventional proteins like casein, soya bean meal, egg protein or meat protein in animal feed. SCP is one of the alternatives that cannot be affected by climate change. SCP has a high protein content containing all the essential amino acids. Microorganisms are an excellent source of SCP because of their rapid growth rate, their ability to use very inexpensive raw materials as carbon sources and their uniquely high efficiency, expressed as grams of protein produced per kilogram of raw material, with which they transform these carbon sources to protein. SCP has many benefits. It is a very fast way of producing protein compared to the production of protein through cultivation of agricultural crops or animal farming. The amino acid profile of many SCP is favourable and very similar to that of fishmeal. SCP can be produced from residual streams from different industries giving the possibility of a cheap production. In addition, SCP production can be performed in bioreactors and does not require agricultural land. Production of SCP may very well fit into the request of a sustainable high-quality alternative to fishmeal since the production can be performed using renewable and sustainable feedstocks such as residual streams from second-generation bioethanol production. The second-generation bioethanol production is predicted to increase in the future, resulting in large volumes of residual and waste streams. These residual streams are commonly used as substrates for biogas production. SCP production is an interesting

A. M. Abdel-Azeem (✉)
Botany Department, Faculty of Science, University of Suez Canal, Ismailia, Egypt
e-mail: ahmed_abdelazeem@science.suez.edu.eg

D. H. Sheir
Chemistry of Natural and Microbial Products Department, Pharmaceutical Industries Division, National Research Centre, Cairo, Egypt

A. M. Abdel-Azeem (ed.), *Recent Developments on Genus Chaetomium*, Fungal Biology, https://doi.org/10.1007/978-3-030-31612-9_13

alternative to biogas production, possibly with a higher economic value. SCP has been found to meet all the requirements for its inclusion as diet supplement for livestock. SCP can replace up to 20–30% of the protein supply by soybean meal without any deleterious effects on growing broiler chicks.

Lignocellulosic biomass presents a readily available feedstock for microbial bioconversion which does not compete with feedstocks used for human food. Lignocellulose is the major structural component of woody plants and non-woody plants and represents a major source of renewable organic matter—a substrate of enormous biotechnological importance. Microorganisms are involved in bioconversion of low-cost carbon feedstocks such as lignocellulose to produce biomass rich in proteins and amino acids. Production of SCP from lignocelluloses is gaining much attention, with the recovery of valuable by-products and simultaneous reduction of the organic load as the chief economic advantages of such processes. This chapter focuses on the bioconversion of lignocellulosic residues into single-cell protein by microorganisms with special reference to *Chaetomium cellulolyticum*.

13.2 Bioconversion of Lignocellulosic Residues into Single-Cell Protein

Cellulose from forestry and agriculture sources constitutes the most plentiful renewable raw material in the planet as potential substrates for SCP production. In nature, cellulose is usually found associated with lignin, hemicellulose, starch, etc. in a complex form. So, if cellulose is to be used as substrate, it must be pretreated chemically or enzymatically in order to remove cellulose as fermentable sugars as mentioned above (Callihan and Clemmer 1979a, b). Currently, the only economical use of lignocellulosic wastes is in mushroom production. Besides the well-known cultivated mushroom *Agaricus bisporus*, there are several important varieties which contain lignocellulolytic enzymes and are grown for food mainly in Asia and Africa. Some are of great economic significance and are produced on an industrial scale. Examples of important ones include *Volvariella* sp., *Lentinus edodes* and *Pleurotus* sp. In the production of industrial cellulose for paper and tissue production, the cost involved in these steps has prevented the generalized production of SCP from cellulose (Callihan and Clemmer 1979a, b). Wood can be also cooked in a medium containing calcium sulphite with excess free sulphur dioxide. Lignin is thus converted to lignosulfonates and hemicellulose is hydrolysed to monosaccharides and may be further converted to furfurols. The amount of free sugars in the spent liquor is found to vary with the type of procedure selected, as various cellulose fibres may be obtained with different degrees of degradation. Spent sulphite liquor has been utilized as a substrate for fermentations since 1909 in Sweden and afterwards in many other parts of the world. The first organism used was *Saccharomyces cerevisiae*, although this organism is unable to utilize pentoses which are found in substantial amounts in this waste product. Later, other organisms better suited for the utilization of all the sugar monomers were selected, namely, *Candida tropicalis* and *Candida utilis*. Yeast produced from sulphite liquor has been used for feeding at war periods, but lost favour in peace time. However in Finland, Peliko process was used for production of baker's yeast from sulphite liquor. The protein content of the fungus *Paecilomyces variotii* is more than 55% (w/w) and has been officially approved as a food in Finland. In 1983, the projected biomass production of the process was estimated to be around 7000 tonnes per year (Oura 1983). Currently, extracellular cellulases are commercially used in cellulose-separating processes. A number of efficient cellulase producers have been reported, but *Trichoderma viride* is a well-known high cellulase-producing organism. Another cellulolytic fungus is *Chaetomium cellulolyticum* (Fig. 13.1) which grows faster and produces 80% more biomass protein than *Trichoderma*. This shows that *C. cellulolyticum* is suitable for SCP production, while *T. viride* is a hyper-producer of extracellular cellulases. The amino acid composition of *C. cellulolyticum* is generally found better than that of *T. viride* and similar to alfalfa and soya meal protein (Bhalla et al. 2007). A cheaper, more amenable SCP substrate of carbohydrate origin is starch which may be obtained from rice, maize and cereals.

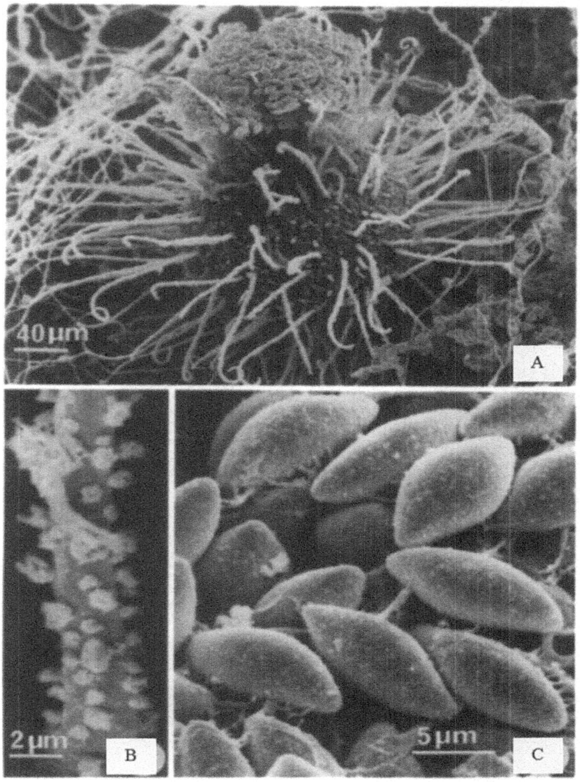

Fig. 13.1 *Chaetomium cellulolyticum*. A. Perithecium. B. Terminal hair. C. Ascospores. (After Ellis 1981)

Lignocellulose has diverse biotechnological potential in the production of value-added products such as biofuels, biofertilizers, animal feed products, biochemicals, biopesticides, biopromoters and biotransformation of the biomass itself into compost or biopulp (Tengerdy and Szakacs 2003). To fully utilize the potential of lignocellulosic biomass, hydrolysis of lignocellulose into fermentable sugars by physical, chemical, physicochemical and biological pretreatment methods is the primary requirement in all applications (Bajpai 2016). During pretreatment, lignin is removed and the porosity of the lignocellulosic materials increased to release the cellulose and hemicellulose sugars. Inhibitors are also formed in the hydrolysates. Complete substrate utilization by the microorganism and inhibitor tolerance are the prerequisites to render lignocellulosic microbial bioconversion processes economically competitive (Hahn-Hägerdal et al. 2007). However, owing to the chemical and structural complexity of lignocellulosic biomass, the sustainable utilization of lignocelluloses is limited until it undergoes pretreatment. This is usually followed by enzymatic hydrolysis, during which oligomeric sugars such as cellulose are broken down to monomeric sugars. The goal of the pretreatment process is to alter the recalcitrant structure of lignocellulosic biomass to increase the availability of degradable carbohydrates present in biomass. A pretreatment method is regarded as an effective method based on a number of features such as a high recovery of all carbohydrates, the production of limited amounts of by-products that are inhibitory to the subsequent hydrolysis and bioconversion processes, minimum energy consumption and cost-effectiveness (Kumar et al. 2009). Many pretreatment methods have been reported which vary from alkali or acid treatment, steam explosion or even X-ray radiation.

13.3 Microorganisms Used for Single-Cell Protein Production

Several fungi, yeast, bacteria and algae have been used for the production of single-cell proteins (Bhalla et al. 1999, 2007; Ukaegbu-Obi 2016). The selection of microorganism depends on several criteria: the growth of microorganism should be fast and a broader range of materials can be considered as suitable substrates. The other criteria may be nutritional, which are energy value, protein content and amino acid balance, and technical, which are type of culture, type of separation and nutritional requirements. The desired microorganisms should be grown on the medium under aseptic conditions. Organisms to be cultured should have the following properties:

1. Should be non-pathogenic to human, animals and plants.
2. Usable as food and feed.
3. Should have good nutritional values.
4. Should not contain toxic compounds.
5. Production cost should be near to the ground.

Some microbial sources of SCP are listed in Table 13.1 (Bhalla et al. 2007; Shipman et al. 1975; Singh et al. 1988, 1991; Singh 1998; Wong and Chan 1980; Anupama and Ravindra 2000; Giec and Skupin 1988; Saliceti-Piazza et al. 1992; Staron 1981; Scerra et al. 1999; Shuler et al. 1979; el-Saadany et al. 1988; Kolani et al. 1996; Guven and Cansunar 1989; Rodriguez et al. 1997; Rhishipal and Philip 1998; Kim and Lebeault 1981; Deibel et al. 1988; Invarson and Morita 1982; Ghanem 1992a, b; Zadrazil and Puniya 1995; Chanda and Chakrabarti 1996; Tannenbaum and Wang 1975; Ekerott and Villadseer 1991; Callihan and Clemmer 1979a, b; Litchfield 1979; Vashista 1989; Mahasneh 1997; Trehan 1993).

Among bacterial species, *Cellulomonas* and *Alcaligenes* are the mostly used bacterial species as a SCP source (Bhalla et al. 2007). Potential phototrophic bacterial strains are recommended for SCP production. Some researchers also suggest use of methanotrophic and other bacterial species for SCP production. The generation time of *Methylophilus methylotrophus* is about 2 h, and this bacterium is used in animal feed and in general produces a more favourable protein composition than yeast or fungi. Therefore, the large quantities of SCP animal feed can be produced using bacteria (Bhalla et al. 2007). Characteristics which make bacteria suitable for this application include rapid growth of bacteria and short generation times of bacteria—almost can double their cell mass in 20 min to 2 h (Bamberg 2000). They are also able to grow on a variety of raw materials which range from carbohydrates such as sugars and starch to gaseous and liquid hydrocarbons which contain methane and petroleum fractions, petrochemicals such as methanol and ethanol, and nitrogen sources which are useful for bacterial growth including ammonia, ammonium salts, nitrates, urea and the organic nitrogen in wastes. It is also suggested to add mineral nutrient supplement to the bacterial culture medium to fulfil deficiency of nutrients which may be absent in natural waters in concentrations sufficient to support growth (Suman et al. 2015).

The use of bacteria is somewhat limited due to the following reasons (Mondal et al. 2012):

1. Poor public acceptance of bacteria as food
2. Small size and difficulty in harvesting
3. High content of nucleic acid on dried weight basis

Many fungal species are used as sources of protein-rich food. Among these, the most popular are yeast species such as *Candida*, *Pichia*, *Hansenula*, *Torulopsis* and *Saccharomyces*. Many other filamentous species are also used as sources of SCP. Actinomycetes and filamentous fungi are reported to produce protein from various substrates. *Fusarium* and *Rhizopus* fungi have been grown as a source of protein food. The inoculum of *Rhizopus arrhizus* or *Aspergillus oryzae* is chosen because of their non-toxic nature. Saprophytic fungi grow on complex organic compounds and convert them into simple structures. High amount of fungal biomass is produced as a result of growth. Mycelial yield varies greatly depending upon organisms and substrates. There are some species of moulds, for example, *Aspergillus fumigatus*, *Aspergillus niger* and *Fusarium graminearum*, which are very dangerous

Table 13.1 List of various bacteria, yeasts, fungi and algae used for SCP production

Taxa	Species	Taxa	Species
Bacteria	*Aeromonas hydrophila*		*Rhizopus chinensis*
	Acinetobacter calcoaceticus		*Trichoderma viride*
	Aeromonas hydrophila		*Thermoleophilum album*
	Alcaligenes eutrophus		*Aspergillus niger* AS 101
	Bacillus sp.		*Aspergillus niger*
	Brevibacterium spp.		*Sporotrichum pulverulentum*
	Cellulomonas sp.		*Chaetomium cellulolyticum*
	Cellulomonas spp.		*Chrysonilia sitophilia*
	Methanomonas methanica		*Fusarium graminearum*
	Methylococcaceae		*Paecilomyces variotii*
	Methylomonas sp.		*Penicillium cyclopium*
	Methylophilus methylotrophus		*Penicillium roqueforti*
	Mycobacterium sp.		*Penicillium camemberti*
	Nocardia sp.		*Scytalidium acidophilum*
	Pseudomonas fluorescens		*Trichoderma album*
	Pseudomonas sp.		*Trichoderma reesei*
	Rhodopseudomonas gelatinosus		White rot fungi
	Rhodopseudomonas sp.	Algae	*Alaria*
	Streptomyces spp.		*Ascophyllum*
	Thermoleophilum album		*Caulerpa racemosa*
Yeast	*Candida guilliermondii*		*Chlorella pyrenoidosa*
	Candida krusei SO1		*Chlorella sorokiniana*
	Candida lipolytica		*Chlorella* sp.
	Candida tropicalis		*Durvillaea antarctica*
	Candida tropicalis ceppo 571		*Ecklonia*
	Candida utilis		*Eisenia*
	Debaryomyces kloekeri		*Fucus*
	Hansenula polymorpha		*Gelidium*
	Kluyveromyces fragilis		*Grateloupia*
	Kluyveromyces marxianus		*Laminaria*
	Marine yeast		*Monostroma*
	Mixed cultures of yeasts		*Nostoc*
	Pichia sp.		*Oedogonium*
	Rhodotorula sp.		*Pelvwtia*
	Saccharomyces sp.		*Porphyra tenera*
	Saccharomyces spp. LK3G		*Porphyra* sp.
	Torulopsis candida		*Rhodymenia* sp.
	Torulopsis methanosorbosa		*Sargassum*
Fungi	*Aspergillus fumigatus*		*Scenedesmus* sp.
	Aspergillus niger		*Spirogyra*
	Aspergillus oryzae		*Spirulina* sp.
	Chaetomium cellulolyticum		*Synechococcus*
	Fusarium graminearum		*Ulva*
	Paecilomyces variotii		*Undaria*

to humans. So such fungi must not be used, or toxicological evaluations should be done before recommending for use as SCP (Weitzel and Winchel 1932). Yeasts are probably the most widely accepted and used microorganism for SCP (Mondal et al. 2012).

Spirulina was grown by people near Lake Chad in Africa and the Aztecs near Lake Texcoco in Mexico since ancient times. They used it as food after drying it. Spirulina is the most widely used algae with 60–72% protein and is also a rich source of vitamins, amino acid, minerals, crude fibre, etc. It is commonly used in supplemented diet for undernourished children. It is a part of diet for sportsmen and is also used in baby foods. Spirulina is recommended for diabetes patients for controlling blood sugar level. It helps in maintaining healthy eyes and skin and is found to be beneficial for lactating mothers. Being a rich source of vitamin A and B, it plays a major role in cosmetic products. Another commonly used SCP is *Chlorella*, single-cell green algae. It contains around 45% protein, 20% carbohydrate, 20% fat and 10% minerals along with fibres and vitamins. It is promoted as a super food and is routinely used to provide health benefits. Apart from being a food supplement it is also used as an alternative medicine in many cases. Biomass obtained from *Chlorella* and *Scenedesmus* is harvested and used as food source by tribal communities in certain parts of the world. Algae are used as food in many different ways and its advantages include simple cultivation, faster growth and rich in protein content (Arora et al. 1991). The production of algae could be limited by certain conditions such as the requirement of warm temperatures and plenty of sunlight in addition to carbon dioxide (Mondal et al. 2012). Another disadvantage associated with using algae as SCP is that digestibility is low with algal cells because of indigestible cell walls (Ware 1977).

For protein production, there are three species most commonly used with a higher commercial value, *Chlorella*, *Spirulina* (*Arthrospira*) and *Dunaliella*, having 55, 65 and 57% protein content. Algae can be used as a food source in many ways (Vashista 1989).

The microorganisms used for SCP production should have reasonable protein content, non-toxic and non-pathogenic to animals and humans. Microorganisms possess the ability to use a range of inexpensive nitrogen and carbon sources and need moderate growth conditions to convert it into valuable product, which is a precondition for SCP production. Some biochemical and cultural features of various groups of microorganism (algae, fungi, yeast and bacteria) for the production of SCP have been summarized in Tables 13.2, 13.3, 13.4 and 13.5 (Bhalla et al. 2009). Table 13.6 presents the comparison of SCP from different organisms (Srividya et al. 2013).

Table 13.2 Characteristics of bacteria for SCP production

Growth rate	Highest
Substrate	A wide range of substrates
pH range	5–7
Cultivation	Bioreactor system
Risk of contamination	High; precautions necessary
Biomass	Sometimes recovery problematic; new improved methods are needed
Protein	80% or more
Amino acid profile	Generally good, a small deficit in S-containing acid
Nucleic acid content	High (8–14%)
Removal of nucleic acids	Necessary
Toxins	Gram-negative bacteria may produce endotoxins
Other features	–

13.4 Cultivation of Microorganisms for Single-Cell Protein Production

Single-cell protein (SCP) production takes six steps as recommended by Ukaegbu-Obi (2016):

1. Screening of microorganisms
2. Selection of raw materials
3. Process engineering and optimization
4. Technology development
5. Economic consideration/process feasibility
6. Safety concerns

SCP is produced using the fermentation process (Chandrani-Wijeyaratne and Tayathilake 2000; Nasseri et al. 2011). Nasseri et al. (2011) summarized seven steps for the fermentation process of single-cell protein production as follows:

1. A pure culture of the selected microorganism that is in the correct physiological state
2. Sterilization of the growth medium which is used for the organism

Table 13.3 Characteristics of yeast for SCP production

Growth rate	Quite high
Substrate	Most substrates except hydrocarbons and CO2
pH range	5–7
Cultivation	Bioreactor system
Risk of contamination	Low
Biomass	Easy by centrifugation
Protein	55–60%
Amino acid profile	Generally good, a small deficit in S-containing acid
Nucleic acid content	High (5–12%)
Removal of nucleic acids	Necessary
Toxins	–
Other features	High B vitamin content

3. A production fermenter which is the equipment used for drawing the culture medium in the steady state
4. Cell separation
5. Collection of cell-free supernatant
6. Product purification
7. Effluent treatment

These processes are done using selected strains of microorganisms. These microorganisms are allowed to multiply on suitable raw materials. Production of SCP involves basic steps of preparation of suitable medium with suitable carbon source, avoiding contamination of the medium and the fermenter, production of microorganisms with desired properties and separation of synthesized biomass and its processing (Soland 2005). Different types of carbon source can be used, such as n-alkenes, gaseous hydrocarbons, methanol, ethanol, renewable sources like carbon dioxide, molasses, polysaccharides, effluents of breweries and other solid substances (Talebnia 2008).

Process development starts with screening of microorganisms. Suitable production strains are obtained from air, water or soil samples or from swabs of biological or inorganic materials and are afterwards optimized by selection, mutation or other genetic methods. Then the technical conditions of cultivation for the optimized and all metabolic pathways and cell structures are determined. Besides, process

Table 13.4 Characteristics of fungi for SCP production

Growth rate	
	Lower than bacteria and yeast
Substrate	
	Limited substrates mostly starchy and cellulosic materials
pH range	
	3–8
Cultivation	
	Bioreactor system
Risk of contamination	
	Low
Biomass	
	Easy for filamentous or pellet forms
Protein	
	50–55%
Amino acid profile	
	Low in S-containing acid
Nucleic acid content	
	High (3–10%)
Removal of nucleic acids	
	Necessary
Toxins	
	Many species produce mycotoxins
Other features	
	Chitin may contain a significant proportion of N content, which is unavailable

engineering and apparatus technology adapt the technical performance of the process for making the production ready for use on the commercial scale. The economic factors—energy and cost—come into play. Environmental protection and safety demands are also considered in the production of SCP in relation to both the product and the process. Finally, safety and the protection of innovation throw up legal and controlled aspects such as product authorizations for particular applications and the legal protection of new process and the microorganisms.

SCP can be produced by fermentation processes. Fermenters vary in size from laboratory scale to industrial scale of several hundred litre capacity. Fermenters are equipped with an aerator, which supplies oxygen to aerobic processes. Also, a stirrer is used to stir the medium. A thermostat is used for controlling the temperature. A pH detector and some other control devices are used which keep all the different parameters required for constant growth (Ferrianti and Fiechter 1983; Sinclair and Cantero 1990). Cost is a major problem for producing and harvesting microbial proteins. Such a production even at a high rate causes dilute solutions usually less than 10% solids. The following are methods available for concentrating the solutions:

- Filtration
- Precipitation
- Centrifugation
- Use of semipermeable membranes

Table 13.5 Characteristics of algae for SCP production

Growth rate	
	Low
Substrate	
	Light, inorganic carbon sources, e.g. CO2
pH range	
	Up to 2.0
Cultivation	
	Open ponds, tanks
Risk of contamination	
	High
Biomass recovery	
	Difficult and costly with unicellular algae
Protein	
	Up to 60%
Amino acid profile	
	Generally good; low in S-containing amino acids
Nucleic acid content	
	Low (4–6%)
Removal of nucleic acids	
	Necessary
Toxins	
	Three types of toxin: —Eendotoxin, neurotoxin, hepatotoxin
Other features	
	Low yield (1–2 g dry wt/l). High chlorophyll content, unsuitable for humans

Table 13.6 Comparison of SCP from different organisms

Percent composition (weight)	Bacteria	Fungi	Algae
True protein	50–83[a]	30–70[a]	40–60[a]
Total nitrogen (protein + nucleic acids)	60–80[a]	35–50[a]	45–65[a]
Lysin	4.3–5.8[a]	6.5–7.8[a]	4.6–7.0[a]
Methionine	2.2–3.0[a]	1.5–1.8[a]	1.4–2.6[a]
Fats/lipids	8–10[a]	5–13[a]	5–10[a]
Carbohydrate	NA	NA	9
Bile pigments and chlorophyll	NA	NA	6
Nucleic acids	15–16[a]	9.70	4–6[a]
Mineral acids	8.6	6.6	7
Amino acids	65	54	NA
Ash	NA	NA	3
Moisture	2.8	4.5–6.0[a]	6.0
Fibre	NA	NA	3

[a]The yield varies with the type of substrate used, the specific organism used and the culture conditions maintained
NA not available

The equipment used for de-watering is expensive and therefore not suitable for small-scale productions and operations. The removal of large amount of water which is necessary to make the material stable for mass storage is not commercially viable. SCP should be dried to 10% moisture or they can be condensed and denatured to prevent spoilage (Sinclair and Cantero 1990). The physiological features of the microorganisms recommend the control of carbon source concentrations, as a limiting substrate, and also sufficient supply of oxygen for the maintenance of balanced growth under an oxidative metabolic pattern. Since microbial growth is a time-dependent process, it applies continuous modifications on all process parameters which affect physiology, but, most markedly, substrate concentration. So, an adequate technology which maintains appropriate growth conditions for an extended period of time must be implemented specifically for obtaining high productivity and yield.

Batch fermentations are not so perfect for biomass production, since the conditions in the reaction medium change with time (Oura 1983). Fed-batch fermentations are better suited for production of biomass, since they involve the control of the carbon source supply through feeding rates. However, with the increase of biomass concentration, the oxygen demand of the culture reaches a level which cannot be met in engineering or economic terms. Fed-batch culture is still in use for baker's yeast production using proven and well-established models (Steinkraus 1986). But, they have not been favoured for the production of SCP on a large industrial scale. Extending a microbial culture by continuous addition of fresh medium with the simultaneous harvesting of a product has been successfully implemented in industrial fermentations for biomass production. Chemostat is the most commonly used principle. This is a perfectly mixed suspension of biomass into which medium is passed at a constant rate and the culture is harvested at the same rate so that the culture volume remains constant. The technical implications of chemostat culture are various and extremely relevant (Cooney 1986). Production periods as long as 6 weeks have been implemented in several yeast and fungi (Forage and Righelato 1979). A common problem of industrial fermentations is the heavy appearance of foam on the head space of the reactor. This foam causes pressurization of the reactor, spillages and contamination hazard. Among the various designs which have been put to effect, the deep-jet fermenter and the airlift fermenter have been used most successfully (Stanbury et al. 2000). Airlift is being used successfully as the configuration of choice for continuous production of SCP. This is presently used in the production of mycoprotein which is the basis for Quorn™ products. The control of major process variables is a critical element of SCP production, from oxygen transfer and substrate and product concentration to the appearance of small amounts of toxic compounds through undesired metabolic processes, which may compromise the quality of the final product.

The biomass from yeast fermentation processes is harvested by continuous centrifugation. Filamentous fungi are harvested by filtration (Solomons 1985). Then the biomass is treated for RNA reduction and dried in steam drums of spray-dryers. Drying is costly but results in a stabilized product with shelf lives of years.

In semisolid fermentations (Adedayo et al. 2011), the insoluble solid substrate is a solid porous matrix that absorbs water with a relatively high water activity and also contains available carbohydrates and mineral nutrients and nitrogen sources. The attraction for this type of culturing method comes from its similarity to the natural way of life for several microorganisms and usage of starchy agricultural wastes makes the whole process more economical.

Submerged culture fermentations have high operating cost and require high capital investment. The cultivation involves many operations which include the following (Andersen et al. 2005):

- Stirring and mixing of a multiphase system
- Transport of oxygen from the gas bubbles through the liquid phase to the microorganisms
- Process of heat transfers from the liquid phase to the surroundings

A special bioreactor is designed for identification of mass and energy transportation phenomena, called U-loop fermenter (Jorgensen 2010). Solid-state fermentation (SSF) involves growth of microorganisms on predominantly insoluble substrate, where there is no free liquid. Thousands of papers have been published on SSF. This technique has been extensively studied describing several types of bioreactor designs, process conditions and microorganisms for the production of different types of value-added products like SCP, feeds, enzymes, ethanol, organic acids, B-complex vitamins, pigments and flavours (Singhania et al. 2009). This process consists of depositing a solid culture substrate, such as rice or wheat bran, on flat beds after inoculating with microorganisms; the substrate is then left in a temperature-controlled room for several days. Growth of fungi is better in SSF; it gives much higher biomass when compared to submerged fermentation. Further, the SSF process is simple and has many advantages over the submerged fermentation (Table 13.7). However, particle size of the substrate, moisture level and C/N ratio are critical factors for SSF (Pandey and Soccol 1998; Tengerdy 1985; Rodriguez-Vazquez et al. 1992; Zadrazil and Puniya 1995; Nigam and Singh 1994).

Liquid-state fermentation is conducted in tanks, which can reach 1001–2500 m^2 (10,770–26,910 ft^2) on a commercial scale. Liquid culture is ideal for the growth of

Table 13.7 Comparison of solid-state and submerged fermentation processes

Parameter	Submerged fermentation	Solid-state fermentation
Substrate condition	Requires continuous agitation and soluble substrate	No agitation required and insoluble polymers as substrate
Moisture	Required in large quantity	Absence of free water
Aerobic condition maintenance	By agitation	By diffusion
Post fermentation waste	Large quantity, hence effluents polluting the environment	Very little, hence non-polluting
Space	Large	Small
Capital investment	Very high	Low
Aseptic conditions	Highly essential	Not required

unicellular organisms such as bacteria or yeasts. In liquid aerobic fermentation, it is important to constantly supply the microorganism with oxygen, which is usually done via stirring of the fermentation media. Synthesis of the desired metabolites requires regulation of temperature, ionic strength and pH, soluble oxygen and control nutrients (Capalbo et al. 2001).

Aeration requirements and foaming characteristics should be well studied before the production of SCP at industrial scale. Protein, RNA and nutritional composition of the product should adhere to recommended parameters. Structural properties of the final product should be well suited for consumption (Scrimshaw and Dillen 1977).

Generally, under conditions of low water activity and presence of intractable solid substrate, fungi show very good growth. Hence, proper growth of fungi in SSF gives much higher concentration of the biomass and higher yield in comparison to submerged fermentation. The advantage of the SSF process is the possibility of efficient utilization of waste as the substrate to produce commercially viable products (Zadrazil and Puniya 1995). The process does not require elaborate prearrangements for preparation of media. The process of SSF initially focussed on enzyme production. But at the moment, there is interest for SCP production due to the dwindling conventional food resources.

13.5 Single-Cell Protein from Lignocellulosic Wastes

13.5.1 Production from Lignocellulosic Wastes

Single-cell protein (SCP) production from various lignocellulosic material has been well documented. Lignocellulosic wastes from different sources have varying composition of hemicellulose, cellulose and lignin. Some sources of lignocellulosic material are listed below (Tanaka and Matsuno 1985; Gupte and Madamwar 1997; Callihan and Clemmer 1979a, b; Dimmling and Seipenbusch 1978):

- Wood (hardwoods and softwoods)
- Grasses
- Leaves
- Wastes from pulp and paper manufacture
- Sugar cane bagasse
- Wheat straw
- Wheat bran
- Rice bran
- Groundnut shell

Based on the dominant component in the waste used, specific fungi can be used for production of biomass. Bacteria can also be grown on wastes or by-products obtained from industrial processes. The biomass thus produced can be harvested and used. SCP from various agro-industrial wastes has been reported from several laboratories (Chaudhary and Sharma 2005; Dimmling and Seipenbusch 1978; Hongpattarakere and Kittikun 1995; Kamel 1979; Israelidis and Coduonis 1982; el-Saadany et al. 1988; Asad et al. 2000; Hashmi et al. 1991; Moo-Young et al. 1992; Paynor et al. 2016). Agro-industrial residues such as sugar cane bagasse (Pessoa 1991, Pessoa et al. 1996), rice straw (Almeida 1991), corncobs (Gonzales-Valdez and Moo-Young 1981) and eucalyptus (Silva 1991; Almeida e Silva et al. 1995), which consist of about 30% hemicellulose, have been hydrolyzed to produce a solution rich in xylose. Other sugar-rich industrial by-products, such as vinasse, spent sulphite liquor and hemicellulose rayon hydrolysates, have also been examined as fermentation media (Bajpai and Bajpai 1986, 1987, 1988; Lo and Moreau 1986). Almeida e Silva et al. (1995) studied microbial protein production by *Paecilomyces variotii* cultivated in eucalyptus hemicellulosic hydrolysate.

Paecilomyces variotii, a fungus frequently found in air and soil in tropical areas, has been used for the production of microbial protein due to its excellent ability to grow in a variety of highly polluting industrial effluents, such as molasses, wood hydrolysates, spent sulphite liquor and vinasse (Romantschuk 1974, Romantschuk and Lehtomaki 1978; Cabib et al. 1983; Castlla et al. 1984; Bajpai and Bajpai 1986, 1987, 1988).

Paecilomyces variotii was the first fungus to be used in an industrial scale for the production of microbial protein. The 'Pekilo' process is a continuous process in which the fungus is grown in sulphite liquor with a production of 10,000 tonnes/year (Romantschuk and Lehtomaki 1978). Spent sulphite liquor is produced during the manufacture of sulphite pulp. *Paecilomyces variotii* has the ability to grow in various complex residual streams from different industries (Almeida e Silva et al. 1995). In the 1970s, the Pekilo process was started in Finland. The protein-rich fungus was approved as animal feed (Romantschuk 1976), although the process is currently not running (Ugalde and Castrillo 2002). The investigation of using *P. variotii* for SCP production has continued, even though the interest has been quite low in the last decades. Pekilo process is designed to operate at unit sizes smaller than SCP plants which use petroleum paraffins and methanol feedstocks. This process also contributes significantly to eliminating problems of water pollution from sulphite spent liquors. The first commercial Pekilo plant went on stream at the United Paper Mills pulp mill at Jämsänkoski. This unit had an annual capacity of 10,000 metric tonnes and produced SCP at significantly reduced costs than petroleum-based plants having an annual capacity of 100,000 tonnes. Although sulphite waste liquor is the first raw material to be used in the Pekilo process, the technology is not restricted to the sulphite pulp industry. Other types of waste, carbohydrates, can also be used. Pekilo SCP is derived mainly from wood sugars.

The development of the Pekilo process began with observations made by Otto Gadd, of the Finnish Pulp and Paper Research Institute, that certain types of microfungi could be cultivated in submerged cultures of spent sulphite liquors. The traditional method of producing fodder yeast from sulphite spent liquor is, of course, well known to pulp manufacturers, particularly in Finland. The prospect of making major improvements in this process by using a fungus seemed particularly promising, given the increasing demand for animal feed and the increased acceptance of

SCP as a substitute for more conventional feeds. The search for the right microfungus started with the testing of more than 300 species. As the cost of research increased, the investment base was broadened; eight major Finnish companies formed a group called SITU for developing Pekilo process on a fully commercial scale. Research continued at the Finnish Pulp and Paper Research Institute. The microfungus finally selected was *Paecilomyces variotii*, which has the following favourable characteristics:

- High crude protein content, i.e. 55–60%, compared to the common micro-fungi level of 25–40%
- Has a satisfactory growth rate under process conditions
- Easily separated from the liquor
- No signs of toxic effects in test animals

Following the selection of the microfungus, the development of Pekilo went from laboratory to bench-scale fermentation. Through fermentation runs with 0.45- and 1-cubic-metre fermenters, it was established that continuous fermentation could be maintained over periods of several weeks without interference from foreign organisms. During this stage of Pekilo's development, it was important to produce enough of it to begin feeding trials at the Finnish Agricultural Research Centre with various animals, mainly pigs, calves and chickens. They proved that Pekilo protein could be used as a partial substitute for soybean meal, fishmeal or skim milk powder. Table 13.8 shows the composition of Pekilo SCP protein. The advantages of Pekilo process over other carbohydrate fermentation processes are presented in Table 13.9.

Pekilo protein obtained from the pilot fermenter was sent for evaluation to the Institute of Veterinary Medicine in Norway and to the National Institute for Research in Dairying in the United Kingdom.

Bajpai and Bajpai (1986, 1987, 1988) investigated SCP production from rayon pulp mill waste, using the fungus *Paecilomyces variotii* and the yeast *Candida*. The prehydrolysate liquor generated from rayon pulp mill on prehydrolysis of wood was acidic in nature, containing about 25 g/L total reducing sugars and having a biological oxygen demand (BOD) of ≈30,000 mg/L. After mild alkali treatment of the liquor, *Candida* species of yeast were grown with a view to produce SCP and to reduce the pollution load. More than 14/L dry yeast could be produced in a 2.6-L fermenter using vegetable oil as an antifoam. The cultivation and recovery of yeast in the prehydrolysate removed 70% BOD. The protein content in yeast was found to vary from 40 to 45% depending upon the species. The repeated fed-batch fermentation gave about 75% higher biomass productivity and there was improvement in sugar utilization. With *Paecilomyces variotii*, 25 g dry wt. biomass/ 1 in the lab. Fermenter was produced with a biomass yield of 96% of the theoretical, with 95% substrate utilization. This resulted in a 70% reduction in the BOD of the prehydrolysate.

Almeida e Silva et al. (1995) used eucalyptus hemicellulose hydrolysate as substrate for production of SCP. The eucalyptus hemicellulose fraction was hydrolyzed by treating eucalyptus woodchips with 1.2% (w/v) sulphuric acid (wood to acid ratio of 1:3) at 150 °C for 120 min. The hydrolysate was used as substrate to grow *Paecilomyces variotii* IOC-3764 in a 7-l fermenter at 30 °C, air 1.5 vvm, stirring speed 400 rpm. The kinetic parameters measured included maximum growth rate (μmax 0.10/h), yield (Yx/s 0.44 g/g) and productivity (Qx 0.26 g/l/h). The amino acid profile of the protein from *Paecilomyces variotii* growth in the eucalyptus hydrolysate was similar to the profile of the protein obtained by the 'Pekilo' process, except for leucine content, which was double the value for the 'Pekilo' process. When compared to the FAO and soybean protein standards, one can see that the protein produced by *Paecilomyces variotii* cultivated in eucalyptus hemicellulose hydrolysate contains all the essential amino acids for animal feed (Table 13.10). Moreover, it has a superior profile when compared with the plant protein.

In a study by Alriksson et al. (2014), sulphite liquor permeate was used as substrate for production of SCP.

Table 13.8 Composition of Pekilo SCP protein

	Percent of dry matter
Protein	55–60
Fat	2–4
Amino acids (mg/kg dry matter)	
Threonine	4.6
Valine	5.1
Cysteine	1.1
Methionine	1.5
Isoleucine	4.3
Leucine	6.9
Tyrosine	3.4
Phenylalanine	3.7
Lysine	6.4
Tryptophan	1.2
Vitamins	
Thiamine	6
Riboflavin	66
Pyridoxine	16
Niacin	488
Pantothenic acid	35
Biotin	2
Folic acid	12

Table 13.9 Advantages of Pekilo process

A 55–60% protein-rich product which also has a favourable amino acid composition
The aseptic operation produces a final product which is microbiologically well defined
The level of impurities originating with the feedstock can be kept very low because of ease of washing in the filter

Table 13.10 Comparison of the amino acid profile of the protein produced by *Paecilomyces variotii* grown in wood hydrolysate with that of other proteins

	IOC 3764[a]	Pekilo[b]	FAO[c]	Soybean[c]	Animal feed[d]
Alanine	7.19	5.80	–	–	–
Valine	6.39	5.03	4.2	5.0	2.70
Glycine	4.99	4–77	–	–	2.43
Isoleucine	5.08	4.18	4.20	4.90	2.57
Leucine	14.38	6.99	4.80	8.00	3.80
Proline	5.98	4.23	–	–	–
Threonine	4.65	4.25	2.80	4.30	1.97
Serine	3.39	4.91	–	–	–
Methionine	1.74	1.73	2.20	1.30	0.72
Phenylalanine	4.31	3.80	2.80	5.30	2.20
Aspartic acid	7.03	8–12	–	–	–
Glutamic acid	11–41	10.35	–	–	–
Tyrosine	5.86	3.36	–	–	–
Lysine	7.34	5.60	4.20	6.60	3–20
Arginine	4.31	6.02	–	–	–
Histidine	2.18	2.15	–	–	–
Cystine	1.17	1.38	2.00	1.60	0.74
Tryptophan	–	–	1.40	1.40	0.60

[a]Values are means of duplicates
[b]Farstad et al. (1975)
[c]Araujo and D'Souza (1986)
[d]Lo and Moreau (1986)

Steen (2014) studied the potential to produce SCP from residual streams from the second-generation bioethanol production. Three different residual streams based on lignocellulosic material (prehydrolysate and stillage of wheat straw and prehydrolysate of spruce) were used and four different microorganisms (*Paecilomyces variotii*, *Cunninghamella echinulata*, *Mortierella isabellina* and *Yarrowia lipolytica*) were evaluated. Pilot-scale cultivation of *P. variotii* on prehydrolysate of wheat straw and on detoxified prehydrolysate of spruce gave promising results with biomass concentrations of 8–10 g/L and with a protein content of around 50%. In addition, the biomass consisted of high levels of β-glucans, about 20%. β-Glucans are an interesting molecule that increasingly is being supplied to fish feed due to their immunostimulatory effect. b-D-Glucans represent part of a group of physiologically active compounds which are generally called 'biological response modifiers'. They are highly conserved carbohydrates forming structural components of cell walls of some plants, bacteria, fungi, yeast and seaweed. Glucan generally represents a group of chemically heterogeneous polysaccharides which exist in various numbers of molecules bound together in several forms of linkage together with several forms and degrees of branching. The high β-glucan content could potentially increase the value of the SCP as an ingredient in fish feed. *Y. lipolytica* was found to grow well on stillage of wheat straw and reached a biomass concentration of 15 g/L with a protein content of over 50% in a pilot-scale experiment. An interesting finding in this study was the utilization of uncharacterized carbon sources within the prehydrolysate and stillage of wheat straw. All the microorganisms, particularly *Y. lipolytica*, were able to utilize a broad range of the carbon sources available within the residual streams.

Several investigations have been carried out using cellulose and hemicellulose waste as a suitable substrate for SCP production (Suman et al. 2015). Many raw materials have been considered as substrate (carbon and energy sources) for SCP production (Nasseri et al. 2011). Further in many cases, these raw materials have been hydrolyzed by physical, chemical and enzymatic methods before use. Various agricultural wastes such as hemicelluloses and cellulose waste from plants have thus been used for the production of SCP (Azzam 1992; Zubi 2005; Ashok et al. 2000; Yakoub Khan and Umar Dahot 2010). These waste products have been converted to biomass, from certain microorganisms. Various forms of organic waste such as cellulose, hemicelluloses and different types of agricultural waste were used in the production of SCP (Adedayo et al. 2011). The degree of SCP production depends on the type of substrate used and also on media composition (Mondal 2006). *Aspergillus terreus* possesses a high protein value and has been used as a better choice for SCP production using inexpensive energy sources like *Eichornia* and banana peel (Jaganmohan et al. 2013). The cladodes of *Opuntia ficus-indica* (cactus pear) were one such lignocellulosic raw material that has potential for production of SCP in arid and semi-arid climate (Gabriel et al. 2014).

Many companies producing SCP including Kanegafuichi (Japan), Liquichimica (Italy) and BP (UK) appeared on the scene. In the United States, less than 15% of the plants producing SCP relied on hydrocarbons as the source of carbons and energy for the microorganisms. Other potential substrates for SCP include citrus wastes, bagasse, sulphite waste liquor, molasses, animal manure, starch and sewage.

With ammonia-pretreated corn stalks as materials and with high-yield cellulase of *Trichoderma reesei* and the feed yeast, Chen et al. (2000a, b) developed a process for production of SCP by fermentation. Two optimized systems of multistrain co-fermentation were set up, the crude protein achieved 18.13 and 21% after 5 days co-fermentation and the cellulose conversion rates were 66.55 and 72%, respectively.

Mixed solid fermentation of *Trichoderma reesei* and *Candida tropicalis* was studied by Wang et al. (2001) for production of SCP from steam explosion maize stalk. Under optimum conditions, the crude protein obtained by mixed fermentation reached 31.82%. The raw cellulose content was reduced by 56.88%.

Zhang et al. (2003) investigated the factors of cellulase and SCP by means of mixed fermentation of *Trichoderma viride* and yeast *Candida utilis* and determined the parameters of mixed solid fermentation.

Chen et al. (1999) extracted the hemicellulose hydrolysate from steam-exploded wheat straw to produce SCP by *Trichosporon cutaneum* 851. In a 2-L auto-fermenter, the biomass concentration of 45 g/L and the productivity of 4.4 g/L/h were obtained in fed-batch fermentation.

Yakoub Khan and Umar Dahot (2010) studied SCP production from rice husk from *Penicillium expansum*. They obtained maximum protein content (30.10%) and

SCP biomass (5.107 g/l) when acid-treated rice husk was supplemented with 1.0% sucrose.

Banerjee et al. (1995) studied the effects of the size of straw and different pre-treatments to protein yield in their research of bioconversion rice straw to SCP by *Neurospora sitophila*. The results showed that the mixed strains of *Rhizopus* and *Trichoderma* had better ability to degrade bagasse in comparison with single strain. The cellulose utilization was about 90%, and the crude protein content in the product exceeded 50% if the rice straw of less than 1 mm was pretreated with 0.15 kg NaOH/kg rice straw.

Ibrahim Rajoka et al. (2004) produced SCP from defatted rice polishings using *Candida utilis* in shake flasks and a 14-L fermenter to optimize fermentation conditions before producing biomass in a 50-L fermenter. Using optimized cultural conditions, specific growth rate, true protein productivity, crude protein productivity, cell mass productivity, substrate consumption rate, cell yield and crude protein yield were found to be 0.224 h^{-1}, 0.94, 1.35, 1.75, 2.12 g l^{-1} h^{-1}, 0.62 g cells g^{-1} substrate utilized and 0.38 g g^{-1}, respectively, in a 50-L fermenter. Maximum values were found to compare favourably with the published data. The biomass protein in the 50-L fermenter contained 22.3% true protein, 27.8% crude protein, 19.2% crude fibre, 9.5% ash, 38.12% carbon, 8.5% cellulose and 0.27% RNA content. The dried biomass showed a gross metabolizable energy value of 2678 kcal kg^{-1} and contained all the essential and non-essential amino acids. Yeast biomass as animal feed may replace expensive feed ingredients presently being used in poultry feed and may improve the economics of feed produced.

Wang and Bian (1999) from the Nanjing Forestry University grew yeast by solid fermentation on substrate of poplar leaves. He found that forestry wastes as poplar leaves could be regarded as materials to produce feed yeast. The ingredients of yeast cultures were analysed. It was found that the total amino acid content increased 92.5%, and the L-cystine which was advantageous to poultry breeding increased by 2–3 times. The yeast cultures were used to cultivate fish, the mix cultures *Parabramis pekinensis* and *Calosoma* grew quickly and the feed coefficient reduced. Furthermore, due to the lower price of the yeast cultures compared to fishing feed, the feed cost reduced and the gross profit increased substantially (Zhao et al. 2000).

Agricultural residue (wheat bran) rich in carbohydrates was used in the fermentation process to produce microbial biomass. Yunus et al. (2015) used two different microorganisms (*Candida utilis* and *Rhizopus oligosporus*) for biomass production. To increase the nutritional contents of wheat bran, they optimized a number of different fermentation parameters such as effect of inoculum size, age of inoculum, incubation period, moisture to substrate ratio and incubation temperature. Maximum yield was obtained at an inoculum size of 10% (v/w), with the age of the inoculum being a 48-h-old culture. A fermentation period of 48 h gave the maximum protein yield and viable counts of yeast cells and mould hyphae. The microorganisms showed good growth at 30 °C. A batch of wheat bran was fermented with *C. utilis* and *R. oligosporus* under optimized conditions. Maximum crude protein yield of 41.02% was obtained compared with the 4.21% crude protein of the non-fermented wheat bran.

Sugar cane bagasse hemicellulosic fraction was hydrolyzed by treatment with 70 mg of sulphuric acid per gram of dry mass at 125 °C for 2 h. The hydrolysate was used as the substrate for growing *Candida langeronii* RLJ Y-019 at 42 °C, initial pH 6.0, agitation at 700 rev/min and aeration at 1.0 and 2.0 v/v/min. The utilization of D-xylose, L-arabinose and acetic acid was delayed due to the presence of D-glucose, but after D-glucose utilization, the other carbon sources were used. The kinetic parameters calculated for both cultivations at 1.0 and 2.0 v/v/min included maximum specific growth rate (μmax) of 0.29 ± 0.01 h^{-1} and 0.43 ± 0.016 h^{-1}, yields (Yx/s) of 0.36 ± 0.012 and 0.40 ± 0.012 gx/gs and productivity (Qx) of 0.81 ± 0.016 and 0.97 < 0.012 g/l/h, respectively, and are compared favourably with the published results obtained with *Candida utilis* and *Geotrichum candidum*. *Candida langeronii* was found superior to *C. utilis* for biomass production from hemicellulose hydrolysate, in that it used L-arabinose and was capable of growth at higher temperatures. The biomass contained 48.2, 1.4, 5.8 and 23.4% of total protein, DNA, RNA and carbohydrate, respectively, and contained all the essential amino acids for animal feed (Nigam 2000).

Wu and Ma (2002) used the mix culture technique to produce SCP utilizing bagasse as the sole carbon source. The results showed that the mixed strains of *Rhizopus* and *Trichoderma* had better ability to degrade bagasse. Liquid fermentation was carried for 108 h at 32 °C, pH 6.0. The crude protein of the dry product was 260.2 g/kg.

The effect of chemical solutions sprayed on sugar cane bagasse pith to produce SCP was examined by Rodriguez-Vazquez and Diaz-Cervantes (1994). They found that the pore size of vessels in pith pretreated with calcium hydroxide and sodium hydroxide increased and the hemicellulose–phenolic compound linkage broke, which made the carbohydrate be easily attacked by microorganisms. It was found to be favourable for fermentation.

Zayed and Mostafa (1992) also found that delignification to bagasse promotes the saccharification by *Aspergillus niger*. Samadi et al. (2016) carried out solid-state fermentation (SSF) to produce SCP from sugar cane bagasse using *Saccharomyces cerevisiae*. The SSF experiments were conducted in a tray bioreactor. The effects of several parameters were studied. These included initial moisture content of substrate, fermentation time, extraction buffer, relative humidity in the bioreactor, bioreactor temperature and pretreatment of substrate. Among the extraction buffers used in this work, carbonate–bicarbonate buffer was the most effective one for protein extraction. Results showed that suitable fermentation conditions were initial substrate moisture content of 70%, fermentation time of 72 h, relative humidity of 85%, bioreactor temperature of 35 °C and pretreatment of substrate using 2% sodium hydroxide solution; at this optimum condition, protein production yield of 13.41% was obtained. The amino acid analysis of the produced protein showed that the product contained almost all of the essential and non-essential amino acids.

SCP production from sugar cane bagasse has been reported (Molina et al. 1984; Sindhu and Sandhu 1980). Molina et al. (1984) treated sugar cane bagasse pith with 1% sodium hydroxide solution at room temperature, at a sodium hydroxide/pith ratio of 10%. They used different contact times and found that the shortest period required for maximum protein production was 24 h at 25 °C. These workers used mixed culture of *Cellulomonas* sp. and *Bacillus subtilis*.

Rodriguez et al. (1993) reported production of *Cellulomonas* SCP with 1% (w/v) bagasse pith. The pith was pretreated with either 0.2 M sodium hydroxide for 1 h at 80 °C or 0.4 M sodium hydroxide for 40 h at 28–30 °C. With these milder pretreatments, they found growth comparable to the one found for the substrate prepared with a more severe treatment. Growth was also comparable with other reports for cellulolytic bacteria grown on pretreated bagasse pith.

Rodriguez and Gallardo (1993) studied the association of *Cellulomonas* sp. with an isolate of *Pseudomonas* sp. for SCP production from bagasse pith. They observed a mutualistic symbiotic relationship during their mixed growth on bagasse pith, the *Cellulomonas* supplying carbon source (glucose produced from bagasse) to the *Pseudomonas* and the latter producing the vitamin supplements required for *Cellulomonas* growth. The metabolic symbiosis allowed the growth of the mixed culture in a minimal medium, without any growth factor supplement. Fed-batch cultivation of the mixed culture showed high biomass production (19.4 g/L).

Perez et al. (2002) while reporting use of sugar cane bagasse in a mineral medium and inoculated with *Candida utilis* for ethanol reported that 57% of the carbon from ethanol was converted to carbon dioxide and 8.7% into biomass. They observed final yeast population of 7 x 10^9 cells/g of dry matter corresponding to 56 mg protein/g dry matter. Perez et al. (2002) concluded that this much protein offers potential for using the protein-enriched bagasse as feed also. These studies clearly show that the sugar cane bagasse or its pith can be upgraded with the production of SCP using the suitable microorganisms on untreated and pretreated substrates. The bagasse is a waste of the sugar industries. Instead of other lignocellulosic materials, its use as substrate for biofuel ethanol production has the following advantages: tackling of a waste and presence of some amounts of soluble sugars which may be assimilated rapidly by the inoculated microorganisms meant for saccharification and/or ethanol fermentation of the substrate. Furthermore, the fermented residual material enriched with microbial cells may find its application as animal feed or its supplement thereof. The latter concept is likely to bring support to the economic constraints regarding the process developments for producing ethanol from lignocellulosic materials in general and from sugar cane bagasse specifically.

Ahlam (2005) and Chaudhary (2008) conducted studies along the same lines and reported isolation, characterization and optimization of microorganisms, both prokaryotic and eukaryotic, which are found useful for saccharifying and fermenting fruits and vegetable wastes and the sugar cane bagasse, respectively. Following maximum yield extraction, the fermented residue is likely to find its application to supplement animal feed with SCP.

Ahmed et al. (2010) reported that sequential culture fermentation by *Arachniotus* sp. at 35 °C for 72 h followed by *Candida utilis* fermentation at 35 °C for 72 h more resulted in higher production of microbial biomass protein. Further, 6% (w/v) corn stover, 0.0075% calcium chloride, 0.005% magnesium sulphate, 0.01% potassium dihydrogen phosphate, C/N ratio of 30:1 and 1% molasses gave higher microbial biomass protein production by the sequential culture fermentation of *Arachniotus* sp. and *C. utilis*. The mixed microbial biomass protein produced in the 75-L fermenter contained 16.41% true protein, 23.51% crude protein, 10.9% crude fibre, 12.11% ash and 0.12% RNA content. The amino acid profile of the final mixed microbial biomass protein showed that it was enriched with essential amino acids. Thus, the potential utilization of corn stover can reduce the cost for growth of these microorganisms and increase microbial biomass protein production by sequential culture fermentation.

Miller and Srinivasan (1983) and El-Nawwi and El-Kader (1996) took alkali-pretreated bagasse as materials and studied the cultivation conditions of SCP and cellulase production with *Aspergillus terreus*. Miller and Srinivasan (1983) analysed the SCP production under conditions of batch, semicontinuous and continuous cultivation. They found that the doubling time of continuous cultivation was short, and the crude protein content was steady at different temperatures. El-Nawwi and El-Kader (1996) achieved 21–28% of SCP content and 11–14.5 g/kg bagasse of SCP yield under 1.5% alkali concentration, pH 4.5, 35 °C fermentation temperature, 4% inoculum and 7-day continuous cultivation in shake flask. Carboxymethyl cellulase and filter paper activity were 0.85–1.2 U/ml and 0.08–0.11 U/ml, respectively, and the enzymatic activities were proportionate with crude protein content in the product. Other than the whole sugar cane bagasse, hemicellulosic hydrolysate of bagasse can also be used to ferment yeast SCP, for the bagasse contains ca. 30–35% hemicellulose.

Nigam (2000) and Pessoa et al. (1996) both studied the microbial protein production and its kinetics with sugar cane bagasse hemicellulosic hydrolysate, and the total protein in biomass product reached 48.2 and 31.3% separately and contained essential amino acids for three animal feed. Nigam (2000) also compared the biomass production of *Candida utilis* and *Candida langeronii* from hemicellulosic hydrolysate and found that *C. langeronii* was superior to *Candida utilis* in that it was able to utilize L-arabinose and was able to grow at higher temperature. Compared to agricultural residues, the lignification degree and cellulose crystallinity are higher, which make it more difficult for organisms to attack wood cellulose, so pretreatment to wood before bioconversion appeared to be more important.

Tong et al. (1995) studied the fermented sawdust to be used as feed. They pretreated the sawdust by heat before fermentation and then inoculated with cellulose degradation organisms. The crude protein content was found to increase after 72-h fermentation. The results of feeding egg-laying chicken using fermented sawdust revealed that it could substitute part of the corn in the day feed, and the laying quotiety and feed to eggs ratio were both petty compared to the contrast group.

Chahal et al. (1981) examined the effects of different pretreatment methods to aspen wood for SCP production with *Chaetomium cellulolyticum* and found that high-pressure steam was superior to atmospheric pressure steam, because high-pressure steam could make wood break to smaller pieces. More complete delignification of wood using sodium chlorite increased the protein composition in the final product to 37.9%, at a specific growth rate of 0.19 h^{-1}, and the cellulose utilization was highest, reaching 90%. The hemicellulose fraction of eucalyptus wood can be easily removed by acid treatment and the hydrolysate is rich in fermentable sugars, mainly xylose, which has been used as a substrate for different bioconversion products.

Almeida e Silva et al. (1995) and Almeida e Silva et al. (2003) conducted studies of bioconversion to SCP with eucalyptus wood hydrolysate and used the response surface methodology for selecting the nutrient level for culturing *Paecilomyces variotii* IOC-3764 in eucalyptus hemicellulosic hydrolysate. Cell biomass concentration obtained 12.06 g/L in medium of 10 g/L rice pollard, 2.0 g/L nitrogen and 1.1 g/L sodium phosphorus acid after 89 h of cultivation.

SCP production from low-cost wastes offers a potential substrate for conversion of low-quality biomass into an improved animal feed and human food (Anupama and Ravindra 2000). The crude protein content was increased from 3.10% in wheat straw to 10.91% in spent wheat straw (Bakshi and Jangar 1991; Bajwa et al. 1991).

Asad et al. (2000) conducted a study using alkali-treated corn cob as a substrate for SCP production by *Arachniotus* sp. The maximum crude protein (18.87%) was obtained after 96 h of continuous agitation at 120 rpm. The optimum culture medium of alkali-treated corn cob (3%) contained (g/100 ml) urea 2, calcium chloride 0.05, magnesium sulphate 0.10 and potassium dihydrogen phosphate 0.3. This biomass product may be used as a protein supplement in the poultry and livestock rations which is presently very costly due to the use of conventional ingredients such as oilseed cakes.

Ghanem (1992a, b) used beet pulp as a medium cultivated with a mixed culture of *Trichoderma reesei* and *Kluyveromyces marxianus*. Bhalla and Joshi (1994) improved the protein content of apple pomace by cultivating cellulolytic filamentous fungi (*Trichoderma* and *Aspergillus niger*) and yeasts (*Saccharomyces cerevisiae*, *Candida utilis* and *C. tropicalis*) in different combinations. The co-culture of *C. utilis* and *A. niger* was found to be the best combination. It resulted in a 200% increase in protein content after only 7 days of solid-state fermentation.

Dhanda et al. (1994) fermented wheat straw with white rot fungi and found that the crude protein content was increased from 3.42 to 6.18%.

Vibha and Sinha (2005) evaluated SCP production of six cellulolytic fungi, viz. *Curvularia lunata*, *T. harzianum*, *Penicillium citrinum*, *Aspergillus flavus*, *A. niger* and *Alternaria alternata,* from pretreated rice stubble. Only *T. harzianum* showed high SCP production with crude protein and biomass, particularly when potassium nitrate was used as a nitrogen source, whereas the lowest protein and biomass were observed in case of *A. niger*.

Shahzad and Rajoka (2011) produced single-cell protein from rice tips as substrate through fermentation with *Aspergillus terreus* and evaluated on broiler chicks to examine its potential as a suitable poultry feed. Chemical analysis and chick assay were performed for evaluation of chemical and biological potential of the biomass. The specific growth rate (l) of the *A. terreus* was 0.451 h − 1. The product coefficients were 0.553 g cell/g substrate utilized, 0.344 g protein/g substrate utilized and 0.622 g protein/g cell mass formation, respectively. Chemical evaluation of biomass showed crude protein 43.7%, true protein 26.60% and crude fibre 11.35% with calorific value as 2730 kcal. Its ash content was 15.20% with 1.01%

calcium, 3.05% phosphorus, 0.64% sodium and 0.98% chloride. Biomass was replaced with soybean meal as 30 and 60% on the basis of protein supply, and the birds' response in terms of weight gain, feed consumption, feed conversion ratio, protein efficiency ratio, feed efficiency and net protein utilization was taken into account. The study showed that microbial biomass produced by *A. terreus* can be replaced up to 30% of the protein supply by soybean meal without any adverse effect on growing broiler chicks.

Rao et al. (2003) produced microbial biomass having 46% crude protein content by an efficient fungal strain, *Penicillium janthinellum* (NCIM St-F-3b). The fungus required a simple medium containing bagasse hemicellulose as carbon source and ammonium sulphate as the nitrogen source. Therefore, bagasse, which is a waste product of the sugar industry, can be efficiently used in microbial biomass protein preparation for animal feed. The microbial biomass contained all the essential amino acids (Table 13.11). The fungal biomass contained high content of lysine, valine and tyrosine but low content of methionine. However content of cysteine and methionine was adequate.

Hu et al. (2015) described a novel technology for SCP production by *Candida utilis* using black liquor produced from the soda pulping process while investigating the effects of culture conditions on the production of microbial biomass and examining the nutritional quality of the SCP. Various parameters were evaluated, and the COD of black liquor, initial pH and nitrogen sources had significant influences on biomass and crude protein production. Maximum values of COD removal rate and crude protein production with 78.78 ± 3.21% and 1.18 ± 0.02 g/L were obtained, respectively, under the optimized condition of black liquor concentration (60%), use of urea (0.5 g/L), initial pH (6.0), temperature (34 °C), agitation speed (180 rpm) and incubation time (36 h). This study provided a potential viable treatment of black liquor and showed a feasible way to make full use of black liquor for the economical production of SCP. The biomass protein contains sorts of potential AA, and the total AA content was up to 39.82%, suggesting that the SCP generated could be useful for poultry feed and the fertilizer industry. A high yield of 1.18 ± 0.02 g SCP could be obtained per kilogram of black liquor under optimized conditions, showing that black liquor has excellent potential as a carbon and energy source for yeasts.

Table 13.11 Essential amino acids (g amino acid/100 g protein) in *Penicillium janthinellum* biomass

Amino acids	*Penicillium janthinellum*	FAO standard
Threonine	3.3	2.8
Lysine	14.0	4.2
Valine	9.0	4.2
Leucine	2.4	4.8
Tyrosine	4.6	2.8
Phenylalanine	2.7	2.8
Methionine	0.3	2.2
Cystine + methionine	2.6	2.2

13.5.2 Economic Aspects

The SCP production products obtained via microbiological synthesis must be competitive with traditional food sources. When estimating costs involved in SCP production, such major factors as the biomass yield, oxygen requirements and cooling should be taken into consideration. They depend on the choice of substrate and also on the choice of the microorganism. All these determine the cost of production and economic feasibility.

For SCP production large-scale fermenters are required. So with high biomass production, high oxygen transfer rates and high respiration rates which in turn increase metabolic heat production and the need of an efficient cooling system ensued. In such a continuous operation for SCP production, the economics of this production must be strongly taken into consideration. The economic factors that should be taken into account are mentioned below (Nasseri et al. 2011):

- Investment
- Energy
- Operating costs
- Waste
- Safety
- Global market

The industrial manufacture of SCP is greatly affected from the economic point of view by the carbon substrate used. The substrate costs are the largest single cost factor (Srividya et al. 2013). Variation in the substrate cost results in the greatest possibility of affecting the total manufacturing cost.

Simplifying the manufacture and purification of raw material can save costs.

Furthermore, the manufacture of raw materials is more economical in larger plants. Factors involved in the raw material costs are site, raw material production, process capacity of the plant and substrate yield.

The energy for compressing air, cooling, sterilizing and drying forms the next most important cost factor. Sites with cheaply available thermal, electrical, fossil or process-derived energy are to be preferred. The capital-dependent costs are determined, by the cost of the apparatus for the process, the capacity of a plant and the capacity conditions. The main variable here is the size of the plant. Small plants can be profitable if they include simplifications of processes and material to a large extent. The large expenditure on apparatus in processes with cheap, simple and unpurified raw materials usually does not pay in comparison with more expensive pure substrates with simpler technology. High productivities in fermentation are compensated by the greater expenditure on energy to obtain these productivities, so that optimum can be determined.

The process costs arising are covered only by the product produced. The absolute value of the product is governed by the amount of product referred to the costs involved and by the product quality. The product quality is poorer for a low-value unpurified product of a rational mini-process with varying composition or one including many additional components than for upgraded products. The upgradation of the product may consist of purification and separation into the components of the microbial biomass. Because of genetic variability, the possibilities of technical control in manufacture and the simplicity of the process, microbial biomass is more suitable for such special products than biomass from plants or animals (Rathoure 2014).

13.6 Conclusion

In this chapter, we discussed most of the microorganisms and agricultural wastes that can be used in the production of SCP. In general, the key considerations for choosing the most suitable waste product for SCP production are: (1) local availability of the particular waste product, (2) pretreatment costs of the waste product before using it in fermentation, (3) the costs of transportation of the waste product, (4) type of microorganism used and (5) SCP concentrations in the final microbial biomass after fermentation.

References

Adedayo MR, Ajiboye EA, Akintunde JK, Odaibo A (2011) SCP: as nutritional enhancer. J Microbiol 2(5):396–409

Ahlam AW (2005) Microbiological conversion of waste fruits and vegetables into ethanol. Ph.D. thesis. University of Punjab, Lahore

Ahmed S, Ahmad F, Hashmi AS (2010) Production of microbial biomass protein by sequential culture fermentation of *Arachniotus* sp. and *Candida utilis*. Pak J Bot 42(2):1225–1234

Almeida A (1991) Produção de proteina microbiana a partir de a partir de hidrolisado hemicelul6sico de Palha-de-arroz. M.Sc. thesis. Universidade Federal de Viqosa, Viqosa

Almeida e Silva JB, de Mancilha IM, Vannetti MCD, Teixeira MA (1995) Microbial protein production by *Paecilomyces variotii* cultivated in eucalyptus hemicellulosic hydrolyzate. Bioresour Technol 82:197–200

Almeida e Silva JB, Lima UA, Taqueda MES (2003) Use of response surface methodology for selection of nutrient levels for culturing *Paecilomyces variotii* in eucalyptus hemicellulosic hydrolyzate. Bioresour Technol 87:45–50

Alriksson B, Hörnberg A, Gudnason AE, Knobloch S, Arnason J, Johansson R (2014) Fish feed from wood. Nordic Wood Biorefinery Conference, Stockholm

Andersen BR, Andersen JB, Jorgensen SB (2005) U-loop reactor modelling for optimization, part 63: estimation of heat loss. J Environ Issues 9:88–90

Anupama, Ravindra P (2000) Value-added food: single cell protein. Biotechnol Adv 18:459–479

Araujo A, D'Souza J (1986) Enzymatic saccharification of pretreated rice straw and biomass production. Biotechnol Bioeng 28:15039

Arora D, Mukerji K, Marth E (1991) Single cell protein in hand book of applied mycology. J Am Sci 18(499):539

Asad MJ, Asghar M, Yaqub M, Shahzad K (2000) Production of single cell protein from delignified corn cob by *Arachniotus* species. Pak J Agri Sci 37:3–4

Ashok RS, Nigam P, Vanete T, Luciana PS (2000) Bio resource technology. J Am Sci 16(8):35

Azzam AM (1992) Production of metabolites, industrial enzymes, amino acids. J Environ Sci Eng 56(67):99

Bajpai P (2016) Pretreatment of lignocellulosic biomass for biofuel production. SpringerBr Gr Chem Sustain. https://doi.org/10.1007/978-981-10-0687-6

Bajpai P, Bajpai PK (1986) Cultivation of yeasts from the Prehydrolysate of a Rayon Pulp Mill Enzyme. Microbiol Technol 8(10):610

Bajpai P, Bajpai PK (1987) Single cell protein production from Rayon Pulp Mill waste by Paecilomyces variotii. J Ferment Technol 65(3):349

Bajpai P, Bajpai PK (1988) Repeated batch fermentation for SCP production from Prehydrolysate of Pulp Mill. Enzyme Microbial Technol 10(5):280. S. N. & Mor

Bajwa MA, Aziz T, Hashmi AS (1991) Production of fungal biomass protein from alkali-treated rice straw by *Arachniotus* sp. J Anim Plant Sci 1(2):79–81

Bakshi MPS, Jangar PN (1991) *Agaricus bisporus* sharvested spent wheat straw as livestock feed Indian. J Anim Sci 61(6):653–654

Bamberg JH (2000) British petroleum and global oil 1950–1975: the challenge of nationalism, vol. 15. Cambridge University Press, Cambridge, UK, pp 445–478

Banerjee UC, Chisti Y, Moo-Young M (1995) Effects of substrate particle size and alkaline pretreatment on protein enrichment by Neurospora sitophila. Resour Conserv Recycl 13(2):139–146

Bhalla TC, Joshi M (1994) Protein enrichment of apple pomace by co-culture of cellulolytic moulds and yeasts. World J Microbiol Biotechnol 10(1):116–117

Bhalla TC, Gajju H, Agrawal HO (1999) Production of single-cell proteins. In: Joshi VK, Pandey AK (eds) Biotechnology: food fermentation, microbiology, biochemistry and technology. Educational Publishers and Distributers, New Delhi, pp 1003–1022

Bhalla TC, Sharma NN, Sharma M (2007) Production of metabolites, industrial enzymes, amino acid, organic acids, antibiotics, vitamins and single cell proteins. National Science Digital Library, India

Bhalla TC, Mehta PK Savitri, Bhatia SK, Pratush A (2009) Microorganism for food and Feed. https://www.researchgate.net/…/303941522

Cabib G, Silva HJ, Giulietti A, Ertola R (1983) The use of sugar cane stillage for single cell protein production. J Chem Technol Biotechnol 33:21–28

Callihan CD, Clemmer JE (1979a) Biomass from cellulosic materials. In: Rose AH (ed) Microbial biomass-economic microbiology. Academic Press, London, pp 271–273

Callihan CD, Clemmer JE (1979b) Biomass from cellulosic materials. In: Rose A (ed) Microbial biomass-economic microbiology, vol 4. Academic Press, New York, pp 271–288

Capalbo FH, Moraes IO, Pelizer MH (2001) Solid-state fermentation of *Bacillus thuringiensis-tolworthi* to control fall armyworm in maize. Electron J Biotechnol 4(2):1–5

Castlla RB, Waehner RS, Giulietti AM (1984) Aerobic microbial treatment of sugarcane stillage by *Candida utilis* and *Paecilomyces variotii* in two step continuous cultures. Biotechnol Lett 121:195–198

Chahal DS, Moo-Young M, Vlach D (1981) Effect of physical and physicochemical pretreatments of wood for SCP production with *Chaetomium*. Biotechnol Bioeng 23:2417–2420

Chanda S, Chakrabarti S (1996) Plant origin liquid waste, a source for single cell protein production by yeast. Bioresour Technol 57:51–54

Chandrani-Wijeyaratne S, Tayathilake AN (2000) Characteristics of two yeast strain (*Candida tropicalis*) isolated from Caryotaurens (Khitul) toddy for single cell protein production. J Natl Sci Found Sri Lanka 28:79–86

Chaudhary N (2008) Designing microbial consortium for ethanol production from sugarcane bagasse. Ph.D. Thesis. University of the Punjab, Lahore

Chaudhary N, Sharma CB (2005) Production of citric acid and single cell protein from agrowaste. Natl Acad Sci Lett 28:189–193

Chen H, Liu J, Li Z (1999) Production of single cell protein by fermentation of extract from hemicellulose autohydrolyzate. Eng Chem Metall 20(4):428–431

Chen Q, Liu J, Cai H (2000a) Bioconversion of natural cellulose materials by multi-strains co-fermentation systems. J Tianjin Univ Commer 20(3):1–6

Chen Q, Liu J, Li T (2000b) Establishment of multi-strain co-fermentation system and bioconversion of maize stalk. Guangzhou Chemical Industry 28(4):27, 69–27, 73

Cooney CL (1986) Continuous culture: a tool for research, development and production. In: Alanl DI, Moo-Young M (eds) Perspectives in biotechnology and applied microbiology. Elsevier Applied Science, London, pp 271–276

Deibel MR, Hiebsch RR, Klein RD (1988) Secreted amylolytic enzymes from *Schwanniomyces occidentalis*: purification by fast protein liquid chromatography (FPLC) and preliminary characterization. Prep Biochem 18(1):77–120

Dhanda S, Kakkar VK, Garcha HS, Makkar GS (1994) Biological treatment of paddy straw and its evaluation through ruminant feeding. Indian J Anim Nutr 11(2):73–79

Dimmling W, Seipenbusch R (1978) Raw materials for the production of SCP. Proc Biochem 131:9–15

el-Saadany R, Khalaf H, el-Manawaty H, Salom F (1988) The production of single cell protein from agricultural wastes by fungi. Acta-Aliment. Acad Sci Hung 17(4):376–377

Ekerott L, Villadseer J (1991) Biotechnology 18:205–231

Ellis DH (1981) Ascocarp morphology and terminal hair ornamentation in thermophilic *Chaetomium* species. Mycologia 73:755–773

El-Nawwi SA, Abd El-Kader A (1996) Production of single-cell protein and cellulase from sugarcane bagasse: effect of culture factors. Biomass Bioenergy 11(4):361–364

Farstad L, Liven E, Fiatlandsmo E, Ness B (1975) Effect of feeding Pekilo single cell protein in various concentrations to growing pigs. Acta Agricultures Scandinavica

Ferrianti MP, Fiechter A (1983) Production and feeding of single cell protein. Applied Science Publications, London, ISBN: 085334231, p 201

Forage AJ, Righelato RC (1979) Biomass from carbohydrates. In: Rose AH (ed) Microbial biomass-economic microbiology. Academic Press, New York, pp 289–292

Gabriel A, Ntuli V, James D (2014) C1Cactus pear biomass, a potential lignocellulose raw material for Single Cell Protein production (SCP). Int J Curr Microbiol Appl Sci 3(7):171–197

Ghanem KM (1992a) Single cell protein production from beet-pulp by mixed culture. Microbiologia 8(1):39–43

Ghanem KM (1992b) Single cell protein production from beet pulp by mixed culture. Qatar Univ Sci J 12:85–88

Giec A, Skupin J (1988) Single cell protein as food and feed. Nahrung 32(3):219–229

Gonzales-Valdez A, Moo-Young M (1981) Production of yeast SCP from corn stover hydrolysate. Biotechnol Lett 3:148–153

Gupte A, Madamwar D (1997) Solid state fermentation of lignocellulosic waste for cellulase and beta-glucosidase production by cocultivation of *Aspergillus ellipticus* and *Aspergillus fumigatus*. Biotechnol Prog 13:166–169

Guven M, Cansunar E (1989) Single cell protein production and reduction of pollutants in waste sulphite liquor. Mikrobiyol Bul 23(4):329–335

Hahn-Hägerdal B, Karhumaa K, Fonseca C, Spencer-Martins I, Gorwa-Grauslund MF (2007) Towards industrial pentose-fermenting yeast strains. Appl Microbiol Biotechnol 74:937–953

Hashmi AS, Batajoo KK, Bajwa MA (1991) Bioconversion of rice straw into protein concentrate with Arachniotus sp. Proceedings of International Symposium on Biotechnology for Energy 16–21 Dec 1969. Faisalabad, pp. 149–155

Hongpattarakere T, Kittikun AH (1995) Optimization of single-cell-protein production from cassava starch using *Schwanniomyces castellii*. Biomed Life Sci 11:607–609

Hu Z, Que Y, Gao Y, Yin Y, Yufen Zhao Y (2015) Using black liquor from the soda pulping process for protein production by *Candida utilis*. Bioresources 10(3):3908–3921

Ibrahim Rajoka M, Kiani MA, Sohail TK (2004) Production of single cell protein from rice polishing using *Candida utilis*. World J Microbiol Biotechnol 20(3):297–301

Invarson KC, Morita H (1982) Single cell protein production by the acid tolerant fungus, *Scytalidium acidophilum* from acid hydrolysates of waste paper. Appl Environ Microbiol 43:643–647

Israelidis CJ, Coduonis MI (1982) Utilisation of agricultural wastes for animal feed and energy. Agr Res 6:243–253

Jaganmohan P, Purushottam B, Prasad SV (2013) Production of SCP with *Aspergillus terrus* using Solid State fermentation. Eur J Biol Sci 5(2):38–45

Jorgensen JB (2010) Systematic model analysis for single cell protein (SCP), production in a U-loop reactor. In: 20th European symposium on computer aided process engineering escape. Am- Eur J Agric Environ Sci 20:79–90

Kamel BS (1979) Utilization of date carbohydrate as substrate in microbial fermentation. Proc Biochem 14:95–100

Kim JH, Lebeault JM (1981) Protein production from whey using *Penicillium cyclopum*: growth parameters and cellular composition. Eur J Appl Microbiol Biotechnol 13:151–154

Kolani S, Delgenes JP, Moletta R, Traore A, Doh A (1996) Optimization of cell yield of *Candida krusei* SO1 and *Saccharomyces* sp. LK3G cultured in sorghum hydrolysate. Biores Technol 57(3):275–281

Kumar P, Barret DM, Delwiche MJ, Stroeve P (2009) Methods for pretreatment of lignocellulosic biomass for efficient hydrolysis and biofuel production. Ind Eng Chem Res 48:3713–3729

Litchfield JH (1979) Production of single cell protein for use in food and feed. In: Peppler HJ, Perlman D (eds) Microbial technology, vol. I, 2nd edn. Academic Press, New York, pp 93–35

Lo SN, Moreau JR (1986) Mixed culture microbial protein from waste sulfite pulping liquor II. Its production on pilot plant scale and use in animal feed. Can J Chem Eng 64:639–646

Mahasneh IA (1997) Production of single cell protein from five strains of *Chlorella* spp. (Chlorophyta). Cytobios 90:153–161

Miller TF, Srinivasan VR (1983) Production of single cell protein by *Aspergillus terreus*. Biotechnol Bioeng 25:1509–1519

Molina OE, Perotti de Gálvez NI, Frigerio CI, Córdoba PR (1984) Planta piloto de Procesos Industrials microbiologicos, PROIMI, Avda Belgrano y Pje Caseros, 4000 S.M. de Tucumán, Argentina

Mondal AK (2006) Production of single cell protein from fruits waste by using *Saccharomyces cerevisiae*. Am J Food Technol 58(117):134

Mondal AK, Sengupta S, Bhowal J, Bhattacharya DK (2012) Utilization of fruit wastes in producing single cell protein. Int J Sci Environ Technol 1:430–438

Moo-Young MY, Chisti U, Vlach D (1992) Fermentativc conversion of cellulosic substrates into microbial protein by *Neurospora gitophila*. Biotech Lett 1–l(9):863–868

Nasseri AT, Rasoul-Amini S, Morowvat MH, Ghasemi Y (2011) Single cell protein: production and process. Am J Food Technol 6(2):103–116

Nigam JN (2000) Cultivation of *Candida langeronii* in sugar cane bagasse hemicellulosic hydrolyzate for the production of single cell protein. World J Microbiol Biotechnol 16:367–158

Nigam P, Singh D (1994) Solid state (substrate) fermentation systems and their applications in biotechnology. J Basic Microbiol 34:405–414

Oura E (1983) Biomass from carbohydrates. In: Rehm H-J, Reed G (eds) Biotechnology, 3rd edn. Verlag Chemie, Weinheim, p 3

Pandey A, Soccol CR (1998) Bioconversion of biomass: a case study of lignocellulosic bioconversion in solid state fermentation. Brazilian Arch Biol Technol 1998(41):379–390

Paynor KA, David ES, Valentino MJG (2016) Endophytic fungi associated with bamboo as possible sources of single cell protein using corn cob as a substrate. Mycosphere 7(2):139–147

Perez JM, Rodriguez F, Alonso MV, Oliet M, Echeverria JM (2002) Characterization of a novolac resin substituting phenol by ammonium lignosulfonate as filler or extender. Bio Res 2:270–283

Pessoa A Jr (1991) Producao de biomassa microbiana a partir de hidrolisado hemicelulosico de cana-de aqficar. M.Sc. thesis. Universidade de Sao Paulo, Sao Paulo

Pessoa A Jr, Mancilha IM, Sato S (1996) Cultivation of *Candida tropicalis* in sugar cane hemicellulosic hydrolyzate for microbial protein production. J Biotechnol 51:83–88

Rao M, Vaema AJ, Deshmukh SS (2003) Production of single cell protein, essential amino acids, and xylanase by *Penivillium janthinellum*. Bio Resour 5:2470–2477
Rathoure AK (2014) In: Harzevili FD, Chen H (eds) Microbial biotechnology, progress and trends. CRC Press, Boca Raton, pp 279–296. https://doi.org/10.1201/b17587-13.,
Rhishipal R, Philip R (1998) Selection of marine yeast for generation of single cell protein from prawn shell wastes. Bioresour Technol 65:255–256
Rodriguez H, Gallardo R (1993) Single cell protein from bagasse pith by a mixed bacterial culture. Acta Biotechnol 13:141–149
Rodriguez H, Alvarez R, Enriquez A (1993) Evaluation of different alkali treatments of bagasse pith for cultivation of *Cellulomonas* sp. World J Microbiol Biotechnol 9:213–215
Rodriguez J, Ferraz A, Nogueira RF, Ferrer I, Esposito E, Duran N (1997) Lignin biodegradation by the ascomycete *Chrysonilia sitophila*. Appl Biochem Biotechnol 62(2–3):233–242
Rodriguez-Vazquez R, Diaz-Cervantes D (1994) Effect of chemical solutions sprayed on sugarcane bagasse pith to production single cell protein: physical and chemical analyses of pith. Bioresour Technol 47:159–164
Rodriguez-Vazquez R, Villanuevaventure G, Riosleal E (1992) Sugar cane bagasse pith dry pretreatment for single cell protein production. Bioresour Technol 39:17–22
Romantschuk H (1974) The Pekilo Process: proteins from spent sulphur liquor. In: Tannenbaunl SR, Wang DIC (eds) Single cell protein, vol. 2. MIT' Press, Cambridge, MA, pp 344–357
Romantschuk H (1976) The Pekilo process: a development project. 6th international symposium on continuous culture of microorganisms. Society of Chemical Industry
Romantschuk H, Lehtomaki M (1978) Operational experiences of first full-scale Pekilo SCP mill application. Process Biochem 3(16–19):29
Saliceti-Piazza L, Dale MC, Moelhman M, Ooks MR, Wankat PC (1992) Free and immobilized yeasts for BOD reduction in dairy wastes: growth on low levels of lactose, lactic acid and glycerol. Annual Meeting, Paper No. 165 k, AlChE; Miami FL, Scerra V, Caridi A, Foti F, Sinatra MC (1999) Influence of dairy *Penicillium* spp. on nutrient content of citrus fruit peel. Anim Feed Sci Technol 78:169–176
Samadi S, Mohammadi M, Najafpour Darzi G (2016) Production of single cell protein from sugarcane bagasse by *Saccharomyces cerevisiae* in tray bioreactor. IJE Trans B: App 29(8):1029–1036
Scerra V, Caridi A, Foti F, Sinatra MC (1999) Influence of dairy *Penicillium* spp. on nutrient content of citrus fruit peel. Anim Feed Sci Technol 78:169–176
Scrimshaw NS, Dillen JC (1977) Single cell protein as food and feed. In: Garattini S, Paglialunga S, Scrimshaw NS (eds) Single cell protein-safety for animal and human feeding. Pergamon Press, Oxford, UK, pp 171–173
Shahzad MA, Rajoka MI (2011) Single cell protein production from *Aspergillus terreus* and its evaluation in broiler chick. Int J Biosci, Biochem Bioinform 1(2):137
Shipman RH, Kao IC, Fan LT (1975) Single cell protein production by photosynthetic bacteria cultivation in agricultural byproducts. Biotechnol Bioeng 17:1561–1570
Shuler ML, Roberts ED, Mitchell DW, Kargi F, Austic RE, Henry A, Vashon R, Seeley HR (1979) Process for aerobic conversion of poultry manure into high protein feed stuff. Biotechnol Bioeng 21:19–38
Silva JBA (1991) Utilizar∗o do hindrolisado hemicelul6sico de eucalipto para produq∗o de protein microbiana. M.Sc. thesis. Universidade Federal de Viqosa, Viscosa
Sinclair CG, Cantero D (1990) Fermentation modelling. In: McNeil B, Harvey LM (eds) Fermentation: a practical approach. IRL Press, Oxford, UK, pp 65–113
Sindhu MS, Sandhu DK (1980) Single-cell protein production by *Trichoderma longibrachiatum* on treated sugarcane bagasse. Biotechnol Bioeng 22:689–692
Singh BD (1998) Biotechnology. Kalyani Publishers, New Delhi, pp 498–510
Singh A, Abidi AB, Darmwal NS, Agrawal AK (1988) Evaluation of chemical pre-treatment for biodegradation of agricultural lignocellulosic wastes by *Aspergillus niger*. MIRCEN J Appl Microb 4:473–479
Singh A, Abidi AB, Agarwal AK, Dharmwal NS (1991) Single cell protein production from *Aspergillus niger* and its evaluation. Zentralbl-Mikrobiol 146(3):181–184

Singhania AK, Soccol CR, Pandey A (2009) Recent advances in solid state fermentation. Biochemist Eng J\(667):789
Soland L (2005) Characterization of liquid mixing and dispersion in a U-loop fermentor. Am-Eur J Agric Environ Sci 67(99):109
Solomons GL (1985) Production of biomass by filamentous fungi. In: Blanch HW, Drew S, DIC W (eds) Comprehensive biotechnology. Pergamon Press, Oxford, UK, pp 483–505
Srividya AR, Vishnuvarthan VJ, Murugappan M, Dahake PG (2013) Single cell protein: a review. Int J Pharmaceut Res Scholars 2:472–485
Stanbury PF, Whitaker A, Hall SJ (2000) Principles of fermentation technology, 2nd edn. Elsevier, Oxford, UK
Staron TJ (1981) Production of proteins by mycological process. Br Pat 1604781
Steen A (2014) Production of single cell protein from residual streams from 2nd generation bio-ethanol production. Master of Science Thesis. Royal Institute of Technology, Sweden
Steinkraus KH (1986) Microbial biomass protein grown on edible substrates: the indigenous fermented foods. In: Moo-Young M, Gregory KF (eds) Microbial biomass protein. Elsevier Applied Science, London, pp 35–45
Suman G, Nupur M, Anuradha S, Pradeep B (2015) Single cell protein production: a review. Int J Curr Microbiol Appl Sci 4:251–262
Talebnia F (2008) Ethanol production from cellulosic biomass by encapsulated Saccharomyces cerevisiae, vol. 334. Ph.D. Thesis. Chalmers University of Technology, Gothenburg, pp 113–145
Tanaka M, Matsuno R (1985) Conversion of lignocellulosic materials to single cell protein, recent developments and problems. Enzym Microb Technol 7:197–206
Tannenbaum SR, Wang DIC (1975) Single cell protein, vol II. MIT Press, Cambridge, MA
Tengerdy RP (1985) Solid state fermentation. Trends Biotechnol 3(4):96–99
Tengerdy RP, Szakacs G (2003) Bioconversion of lignocellulose in solid substrate fermentation. Biochem Eng J 13:169–179
Tong Y, Wang X, Song S (1995) Exploitation study on sawdust used to feed. Exploit Farm Produce 10:20–21
Trehan K (1993) Biotechnology. Wiley Eastern Limited, New Delhi, pp 79–88
Ugalde UO, Castrillo JI (2002) Single cell proteins from fungi and yeasts. Appl Myco Biotechnol 186:123–149
Ukaegbu-Obi KM (2016) Single Cell protein: a resort to global protein challenge and waste management. J Microbiol Microb Technol 1(1):5
Vashista BR (1989) Botany for degree students-Algae. S. Chand and Co., Ltd., New Delhi, pp 503–514
Vibha, Sinha A (2005) Production of soluble crude protein using cellulolytic fungi on rice stubble as substrate under waste program management. Microbiol 33(3):147–149
Wang C, Bian Q (1999) Study on improving feed yeast quality and action efficiency. Feed Study 189:6–9
Wang C, Ding Y, Xiao C (2001) Production of high enzyme activity SCP from cellulose material. Ind Microbiol 31(1):30–33
Ware SA (1977) Single cell protein and other food recovery technologies from wastes. Municipal Environmental Research Laboratory Office of Research and Development, U.S. Environmental Protection Agency, Cincinnati
Weitzel W, Winchel M (1932) The yeast its nutritive and therapeutic value. Verlag Rothgiese und Diesing, Berlin
Wong PK, Chan K (1980) Algal single cell protein production from sewage effluents with high salinity. Experientia 36(9):1065–1066
Wu Q, Ma L (2002) Fermentation production of single cell protein by utilizing bagasse as the sole carbon source. J Hubei Agri Coll 22(2):150–152
Yakoub Khan M, Umar Dahot M (2010) Effect of various agriculture wastes and pure sugars on the production of single cell protein by Penicillium Expansum. World Appl Sci J 8:80–84
Yunus F, Nadeem M, Rashid F (2015) Single-cell protein production through microbial conversion of lignocellulosic residue (wheat bran) for animal feed. J Inst Brew 121(4):553–555

Zadrazil F, Puniya AK (1995) Studies on effect of particle size on solid state fermentation of sugarcane bagasse into animal feed using white rot fungi. Bioresour Technol 54:85–87
Zayed G, Mostafa N (1992) Studies on the production and kinetic aspects of single cell protein from sugar cane bagasse saccharified by *Aspergillus niger*. Biomass Bioenergy 3(5):363–367
Zhang D, Zhang R, Zhang T (2003) Preparation of cellulase and SCP by mixed fermentation of *Trichoderma viride* and *Candida utilis*. J Innermongolia Polytech Univ 22(3):180–185
Zhao L, Wang S, Ye H (2000) Application study on fish cultivating with yeast product from agricultural and forestry wastes. Feed Study 5:11–13
Zubi W (2005) Production of single cell protein from base hydrolyzed of date extract byproduct by the fungus *Fusarium graminearum*. M.Sc. Thesis, vol 19. Garyounis University, Benghazi, pp 167–200

Chapter 14
Light, Electromagnetic Spectrum, and Photostimulation of Microorganisms with Special Reference to *Chaetomium*

Abdelghafar M. Abu-Elsaoud and Ahmed M. Abdel-Azeem

14.1 The Nature of Light

Light is the ultimate source of energy for most of life on earth. Various living organisms can capture light energy and use it, e.g. plants and autotrophs can capture light energy and convert it through photosynthesis. However, light is more than the driver of energy-demanding metabolic activities; it also reports on the state of the environment through its quality (the balance of photons of different wavelengths), intensity (energy flux), and interaction with other environmental factors (Jones 2013).

The light behaviour when it travels through space and when it interacts with matter plays a central role in the two main paradigms of twentieth-century physics: relativity and quantum physics. It is also important to understand the behaviour and function of organisms (Björn 2015).

The developmental response of an organism to information in light, which may be its quantity, quality (i.e. wavelengths present) and direction or the relative length of day and night (photoperiod), is defined as *photomorphogenesis*. In order to respond to light, organisms must possess *photoreceptors*, molecules that absorb light and set in motion a cascade of events leading to biological responses (Jones 2013). The use of light wavelengths to activate biological process is called *photostimulation* and the light type used is the photostimulators.

A. M. Abu-Elsaoud (✉) · A. M. Abdel-Azeem
Botany Department, Faculty of Science, Suez Canal University, Ismailia, Egypt
e-mail: Abuelsaoud@science.suez.edu.eg; ahmed_abdelazeem@science.suez.edu.eg

A. M. Abdel-Azeem (ed.), *Recent Developments on Genus Chaetomium*,
Fungal Biology, https://doi.org/10.1007/978-3-030-31612-9_14

14.1.1 Electromagnetic Spectrum

The sun is the fundamental source of all non-nuclear energy on earth. The solar energy is the product of the sun's nuclear fusion, and the total annual solar energy absorbed by the earth's atmosphere, oceans and landmasses amounts to some 5.62 × 1024 joules, of which photosynthesis captures 3.16 × 1021 joules per year (Table 14.1). Visible light is just one part of the electromagnetic spectrum, which stretches from γ- and X-rays at one extreme through to radiowaves at the other. Light has the properties simultaneously of a wave and a particle. In simple terms it can be thought of as individual packets of energy or quanta that move in waves. A quantum of light energy is called a **photon**. The wavelength (λ, the Greek letter lambda) of visible light is usually expressed in nanometres (nm). The visible spectrum ranges from a wavelength of about 380 nm (violet) through to 760 nm (far red) (Fig. 14.1). Equation 14.1 expresses the relationship between wavelength (in meters) and frequency (υ, Greek letter nu; units = s^{-1}) and speed of light (c, units = $m\ s^{-1}$). Light has two main properties – **particles** and **waves**. The particle and wave properties of light are well demonstrated by a modification of Young's double-slit experiment (Jones 2013).

Relationship between wavelength, frequency, and light speed

$$c = \upsilon\lambda \tag{14.1}$$

Relationship between wavelength or frequency of electromagnetic radiations and energy

$$E = hc / \lambda \tag{14.2}$$

Table 14.1 Fate of solar power reaching earth (Jones 2013)

Global solar power balance	Amount in terawatts[a]
Solar power input[b]	178,000
Reflected to space immediately	53,000
Absorbed and then reflected as heat	82,000
Used to evaporate water	40,000
Captured by photosynthesis (net primary productivity)[c]	100
Total power used by human society	
In 2005	13
Projected use in 2100	46
Total used for food	0.6

[a]Watt is the unit of power and is related to the joule, the unit of energy, by watts = joules per unit time. A terawatt is 1012 watts and is equal to 1012 joules s^{-1}

[b]Total solar energy input per year = 5.62 Å~1012 terawatts (5.62 Å~1024 joules)

[c]Total solar energy per year captured by photosynthetic organisms = 3.16 Å~109 terawatts (3.16 Å~1021 joules)

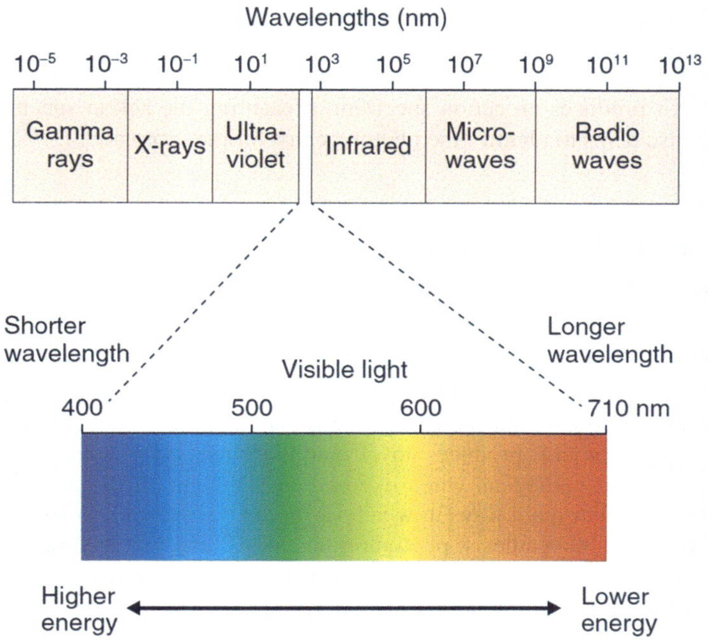

Fig. 14.1 The electromagnetic spectrum with the range from 400 to 710 nm expanded to show the colours of the visible wavelengths. The limit perception can extend beyond this range, as far as 380 nm at the blue and 760 nm at the red end. Note that energy is given in *J* mol^{-1} (Jones 2013)

where c = speed of light (approximately 300×10^6 m s^{-1}) and h = Planck's constant (4.14×10^{-15} eV.s). It follows that the energy of a given wavelength of light (in the commonly expressed unit of nm) is: $E = 1240/\lambda_{\nu\mu}$.

14.2 Photobiology: Interaction of Light with Living Organisms

The science of interaction between light wavelengths and life is called **photobiology**. In order for organisms to respond to light, organisms must have **photoreceptors**, molecules that absorb light and set in motion a cascade of events leading to biological responses (Jones 2013). The use of light wavelengths to activate biological process is called **photostimulation** and the light type used is the **photostimulators**. Light-absorbing photoreceptor molecules allow the organism to monitor environmental rhythms and fluctuations and to adjust its physiology and metabolism appropriately.

In humans and other animals, rhodopsin is one of the photoreceptors for vision. Plants and microorganisms have a number of different photoreceptors. These include the photosynthetic pigments, phytochromes, and cryptochromes and phototropins. Each photoreceptor has a characteristic absorption spectrum. The wavelengths of light that are absorbed by a photoreceptor activate specific responses. Plotting the intensity of a particular physiological response against the wavelengths that trigger it produces an action spectrum. Measuring the action spectrum for a photoresponse helps to identify the photoreceptor for the response.

14.3 Light, Electromagnetic Radiations, and Photostimulation

Electromagnetic radiations (EMR) can enhance production of bioactive secondary metabolites of actinomycetes; these include microwaves, ultraviolet light, visible light, and lasers. For example, actinomycete strains isolated from a volcanic cave in western Canada could produce novel antimicrobial compounds against six multidrug-resistant pathogens when exposed to UV light (Rule and Cheeptham 2013). There are four main ways in which EMR can be essential for life including: thermal effects, photosynthesis, photomorphogenesis, and mutagenesis. EMR has properties of both waves (where it has a wavelength) and of particles (where energy is transferred as quanta or photons).

14.3.1 Light Polarization

Light waves are transverse, i.e. the oscillation is perpendicular to the direction of wave propagation and the direction of the light (this is in contrast to sound waves, in which particles vibrate in the line of wave propagation). In the case of light, there are no vibrating particles but a variation in electric and magnetic fields. The electric and magnetic fields are both perpendicular to the direction of propagation and also perpendicular to one another. When the electric fields of all the components of a light beam are parallel, the beam is said to be plane-polarized (Björn 2015). The plane of polarization is the plane that contains both the electrical field direction and the line of propagation. If we add two beams which travel in the same direction and are both plane-polarized and have the same phase (i.e. the waves are in step) but different planes of polarization, the resulting light is also plane-polarized with its plane of polarization at an intermediate angle.

Light can also be circularly polarized, in which case the electrical field direction spirals along the line of propagation. Since such a spiral can be left- or right-handed, there are two kinds of circular polarization, left-handed and right-handed (Fig. 14.2). Circularly polarized light can be regarded as the sum of two equally strong plane-polarized components with right angles between the planes of polarizations and a 90Åã phase difference between the components. On the other hand, plane-polarized light can be regarded as a sum of equally strong left- and right-handed components of circularly polarized light (Björn 2015).

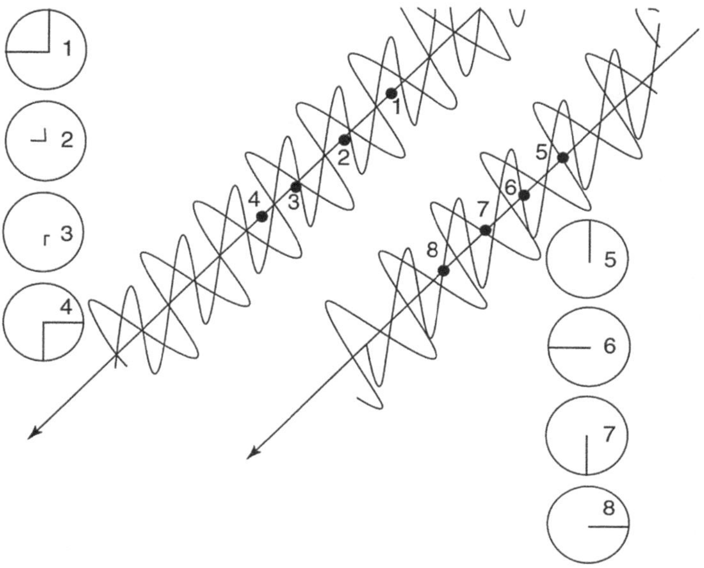

Fig. 14.2 In the upper left part of the figure, a plane-polarized light beam, composed of one vertically and one horizontally polarized component, is depicted in perspective and also "head-on" at different points (or at one point at different moments). Numbered points in the perspective drawing correspond to the numbers on the "head-on" drawings. Only the electric components of the electromagnetic fields are shown (wavy lines in the perspective drawing, straight lines in the "head-on" drawings). In the lower right part of the drawing, the same is shown for a circularly polarized beam

Natural light, such as direct sunlight, is often almost unpolarized, i.e. a random mixture of all possible polarizations. After reflection in a water surface, the light becomes partially plane-polarized. Skylight is a mixture of circularly and plane-polarized light, which we call elliptically polarized light. We cannot directly perceive the polarization of the light we see. Insects do and often use the polarization of skylight as an aid in their orientation. Plants in many cases react differently to plane-polarized light depending on its plane of polarization. This holds for chloroplast orientation in seed plants, mosses, and green algae and also for growth of fern gametophytes.

14.4 Generation and Control of Light for Photostimulation of Microorganisms

Almost all the natural light at the surface of the earth comes from the sun (this holds, of course, also for moonlight). The sun, on the whole, radiates as a glowing black-body at a temperature of about 6000 K. We have already mentioned the absence of some wavelength components from sunlight due to absorption in the outer cooler layers of the sun. Sunlight is further modified by the earth's atmosphere before it reaches ground level.

14.4.1 Lasers

Laser is an acronym for light amplification by stimulated emission of radiation. Stimulated emission occurs when a photon causes a molecule in an excited state to emit a second photon. Stimulated emission as such requires no special equipment. It occurs regularly when photons of the proper energy encounter excited molecules. However, as a rule, excited molecules are very rare compared to molecules in the ground state. To get light amplification by stimulated emission, we must have more stimulated emission than light absorption, which means that we must have more molecules in the proper excited state than in the ground state. This can be achieved in different ways but never by "direct lift" from the ground state by absorption of light. Various lasers employ indirect "optical pumping", electrical energy, or chemical reactions. For a laser to work, photon losses must also be minimized by a suitable optical configuration, often involving mirrors.

14.4.1.1 Laser Light Has some Unusual Properties

1. Laser is coherent in the sense that the light constitutes very long wave trains, contrary to ordinary light, where each photon is regarded as a limited wave packet independent of other photons.
2. Laser light can be made very collimated (all rays very parallel).
3. Laser light is usually very monochromatic (very narrow spectral bandwidth) or consists of a small number of such very narrow bands.
4. Laser light may be (but is not necessarily) plane-polarized.
5. The light from some types of lasers is given off in extremely short pulses of extremely high power (energy per time unit). However, this does not hold for all lasers. Some lasers emit light continuously and have very feeble power.

Even lasers of low power, such as the helium–neon laser, should be handled with some caution. This is because the beam is so narrow, parallel, and monochromatic that if it hits your eye all its power will be focused onto a very small area of your retina and blind that particular spot. One kind of laser that is in everyday use is the continuous helium–neon laser, emitting at 632.8 nm and a few infrared wavelengths. Dye lasers are advantageous in many cases because the wavelength can be selected over a wide range (within the fluorescence band of the dye used, and the dye can be changed if necessary). You may sometimes encounter a YAG laser. YAG is the acronym for yttrium aluminum garnet, containing trivalent neodymium ions in Y 3 Al 5 O 12. They are very powerful emitters of infrared radiation of 1060 nm wavelength. For photobiological purposes, they are sometimes combined with frequency doublers made of potassium phosphate crystals, so that green light of 530 nm wavelengths is obtained. The wavelength can be further changed either by letting the light undergo Raman scattering or by using it as a power source for a dye laser. Diode lasers are photodiodes emitting coherent light. They are now the most common lasers, used in CD players and other optical readout devices and laser pointers. They are available from 370 nm in the UV-A band to the long-wavelength red.

14.4.2 Light-Emitting Diodes

Light-emitting diodes (LEDs) are used in applications where very strong light is not needed, for instance, as indicator their red light is not easy to efficiently separate from the chlorophyll fluorescence. Now sufficiently intense blue-emitting diodes are also available. LEDs of several spectral emission types are presently manufactured: ultraviolet A, B, and C and blue, green, yellow, red, and infrared. It should be noted that they are not monochromatic light sources and especially the shortwave rated LEDs often have a broadband emission of longer wavelength than the nominal emission. For some types, the emission spectrum changes with operating current. LEDs are powered by a low-voltage source (e.g. a 1.5 V battery; some types need up to 5 V) in series with a resistor limiting the current to the rated value. Proper polarity should be observed.

Traditional LEDs contain inorganic semiconductors such as GaN, InGaN, SiC, and GaAsP (Table 14.2). Very recently, several laboratories and companies have started to develop organic light-emitting diodes (OLEDs), which will probably widen the range of spectral types available. Roithner also markets a range of infrared-emitting diodes with emission peak wavelengths greater than >4.5 μm. A LED emitting at 210 nm has been constructed (Taniyasu et al. 2006) but as of this writing is not yet commercially available. An interesting new development is the construction of a LED that can generate a single photon at a time (Yuan et al. 2002).

Table 14.2 Examples of LEDs and where to obtain them

Peak wavelength (nm)	Semiconductors	Company and URL
237–365	AlGaN/GaN	Sensor Electronic Technology, Inc.
		www.s-et.com
370–390	GaN	Nichia America
460	GaN	www.nichia.co.jp/en
470, 505, 525	SiC/GaP	LEDtronics
574, 595, 611	InCaAlP	www.ledtronics.com
630	GaAsP/GaP	
660	GaAlAs/G	
	AlAs	
660, 700, 720	GaAlAsP	Roithner Lasertechnik
780, 810, 905	GaAlAsP	http://www.roithner-laser.com/
375–1550	InGaN	Epitex
	AlGaAs/AlGaAs	http://www.epitex.com/
240–415	InGaAsP	Spectrecology
		http://www.spectrecology.com/
		(All accessed Aug. 17, 2013)

14.4.3 Ultraviolet Radiations

Short-wavelength UV light exhibits more quantum properties than its visible and infrared counterparts. Ultraviolet light is arbitrarily broken down into three bands, according to its anecdotal effects. UV-A is the least harmful and most commonly found type of UV light, because it has the least energy. UV-A light is often called black light and is used for its relative harmlessness and its ability to cause fluorescent materials to emit visible light, thus appearing to glow in the dark. Most phototherapy and tanning booths use UV-A lamps.

UV-B is typically the most destructive form of UV light, because it has enough energy to damage biological tissues, yet not quite enough to be completely absorbed by the atmosphere. UV-B is known to cause skin cancer. Since the atmosphere blocks most of the extra-terrestrial UV-B light, a small change in the ozone layer could dramatically increase the danger of skin cancer. Short-wavelength UV-C is almost completely absorbed in air within a few hundred meters. When UV-C photons collide with oxygen atoms, the energy exchange causes the formation of ozone. UV-C is almost never observed in nature, since it is absorbed so quickly. Germicidal UV-C lamps are often used to purify air and water, because of their ability to kill bacteria.

Solar radiation is essential to life on earth, but its UV component may also damage both living organisms and nonliving matter. UV radiation is usually divided into three wavelength bands: UV-A (315–400 nm), UV-B (280–315 nm) and UV-C (100–280 nm). UV-C radiation is potentially the most damaging, but is completely filtered out by the earth's atmosphere and does not reach the surface. The earth's surface is also largely protected from the most damaging short-wavelength UV-B radiation due to absorption by stratospheric ozone. UV-A radiation passes through the atmosphere with little attenuation and thus is the largest component of ground-level solar UV radiation. Although generally less harmful than UV-B radiation, UV-A radiation has important effects on tropospheric chemistry, air quality, and aquatic and soil processes, as well as being mutagenic and causing immune suppression in humans (Damian et al. 2011). Solar UV radiation, in particular UV-B, can be a positive regulator of plant defence systems against a broad spectrum of insect pests and pathogenic microorganisms (Williamson et al. 2014). UV radiations can have either negative or positive effects on microorganisms especially UV-A and UV-B (Fig. 14.3).

14.5 Photostimulation of Microorganisms

From previous electromagnetic radiations including lasers, light-emitting diodes, ultraviolet, etc., these can be used as photostimulators, as they can enhance bioprocesses in microorganisms. The low-intensity Nd-YAG laser radiation stimulates actinomycete individuals, such as *Streptomyces fradiae* isolated from cholesterol-rich materials to increase cholesterol decomposing activity (Ouf et al. 2012).

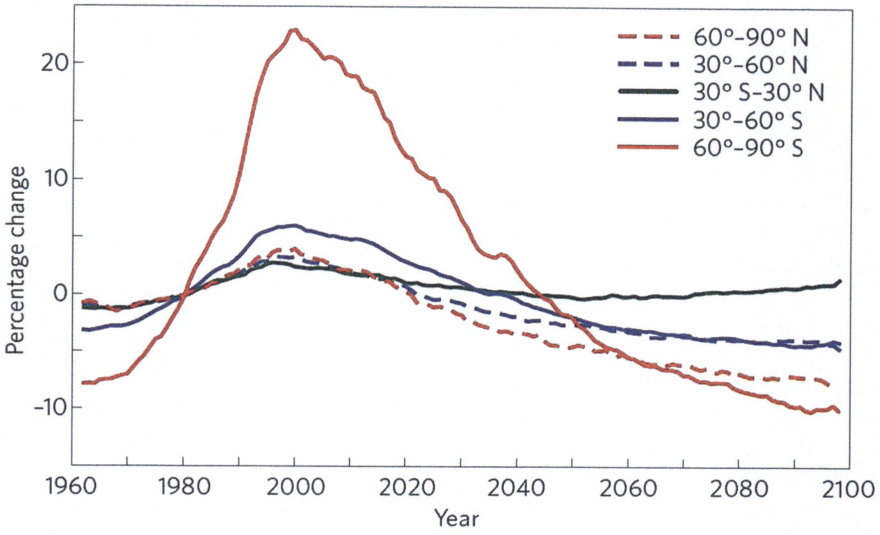

Fig. 14.3 Observed (pre-2010) and projected changes in annual mean erythemal (sunburning) clear-sky UV-B radiation at earth's surface, relative to 1980, for different latitude bands. (McKenzie et al. 2011; Bais et al. 2015; Williamson et al. 2014)

The low-power electromagnetic radiation is an emerging technology for the stimulation of cell activity. Low-intensity laser radiation with a wavelength of 400–500 nm accelerated cell division and enhanced protein synthesis in various microorganisms and stimulated the cholesterol decomposition of *Streptomyces fradiae* (Yew et al. 1982).

There is a large body of data about the utilization of laser radiation to inhibit the growth of microorganisms (Keates et al. 1988; Maisch et al. 2005; Sharma et al. 2008). However, there are some quantitative studies with organisms of different complexity improving the stimulating action of low-intensity laser radiation. The low-intensity Nd-YAG laser radiation stimulates actinomycete individuals, such as *Streptomyces fradiae* isolated from cholesterol-rich materials to increase cholesterol decomposing activity (Ouf et al. 2012). The low-power electromagnetic radiation is an emerging technology for the stimulation of cell activity. Low-intensity laser radiation with a wavelength of 400–500 nm and about 600 nm has brought about accelerated cell division and enhanced protein synthesis in various microorganisms. For example, our data indicate the importance of the photosensitizer in enhancement of laser radiation to stimulate cholesterol decomposition of *Streptomyces fradiae* (Yew et al. 1982). In a photostimulation study on endophytic fungi from medicinal plants, 13 species out of 22 entophytic fungi were screened for production of AgNPs, and photostimulation was carried out by red polarized laser and ultraviolet radiations. Reaction conditions such as silver nitrate concentration, pH, temperature, and efficiency of photostimulation using monochromatic red polarized light and UV radiations were optimized and assessed for high production of AgNPs (Abu-Elsaoud et al. 2015). High concentrations of AgNPs were produced by *Chaetomium globosum* and *Trichoderma viride* recovered from *Tanacetum sinaicum* and *Chiliadenus montanus*, respectively. Both *C. globosum* and *T. viride* showed significantly different response to photostimulation by either red polarized or red LED light. *T. viride* showed promising results and significant increase in AgNP production after photostimulation by monochromatic red polarized light and red light-emitting diodes (rLEDs) (Abu-Elsaoud et al. 2015).

The action of low-intensity laser radiation on *Escherichia coli* was reviewed by Tiphlova and Karu (1991), dealing with the quantitative laws of monochromatic visible light action on the cellular level, as well as with the primary photoacceptors and possible light signal transduction chains in bacterial cell *Escherichia coli* WP2 trp-. The existence of certain parameters of light (dose, intensity, pulse repetition rate, monochromaticity within the absorption bandwidth of biomolecules) for stimulation of *E. coli* WP2 metabolism has been established. The sensitivity of bacteria to irradiation with monochromatic visible light has been proposed to be a delta pH-dependent genetic process (Tiphlova and Karu 1991).

The activity of phenol hydroxylase in *Candida tropicalis* was promoted after irradiation with a He-Ne laser. In addition, the cell growth and intrinsic phenol biodegradation kinetics of mutant strain CTM 2 in batch cultures were also described by Haldane's kinetic equation with a wide range of initial phenol concentrations from 0 to 2600 mg liter^{-1}. The specific growth and degradation rates further demonstrated that the CTM 2 mutant strain possessed a higher capacity to resist phenol toxicity than wild *C. tropicalis* did.

In a study on *Phellinus igniarius* mutant screened through low-power He-Ne laser and ultraviolet (UV) induction, the mutant was then used to improve the production of endo-polysaccharides. These endo-polysaccharides exhibited stronger antioxidant activities in vitro, contained stronger hydroxyl radical scavenging capacity and showed higher Trolox equivalent antioxidant capacity (TEAC) (195.43 μmol Trolox/g sample) and ferric-reducing ability of plasma (FRAP) (20.57 μmol Fe2+/g sample) values compared with those of the CK. Therefore, the mutant screening through low-power He-Ne laser and UV induction could be an efficient and practical method for the development of the *Phellinus* strains and thus could improve the production and antioxidant activities of their endo-polysaccharides (Zhang et al. 2016).

The effect of enhanced levels of ultraviolet radiations especially UVA + UV + B on some aeromycobiota of Ismailia region, Egypt, was evaluated in an in vitro experiment (unpublished data). Two fungal species were tested – *Drechslera* sp. and *Paecilomyces* sp. Ultraviolet radiations (UV-A, UV-B) affected conidial structure (size) and biochemical consequences of both *Drechslera* sp. and *Paecilomyces* sp., and the level of UV-absorbing compound in *Paecilomyces* increased significantly after UV-A and UV-B irradiation when compared to control group, while for *Drechslera*, it showed a non-significant difference in UV-absorbing compounds with enhanced levels of UV radiations (Figs. 14.4 and 14.5). In addition to enhanced levels of mycosporine-like amino acids (MAAs).

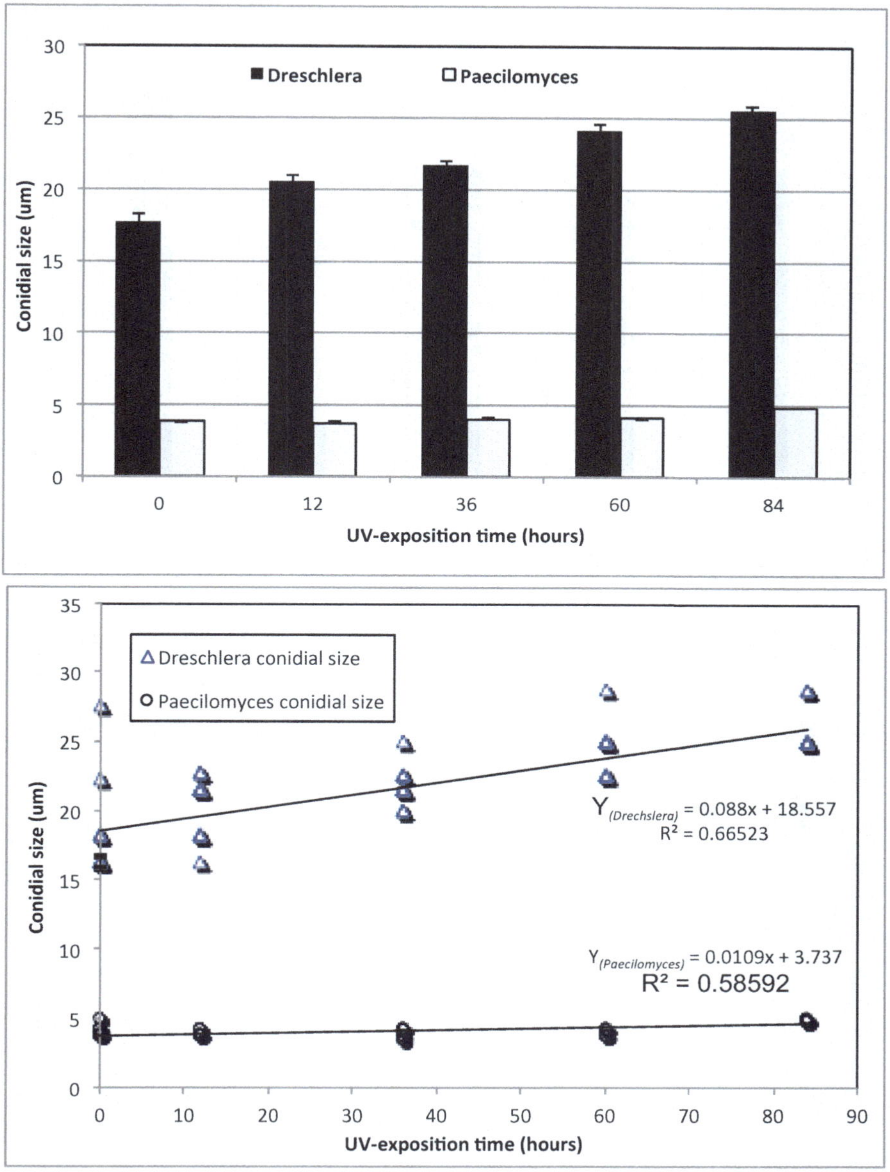

Fig. 14.4 Enhanced levels of ultraviolet radiations (UV-A, UV-B) increase the conidial size (μm) of *Drechslera* sp. and *Paecilomyces* sp. (Abu-Elsaoud and Abdel-Azeem unpublished data)

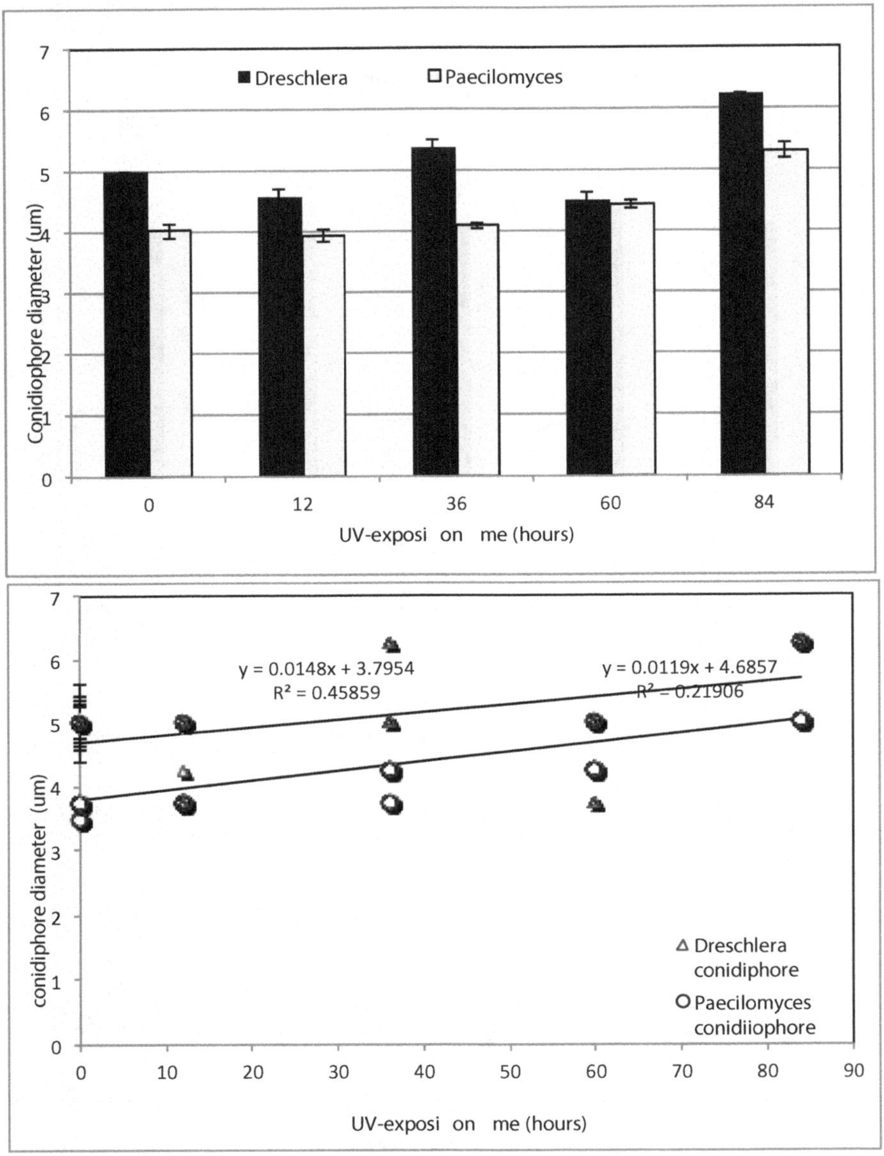

Fig. 14.5 Enhanced levels of ultraviolet radiations (UV-A, UV-B) increase the conidiophore diameter (μm) of *Drechslera* sp. and *Paecilomyces* sp. (Abu-Elsaoud and Abdel-Azeem unpublished data)

A study concerning assessment of two different photostimulation methods for enhancing the production of silver myco-nanoparticles based on NADPH-dependent nitrate was carried out using different irradiation doses (1, 5 and 30 s) of red polarized laser and red light-emitting diodes (r-LED) (Abu-Elsaoud et al. 2015). Monochromatic red polarized light with wavelength 632.8 nm induced a significant increase in the biosynthesis if AgNPs from *Trichoderma viride* cell extract based on both colour changes and UV-visible spectrophotometry. In *Chaetomium globosum*, monochromatic red polarized light did not affect AgNP biosynthesis significantly, and the relation with increasing doses of polarized light was negative, i.e. showing decreased biosynthesis of AgNPs with increasing doses of monochromatic red polarized light (Table 14.3). The differences in AgNP biosynthesis in both species were found to be significant at different wavelengths of UV-visible spectrophotometry. Red light-emitting diodes of wavelength ranging from 620 to 660 nm and maximum wavelength of 635 nm induced a significant increase in the AgNP biosynthesis especially for *Trichoderma viride*. Doses of 1-, 5- and 30-minute irradiation with

Table 14.3 Photostimulation of microorganisms with electromagnetic spectrum

EM radiation	Wavelength (λ; nm)	Photostimulation action	Microorganism	Reference
Nd-YAG laser	500–600	Accelerated cell division Enhanced protein synthesis Stimulated cholesterol degradation	Actinomycetes: *Streptomyces fradiae*	Ouf et al. (2012)
R-polarized laser	632.8	Improved biosynthesis of silver nanoparticles	Fungi: *Trichoderma viride* *Chaetomium globosum*	Abu-Elsaoud et al. (2015)
Red LED	600–700	Improved production of silver nanoparticles Increased activity of exogenous nitrase	Fungi: *Trichoderma viride* *Chaetomium globosum*	Abu-Elsaoud et al. (2015)
He-Ne laser				
UV-A + UV-B	320–400 280–320	Changed conidial structure Increased UV-absorbing compounds Increased mycosporine-like amino acids (MAAs)	Fungi: *Drechslera* sp. *Paecilomyces* sp.	(unpublished data) Abu-Elsaoud et al. (n.d.)
He-Ne laser	400–500, 600	Accelerated cell division Enhanced protein synthesis	*Various microorganisms* *Streptomyces fradiae*	
He-Ne laser		Increased activity of phenol hydroxylase	*Candida tropicalis*	Jiang et al. (2007)
He-Ne laser and UV		Enhanced production/ antioxidant activities of endo-polysaccharides	*Phellinus igniarius*	Zhang et al. (2016)

red LED light induced increase in the AgNP biosynthesis, while *C. globosum* doses of 1 minute had a promising effect on AgNP biosynthesis (Abu-Elsaoud et al. 2015).

On the other hand, ultraviolet radiations ranging from 200 to 400 nm are classified into three different levels: UV-A (320–400 nm), UV-B (280–320 nm) and UV-C (200–280 nm). The inactivation of microorganisms by ultraviolet (UV) light has been known for a long time and is used in many applications (Qualls and Johnson 1983; Zemke et al. 1990; Rames et al. 1997; Warriner et al. 2000; Lin and Blatchley 2001; Favier et al. 2001). It is known that sterilization efficiency of UV systems depends on different parameters like wavelength range and microbial concentration (Bourrouet et al. 2001). A wavelength ranging between 200 and 300 nm that corresponds to peak absorption of DNA is effective and the absorption of UV light by the DNA molecule causes death of microorganisms (Ishida et al. 1991). This is important for sterilization. All cells (both eukaryotic and prokaryotic) can repair their DNA defects. Photoreactivation induction by 350–600 nm lights is involved in these repair mechanisms (Favier et al. 2001; Schoenen and Kolch 1992).

Endophytic fungi are symbiotically associated biota of living plant tissues with symptomless disease to their host (Petrini 1991) and are not host specific (Cohen 2006). In the last few decades, scientists concentrate in their investigation to bioprospect natural chemical biological compounds, especially in extremes or hot biodiverse environments (Abdel-Azeem et al. 2012). Active metabolites from biological origin are produced by a large number of fungal species and most bioprospecting strategies were limited to some ecological groups of fungal species in Egypt (Abdel-Azeem 2010).

Nanotechnology has recently become one of the most active research fields in biology, chemistry, physics, mathematics, technology and engineering which are integrated to explore benefits of the nano-world towards the betterment of the society (Koopmans and Aggeli 2010). The dimension of matter important in nanoscience and nanotechnology is typically on the 0.2 nm to 100 nm scale (nanoscale). The properties of materials change as their size approaches the nanoscale (Eustis and El-Sayed 2006). Research in bio-nanotechnology has shown to provide reliable, eco-friendly processes for synthesis of noble nanomaterial. Biosynthesis of nanoparticles using fungal species has been targeted by many research projects (Ahmad et al. 2003; Hemath Naveen et al. 2010; Vahabi et al. 2011; San et al. 2012).

The applications of silver nanoparticles on different fields were of great importance and attracted attention of many researchers (Galdiero et al. 2011), e.g. applications in food technology, agriculture, biomedicine, and environmental technology (Rogers et al. 2008; Lu et al. 2008; Kumar et al. 2010; Baram-Pinto et al. 2010; Lara et al. 2010; Nameirakpam et al. 2012; Devi et al. 2013; Devi and Joshi 2014; Mustafa et al. 2013; Singh et al. 2015).

Recently the application of myco-nanotechnology for production of nanoparticles, as an alternative to chemical and physical ones, increased in the last decade (Alghuthaymi et al. 2015) even with some considerations concerning the environment (Mustafa et al. 2013). Endophytic anamorphic Ascomycota have been used for green synthesis of AgNPs during the last decade worldwide (Verma et al. 2010, Raheman et al. 2011; Nameirakpam et al. 2012; Devi et al. 2013; Devi and Joshi 2014; Qian 2013).

In Egypt very few studies were carried on the synthesis of AgNPs by endophytic fungi (Mustafa et al. 2013). Moreover, green synthesis optimizations combined with photostimulation studies were never reported up to our knowledge. In our continuous programme to find new biologically active natural products from Egyptian endophytic fungi, our teamwork with Suez Canal University screened various endophytic fungal species representing various selected habitats in Egypt especially arid Sinai. The aim of this work is directed to survey extracellular synthesis of AgNPs by endophytic fungi isolated from six wild medicinally important plants in Wadi Al-Arbaeen, Saint Katherine Protectorate and arid Sinai, Egypt, in addition to optimize the reaction conditions and study the efficiency of photostimulation providing the maximum possible biosynthesis of AgNPs.

References

Abdel-Azeem AM (2010) The history, fungal biodiversity, conservation, and future perspectives for mycology in Egypt. IMA Fungus 1:123

Abdel-Azeem AM, Salem FM, Mohamed HM, Rashad HM, Mohamed RM, Khalil WF (2012) Bioprospecting of Egyptian fungi for ecological and medical applications. Proc. 8th Meet. TWAS-ARO, vol. 31. In: Bibilotheca Alexandrina

Abu-Elsaoud AM, Abdel-Azeem AM, Elian M, Loutfy A (n.d.) Mycosporine like amino acids and UV-absorbing compounds production in fungi under enhanced levels of ultraviolet radiations. 2019: unpublished data

Abu-Elsaoud AM, Abdel-Azeem AM, Mousa SA, Hassan SS (2015) Biosynthesis, optimisation and photostimulation of α-NADPH-dependent nitrate reductase-mediated silver nanoparticles by Egyptian endophytic fungi. Adv Environ Biol 9(24):259–269

Ahmad A, Mukherjee P, Senapati S, Mandal D, Khan MI, Kumar R et al (2003) Extracellular biosynthesis of silver nanoparticles using the fungus *Fusarium oxysporum*. Colloids Surf B Biointerfaces 28:313–318. https://doi.org/10.1016/S0927-7765(02)00174-1

Alghuthaymi MA, Almoammar H, Rai M, Said-Galiev E, Abd-Elsalam KA (2015) Myconanoparticles: synthesis and their role in phytopathogens management. Biotechnol Biotechnol Equip 29:221–236. https://doi.org/10.1080/13102818.2015.1008194

Bais AF, McKenzie RL, Bernhard G, Aucamp PJ, Ilyas M, Madronich S, Tourpali K (2015) Ozone depletion and climate change: impacts on UV radiation. Photochem Photobiol Sci 14(1):19–52. https://doi.org/10.1039/c4pp90032d

Baram-Pinto D, Shukla S, Gedanken A, Sarid R (2010) Inhibition of HSV-1 attachment, entry, and cell-to-cell spread by functionalized multivalent gold nanoparticles. Small Weinh Bergstr Ger 6:1044–1050. https://doi.org/10.1002/smll.200902384

Björn LO (ed) (2015) Photobiology. Springer, New York, NY. https://doi.org/10.1007/978-1-4939-1468-5

Bourrouet A, García J, Mujeriego R, Peñuelas G (2001) Faecal bacteria and bacteriophage inactivation in a full-scale UV disinfection system used for wastewater reclamation. Water Sci Technol J Int Assoc Water Pollut Res 43:187–194

Damian DL, Matthews YJ, Phan TA, Halliday GM (2011) An action spectrum for ultraviolet radiation-induced immunosuppression in humans. Br J Dermatol 164(3):657–659. https://doi.org/10.1111/j.1365-2133.2010.10161.x

Devi LS, Bareh DA, Joshi SR (2013) Studies on biosynthesis of antimicrobial silver nanoparticles using endophytic fungi isolated from the ethno-medicinal plant *Gloriosa superba* L. Proc Natl Acad Sci India Sect B Biol Sci 84:1091–1099. https://doi.org/10.1007/s40011-013-0185-7

Devi LS, Joshi SR (2014) Evaluation of the antimicrobial potency of silver nanoparticles biosynthesized by using an endophytic fungus, *Cryptosporiopsis ericae* PS4. J Microbiol Seoul Korea 52:667–674. https://doi.org/10.1007/s12275-014-4113-1

Eustis S, El-Sayed MA (2006) Why gold nanoparticles are more precious than pretty gold: Noble metal surface plasmon resonance and its enhancement of the radiative and nonradiative properties of nanocrystals of different shapes. Chem Soc Rev 35:209–217. https://doi.org/10.1039/B514191E

Favier GL, Escudero ME, de Guzman AM (2001) Effect of chlorine, sodium chloride, trisodium phosphate, and ultraviolet radiation on the reduction of *Yersinia enterocolitica* and mesophilic aerobic bacteria from eggshell surface. J Food Prot 64:1621–1623. https://doi.org/10.4315/0362-028x-64.10.1621

Galdiero S, Falanga A, Vitiello M, Cantisani M, Marra V, Galdiero M (2011) Silver nanoparticles as potential antiviral agents. Molecules 16:8894–8918. https://doi.org/10.3390/molecules16108894

Hemath Naveen KS, Kumar G, Karthik L, Bhaskara Rao KV (2010) Extracellular biosynthesis of silver nanoparticles using the filamentous fungus *Penicillium* sp. Arch Appl Sci Res 2:161–167

Ishida H, Nahara Y, Tamamoto M, Hamada T (1991) The fungicidal effect of ultraviolet light on impression materials. J Prosthet Dent 65:532–535. https://doi.org/10.1016/0022-3913(91)90295-8

Jiang Y, Wen J, Jia X, Caiyin Q, Hu Z (2007) Mutation of *Candida tropicalis* by irradiation with a He-Ne laser to increase its ability to degrade phenol. Appl Environ Microbiol 73(1):226–231. https://doi.org/10.1128/AEM.00677-06

Jones RL (ed) (2013) The molecular life of plants. Wiley-Blackwell, Chichester, West Sussex\ Hoboken, NJ

Keates RH, Drago PC, Rothchild EJ (1988) Effect of excimer laser on microbiological organisms. Ophthalmic Surg 19(10):715–718

Koopmans RJ, Aggeli A (2010) Nanobiotechnology—quo vadis? Curr Opin Microbiol 13:327–334. https://doi.org/10.1016/j.mib.2010.01.012

Kumar V, Yadav SC, Yadav SK (2010) *Syzygium cumini* leaf and seed extract mediated biosynthesis of silver nanoparticles and their characterization. J Chem Technol Biotechnol 85:1301–1309. https://doi.org/10.1002/jctb.2427

Lara HH, Ixtepan-Turrent L, Garza-Treviño EN, Rodriguez-Padilla C (2010) Research PVP-coated silver nanoparticles block the transmission of cell-free and cellassociated HIV-1 in human cervical culture. J Nanobiotechnol 13:8. https://doi.org/10.1186/1477-3155-8-15

Lin LS, Blatchley ER (2001) UV dose distribution characterization using fractal concepts for system performance evaluation. Water Sci Technol J Int Assoc Water Pollut Res 43:181–188

Lu L, Sun RW-Y, Chen R, Hui C-K, Ho C-M, Luk JM et al (2008) Silver nanoparticles inhibit hepatitis B virus replication. Antivir Ther 13:253–262

Maisch T, Bosl C, Szeimies R-M, Lehn N, Abels C (2005) Photodynamic effects of novel XF porphyrin derivatives on prokaryotic and eukaryotic cells. Antimicrob Agents Chemother 49(4):1542–1552. https://doi.org/10.1128/AAC.49.4.1542-1552.2005

McKenzie RL, Aucamp PJ, Bais AF, Björn LO, Ilyas M, Madronich S (2011) Ozone depletion and climate change: impacts on UV radiation. Photochem Photobiol Sci 10(2):182–198. https://doi.org/10.1039/c0pp90034f

Mustafa G, Tahir H, Sultan M, Akhtar N (2013) Synthesis and characterization of cupric oxide (CuO) nanoparticles and their application for the removal of dyes. Afr J Biotechnol 12:6650–6660. https://doi.org/10.5897/AJB2013.13058

Nameirakpam ND, Dheeban SP, Sutha S (2012) Biomimetic synthesis of silver nanoparticles from an endophytic fungus and their antimicrobial efficacy. Int J Biomed Adv Res 3:409–415. https://doi.org/10.7439/ijbar.v3i5.365.

Qian H, Pretzer LA, Velazquez JC, Zhao Z, Wong MS (2013) Gold nanoparticles for cleaning contaminated water. J Chem Technol Biotechnol 88:735–741. https://doi.org/10.1002/jctb.4030

Qualls RG, Johnson JD (1983) Bioassay and dose measurement in UV disinfection. Appl Environ Microbiol 45:872–877

Ouf SA, Alsarrani AQ, Al-Adly AA, Ibrahim MK (2012) Evaluation of low-intensity laser radiation on stimulating the cholesterol degrading activity: part I. microorganisms isolated from cholesterol-rich materials. Saudi J Biol Sci 19(2):185–193. https://doi.org/10.1016/j.sjbs.2011.12.006

Petrini O (1991) Fungal endophytes of tree leaves. In: Andrews JH, Hirano SS (eds) Microbial Ecology Leaves. Springer, New York, pp 179–197

Rames J, Chaloupecký V, Sojková N, Bencko V (1997) An attempt to demonstrate the increased resistance of selected bacterial strains during repeated exposure to UV radiation at 254 nm. Cent Eur J Public Health 5:30–31

Raheman F, Deshmukh S, Ingle A, Gade A, Rai M (2011) Silver nanoparticles: novel antimicrobial agent synthesized from an endophytic fungus *Pestalotia* sp. isolated from leaves of *Syzygium cumini* (L). Nano Biomed Eng 3. https://doi.org/10.5101/nbe.v3i3.p174-178

Rogers JV, Parkinson CV, Choi YW, Speshock JL, Hussain SM (2008) A preliminary assessment of silver nanoparticle inhibition of Monkeypox virus plaque formation. Nanoscale Res Lett 3:129–133. https://doi.org/10.1007/s11671-008-9128-2

Rule D, Cheeptham N (2013) The effects of UV light on the antimicrobial activities of cave actinomycetes. Int J Speleol 42:147–153. https://doi.org/10.5038/1827-806X.42.2.7

San Y, Chan M, Don M, Selatan SP (2012) Instantaneous biosynthesis of silver nanoparticles by selected macro fungi. Aust J Basic Appl Sci 6:86–88

Schoenen D, Kolch A (1992) Photoreactivation of E. coli depending on light intensity after UV irradiation. Zentralblatt Hyg Umweltmed Int J Hyg Environ Med 192:565–570

Sharma M, Visai L, Bragheri F, Cristiani I, Gupta PK, Speziale P (2008) Toluidine blue-mediated photodynamic effects on staphylococcal biofilms. Antimicrob Agents Chemother 52(1):299–305. https://doi.org/10.1128/AAC.00988-07

Singh AK, Rathod V, Dattu Singh SN, Kulkarni P, Mathew J, ul Haq M (2015) Bioactive silver nanoparticles from endophytic fungus fusarium sp. isolated from an Ethanomedicinal plant *Withania somnifera* (Ashwagandha) and its antibacterial activity. Int J Nanomater Biostructures 5(1):15–19

Taniyasu Y, Kasu M, Makimoto T (2006) An aluminium nitride light-emitting diode with a wavelength of 210 nanometres. Nature 441(7091):325–328. https://doi.org/10.1038/nature04760

Tiphlova O, Karu T (1991) Action of low-intensity laser radiation on *Escherichia coli*. Crit Rev Biomed Eng 18(6):387–412

Vahabi K, Mansoori GA, Karimi S (2011) Biosynthesis of silver nanoparticles by fungus *Trichoderma reesei* (a route for large-scale production of AgNPs). Insciences J 1:65–79

Verma VC, Kharwar RN, Gange AC (2010) Biosynthesis of antimicrobial silver nanoparticles by the endophytic fungus *Aspergillus clavatus*. Nanomed 5:33–40. https://doi.org/10.2217/nnm.09.77

Warriner K, Rysstad G, Murden A, Rumsby P, Thomas D, Waites WM (2000) Inactivation of *Bacillus subtilis* spores on aluminum and polyethylene preformed cartons by UV-excimer laser irradiation. J Food Prot 63:753–757. https://doi.org/10.4315/0362-028x-63.6.753

Williamson CE, Zepp RG, Lucas RM, Madronich S, Austin AT, Ballaré CL, Norval M et al (2014) Solar ultraviolet radiation in a changing climate. Nat Clim Chang 4(6):434–441. https://doi.org/10.1038/nclimate2225

Yew DT, Lam ST, Chan YW (1982) Effect of low dose laser on the chorioallantoic culture of retinal pigment cells. Acta Radiol Oncol 21(6):433–438

Yuan Z, Kardynal BE, Mark Stevenson R, Shields AJ, Lobo CJ, Cooper K, Beattie NS, Ritchie DA, Pepper M (2002) Electrically driven single-photon source. Science 295(5552):102–105. https://doi.org/10.1126/science.1066790

Zemke V, Podgorsek L, Schoenen D (1990) Ultraviolet disinfection of drinking water. 1. Communication: inactivation of *E. coli* and coliform bacteria. Zentralblatt Hyg Umweltmed Int J Hyg Environ Med 190:51–61

Zhang H-N, Ma H-L, Zhou C-S, Yang Y, Yin X-L, Yan J-K (2016) Enhanced production and antioxidant activity of endo-polysaccharides from *Phellinus igniarius* mutants screened by low power He-Ne laser and ultraviolet induction. Bioact Carbohydr Diet Fibre 15:30. https://doi.org/10.1016/j.bcdf.2016.11.006

Chapter 15
Chaetomium as Potential Soft Rot Degrader of Woody and Papery Cultural Heritage

Ahmed M. Abdel-Azeem, Robert A. Blanchette, and Benjamin W. Held

15.1 Introduction

Wood deterioration is an essential process in the environment that recycles complex organic matter and is an integral component of life. These processes, however, also destroy historic wood that has been used as shelter, utility, and art resulting in a loss of valuable cultural properties from archaeological sites. Woods with natural resistance to microbial degradation were often used in ancient times for an application where wood was in contact with the ground, for shipbuilding and for other uses (Meiggs 1982). These extractive-rich woods helped to preserve the wood and resist microbial attack but even the most resistant woods are not immune from decomposition. Wood that persists for long periods of time is usually protected by an environment that limits microbial activity. These special conditions may allow wood to survive centuries or even thousands of years, but even in the most extreme environments, some physical and chemical modification of wood from biodeterioration takes place. What type of deterioration occurs and how these processes impact the wood are important questions that need consideration if wooden cultural properties are to be studied and properly preserved. Since there are relatively few wooden objects surviving from past civilizations, they are extremely valuable resources that deserve careful attention.

Chaetomium is a genus of filamentous fungi (phylum Ascomycota, class Sordariomycetes) encompassing species that typically possess densely setose, ovoid to pyriform ostiolate ascomata, clavate asci, and pigmented, one-celled ascospores (Samson et al. 2000). Species of *Chaetomium* are important in the decomposition of cellulose-rich materials.

A. M. Abdel-Azeem (✉)
Botany Department, Faculty of Science, Suez Canal University, Ismailia, Egypt
e-mail: Ahmed_abdelazeem@science.suez.edu.eg

R. A. Blanchette · B. W. Held
Department of Plant Pathology, University of Minnesota,
St. Paul, Minnesota, United States of America

A. M. Abdel-Azeem (ed.), *Recent Developments on Genus Chaetomium*,
Fungal Biology, https://doi.org/10.1007/978-3-030-31612-9_15

In this chapter we will discuss the different ways of wood degradation caused by fungi and also describe specific degradation found in archaeological wood from a variety of different terrestrial with special reference to genus *Chaetomium*.

15.2 Fungi in the Museum Environment

Contamination of pieces of art presented in exhibition rooms or stored in depots and their spoilage by fungi is not exceptional but rather frequent in old and newly built museums (Allsopp et al. 2004; Nittèrus 2000; Capitelli et al. 2009; Mesquita et al. 2009; Pangallo et al. 2007; Koestler et al. 2003). It is well known to mycologists that fungi are able to inhabit, to alter, and to degrade all types of organic and inorganic materials. However, most conservators and museum curators are not aware of this enormous deteriorative potential.

Historically, pieces of art were made of all types of organic materials, and these materials are again used for an authentic restoration or conservation of the objects in recent times: Paint was made of mineral pigments bound with organic binders such as egg yolk, casein, linseed, poppy seed, hempseed oil, Chinese wood oil, or different resins. Linen canvas clamped on wooden frames serves as painting ground and was often primed with rabbit skin glue before painting. Gold leaf on precious wooden or stucco frames was applied using organic glues, linseed or turpentine oil. Historic glues were based on cellulose or rabbit skin. Sculptures and other art objects carry décors made of textiles, leather, straw, clay, natural hair, or feathers. The most precious documents of humankind are books and scrolls made of paper, papyrus, and parchment. Because of the tremendous diversity of exo-enzymes produced by fungi – cellulases, glucanases, laccases, phenolases, keratinases, monooxygenases, and many more – and their remarkable ability to grow at low aw values, the preservation of museum objects is inevitably connected with prevention of mold, monitoring of fungi, and treatment of fungi on contaminated objects.

A compilation of fungi frequently occurring on paper, paintings, and other materials in museum is given in Table 15.1 after Sterflinger (2010). The data are based on more than 20 studies carried out in Austrian museums since 2000 and on data collected from different case studies in the literature. Most fungi playing a role in the deterioration of cultural heritage phylogenetically belong to Euascomycetes; hemiascomycetes – yeasts – are rarely isolated from art objects. Teleomorphs are rarely found and the only teleomorph genera that frequently occur are *Chaetomium* – mostly on paper, wood, and feathers – and *Eurotium*, in environments with low aw values. Occurrence of basidiomycetes is restricted to wood degradation in churches or other protected historical monuments. Zygomycetes are frequently isolated from pieces of art but in most cases they can be regarded as transients not being really established on the objects.

Fungal growth on objects of cultural heritage often causes a serious aesthetical spoiling due to colony formation and fungal pigments (Sterflinger et al. 1999). Moreover, fungi degrade materials and thus affect objects substantially: the

Table 15.1 Most recovered fungi from museums and on materials objects of arts[a,b,c,d,e]

Substrate	Genus
Paintings: oil, water color, acrylic	*Alternaria sp., Aspergillus flavus, Aspergillus* sect. *niger, A. sydowii, A. versicolor, Aureobasidium pullulans, Chaetomium funicola, Cladosporium herbarum, C. cladosporioides, Eurotium chevalieri, E. rubrum, Fusarium* sp., *Mucor* sp., *Penicillium chrysogenum, P. citrinum, P. decumbens*, and many other species of the genus
Paper (laid paper, wood pulp paper) and cellulose textiles (cotton, linen)	*Alternaria* sp., *Aspergillus clavatus, A. flavus, A. glaucus, A. terreus, A. repens, A. ruber, A. fumigatus, A. ochraceus, A. nidulans, Aspergillus sect. niger, Botrytis cinerea, Chaetomium globosum, C. elatum, C. indicum, Eurotium amstelodami, Fusarium* sp., *Mucor* sp., *Paecilomyces variotii, Penicillium chrysogenum, P. funiculosum, P. purpurogenum, P. rubrum, P. variabile, P. spinulosum, P. fellutatum, P. frequentans, P. citrinum, Pichia guilliermondii, Rhizopus oryzae, Stachybotrys chartarum, Toxicocladosporium irritans, Trichoderma harzianum, T. viride, Stemphylium* sp., *Ulocladium* sp.
Parchment	*Cladosporium cladosporioides, Epicoccum nigrum, Phlebiopsis gigantea, Penicillium chrysogenum, Thanatephorus cucumeris*
Keratinous substrates (leather, wool, feathers, fur, hair)	*Absidia glauca, A. cylindrospora, A. spinosa, Acremonium* sp., *Alternaria alternata, Aspergillus sydowii, A. candidus, A. clavatus, A. carneus, A. foetidus, A. flavus, A. fumigatus*, and many other species of the genus *Arthroderma* sp., *Aureobasidium pullulans, Chaetomium globosum, Chrysosporium* sp., *Coniosporium* sp., *Cladosporium cladosporioides, Cunninghamella echinulata, C. elegans, Epicoccum nigrum, Emericella* sp., *Geotrichum candidum, Mucor* sp., *Penicillium brevicompactum, Penicillium chrysogenum,* and many other species of the genus, *Phoma medicaginis, Scopulariopsis* sp., *Stachybotrys chartarum, Trichophyton* sp., *Rhizopus* sp.
Archeological findings: bones, ceramics	Archaeological findings often carry a large load of spores, in case of contamination the diversity on the objects reflects the diversity of the respective soil

The identification of the fungi was carried out based on morphology and/or sequencing of the ITS1, 5.8 S, ITS2 region with subsequent homology search using the BLAST algorithm [http://www.ncbi.nlm.nih.gov/BLAST/]

[a]Unpublished data Sterflinger/ACBR

[b]Mesquita et al. (2009)

[c]Meier and Petersen (2006)

[d]Blyskal (2009)

[e]Pangallo et al. (2009)

enzymatic degradation of organic binders causes reduction or even loss of paint layers. Fungi penetrate cracks and migrate underneath paint layers thus causing detachment. In paper conservation fungi are a special problem due to their ability to excrete cellulases. Lignin-degrading fungi are rarely observed in indoor environments, but considerable damage can be caused by the cellulose degraders *Serpula lacrymans* or *Coniophora puteana* in churches and other objects of cultural heritage if wooden altars or the roof structures are attacked. Also originals and museum reconstructions of historical buildings are considerably damaged by *S. lacrymans* (Bech-Andersen and Elborne 2004).

The development of fungi in museums is to a large extent determined by the indoor climate, the amount of available nutrients – from the atmosphere and from the materials themselves – and also the cleaning intervals in the museum. The indoor climate as indicated by temperature, relative humidity, and specific humidity is the most important factor for fungal growth. It is also closely related to the building's physical properties, especially the thermal insulation and the tendency to generate condensate from warm indoor air on cold walls of the building envelope (Camuffo 1998). Depending on the climate in the museum or storage rooms, the fungal diversity is restricted to few xerophilic and xerotolerant species such as *Eurotium* sp., *Aspergillus* sp., or *Wallemia* sp. Only in storage rooms where the humidity is raised to more than 70% for a period of several weeks or month is a high fungal diversity able to establish. The climate ranges allowing fungal spores to germinate and that restrict the growth of the fungal mycelium are shown in the isopleth systems by Sedlbauer and Krus (2003). The authors also show that hygroscopic materials support the growth of fungi at low relative humidity and that the water demand depends on the biodegradability of the substrate. The objects influence the development of the fungal community by their chemical composition and biodegradability for species with different exo-enzymes.

In museums the range of 55% Relative Humidity (RH) is generally regarded as the borderline for fungal growth and thus climate control is adjusted below this value. In fact, fungi that are able to survive at a relative humidity of 55% are rare and restricted to extreme environments such as hot and cold deserts. So why do mesophilic hyphomycetes frequently occur in museums? All museums in the world measure temperature and humidity in storage and exhibition rooms by means of modern data logger, data writers, or simple hygro- and thermometers. However, the way and location of climate measurements are often insufficient to reflect the real climate and to detect different climatic zones in the building. In his book on microclimate in museums, Camuffo (1998) illustrates the complexity of climate monitoring that cannot just be monitored by a single data logger in the middle of a museum room. The influence of airstream through doors, warming by sunlight, and daily changes of temperature gradients as well as the isolation and exposition of the building envelope have to be considered as important factors. In fact, fungal growth mostly happens between shelves with little aeration or near to walls with temperatures below the dew point. Micro-niches are often also created by wrapping of single objects into plastic foils or extremely tight boxes not allowing an exchange of air and vapor.

The mycobiota in museums is also influenced by the atmospheric particulate matter carrying carbonates, minerals, and others. Gysels et al. (2004) have shown in a study on the Royal Museum of Fine Art in Antwerp that the indoor aerosols were largely determined by the outdoor atmosphere and the outdoor sources of organic and inorganic pollutants. Fungi are well able to degrade different types of organic pollutants including polycyclic aromatic hydrocarbons (PAHs). Therefore, the fungal diversity on monuments in an urban environment was found to be much higher than in a rural environment of the climatic zone (Sterflinger and Prillinger 2001).

As a common species of *Chaetomium*, *Chaetomium globosum* can utilize a variety of carbohydrates as sources of carbon (Table 15.2) for the listing of the sources based on testing conducted at Health Canada.

Table 15.2 Carbohydrate utilization used for taxonomic identification of *C. globosum* strain ATCC 6205, based on the RapID™ YEAST PLUS system

Carbohydrate[a]	Result
Glucose	–
Maltose	–
Sucrose	–
Trehalose	–
Raffinose	–
Fatty acid ester	–
p-Nitrophenyl-N-acetyl-β,D-galactosaminide	+
p-Nitrophenyl-α,D-glucoside	–
p-Nitrophenyl-β,D-glucoside	+
o-Nitrophenyl-β,D-galactoside	–
p-Nitrophenyl-α,D-galactoside	+
p-Nitrophenyl-β,D-fucoside	–
p-Nitrophenyl phosphate	+
p-Nitrophenyl phosphorylcholine	–
Urea	+
Proline β-naphthylamide	+
Histidine β-naphthylamide	+
Leucyl-glycine β-naphthylamide	–

[a]The Thermo Scientific™ RapID™ YEAST PLUS System was used to evaluate sugar consumption of *C. globosum* strain ATCC 6205. *C. globosum* strain ATCC 6205 was inoculated into wells containing various carbon sources and incubated at 28 °C. The wells were evaluated after 48 hours

"+" means that the test result was positive and "–" means that the test result was negative

Prokhorov and Linnik (2011) reported *C. globosum* growth studies on wort agar, or mineral medium, and either glucose, saccharose, mannite, lactose, amylum or cellulose as carbon sources. For all carbon sources, fungal colony sizes increased on all media beyond three days and ascocarps formed on all media. In the environment, *C. globosum* uses natural sources such as cellulose in straw (Harper and Lynch 1985), the wood of the European beech (*Fagus sylvatica*) (Mohtashamipur and Norpoth 1990), heartwood and sapwood from the brazilwood tree (*Caesalpinia echinata*) (Silva et al. 2007), and fungal cell wall (*Pythium ultimum*) (Inglis and Kawchuk 2002). It is also capable of utilizing keratin in feathers (Kaul and Sumbali 1999) and collagen in leather (Strzelczyk et al. 1989).

15.3 Structural and Chemical Features of Wood

Wood consists of an orderly arrangement of cells with walls composed of varying amounts of cellulose, hemicellulose, and lignin. The great diversity of woody plants is reflected in the varied morphology and chemical composition of their wood. Typically, two general groups, *hardwoods* (angiosperms) and *softwoods* (gymno-

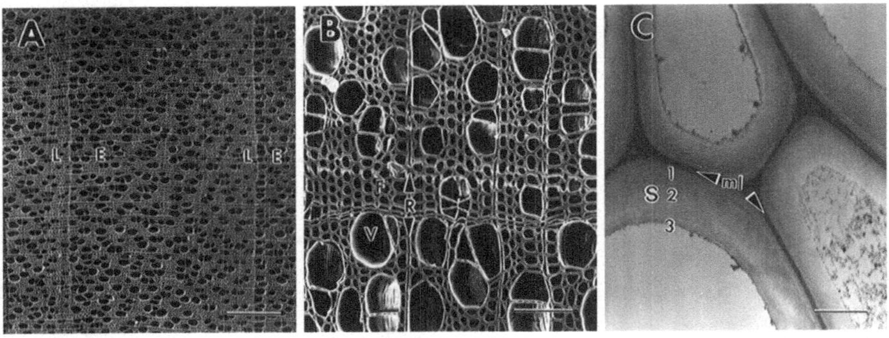

Fig. 15.1 Cell structure of an angiosperm. (**a** and **b**) Sections of a diffuse porous hardwood showing earlywood (E) and latewood (L). The wood consists of vessels (V), fibers (F), and ray parenchyma cells (R). (**c**) Cell walls with secondary wall layers (S1, S2, and S3) and middle lamellae (ml). Transverse sections. A and B SEM, C TEM. Bar = 500 μm in A, 100 μm in B, and 2 μm in C (Blanchette 2000)

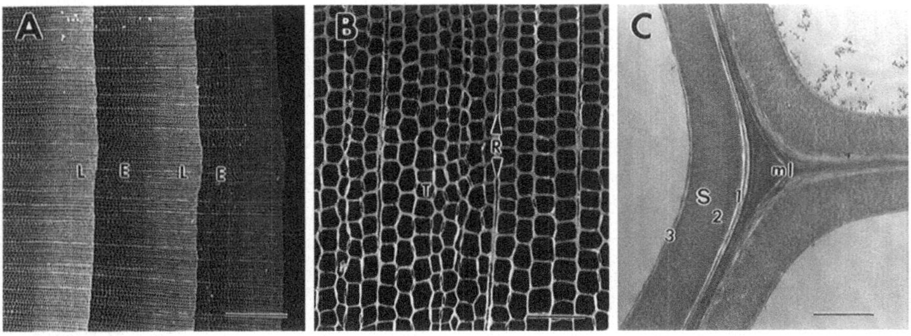

Fig. 15.2 Cell structure of a gymnosperm. (**a** and **b**) Thin-walled earlywood or springwood (E) and thick-walled latewood (L) tracheids. (**b**) Zone of earlywood showing tracheids (T) and ray parenchyma cells (R). (**c**) A group of tracheid cell walls showing secondary wall layers (S1, S2, and S3) and middle lamella (ml). Transverse sections. A and B SEM, C TEM. Bar = 500 μm in A, 100 μm in B, and 2 μm in C (Blanchette 2000)

sperms), can be easily separated. Hardwoods have pores or vessel elements that occur among fiber and parenchyma cells (Fig. 15.1). Cellulose content ranges from 40% to 50% with 15–25% lignin and 15–25% hemicellulose. The remaining components consist of various extracellular compounds.

Softwoods are composed of overlapping tracheids, connected by bordered pit apertures, and parenchyma cells and, in some cases, resin canals (Fig. 15.2). Greater concentrations of lignin, about 5–10% more than hardwoods, are found in softwoods and about the same amount of cellulose 40–50%. Less hemicellulose may be found in softwoods than hardwoods. The chemical composition of softwoods is also different from hardwoods with different types of lignin (primarily guaiacyl propane units), hemicelluloses (mannose is the most common constituent), and wood extractives (different terpenes, fatty acids, etc.).

15.4 Degradation of Wood by Fungi

Microbes that degrade wood produce extracellular enzymes that break down the woody cell wall. Growth characteristics of the microorganism in wood and the type of degradative system produced results in different decay patterns being produced (Blanchette 1995, 1998). Depending on the type of decay, different physical, chemical, and morphological changes occur in wood. These decay processes have been well characterized and provide useful insights to elucidate deterioration in archaeological woods (Table 15.3). A review of decay patterns produced by different fungi suggests that three categories can be used to separate the types of decay produced in wood. Names for these categories are based on visual characteristics of the advanced decay.

Two major groups of decay produced by fungi taxonomically classified in the subdivision *Basidiomycota* are *white* and *brown* rot fungi. *White rot fungi* can degrade all cell wall components, including lignin. They often cause a bleaching of normal wood coloration. Their ability to metabolize large amounts of lignin in wood is unique among microorganisms. The thousands of species that cause white rots are a heterogeneous group that may degrade greater or lesser amounts of a specific cell wall component. Some species preferentially remove lignin from wood leaving pockets of white, degraded cells that consist entirely of cellulose, while others degrade lignin and cellulose simultaneously (Blanchette 1991). Degradation is usually localized to cells colonized by fungal hyphae and substantial amounts of undecayed wood remains. A progressive erosion of the cell wall occurs when components are degraded simultaneously (Fig. 15.3) or a diffuse attack of lignin may occur by species that preferentially remove lignin (Blanchette 1991). Strength losses are not significant until late stages of decay (Cowling 1961; Zabel and Morrell 1992).

Table 15.3 Microbes that cause wood degradation (Blanchette 2000)

Microorganism	Wood components utilized	Decay characteristics
Fungi		
White-rot	All cell wall components, some species preferentially attack lignin	Progressive erosion of all cell wall layers. Middle lamella is degraded
Brown rot	Carbohydrates, some lignin modification	Diffuse depolymerization of cellulose
Soft-rot	Carbohydrates, some lignin modification	Type 1 – cavities form in secondary wall Type 2 – progressive erosion of secondary walls, but middle lamella is not degraded
Bacteria		
Erosion	Carbohydrates, extent of lignin modification not known	Erosion troughs leaving large quantities of residual wall material
Tunneling	Carbohydrates and some lignin	Minute tunnels in secondary walls and middle lamella
Cavitation	Carbohydrates, extent of lignin modification not known	Cavities form in secondary wall leaving residual wall material
Scavengers	Primary organisms: pit membranes but no cell wall decay. Secondary organisms: utilizes modified cell wall components	Primary organisms penetrate pits and degrade wood extractives. Secondary bacteria scavenge altered wood components

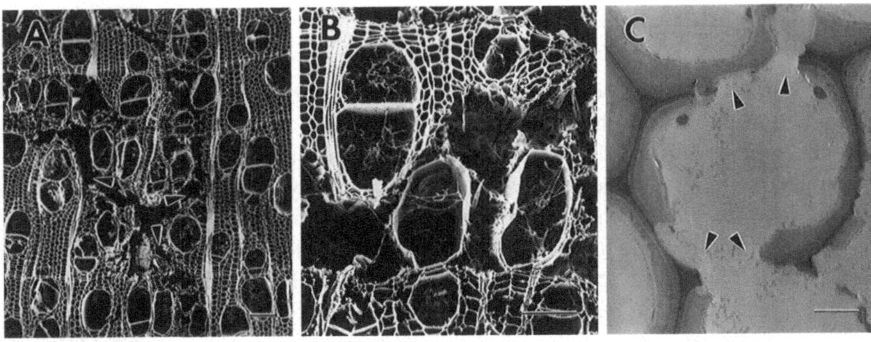

Fig. 15.3 Nonselective attack of wood by a white rot fungus. (**a** and **b**) All cell wall components are degraded resulting in voids within the degraded wood. (**c**) A progressive attack of all cell wall components causing a localized erosion of the secondary wall layers and middle lamellae (arrowheads). Transverse sections. A and B SEM, C TEM. Bar = 100 μm in A and B, 5 μm in C (Blanchette 2000)

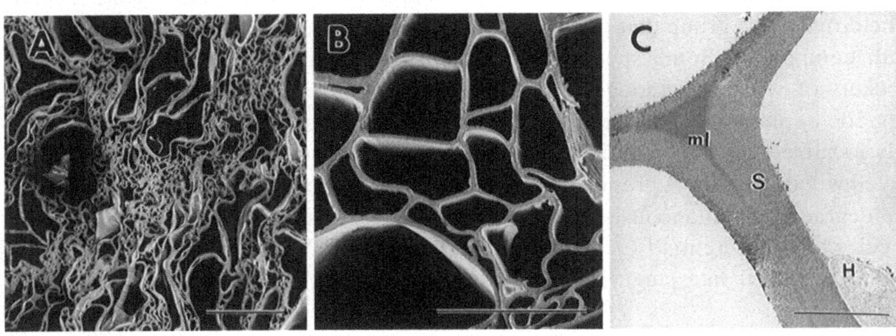

Fig. 15.4 Decay of wood by brown rot fungi. (**a**) Degradation of cellulose in woody cell walls leaves a residual network of lignin. Cell walls collapse and appear distorted. (**b**) Degraded cells showing walls that are porous and fragile. (**c**) Ultrastructure of a cell consisting primarily of residual lignin that has little binding strength and is loosely held together. H hypha, S secondary wall, ml middle lamellae (Blanchette 2000)

White rot fungi are common parasites of heartwood in living trees and are aggressive decomposers of woody debris in forest ecosystems (Blanchette 1991; Rayner and Boddy 1988).

Brown rot fungi depolymerase cellulose rapidly during incipient stages of wood colonization. Considerable losses in wood strength occur very early in the decay process, often before decay characteristics are visually evident (Wilcox 1968). Cell wall carbohydrates are degraded extensively during decay leaving a modified, lignin-rich substrate (Fig. 15.4). The residual wood is brown and often cracks into cubical pieces when dry. Brown rot fungi commonly cause decay of timber in buildings, and these fungi have had serious impact on ancient and historic buildings (Jennings and Bravery 1991). One of the most destructive brown rot fungi is *Serpula lacrymans* (previously called *Merulius lacrymans*). This fungus has become well adapted to attacking timber in service and can spread rapidly on wood and traverse nonnutritional surfaces. Aerial mycelia differentiate into thick strands that allow the fungus to invade new substrates. The mycelial strands also act to transport moisture and nutrients to considerable distances (Jennings 1991). Commonly, this type of decay has been referred to as dry rot. This term, apparently first used to describe any deterioration of dead wood or wood in service (Britton 1875), is misleading because moisture must be present for the decay to occur.

15.5 Wood Cell Wall Degradation by Soft Rot Fungi

Soft rot is caused by a diverse range of microfungi from the ascomycetes (teleomorphic and anamorphic) and has ubiquitous distribution throughout the world. Soft rot can have severe economic consequences since several of the species involved attack wooden constructions including preservative-treated wood under terrestrial (Fig. 15.5a) and aquatic (marine and sweet water) situations (Gersonde and Kerner-Gang 1976; Leightley 1977; Leightley and Eaton 1978; Zabel et al. 1985; Daniel and Nilsson 1989). For example, soft rot has been documented as one of the major forms of attack of preservative-treated (i.e., with waterborne preservatives like copper–chrome–arsenic and creosote) utility poles (Gersonde and Kerner-Gang 1976; Henningsson and Nilsson 1976; Zabel et al. 1985). Visually, soft-rotted wood becomes gray and even black in advanced stages of decay due to colonization by fungi hyphae in the wood. The surface of soft-rotted wood also tends to crack in a similar way as brown rot. The term "soft rot" was originally used to describe the decay from waterlogged wood, which was soft to touch (Savory 1954). However, this definition is not correct, and wood from terrestrial environments can be very heavily degraded but still remain hard as, for example, seen with utility poles. When such poles are strained, typical brash fractures are produced in contrast to that of brown rot, which produces brittle fractures due to depolymerization of the cellulose. Thus, soft rot can give rise to significant reductions in the strength of wood at comparable small weight losses although not to the level of brown rot.

Soft rot decay of wood is strongly affected by both lignin type and concentration in contrast to brown rot fungi and more significantly than white rot decay. Thus, softwoods (e.g., spruce and pine) show greater resistance to attack than hardwoods especially low lignin hardwoods (e.g., aspen and birch beech). In addition, at the cellular level in hardwoods, the variation in microdistribution of syringyl and guaiacyl lignin that occurs between cell types (vessels/fibers/ parenchyma) very often creates variations in decay morphology (Daniel and Nilsson 1998).

15.5.1 Morphological Aspects of Cell Wall Decay by Soft Rot Fungi

Soft rot produces two distinct forms of attack of wood decay known as Type I in which characteristic cavities are produced in cell walls and Type II where hyphae localized in the cell lumina cause cell wall erosion (i.e., cell wall thinning) (Figs. 15.5, 15.6, and 15.8) (Table 15.4). In particular, cavity formation is very

Fig. 15.5 Soft rot decay of wood. (**a**) Utility pole with outer regions degraded by soft rot. (**b**, c) Soft rot Type I with cavities orientated along the cellulose microfibrils in the S2 layer. (**d**, **e**) Transverse sections of pine at early (cavities) and late stages of soft rot Type I. At late stages, only the middle lamellae and S3 layers remain in softwoods. Transverse sections of pine (**f**) and birch (**g**) showing soft rot cell wall erosion (Type II). Ultimately, only the middle lamellae remain (Daniel 2016)

diagnostic for soft rot decay although many of the soft rot fungi cause both decay types simultaneously, sometimes in the same cells. A few species only cause Type II attack (Nilsson 1973). Type I decay results in the formation of characteristic cavities produced by hyphae within secondary cell walls that align along the cellulose microfibrils (Figs 15.5b–d), their presence best observed using longitudinal sections under polarized light (Fig. 15.5c). Type I decay is initiated by specialized microhyphae (~0.5 mm) or fine penetration hyphae that grow from the cell lumen into the S2 layer of wood cell walls (Leightley 1977; Crossley 1979; Hale and Eaton 1985a, b; Daniel and Nilsson 1989).

Such hyphae can penetrate directly through adjacent cells without forming a cavity and expand again to the native size in the adjacent lumen (i.e., rather like blue-stain fungi). It is only when the hypha reorientates itself in the S2 layer by forming either a T-branch or L-bend that cavity formation is initiated. When orientated along the

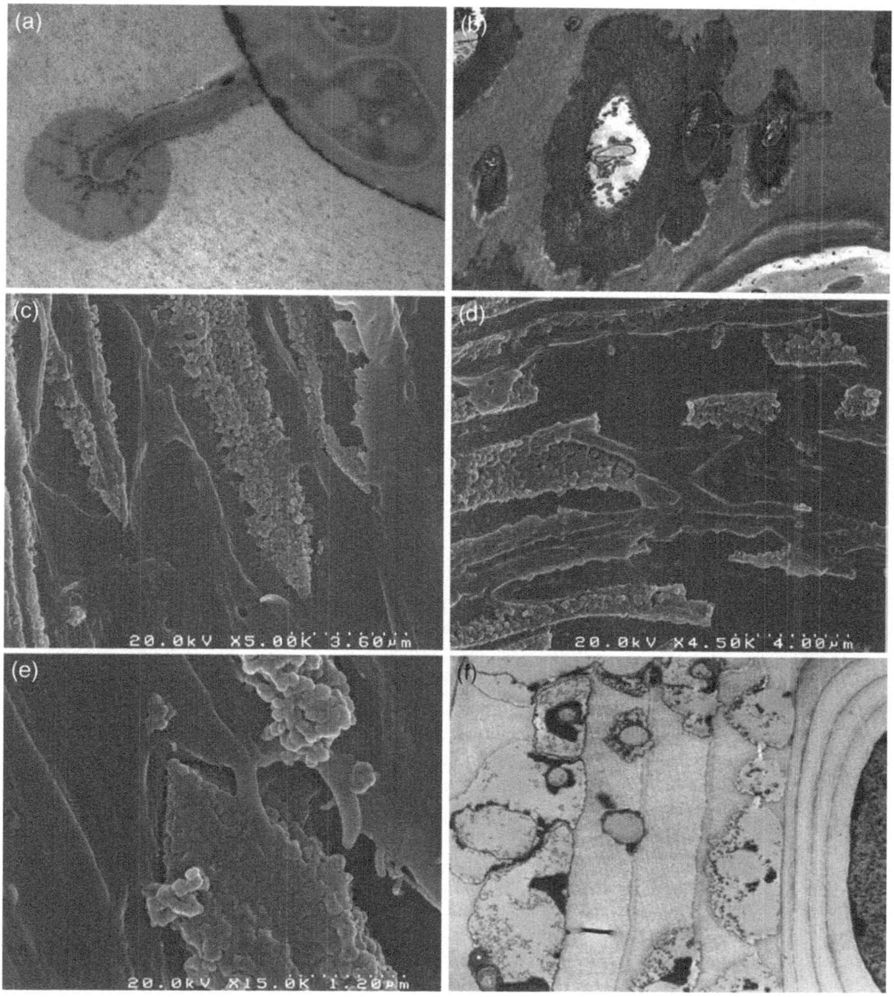

Fig. 15.6 Soft rot Type I. TEM sections through birch (**a**), pine (**b**), and *H. foetidum* (**f**) wood degraded by soft rot Type I. Sites of Tbranching, multiple Tbranching, and cavity formation in the S2 layer (**a**, **b**). (**c**–**e**) SEM images of soft rot cavities in S2 (**c**, **e**) and S1 layers (**d**). Hyphae are covered in melanin deposits and lignin breakdown products. Multilayered fiber of *H. foetidum* showing effect of lignin on soft rot attack with halfmoon-shaped cavities produced against thin concentric layers containing high lignin in the S2 layer (**f**) (Daniel (2016)

cellulose microfibrils in the cell wall, the hypha stops growing and produces a cavity (Hale and Eaton 1985a, b; Daniel and Nilsson 1998). When the cavity has enlarged, growth can be resumed by the hypha either from one end (L-bending) or from both ends (T-branching) simultaneously. After the cavities have enlarged, the process repeats itself and chains of cavities are produced (Figs 15.5b, c). The process of cavity formation has been very well documented and the size and shape of the cavities are known to vary with fungal species and wood type (Daniel and Nilsson 1998). T-branching is not limited to single branches and soft rot fungi are frequently known

Table 15.4 Classification of morphological effects of fungal attack of wood cell walls (Daniel 2016)

Decay type	Morphological changes of wood cell walls	Cell wall components attacked	Taxonomic grouping	Typical examples
White rot (simultaneous)	Hyphal bore holes enlarge Cell wall thinning from lumen Middle lamellae degraded Development of cavities in some species	Cellulose, lignin, hemicelluloses Extractives	Basidiomycetes Higher ascomycetes	*T. versicolor*[a] *Heterobasidium annosus* *Xylaria polymorpha*[b] *Daldinia concentrica*[b]
White rot (preferential)	Hyphal bore holes enlarge Cell wall attack from lumen	Hemicelluloses, lignin Extractives	Basidiomycetes	*Ceriporiopsis subvermispora* *Heterobasidion annosum* *Phl. radiata Cel 26*
	Middle lamellae degraded Fiber defibration and separation			
Brown rot	Rapid attack of cell walls Cell wall attack from lumen	Depolymerization of cellulose, hemicelluloses; lignin modified	Basidiomycetes	*C. puteana*[a] *Ol. (Postia) placenta*[a]
Soft rot Type I	Hyphal bore holes remain small Longitudinal cavities Middle lamellae remain	Cellulose, hemicelluloses lignin modified/ degraded	Ascomycetes Fungi imperfecti	*Chaetomium globosum*[a] *Phialophora mutabilis*
Soft rot Type I, *diffuse type*	Hyphal bore holes remain small Coalescence of longitudinal cavities	Cellulose, hemicelluloses lignin modified/ degraded	Ascomycetes	*Phi. dimorphospora* *C. globosum*
	Middle lamellae remain		Fungi imperfecti	*Bispora betulina*
Soft rot Type II	Hyphal bore holes remain small; cell wall thinning from lumen; middle lamella remains	Cellulose, hemicelluloses lignin modified/ degraded	Ascomycetes Fungi imperfecti	*C. globosum* *Phialop. mutabilis*
Blue stain fungi	Fine bore holes Small cavities/erosion troughs	Primarily nonlignified cells (ray parenchyma), extractives	Ascomycetes Fungi imperfecti	*Ophiostoma piceae* *L. theobromae*
Mold fungi	Growth in surface regions	Soluble sugars, extractives	Ascomycetes Fungi imperfecti Zygomycetes	*C. globosum* *Penicillium brevicompactum* *Aspergillus versicolor* *Rhizopus* spp.

[a]*Frequently used as test fungi for assessing new types of wood protection – for example, preservatives, wood modification*
[b]*Higher ascomycetes cause white rot but are not known to degrade middle lamella regions*

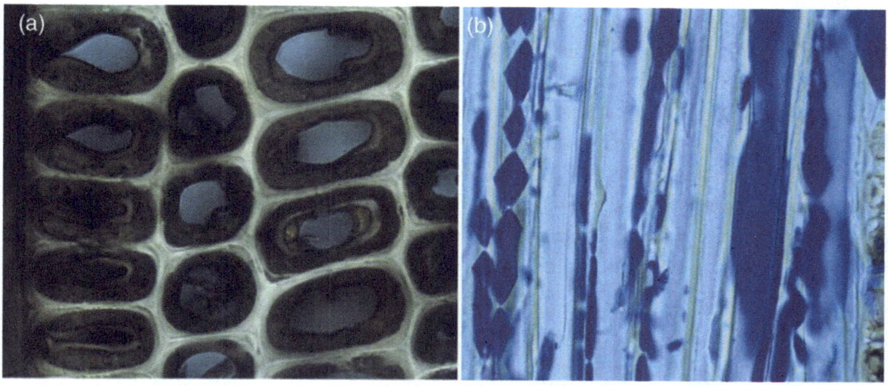

Fig. 15.7 Diffuse soft rot Type I decay of pine and birchwood by *Phi. dimorphospora*. (**a**) Soft rot hyphae within the secondary cell walls (S2) with diffuse widespread decay not restricted to cavities. (**b**) Cavities (stained blue to right of photo) are often very large and most of the secondary wall degraded. Note: Subpart (**a**) is not stained, but colored black through melanized hyphae and melanin secretions (Daniel 2016)

to develop multi-T-branching in cell walls (Fig. 15.7b) as observed with some white rot basidiomycetes. The fact that cavity formation always follows the orientation of the cellulose microfibrils is shown by the circular orientation of hyphae that invade pit chamber walls and between pits (Khalili et al. 2000). Cavities are most often observed first in the thick latewood cells (Fig. 15.5d, e) of wood spreading thereafter to earlywood. The relationship here is thought to be related to the size of the hyphae as cavities are rarely observed in the S1 and never in the S3 layer (i.e., it is too thin and has high lignin in softwoods), although in advanced stages of decay the S1 and S3 are degraded through hyphal attack from the adjoining S2 layer.

The effect of lignin content and lignin type is shown at the cellular level in advanced stages of attack where only the high lignin middle lamellae remain in hard-and softwoods (Fig. 15.5e). In addition, the hyphae and cavities often have darkly staining materials closely associated consisting of lignin remnants and melanin, the latter directly attached to hyphae or extracellular within the cavities (Figs 15.5e, 15.6c–f, and 15.7a). A similar effect can also be noted in fibers showing very thin concentric layers containing higher lignin levels than the surrounding cell wall (e.g., *Homalium foetidum*; Fig. 15.6f). Here, "half-moon" cavities are produced with flat axis orientated along the more lignified layer. That soft rot fungi need to orientate along the cellulose to cause cavity formation suggests it is an adaptation to overcome the effect of lignin coating the cellulose at the macromolecular level. Since the effect is shown in soft and hardwoods with different lignin types and content, it is suggested that it is the cellulose orientation that is most important. This is shown quite easily by slight delignification of wood, which enhances not only its susceptibility to soft rot but changes the nature of the decay process by modification or loss of cavity formation.

Type II soft rot is very similar to simultaneous white rot attack and can result in a complete removal of the secondary cell walls but in contrast the middle lamellae remain (Fig. 15.5f, g). This decay type is very frequently observed in low-lignin hardwood (e.g., aspen and birch) especially under high moisture situations.

A third decay form known as "diffuse cavity formation" has been described in which attack is similar to Type I during the initial stages but after cavity formation solubilization of the polysaccharides is more widespread and diffuse in the S2 layer and more like brown rot attack (Fig. 15.7a, b). The decay type is best observed in low-lignin-containing hardwoods (e.g., birch) although it does occur in softwoods. Typical soft rot species producing this decay pattern include *Phialocephala dimorphospora* (Anagnost et al. 1994). Blanchette (2000) concluded that Type 1 is characterized by longitudinal cavities formed within the secondary wall of wood cells and Type 2 used to describe an erosion of the entire secondary wall (Fig. 15.8).

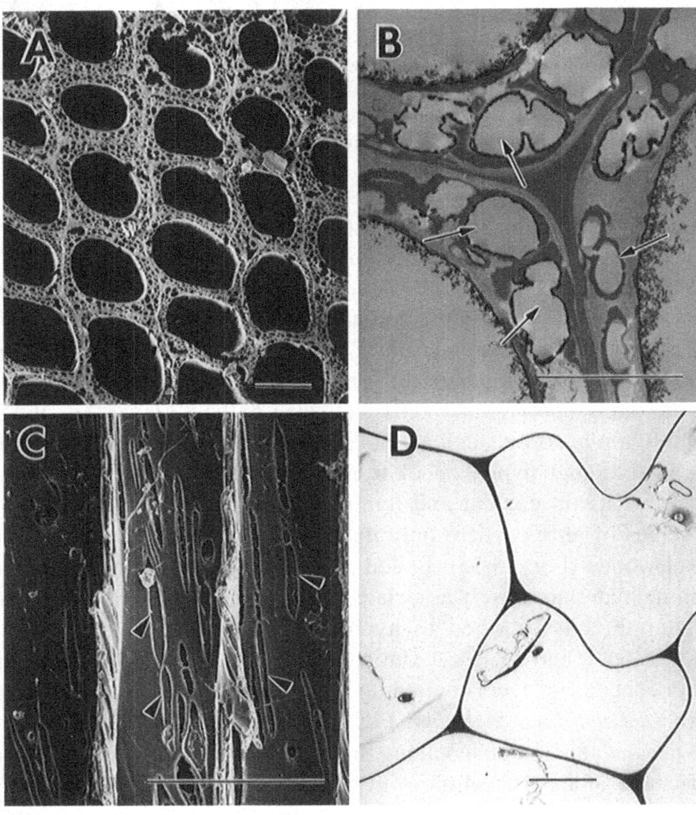

Fig. 15.8 Soft rot attack of wood. (**a–c**). Type I attack forms chains of cavities within the secondary wall (arrows). In advanced stages of decay, cell walls contain numerous cavities that often coalesce together (arrows). In some cells, the chains of cavities are visible from the cell lumina (arrowheads) where holes in the wall have been exposed. (**d**) Type II form of attack showing an erosion of the secondary wall but no degradation of the middle lamella. In advanced decay, the secondary walls are completely degraded and only the middle lamellae remain. A, B, and D transverse sections, C radial section. A and C SEM, B and D TEM. Bar = 50 μm in A and C, 5 μm in B and D (Blanchette 2000)

15.6 Key to Identification of Wood Decays Based on Light Microscopic Features (Anagnost 1998)

la Erosion channels on the lumen surface present 2
b Erosion channels absent from the lumen surface 5
2a Erosion channels rounded with V-notches, hardwoods or softwoods 3
b Erosion channels angular with V-notches, in hardwoods only 4
3a Erosion appears as elongate channels beneath lumen hyphae; channels may be numerous; pit erosion is rounded or oval; bore holes and pit erosion are common; some cell separation may be evident; cavities in the S2 are rare; if present, cavities not in chains, penetrating bore hole is wider than hyphae ....
White rot (simultaneous)
b Cell separation is common; erosion channels are sometimes present ....
White rot (selective delignification)
4a Bore holes, if present, are smaller in diameter than penetrating hyphae; pit erosion is angular often forming diamond-shaped erosion or lacking ....
Soft rot (type 2)
b Erosion is accompanied by cavities within the cell wall; pit erosion is angular (diamond-shaped) or lacking ....
Soft rot (type 1 and 2)
5a Bore holes often lacking or rare (except in some cases; if present, bore hole diameters are the same as (early stages) or slightly wider than (later stages; up to 4*um*) associated hyphae); brown discoloration or shrinkage cracks may be evident; birefringence in polarized light may be lacking; loss of cell shape and wall thickness (incrosssection);rayparenchymadestroyedinearlystages.................................
Brown rot
b Bore holes initially smaller than hyphae. 6
6a Bore holes initially small, but enlarge to become much larger than penetrating hyphae; erosion channels may develop; cell separation sometimes evident; cavities in S2 are rare; if present, hyphae t- or I-branching (sometimes multiple branching occurs) to form very narrow cavities within the S2 layer which extend and widen; initial bore holes widen obviously.........................*White rot*
b Bore holes remain smaller than penetrating hyphae 7
7a Cavities are present within the S2 layer arising from t- or I-branching from transverse bore holes 8
b Narrow bore holes may be present; hyphae in cell lumens, often primarily in ray cells; wood may be discolored .... *Early decay, stain or mold*
8a In hardwoods and softwoods, hyphae form diamond-shaped cavities within the S2 layer .... *Soft rot* (type 1 regular)
b In softwoods only, hyphae form individual cavities that develop into diffuse cavities within the S2 cell wall; cavities appear as patches of S2 destruction at later stages with differential interference contrast microscopy ....
Soft rot (type 1 diffuse)

15.7 Soft Rot Fungi: Enzymes Involved and Some Biochemical Aspects of Decay

Numerous studies have been conducted to understand the biochemical/chemical mechanisms involved in fungal decay of lignocellulose, the majority aimed at biotechnological goals rather than using the understanding to produce better protective measures of wood in service. In principle the majority of these studies have been carried out using liquid cultures of fungal monocultures together with lignocellulose in various forms (e.g., as particles, sawdust, or flakes), purified cellulose/hemicelluloses, or lignin monomers (e.g., synthetic/natural). While these studies give information on the types of enzymes that may be produced under various physiological conditions (e.g., temperature, pH, and shaking/static), it may not reflect the situation that occurs in wood substrates under native conditions, but rather the potential ability of the fungi involved. For example, few studies have actually been involved in measuring enzyme activities in wood undergoing decay because of the difficulties in extracting sufficient amounts of "active proteins" in order to carry out enzymatic assays. Notable exceptions include the studies by Daniel et al. (1992, 1994) on the extraction of lignin-degrading enzyme (LiP, MnP, laccase) produced by *Pha. chrysosporium*, *T. versicolor*, and *O. mucida* in birchwood. A further problem concerns the sensitivity of the enzyme assays, which may not be sufficiently sensitive to detect the minor amount of proteins that can be extracted even in highly degraded wood materials. For example, extraction of proteins from whiterotted wood in which profuse hyphae growth is often seen is easier than from either brown or softrotted wood. In recent years, a variety of genomic, transcriptome, and secretome analytical approaches have been developed allowing for profile overviews of the enzyme systems available or employed by various fungi when grown on purified wood components (e.g., cellulose) (Martinez et al. 2009). This approach gives a more in-depth view of the enzymes or enzyme systems potentially involved, and the changes in enzyme profiles over time during wood decay can be followed. Evidence for the upregulation of a gene does not however mean it is actually involved in decay but rather that it has a potential. Possibly one of the most important approaches although indirect for proving the involvement of enzymes at sites of wood decay is by using antibodies produced against the purified proteins in microscopy assays. The easiest approach is by indirect labeling and the use of secondary antibodies whereby the sites of primary antibody labeling of enzymes in situ are recognized by the secondary antibodies carrying a fluorescence or gold label, which can then be visualized using fluorescence/confocal microscopy or electron microscopy. These studies were very important in proving the extracellular secretion of enzymes involved in decay and their remote distribution from hyphae during the wood decay process (Daniel et al. 1989, 1990, 1994, b, 2007; Ruel et al. 1990; Srinivasan et al. 1995). A wide variety of enzymes may be directly involved in hydrolytic activities (e.g., cellulases, hemicellulases, and pectinases) or act as oxidases in the production of oxidants that indirectly affect wood components (e.g., OH radicals from H_2O_2, pyranose 2oxidase, and glucose 1oxidase). A further important characteristic concerns the nature of the enzymes and whether they are associated with extracellular matrix materials such as slime and other polysaccharidebased materials (Daniel 2014) (Fig. 15.8).

Compared with white and brown rot fungi, much less is known about the degradative enzyme systems produced by ascomycetes during soft rot attack of wood in which cavity and erosion decay occur. Cavity formation in hardwoods is easier as shown by the larger number of cavities formed in a fixed period of time. Currently, there are no indications that enzymatic systems used in cavity formation differ from those used in cell wall erosion. The most important requirement for soft rot cavity formation is the alignment of hyphae with the cellulose microfibrils, the alignment probably inducing the secretion of cell wall-degrading cellulases. This indicates that the enzyme systems involved are likely to be closely associated with the fungal hyphae and initially at least present on the hyphal surfaces (Fig. 15.6a–f). The fact that the bioconical cavities are formed by hyphae aligned with the cellulose microfibrils in wood cell walls and not by the thin penetration (i.e., bore hyphae) traversing cell walls also suggests involvement of different or the absence of enzyme systems. Studies indicate that most cellulolytic microfungi cause some erosion of hardwood cell walls even those essentially regarded as typical mold fungi (e.g., *Aspergillus*, *Penicillium*, *Trichoderma* spp.); however, their ability to cause the same attack on softwoods is greatly limited. Frequently, *Trichoderma* spp. are reported as a typical soft rot fungi of wood and thus used as an example for comparison with white and brown rot fungi. However, as indicated earlier, *Trichoderma* cannot be really classified as a true soft rot fungus as it has not been unequivocally confirmed to produce cavities in the secondary cell walls of softwoods.

Few studies have been conducted on the chemical changes in wood following soft rot attack. It is clear, however, that since soft rot fungi have an inability to degrade the middle lamella regions in wood, their ability to degrade lignin is more limited than that of white rot fungi, and thus an increase in the relative lignin content in wood due to preferential removal of polysaccharides is expected. Lignin degradation has been reported, but it seems related to the fungal species involved. Chemical analyses of lignin from beechwood degraded by soft rot fungi have shown a lower methoxyl content and greater acid solubility than in undegraded wood (Levi and Preston 1965). Recent studies on birchwood degraded by a range of fungi imperfecti and ascomycetes have shown decay characterized by lower lignin (Klason) loss compared to white rot and much lower alkali solubility compared with brown rot (Worrall et al. 1997). Lignin peroxidase has been isolated and purified from the ascomycete *Chrysonilia sitophila* (Durán et al. 1987; Rodríguez et al. 1997), although the fungus has not been shown to degrade wood. Similarly, a range of phenolic and lignin-related compounds have been shown degraded by soft rot fungi (Haider and Trojanowski 1975; Bugos et al. 1988), but this does not confirm they degrade lignin in wood or cause wood decay. A number of thermophilic ascomycetes (Machuca et al. 1998) and some heat-tolerant soft rot fungi like *Talaromyces thermophilus* and *Thielavia terrestris* are also weakly ligninolytic, but their true effect of wood is limited (Dix and Webster 1995). Observations show extensive removal of wood cell wall materials surrounding cavities produced in the secondary walls (S2, S1 layers) of both hard and softwoods in advanced stages (Figs 15.5d, e and 15.6a, b, f). However, in both wood species and depending on soft rot fungus, there are generally always large amounts of residual electron materials surrounding and attached to the hyphae (e.g., Fig. 15.6f) (Daniel and Nilsson 1989).

These materials represent partially degraded lignin remaining after preferential cellulose/hemicellulose removal as well as melanin.

The ability of soft fungi to preferentially remove carbohydrates and leave lignin in cavities and in the cell wall during erosion decay indicates the involvement of a diffusive cellulase/hemicellulase system. This is consistent with several large screening studies carried out on soft rot fungi from terrestrial, marine, and freshwater situations showing enzymatic "clearing" (i.e., decay) of ball mill cellulose/cellulose agar and hemicellulose (xylan) agar during growth of these microfungi (Nilsson 1973; Bucher et al. 2004; Duncan et al. 2006, 2008; Simonis et al. 2008). These enzyme–agar studies showed that clearing zones could occur a considerable distance from the fungal hyphae indicating a highly diffusible enzymatic system that varies with fungal species. The fact that cavity formation and erosion decay is a frequent form of decay in aquatic situations emphasizes not only the success of this form of decay compared to higher fungi, but also that the hydrated cell wall may be advantageous for the enzymatic systems involved. Very little research has been conducted on the cellulolytic and hemicellulolytic systems on true soft rot fungi and few studies on the types of enzymes involved. Screening studies indicate, however, the effective endoglucanase activities, which is consistent with the alignment of hyphae with the cellulose microfibrils in wood cell walls with cavity formation. Soft rot fungi producing cavities and erosion in preservative-treated wood cell walls also have an effective system for immobilization of heavy metals in addition to causing decay.

For an overview of carbohydrate-degrading enzymes produced by wood decay fungi, readers should consult the CAZy database (www.cazy.org) and for fungal oxidoreductases (i.e., lignin-degrading enzymes) the FOLy (Fungal Oxidative Lignin Enzymes) database (https://foly-db.esil.univ-mrs.fr/) (Levasseur et al. 2008) and the more recently updated CAZy database that includes auxiliary activities and covers redox enzymes (http://www.cazy.org/Auxiliary-Activities.html). The aim of the databases is to provide an overview of cellulose- and lignin-degrading enzymes for biotechnical applications.

15.8 *Chaetomium* and Egyptian Papyrus

In 1973 Kowalik and Sadurska studied the microbiota of papyrus from samples of Cairo museums. Different fungi imperfecti, ascomycetes, and actinomycetes were isolated from samples of papyrus of Cairo museums. They recovered many fungi such as *Alternaria geophila, Botryodiplodia theobromae, Emericellopsis minima, Fusarium lactis, Helminthosporium sativum, Spondylocladium australe*, some species of the genus *Chaetomium*, and some actinobacteria related to genus *Streptomyces* which seemed to be specific for papyrus and/or for Egyptian climatic conditions. The genus *Chaetomium* and *Emericellopsis* may play a great role in decomposition of basic polymers of papyrus. Considering the nitrogen source of microbiota, it can be observed that papyrus-destroying microorganisms preferred ammonium to nitrate ions. It was found that papyrus-decomposing microorganisms may grow equally well at 24–26 °C at 30 °C and some fungi are growing even at 42 °C. The members of the genus *Penicillium*, which prefer rather low temperatures and are frequent inhabitants of paper, were isolated from papyrus only once. By using the method of paper sheets damped with a 10 percent ethyl alcohol solution of thymol, pentachlorophenol, dichlorophene, and p-chloro-m-cresol as microbiocides, it could be concluded that only the last fungicide may assure the protection of papyrus. They recovered 12 species of *Chaetomium*; they were the following: *C. angustum, C. atrobrunneum, C. bostrychodes, C. cochliodes, C. elatum, C. fusiforme, C. globosum, C. indicum, C. ochraceum, C. olivaceum, C. trilaterale*, and *C. turgidopilosum.*

15.9 *Chaetomium* and Egyptian Archeological Wood

In 2019 Abdel-Azeem et al. studied the assessment of biodegradation in ancient archaeological wood from the Middle Cemetery at Abydos, Egypt. Abydos is a large, complex archaeological site located approximately 500 km south of Cairo in Upper Egypt. The site has served as a cemetery for thousands of years and is where most of the Early Dynastic royal tombs are located. North Abydos includes the Middle Cemetery and the North Cemetery, which are separated from each other by a wadi. The Middle Cemetery was the burial ground for important Sixth Dynasty (2407–2260 BC) officials and over time for thousands of elite and non-elite individuals as well. Excavations at the core area of the Old Kingdom mortuary landscape have revealed many culturally important wooden objects but these are often found with extensive deterioration that can compromise their preservation. The objectives of this study were to characterize the biodegradation that has taken place in excavated wooden objects, elucidate the type of wood degradation present, obtain information on soil properties at the site, and identify fungi currently associated with the wood and soils. Light and scanning electron microscopy studies were used to observe the micromorphological characteristics of the wood, and culturing on different media was done to isolate fungi. Identification of the fungi was done by examining morphological characteristics and extracting rDNA from pure cultures and sequencing the ITS region. Wooden objects, made from *Cedrus*, *Juniperus* and *Acacia* as well as several unidentified hardwoods, were found with extensive degradation and were exceedingly fragile. Termite damage was evident and frass from the subterranean termites along with sand particles were present in most woods. Evidence of soft rot attack was found in sections of wood that remained. Fungi isolated from wood and soils were identified as species of *Aspergillus*, *Chaetomium*, *Cladosporium*, *Fusarium*, *Penicillium*, *Stemphylium, Talaromyces,* and *Trichoderma*. Results provide important information on the current condition of the wood and gives insights to the identity of the fungi in wood and soils at the site. These results provide needed information to help develop conservation plans to preserve these degraded and fragile wooden objects.

Light microscopy (Fig. 15.9a) of wood sections revealed cavities within the secondary wall of tracheids that was characteristic of Type I soft rot. Advanced stages

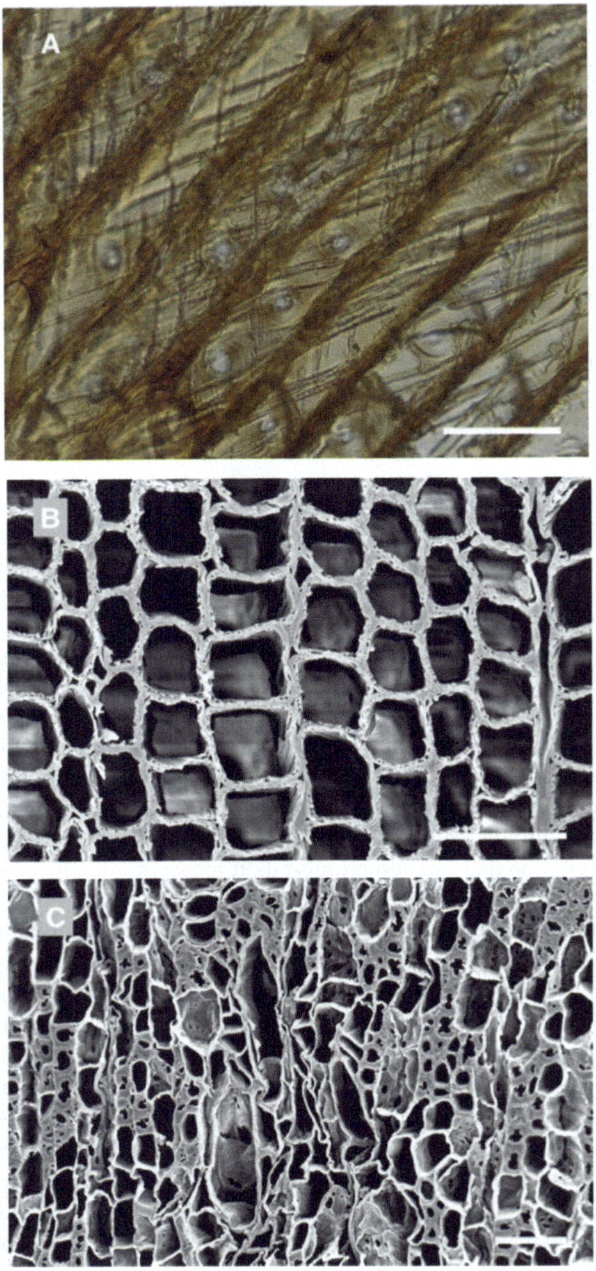

Fig. 15.9 showed the micromorphology of decayed wood. Tangential section (**a**) and scanning electron micrographs of transverse sections (**b** and **c**) from wooden objects with wood degradation. **a**) Type I soft rot in wood cells of Cedrus showing cavities that form inside the tracheid secondary walls. Soft rot cavities spiral within the cells. **b**) Small cavities are evident within the cell walls of the tracheids. **c**) An unknown hardwood with Type I soft rot and cavities within the secondary cell walls and also Type II soft rot causing an erosion of the entire secondary wall. Only a fragile middle lamella remains in some cells. Weakened cells appear distorted and some have collapsed. Scale bar = 50 μm (Abdel-Azeem et al. 2019)

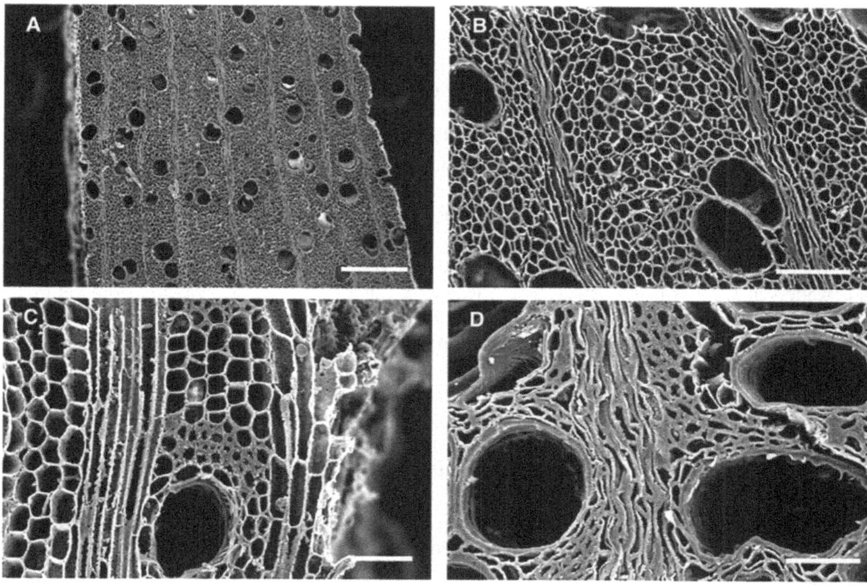

Fig. 15.10 showing scanning electron micrographs of transverse sections of degraded wood. **a** and **b**) Cell walls degraded by Type II soft rot from AMC 2013 field coffin sample 7. Secondary cell walls are degraded with most cells having only the middle lamella remaining. Although the wood structure can still be seen, the thin cell walls remaining are weak and fragile. **c**) Wood from coffin (AMC2013, Unit 28, Feat. 14, Burial 26.4) showing cells with Type II soft rot with degraded secondary walls and a thin middle lamella remaining as well as some cells with less severe attack that have some of the fiber secondary walls still present. **d**) Wood section from a box found in the Weni tomb (AMC01 level 1). The degraded cell walls caused by the soft rot have little strength and often break or collapse as seen in this micrograph. Scale bar in A = 250 μ and B, C, and D = 50 μ (Abdel-Azeem et al. 2019)

of decay were present and many soft rot cavities were seen in a spiral pattern within the secondary walls. In transverse sections, small cavities were seen within tracheids (Fig. 15.9a). Wood from objects that had been made of various hardwoods also had evidence of soft rot. Some cavities within secondary walls were present but more commonly observed was a Type II soft rot where the secondary cell walls were eroded (Figs. 15.9 and 15.10). In some cells with advanced decay, the entire secondary cell wall was removed leaving only the middle lamella.

The altered cells were often collapsed and distorted. Sections from other wooden objects showed similar patterns of degradation with Type II soft rot in objects made from hardwood (Fig. 15.9). The secondary walls were degraded leaving only a weak and fragile framework of the middle lamella and cell walls were often fractured and collapsed (Fig. 15.10).

References

Abdel-Azeem AM, Held BW, Richards JE, Davis SL, Blanchette RA (2019) Assessment of biodegradation in ancient archaeological wood from the Middle Cemetery at Abydos, Egypt. PLoS One 4(3):e0213753. https://doi.org/10.1371/journal.pone.0213753

Allsopp D, Seral K, Gaylarde C (2004) Introduction to biodeterioration. Cambridge University Press, Cambridge, p 237

Anagnost SE, Worrall JJ, Wang CK (1994) Diffuse cavity formation in soft rot of pine. Wood Sci Technol 28:199–208

Anagnost SE, (1998) Light Microscopic Diagnosis of Wood Decay. IAWA J. 19, 141–167

Bech-Andersen J, Elborne SA (2004) The true dry rot fungus (*Serpula lacrymans*) from nature to houses. In: Dradacky M (ed) European research on cultural heritage state of the art studies, vol. 2. Institute of Theoretical and Applied Mechanics Academy of Sciences of the Czech Republic, Prague, pp 445–448

Blanchette RA (1991) Delignification by wood-decay fungi. Annu Rev Phytopathol 29:381–398

Blanchette RA (1995) Biodeterioration of archaeological wood. CAB Biodeterior Abstr 9:113–127

Blanchette RA (1998) A guide to wood deterioration caused by fungi and insects. In: Dardes K, Rothe A (eds) The conservation of panel paintings. Getty Conservation Institute, Los Angeles, pp 55–68

Blanchette RA (2000) A review of microbial deterioration found in archaeological wood from different environments. Int Biodeterior Biodegrad 46:189–204

Blyskal B (2009) Fungi utilizing keratinous substrates. Int Biodeterior Biodegrad 63:631–653

Britton TA (1875) A treatise on the origin, progress, prevention, and cure of dry rot in timber. E. and F.N. Spon, London, p 311

Bucher VVC, Hyde KD, Pointing SB, Reddy CA (2004) Production of wood decay enzymes, mass loss and lignin solubilization in wood by marine ascomycetes and their ana morphs. Fungal Divers 15:1–14

Bugos RC, Sutherland JB, Adler JH (1988) Phenolic compound utilization by the soft rot fungus *Lecythophora hoffmannii*. Appl Environ Microbiol 54:1882–1885

Camuffo D (1998) Microclimate for cultural heritage. Elsevier, Amsterdam, p 415

Capitelli F, Fermo P, Vechi R, Piazzalunga A, Valli G, Zanardini E, Sorlini C (2009) Chemical–physical and micro-biological measurements for indoor air qualità assessment at the ca'Granda historical archive, Milan (Italy). Water Air Soil Pollut 201:109–120

Cowling EB (1961) Comparative biochemistry of the decay of sweet gum by white-rot fungi. US Department of Agriculture Technical Bulletin 1258:79

Crossley A (1979) The use of electron microscopy to compare wood decay mechanisms. PhD thesis. Imperial College of Science and Technology, UK

Daniel G (2016) Fungal degradation of wood cell walls. In: Secondary xylem biology. Elsevier, Amsterdam, pp 131–167

Daniel G, Nilsson T (1998) Developments in the study of soft rot and bacterial decay. In: Bruce A, Palfreyman JW (eds) Forest products biotechnology. Taylor and Francis, London, p 326

Daniel G (2014) Fungal and bacterial biodeterioration: white rots, brown rots, soft rots and bac teria. In: Schultz TP, Goodell B, Nicholas DD (eds) Deterioration and protection of Sustainable Biomaterials. American Chemical Society, Washington, DC, pp 23–58

Daniel G, Nilsson T (1989) Interactions between soft rot fungi and CCA preservatives in *Betula verrucosa*. J Inst Wood Sci 11:162–171

Daniel G, Pettersson B, Nilsson T, Volc J (1990) Use of immunogold cytochemistry to detect Mn (II)dependent and lignin peroxidases in wood degraded by the white rot fungi *Phanerochaete chrysosporium* and *Lentinula edodes*. Can J Bot 68:920–933

Daniel G, Volc J, Filonova L, Plíhal O, Kubátová E, Halada P (2007) Characteristics of *Gloeophyllum trabeum* alcohol oxidase, an extracellular source of H_2O_2 in brown rot decay of wood. Appl Environ Microbiol 73:6241–6253

Daniel G, Volc J, Kubatova E (1994) Pyranose oxidase, a major source of H_2O_2 during wood degradation by *Phanerochaete chrysosporium*, *Trametes versicolor*, and *Oudemansiella mucida*. Appl Environ Microbiol 60:2524–2532

Daniel G, Volc J, Nilsson T (1992) Soft rot and multiple Tbranching by the basidiomycete *Oudemansiella mucida*. Mycol Res 96:49–54

Dix NJ, Webster J (1995) Colonization and decay of wood. In: Dix NJ, Webster J (eds) Fungal Ecology. Springer, Wallington, pp 145–171

Duncan SM, Farrell RL, Thwaites JM, Held BW, Arenz BE, Jurgens JA, Blanchette RA (2006) Endoglucanase producing fungi isolated from Cape Evans historic expedition hut on Ross Island. Antarctica. Environ Microbiol 8:1212–1219

Duncan SM, Minasaki R, Farrell RL, Thwaites JM, Held BW, Arenz BE, Jurgens JA, Blanchette RA (2008) Screening fungi isolated from historic Discovery Hut on Ross Island, Antarctica for cellulose degradation. Antarct Sci 20:463–470

Durán N, Ferrer I, Rodríguez J (1987) Ligninases from *Chrysonilia sitophila* (TFB27441 strain). Appl Biochem Biotechol 16:157–167

Gersonde M, Kerner Gang W (1976) Review of information available for development of method for testing wood preservatives with soft rot fungi. Int Biodeterior Bull 12:5–13

Gysels K, Delalieeux F, Deutsch F, Van Gieken R, Camuffo D, Bernardi A, Sturaro G, Busse HJ, Wieser M (2004) Indoor environment and conservation in the Royal Museum of Fine Arts. J Cult Herit 5:221–230

Haider K, Trojanowski J (1975) Decomposition of specifically ^{14}Clabelled phenols and dehydropolymers of coniferyl alcohol as models for lignin degradation by soft and white rot fungi. Arch Microbiol 105:33–41

Hale MD, Eaton RA (1985a) The ultrastructure of soft rot fungi. II. Cavity forming hyphae in wood cell walls. Mycologia 77:594–605

Hale MD, Eaton RA (1985b) The ultrastructure of soft rot fungi. I. Fine hyphae in wood cell walls. Mycologia 77:447–463

Harper SHT, Lynch JM (1985) Colonization and decomposition of straw by fungi. Trans Br Mycol Soc 85:7–655

Henningsson B, Nilsson T (1976) Some aspects on microflora and the decomposition on preservativetreated wood in ground contact. Mater Org 3:307–318

Inglis GD, Kawchuk LM (2002) Comparative degradation of oomycete, ascomycete, and basidiomycete cell walls by mycoparasitic and biocontrol fungi. Can J Microbiol 48:60–70

Jennings DH (1991) The physiology and biochemistry of the vegetative mycelium. In: Jennings DH, Bravery AF (eds) *Serpula lacrymans*: fundamental biology and control strategies. Wiley, New York, pp 55–79

Jennings DH, Bravery AF (1991) Serpula lacrymans: fundamental biology and control strategies. Wiley, New York, p 217

Kaul S, Sumbali G (1999) Production of extracellular keratinases by keratinophilic fungal species inhabiting feathers of living poultry birds (*Gallus domesticus*). Mycopathologia 146:6–19

Khalili S, Daniel G, Nilsson T (2000) Use of soft rot fungi for studies on the microstructure of kapok (*Ceiba pentandra* (L.) Gaertn.) fibre cell walls. Holzforschung 54:229–233

Koestler RJ, Koestler VH, Charola AE, Nieto Fernandez FE (eds) (2003) Art, biology and conservation: biodeterioration of works of art. The Metropolitan Museum of Art, New York

Kowalik R, Sadurska I (1973) Microflora of papyrus from samples of Cairo Museum. Stud Conserv 18:1–24

Leightley L, Eaton R (1978) Mechanisms of decay of timber by aquatic microorganisms. Proceedings of the Annual Convention of British Wood Preserving Association. British Wood Preserving Association, Cambridge, pp 221–246

Leightley LE (1977) Mechanisms of timber degradation by aquatic fungi. PhD thesis. Portmouth University, Portsmouth

Levasseur A, Piumi F, Coutinho PM, Rancurel C, Asther M, Delattre M, Henrissat B, Pontarotti P, Asther M, Record E (2008) FOLy: an integrated database for the classification and functional annotation of fungal oxidoreductases potentially involved in the degradation of lignin and related aromatic compounds. Fungal Genet Biol 45:638–645

Levi MP, Preston RD (1965) A chemical and microscopic examination of the action of the softrot fungus *Chaetomium globosum* on beechwood (*Fagus sylvaticus*). Holzforschung 19:183–190

Machuca A, Aoyama H, Durán N (1998) Production and characterization of thermostable phenol oxidases of the ascomycete *Thermoascus aurantiacus*. Biotechnol Appl Biochem 27:217–223

Martinez D, Challacombe J, Morgenstern I, Hibbett D, Schmoll M, Kubicek CP, Ferreira P, Ruiz-Duenas FJ, Martinez AT, Kersten P et al. (2009) Genome, transcriptome, and secretome analysis of wood decay fungus *Postia placenta* supports unique mechanisms of lignocellulose conversion. Proc Natl Acad Sci USA 106:1954–1959

Meier C, Petersen K (2006) Schimmelpilze auf Papier – Ein Handbuch fü r Restauratoren. Der Andere Verlag, 198 pp. Manoharachary C, Reddy PJM, Prabhakar B, Mohan KC

Meiggs R (1982) Trees and timber in the ancient Mediterranean world. Oxford University Press, London, p 553

Mesquita N, Portugal A, Videira S, Rodrı́guez-Echeverrı́a S, Bandeira AML, Santos MJA, Freitas H (2009) Fungal diversity in ancient documents. A case study on the Archive of the University of Coimbra. Int Biodeterior Biodegrad 63:626–629

Mohtashamipur E, Norpoth K (1990) Release of mutagens after chemical or microbial degradation of beech wood lignin. Toxicol Lett 51:277–285

Nilsson T (1973) Studies on wood degradation and cellulolytic activity of microfungi. Stud For Suec 104:1–40

Nittèrus M (2000) Fungi in archives and libraries, a literary survey. Restaurator 21:25–40

Pangallo D, Chovanova K, Simonovicova A, Ferianc P (2009) Investigation of microbial community isolated from indoor artworks and their environment: identification, biodegradative abilities, and DNA typing. Can J Microbiol 55:277–287

Pangallo D, Simonovicová A, Chovanová K, Ferianc P (2007) Wooden art objects and the museum environment: identification and biodeteriorative characteristics of isolated micro flora. Lett Appl Microbiol 45:78–94

Prokhorov V, Linnik M (2011) Morphological, cultural, and biodestructive peculiarities of *Chaetomium* species. Mosc Univ Biol Sci Bull 66:95–101

Rayner ADM, Boddy L (1988) Fungal decomposition of wood: its biology and ecology. Wiley, New York, p 587

Rodrı́guez J, Ferraz A, Nogueira RF, Ferrer I, Esposito E, Durán N (1997) Lignin biodegradation by the ascomycete *Chrysonilia sitophila*. Appl Biochem Biotechnol 62:233–242

Ruel K, Odier E, Joseleau Jp (1990) Immunocytochemical observation of fungal enzymes during degradation of wood cell walls by *Phanerochaete chrysosporium*. In: Kirk TK, Chang HM (eds) Biotechnology in pulp and paper manufacture: applications and Fundamental Investigations. ButterworthHeinemann, Boston, pp 83–97

Samson RA, Hoekstra ES, Frisvad JC, Filtenborg O (2000) Introduction to a food and airborne Fungi, 6th edn. Centraalbureau voor Schimmelcultures, Utrecht

Savory JG (1954) Breakdown of timber by Ascomycetes and Fungi Imperfecti. Ann Appl Biol 41:336–347

Sedlbauer K, Krus M (2003) Schimmelpilz aus bauphysika- lischer Sicht. Fraunhofer-Institut fü r Bauphysik, Holzkirchen, p 45

Silva CAd, Monteiro MBB, Brazolin S, Lopez GAC, Richter A, Braga MR (2007) Biodeterioration of brazilwood *Caesalpinia echinata* Lam. (Leguminosae – Caesalpinioideae) by rot fungi and termintes. Int Biodeterior Biodegrad 60:8–285

Simonis JL, Raja HA, Shearer CA (2008) Extracellular enzymes and soft rot decay: are ascomycetes important degraders in fresh water. Fungal Divers 31:135–146

Srinivasan C, Dsouza TM, Boominathan K, Reddy CA (1995) Demonstration of laccase in the white rot basidiomycete *Phanerochaete chrysosporium* BKMF1767. Appl Environ Microbiol 61:4274–4277

Sterflinger K (2010) Fungi: their role in deterioration of cultural heritage. Fungal Biol Rev 24:47–55

Sterflinger K, Krumbein WE, Rullkö tter J (1999) Patination of marble sandstone and granite by microbial communities. Z Dtsch Geol Ges 150:299–311

Sterflinger K, Prillinger H (2001) Molecular taxonomy and biodiversity of rock fungal communities in an urban environment (Vienna, Austria). Antonie Van Leeuwenhoek 80:275–286

Strzelczyk AB, Kuroczkin J, Krumbein WE (1989) Studies on the microbial degradation of ancient leather bookbindings. Part 2. Int Biodeterior 25(1):39–47

Wilcox WW (1968) Changes in wood microstructure through progressive stages of decay. U.S. For Service Prod Lab Res Pap FPL-70

Worrall JJ, Anagnost SE, Zabel RA (1997) Comparison of wood decay among diverse lignicolous fungi. Mycologia 89:199–219

Zabel RA, Lombard FF, Wang CJK, Terracina FC (1985) Fungi associated with decay in treated southern pine utility poles in the eastern United States. Wood Fiber Sci 17:75–91

Zabel RA, Morrell JJ (1992) Wood Microbiology, Decay and its Prevention. Academic Press, Orlando, 476 pp

Chapter 16
Thermophilic *Chaetomium* in Biotechnology

Ahmed M. Abdel-Azeem, Shimaa A. El-Mansy, Hebat Allah H. Abo Nahas, Mariam K. Mousa, Hossam E. Gandal, Salma E. Hamdy, Mahmoud M. El-Ansary, and Mohamed A. Abdel-Azeem

16.1 Introduction

Enzymes from thermophilic fungi have been receiving considerable attention chiefly because of their potential to catalyze reactions at elevated temperatures in various industrial processes. These fungi produce an array of thermostable enzymes, including cellulases, amylases, proteases, phosphatases (including phytase), xylanases, lipases, and several other enzymes that are useful in industries like food and feed, textiles, detergents, leather, dairy, and pharmaceuticals. Depending on their intended use, the industrial enzyme market can be classified into three different sections: (1) technical enzymes, (2) food enzymes, and (3) animal feed enzymes. The largest section is of technical enzymes, where enzymes used for detergents and the pulp and paper industry constitute 52% of the total world market (Business Communications Company 2004). The leading enzymes in this section are hydrolytic enzymes, such as proteases and amylases, which comprise 20% and 25% of the total market, respectively (Business Communications Company 2004).

Chaetomium is regarded as one of the relatively rich genera of *Ascomycetes* by accommodating more than 100 species inhabiting ecologically diverse habitats including soil, composting materials, decomposing wood, herbivore dung, stored seeds, etc. Since introduced as a genus, by Kunze and Schmidt (1817) and typified with *Chaetomium globosum*, further species were added and several taxonomic revisions were tried. Since the establishment of the genus, more than 300 species

A. M. Abdel-Azeem (✉) · M. K. Mousa · H. E. Gandal · S. E. Hamdy · M. M. El-Ansary
Botany Department, Faculty of Science, University of Suez Canal, Ismailia, Egypt
e-mail: ahmed_abdelazeem@science.suez.edu.eg

S. A. El-Mansy · H. A. H. Abo Nahas
Zoology Department, Faculty of Science, University of Suez Canal, Ismailia, Egypt

M. A. Abdel-Azeem
Pharmacognosy Department, Faculty of Pharmacy, University of Suez Canal, Ismailia, Egypt

A. M. Abdel-Azeem (ed.), *Recent Developments on Genus Chaetomium*,
Fungal Biology, https://doi.org/10.1007/978-3-030-31612-9_16

have been described, many of which were synonymized/excluded and only 163 species were accepted (Doveri 2013).

16.2 Thermophiles and Thermostable Enzymes

16.2.1 Advantages of Thermostable Enzymes

The thermophilic biocatalysts are important not only due to their thermostabilities but also because they are more resistant to denaturing agents and tolerant to higher solute (reactants) concentrations. A higher degree of thermophilicity of an organism does not necessarily imply that pure enzymes derived from such organisms will always be very thermostable. The fact remains, however, that one has a better chance of finding more thermostable or more chemoresistant enzymes in thermophiles than in mesophiles. High productivity is also expected from thermophiles. According to Arrhenius law, an increase in temperature speeds up chemical and enzymatic reactions, and therefore, microbial growth and product formation. Productivity is unfortunately given in terms of yield coefficients only, and not specific product formation rates, a parameter that allows easy and direct comparison between organisms (Sonnleitner and Fiechter 1983).

The benefits of using enzymes as catalysts in industrial processes lie in their specificity and efficiency, leading to the production of by-products, less toxic wastes, and reduced handling problems. The main disadvantages of using enzyme reside in stability problems and high cost. The latter partially depends on the former, since the frequency of replacement of enzyme in a bio-reactor (and therefore total production costs) is stability-dependent. Production cost itself is generally high due to low yields of enzyme per unit biomass, the expense of extraction and concentration and/or purification, losses of activity during purification and storage, and the handling-stability of the enzyme again being a factor (Kristjansson 1989; Coolbear et al. 1992).

The use of thermostable enzymes reduces stability problems and in doing so alleviates some of the expense of production and replacement in a bioreactor. The stability of enzymes from thermophiles should lead to higher recoveries at ambient temperatures than is possible for mesophiles. The low activity of extremely thermophilic enzymes at ambient temperatures eases handling and storage problems, and the comparative molecular inflexibility that results in this inactivity at lower temperature has been suggested to lower the immune response to such proteins, thus reducing potential health risks (Coolbear et al. 1992). Besides higher thermostability, the other expected advantages of thermophilic enzymes are increased chemoresistance, a longer useful shelf life, and less contamination problems (Sonnleitner and Fiechter 1983) (Table 16.1).

Many of the enzymes presently used in industrial processes are quite thermostable even though they originate from mesophilic bacteria or fungi. Thermostability

Table 16.1 Main advantages of thermostable enzymes in industrial processes

S. no.	Property	Advantage in process
1	Thermostability	Tolerate high temp., last longer
2	High optimum temperature	Little activity at low temperature, long shelf life
3	Resistance to denaturing agents	Tolerate organic solvents, high and low pH
4	General robustness	Tolerate harsh purification, gives better yield
5	Genes can be cloned in *E. coli*	A heating step makes purification easier
6	Chemical reaction rates	Diffusion and other chemical processes are accelerated
7	Solubility	Higher concentrations of poorly soluble compounds are possible
8	Viscosity	Decreases; mixing and pumping can be accelerated; mass transfer rate increases
9	Microbial contamination	Growth of all pathogens and most environmental mesophiles prevented
10	Biological activity in raw materials	Heating kills most interfering enzymes or microbial activities

in enzymes from mesophiles is, however, the exception, not the rule. Since industrially useful enzymes must usually be thermostable, this characteristic is of primary importance in screening programs for enzymes. This is also important because the intrinsic basis underlying the thermostability of thermophilic enzymes is yet to be revealed and so engineering this characteristic into less thermophilic enzymes is not possible at this time. Successful cloning and expression of genes encoding hyperthermophilic enzymes in mesophilic hosts has improved the availability of high-temperature biocatalysts (Adams and Kelly 1998).

16.2.2 List of Thermophilic Chaetomium *Species According to Salar (2018)*

Chaetomium britannicum Ames (1963)

Description: Growth of the fungus and production of perithecia take place at 47 °C. Ascomata ovoid to vase-shaped. Terminal and lateral hairs are very slender, grayish, and straight to undulate. Asci club-shaped, eight-spored. Ascospores brown, large, 19–24 × 11–14 μm, irregularly oval, rounded on the ends, and have a single apical germ pore.

Habitat diversity: Mushroom compost and soil.

Geographic distribution: United Kingdom.

Description: Based on Ames (1963).

Chaetomium mesopotamicum Abdullah and Zora (1993)

Description: This is a newly described species and has a growth temperature ranging from 30 °C to 52 °C. It differs from *Chaetomium thermophilum* La Touche and *C. virginicum* Ames by its asci, which are clavate and possess long highly branched terminal hairs. Ascospores globose to ovoid, olive to brown, 5.5–7.8 × 5.2–6.3 μm, provided with one apical germ pore.

Habitat diversity: Date palm plantation.

Geographic distribution: Iraq.

Description: Based on Abdullah and Zora (1993).

Chaetomium senegalensis Ames (1963)

Description: Colonies on YpSs at 45 °C appear white with a daily growth rate of 6–7 mm. Yellow exudate in mature colonies appears after 18–20 days. Ascomata dark, spherical or ovate, ostiolate, 110–230 μm in diameter, with a dark brown or black wall of angular, flattened, 6–10 μm cells. Ascomatal hairs narrow, delicate, often branched with bulbous base and tapering ends, punctulate or verrucose, 1.5–2 μm in width. Asci fasciculate, narrow, cylindrical, eight-spored, 50–70 × 7–9 μm. Ascospores uniseriate, ovate or clavate, bilaterally flattened, dark greenish, brown when mature, 6.5–12.5 × 4–7 μm, teardrop-shaped with a subapical germ pore.

Habitat diversity: On plant remains, seeds of Capsicum annuum, soil, and decomposing wheat straw.

Geographic distribution: Senegal, Netherlands, Kuwait, Iran, and India.

Description: Based on Aneja and Kumar (1994).

Chaetomium thermophile var. ***coprophile*** Cooney and Emerson (1964).

Description: On YpSs agar at 45 °C, the colonies appear white at first, producing typical concentric rings of growth; the rings of dark brown perithecia are separated by narrow zones of whitish hyphae. Perithecia superficial, more or less gregarious, globose or subglobose, 75–150 μm in diameter, ostiolate, but the ostiole is not easily observed, as the perithecia are densely clothed with dark, dichotomously branched hairs. Hairs smooth, 40–140 × 4.5–7 μm. Asci short-stalked, cylindrical, produced in a basal tuft, bearing eight spores in a linear row, 45–55 × 6–8 μm. Ascospores dark or olive brown, globose or subglobose, 7–9 μm in diameter.

Habitat diversity: Decomposing wheat straw, horse dung, mushroom compost, vegetable detritus, and soil.

Geographic distribution: United States, United Kingdom, India, Netherlands, and Ghana.

Description: Based on our isolate IMI 348710.

Chaetomium thermophile var. ***dissitum*** Cooney and Emerson (1964)

Description: On YpSs agar at 45 °C, the colonies appear colorless at first and then become dark brown. Perithecia initially appear in about 2 days as small, colorless knots on the hyphae. Perithecia superficial; gray, gray green, to dark brown; produced in diffuse manner; scattered; ostiolate but the opening is not easily observed as the perithecia are densely clothed with dense terminal hairs; mature perithecia 60–175 μm in diameter, usually globose, although subglobose or oval forms are also quite common. Asci short-stalked, cylindrical, eight-spored, arranged in a single row, 50–60 × 7–8 μm. Ascospores olive brown in color, globose or subglobose, smoothwalled, 6.5–9.5 μm in diameter, with a single apiculus.

Habitat diversity: Nesting materials, decomposing wheat straw, mushroom compost, and soil.

Geographic distribution: United States, United Kingdom, India, Netherlands, and Ghana.

Description: Based on our isolate maintained in our culture collection under accession number T-132.

Chaetomium virginicum Ames (1963)

Description: Colonies appear rich brown in color at 45 °C, hyphae 2–4 μm wide, often anastomose to form a loose network. Perithecia brown, globose, attached to the substratum with undifferentiated rhizoids, terminal hairs cover the entire perithecium, giving it the appearance of a tumbleweed in miniature. Hairs dense, granular, intricate, irregularly branched, or suggestive of dichotomous branching. Asci stalked, long, cylindrical, eight-spored, 70 × 10 μm. Ascospores unistichous, light yellow brown to pale brown, almond-shaped, 8–11 μm.

Habitat diversity: On decomposing leaves.
Geographic distribution: United States.
Description: Based on Ames (1963).

16.3 Enzymes of Thermophilic Chaetomia

16.3.1 Cellulase

The cellulase system in fungi is considered to comprise three hydrolytic enzymes: (i) the endo-(1,4)-β-D-glucanase (synonyms: endoglucanase, endocellulase, and carboxymethyl cellulase [EC 3.2.1.4]), which cleaves β-linkages at random, commonly in the amorphous parts of cellulose; (ii) the exo-(1,4)-β-D-glucanase (synonyms: cellobiohydrolase, exocellulase, microcrystalline cellulase, and Avicelase [EC 3.2.1.91]), which releases cellobiose from either the nonreducing or the reducing end, generally from the crystalline parts of cellulose; and (iii) the β-glucosidase (synonym: cellobiase [EC 3.2.1.21]), which releases glucose from cellobiose and short-chain cellooligosac-charides (33). Although β-glucosidase has no direct action on cellulose, it is regarded as a component of cellulase system because it stimulates cellulose hydrolysis (Bhat and Bhat 1997).

Purified thermophilic fungal cellulases have been characterized in terms of their molecular weight, optimal pH, optimal temperature, thermostability, and glycosylation. Usually, thermophilic fungal cellulases are single polypeptides although it has been reported that some beta-glucosidases are dimeric (Mamma et al. 2004). The molecular weight of thermophilic fungal cellulases spans a wide range (30–250 kDa) with different carbohydrate contents (2–50%). Optimal pH and temperature are similar for the majority of the purified cellulases from thermophilic fungi. Thermophilic fungal cellulases are active in the pH range 4.0–7.0 and have a high temperature maximum at 50–80 °C for activity (Li et al. 2011). In addition, they exhibit remarkable thermal stability and are stable at 60 °C with longer half-lives at 70, 80, and 90 °C than those from other fungi.

16.3.2 Structure of Thermophilic Fungal Cellulases

Primary Structure A common characteristic of cellulases is their modular structure. Typically, endocellulases and cellobiohydrolases are composed of four domains or regions (Fig. 16.1): a signal peptide that mediates secretion, a cellulose-binding domain (CBD) for anchorage to the substrate, a hinge region (linker) rich in Ser, Thr, and Pro residues, and a catalytic domain (CD) responsible for the hydrolysis of the substrate. The mature proteins are *O*- and *N*-glycosylated in the hinge region and the CDs, respectively. The effect of the glycosylation sites in the hinge

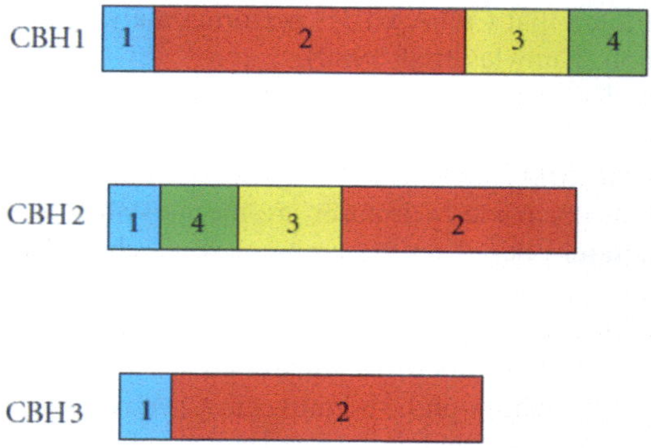

Fig. 16.1 Domain organization of cellobiohydrolases CBH1 (AY861347), CBH2 (AY861348), and CBH3 of *C. thermophilum* (DQ085790). 1: signal peptide region, 2: catalytic domain, 3: hinge region, and 4: cellulose-binding domain. (After Li et al. 2011)

region is not clear yet but they may play a role in the flexibility and disorder of the linker (Beckham et al. 2010). Variations between cellulases within the same mechanistic class have been observed. An example is illustrated by *T. emersonii* CBHII, which is characterized by a modular structure (Murray et al. 2003) whereas CBH1 from the same fungus consists solely of a catalytic domain (Grassick et al. 2004). Similarly, *Chaetomium thermophilum* CBH1 and CBH2 consist of a typical CBD, a linker, and a catalytic domain. In contrast, CBH3 only comprises a catalytic domain and lacks a CBD and a hinge region (Li et al. 2009).

Fungal CBDs are composed of less than 40 amino acid residues, and they interact with cellulose through a flat or platform-like hydrophobic binding site formed by three conserved aromatic residues. The binding site is thought to be complementary to the flat surfaces presented by cellulose crystals (Hashimoto 2006; Shoseyov et al. 2006). The (110) faces of the cellulose crystalline microfibrils have been proposed as the putative CBD binding site (Dagel et al. 2011). With this arrangement, the glucopyranoside rings of cellulose are expected to be fully exposed and available for hydrophobic interactions.

The CAZy (see http://www.cazypedia.org/index.php/) classification shows that the thermophilic soil-dwelling fungus *Chaetomium thermophilum* possesses at least seven β-glucosidase/cellulase enzymes from GH families 3, 6, and 7 (Cantarel et al. 2009). The cellobiohydrolases from the GH6 family of enzymes are processive enzymes that remove the disaccharide cellobiose in a processive manner, acting from the nonreducing end of cellulosic polysaccharides (Rouvinen et al. 1990; Koivula et al. 1996). This disaccharide liberating activity is believed to reflect the structural properties of the cellulose molecule. Each β-1,4-glucose moiety is rotated 180° relative to the next along the axis of the chain. Thus, only every second glycosidic linkage is presented to the catalytic machinery in the correct orientation to allow bond hydrolysis. The catalytic core module of the Cel6A cellobiohydrolases has been shown to form a modified TIM-barrel structure (Rouvinen et al. 1990) in which a buried cleft spanning the C-terminal region of the domain and capable of receiving β-1,4-glucose polysaccharides of various lengths is thought to contain the catalytic active site (Koivula et al. 1996; Varrot et al. 1999). It has been demonstrated that Cel6A CBHs perform catalysis with inversion of anomeric configuration (Knowles et al. 1988).

By now, many glycoside hydrolases have been isolated from C. thermophilum, such as a GH55 β-1,3-glucanase (Papageorgiou and Li 2015), a β-glucosidase (Xu et al. 2011), and a cellobiohydrolase II (Wang et al. 2013). Generally, C. thermophilum glycoside hydrolases are thermostable and have a high optimal reaction temperature based on the previous researches. Thermostable enzymes have potential advantages in lignocelluloses conversation, on account of effectively improving hydrolysis efficiency and reducing the possible contamination at high temperature in industrial processes (Huy et al. 2016).

The thermophilic ascomycete *C. thermophilum* also awakes interests for biotechnological applications. In order to determine the enzymatic properties, these fungus glycosyl hydrolases-encoding genes have been cloned and expressed in *P. pastoris* and *T. reesei*. Cellobiohydrolase II (CBHII)-encoding gene was subject to in vitro directed evolution (Wang et al. 2012) aiming the thermostability increasing. In addition to an enhanced thermal stability, the two CBHII mutant versions produced in *P. pastoris* also presented higher optima temperature and pH values (60 °C, pH 5–6) in comparison to the wild-type enzyme (50 °C and pH 4). Not only thermostability accounts for *C. thermophilum* enzymes' advantages but also Voutilainen et al. (2008) reported that the acidic cellobiohydrolase Cel7A produced in *P. pastoris* was more thermostable (by 10 °C) and more active (by fourfold) in the hydrolysis of microcrystalline cellulose when compared to *T. reesei* Cel7A, thus representing a competitive choice for industrial purposes. Another interesting example of such a versatility is xylanase Xyn11A which produced by *T. reesei* (Mäntylä et al. 2007). Experiments showed that pH 7.0 and at 70 °C of this enzyme could be commercially feasible for industrial-scale bleaching of kraft pulp at high temperatures in comparison with other enzymes produced by other taxa.

16.3.3 Xylanases

Next to cellulose, xylan is the most abundant structural polysaccharide in nature. Its complete degradation requires the cooperative action of a variety of hydrolytic enzymes: the endoxylanases (EC 3.2.1.8), which randomly cleave 13–1,4-linked xylose (the xylan backbone); the 13-xylosidases (EC 3.2.1.37), which hydrolyze xylooligomers; and the different side-branch-splitting enzymes, for example, a-glucuronidase and a-arabinosidase, acetylxylan esterase, and acetyl esterase, which liberate other sugars (glucuronic acid arabinose) that are attached as branches to the backbone (Biely 1985). Xylanases of thermophilic fungi are receiving considerable attention because of their application in biobleaching of pulp in the paper industry, wherein the enzymatic removal of xylan from lignin-carbohydrate complexes facilitates the leaching of lignin from the fiber cell wall, obviating the need for chlorine for pulp bleaching in the brightening process. They also have applications in the pretreatment of animal feed to improve its digestibility.

A variety of materials have been used for induction of xylanases: pure xylan (Gomes et al. 1993; Purkarthofer et al. 1993) and xylan-rich natural substrates, such as sawdust, corn cob, wheat bran, sugar beet pulp, and sugarcane bagasse (Maheshwari et al. 2000). Paper of inferior quality was an excellent carbon source and inducer for xylanase in *Thermoascus aurantiacus*, *Humicola lanuginosa*, and *Paecilomyces varioti* (Maheshwari et al. 2000). In *Melanocarpus albomyces* and *Thermomyces lanuginosus*, xylose, the pentosan unit of xylan, could also induce xylanase. Xylanases are often coinduced with cellulases by pure cellulose, as in *T. aurantiacus*, *Chaetomium thermophile* var. *coprophile*, and H. insolens (Maheshwari et al. 2000).

The xylanases from these fungi possess optimum temperatures between 60 and 80 °C and are very stable in this range. These enzymes are usually glycoproteins and most show highest activity at an acid pH (4.5–6.5). They exist in a multiplicity of forms and the majority exhibit variable MWs in the range 6–38 kDa. Many endoxylanases from thermophiles have some degree of structural homology with those from mesophiles. A number of authors have tried to explain the thermostability observed in enzymes from thermophiles in terms of extra disulfide bridges, an N-terminal proline residue causing a reduction in conformational freedom, salt bridges, and presence of hydrophobic side-chains (Turunen et al. 2001). Hakulinen et al. (2003) describe also some minor modifications responsible for the increased thermal stability of xylanases: (1) higher Thr/Ser ratio; (2) increased number of charged residues, especially Arg, resulting in enhanced polar interactions; and (3) improved stabilization of secondary structures involving a higher number of residues in the β-strands and stabilization of the α-helix region.

In 2003, Hakulinen et al. found that the crystal structures of thermophilic xylanases isolated from *Chaetomium thermophilum* determined at 1.75 Å and compared with other 12 xylanases. The enzymes have the overall fold typical to family 11 xylanases with two highly twisted β-sheets forming a large cleft. The comparison of 12 crystal structures of family 11 xylanases from both mesophilic and thermophilic organisms showed that the structures of different xylanases are very similar. The sequence identity differences correlated well with the structural differences. Several minor modifications appeared to be responsible for the increased thermal stability of family 11 xylanases: (a) higher Thr:Ser ratio, (b) increased number of charged residues, especially Arg, resulting in enhanced polar interactions, and (c) improved stabilization of secondary structures involved the higher number of residues in the β-strands and stabilization of the α-helix region. Some members of family 11 xylanases have a unique strategy to improve their stability, such as a higher number of ion pairs or aromatic residues on protein surface, a more compact structure, a tighter packing, and insertions at some regions resulting in enhanced interactions.

Hakulinen et al. (2003) found that the overall structure of xylanase from C. thermophilum (CTX) was dominated by one α-helix and two strongly twisted β-sheets, which were packed against each other. This is the protein fold of family 11 xylanases. According to Törrönen et al. (1992), the shape of the molecule resembles a right hand: two β-sheets and the α-helix form fingers and a palm, a long loop between the B7 and B8 strands forms a thumb, and a loop between the B6 and B9 strands forms a cord (Fig. 16.2a). The final model of CTX contained residues 1–191 for both molecules in the asymmetric unit (labeled A and B). The first residue, glutamine, was deaminated and cyclized to pyrrolidone carboxylic acid. When the ESI mass spectrum of CTX was measured, the unique molecular masses, 21,479 Da and 21,682 Da, were obtained. Assuming that CTX contains 196 residues, the calculated molecular mass would be 21,478 Da, which agrees well with the lower mass obtained. The difference between the two obtained masses was 203 Da corresponding to one N-acetyl-glucosamine (GlcNAc). There is one potential N-glycosylation site (Asn62) in the sequence of CTX, but there was no clear sign of glycosylation in the electron density map. It is possible that only the protein molecules without GlcNAcs had been crystallized or that the GlcNAc is disordered.

According to the mass spectrum, approximately 20% of the material did not contain GlcNAc or alternatively, the GlcNAc had been lost. In the crystal structure, a glycerol molecule was located in the active site of molecule A, but was not observed in molecule B. The cryoprotectant soaking solution was most likely the source of glycerol, which was packed against Trp19 by stacking interactions and was hydrogen-bonded to carboxyl group of Pro127. In addition, Arg123 had two conformations in molecule A and in one of the conformations the guadinine group of the Arg was located toward the hydroxyl group of the glycerol. The rms deviation between the A and B molecules of CTX was 0.8 A°. The crystal structure showed four sulfate ions and a calcium ion in the asymmetric unit. The calcium ion was located between molecules A and B exactly on the noncrystallographic axis. The calcium ion interacted with side chains Oc of Thr10 from molecule A and B, both of which clearly had two conformations in the electron density map. Two of the sulfate ions were located exactly on the crystallographic axes. In addition to these two sulfate ions, which are attached to Argresidues A27 and B27, there are two other sulfates, which are attached to Arg residues A68 and B68. Due to the crystal

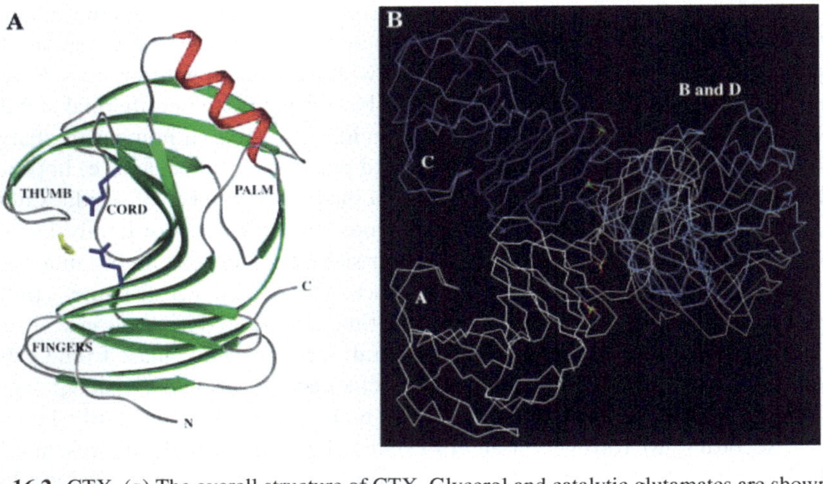

Fig. 16.2 CTX. (**a**) The overall structure of CTX. Glycerol and catalytic glutamates are shown in the active site. (**b**) A tetrameric assembly with sulfate ions. Molecules A and B are shown in white and symmetry molecules C and D in blue. (Hakulinen et al. 2003)

packing, the enzyme resembles a tetrameric assembly (Fig. 16.2b). Four sulfate ions link molecule A to symmetry molecule D and correspondingly molecule B to symmetry molecule C. However, according to the dynamic light scattering measurements, the protein was a monomer. Therefore, the sulfate ions from the crystallization solution might have been involved in this "tetramerization" process. It is possible that tetramers were assembled first and their stacking then led to crystal formation in the high salt concentration.

16.4 Glucoamylases

The application of glucoamylases (E.C. 3.2.1.3, glucan 1,4-α-glucosidase) in starch saccharification lies in sugar industry due to its ability to release glucose as the major end product, which is used in food, beverage, ethanol, amino acids, and organic acids (Kumar and Satyanarayana 2009). Glucoamylase is one of the enzymes of worldwide interest in starch saccharification to yield glucose for use in food and fermentation industries. Glucoamylase is one of the high demand commercial biocatalysts in food industry, which is required in higher tonnage than almost any other enzyme (Reilly 1999; Ford 1999).

Industrially glucoamylases are produced from filamentous fungi, *Aspergillus* and *Rhizopus* spp. via submerged fermentation using production medium with a typical concentration of 20% corn and 2.5% of corn steep liquor at 30–35 °C (Nigam and Singh 1995). The fungal glucoamylases being optimally active at around pH 4.0–4.5, the saccharification is essentially carried out under acidic conditions at 60 °C for 3–4 days to achieve a final yield of 98% glucose (Manjunath et al. 1983; Crabb and Mitchinson 1997). Glucoamylase is one of the high demand commercial biocatalysts in food industry, which is required in higher tonnage than almost any other enzyme (Reilly 1999). A number of glucoamylase production media for a variety of microbes have been optimized at laboratory scale including liquid, solid as well as aqueous two-phase systems. Different fermentation vessels, from shake flasks to bioreactors, and strategies including batch, fed-batch, and continuous fermentations have been employed (Uma Maheswar Rao et al. 2011). Glucoamylase production by *Thermomyces lanuginosus* was found to be 2.5-fold higher in shake flasks compared to static cultures (Uma Maheswar Rao et al. 2011).

Chen et al. (2005) purified a thermostable extracellular glucoamylase (exo-1, 4-a-D-glucanohydrolase, E.C.3.2.1.3) from the culture supernatant of a thermophilic fungus *Chaetomium thermophilum* to sodium dodecyl sulfate-polyacrylamide gel electrophoresis (SDSPAGE) homogeneity by using ammonium sulfate fraction, DEAE-Sepharose Fast Flow chromatography, and Phenyl-Sepharose Fast Flow chromatography. SDS-PAGE of the purified enzyme showed a single protein band of molecular weight 64 kDa. The glucoamylase exhibited optimum catalytic activity at pH 4.0 and 65 °C. It was thermostable at 50 °C and 60 °C, and retained 50% activity after 60 min at 65 °C. The half-life of the enzyme at 70 °C was 20 min. N-terminal amino acid sequencing (15 residues) was AVDSYIERETPIAWN. Different metal ions showed different effects on the glucoamylase activity of *C. thermophi-*

lum. Ca^{2+}, Mg^{2+}, Na^{+}, and K^{+} enhanced the enzyme activity, whereas Fe^{2+}, Ag^{+}, and Hg^{2+} cause obvious inhibition. These properties make it applicable to other biotechnological purposes.

16.5 Cellobiohydrolase

Thermoactive and thermostable cellulases, in general, have favorable effects on the resistance of adverse conditions, including high salt concentrations and extreme pHs. In particular, enzymes with such excellent properties are desired to effectively enhance hydrolysis efficiency at elevated temperatures while simultaneously reducing microbial contamination in industrial processes (Alponti et al. 2016). Therefore, it is essential to explore thermoactive enzymes with considerable thermostability. *Chaetomium thermophilum* produces multiple thermostable cellulases with high efficiencies (Bock et al. 2014), such as a β-1,4-endoglucanase CTendo45 (Chen et al. 2018) and two cellobiohydrolases CtCel6 (Zhou et al. 2017) and CtCBH1 (Lee et al. 2017). According to the classification of the Carbohydrate-Active Enzyme (CAZy) database (Cantarel et al. 2009), cellobiohydrolases are mainly assigned into two glycoside hydrolase families (GH6 and GH7). Besides, the GH6 cellobiohydrolase is extensively considered to act processively from the nonreducing terminal of cellulose chains to release disaccharide cellobioses with an atypical single-displacement mechanism (Thompson et al. 2012). However, although GH6 cellobiohydrolases possess excellent activity and high thermostability, there is certainly room for characteristic improvement (Baramee et al. 2017).

Han et al. (2018) studied a thermostable cellobiohydrolase CtCel6 from *Chaetomium thermophilum* with high hydrolytic activity was employed to construct mutants to further enhance catalytic activity and thermostability. Based on structural analysis of the corresponding homologous model (Fig. 16.3a), four conserved and noncatalytic residues around the substrate-binding site in buried cleft were selected for site-directed mutagenesis (Fig. 16.3b). These recombinant enzymes were successfully expressed using the yeast *Pichia pastoris* and purified to determine the biochemical properties. The wild-type and mutant cellobiohydrolases shared a similar pattern of the optimum reaction condition at pH 5 and 70 °C, which could be attributed to the inapparent conformational rearrangement caused by residue substitutions (Xie et al. 2014).

In their study, the mutant Y119F increased the catalytic activity 1.82-, 1.65-, and 1.43-fold against β-D-glucan, phosphoric acid swollen cellulose (PASC), and carboxymethylcellulose sodium (CMC-Na), respectively. In addition, S131 W effectively enhanced the enzyme's heat resistance to elevated temperatures. The half-life (t1/2) of this mutant enzyme was increased 1.42- and 2.40-fold at 80 °C and 90 °C, respectively, compared to the wild-type. This study offers initial insight into the biological function of the conserved and noncatalytic residues of thermostable cellobiohydrolases and provides a valid approach to the improvement of enzyme redesign proposal.

16.6 Superoxide Dismutase

A thermostable superoxide dismutase (SOD) from the culture supernatant of a thermophilic fungus *Chaetomium thermophilum* strain CT2 was purified to homogeneity by fractional ammonium sulfate precipitation, ion-exchange chromatography on DEAE-sepharose, and phenyl-sepharose hydrophobic interaction chromatography (Guo et al. 2008). The pure SOD had a specific activity of 115.77 U/mg of protein and was purified 7.49-fold, with a yield of 14.4%. The molecular mass of a single band of the enzyme was estimated to be 23.5 kDa, using sodium dodecyl sulfate-polyacrylamide gel electrophoresis. Using gel filtration on Sephacryl S-100, the molecular mass was estimated to be 94.4 kDa, indicating that this enzyme was composed of four identical subunits of 23.5 kDa each. The SOD was found to be inhibited by NaN_3, but not by KCN and H_2O_2. Atomic absorption spectrophotometric

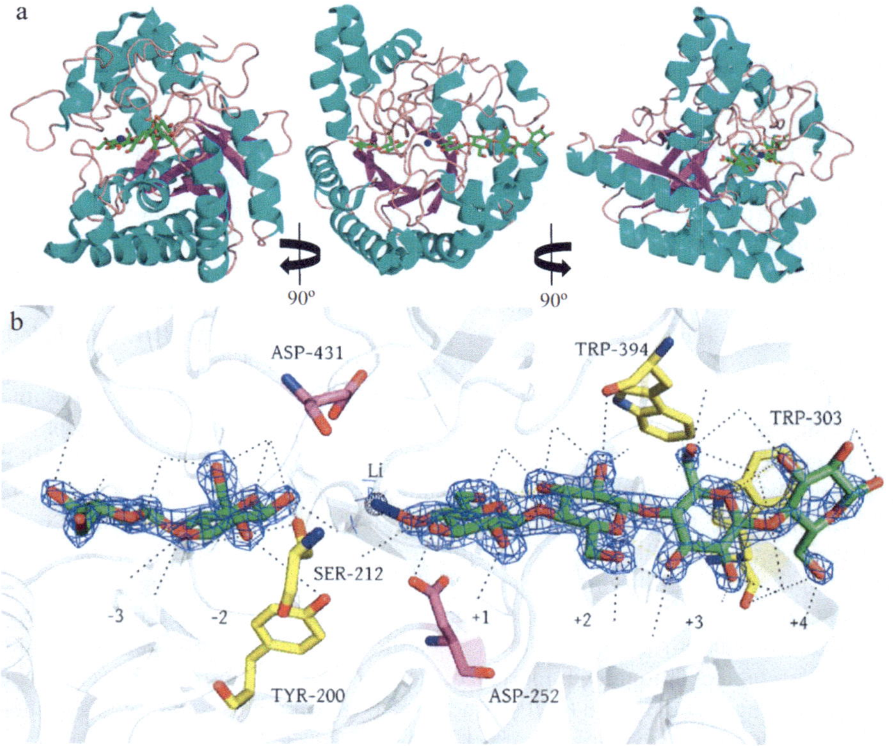

Fig. 16.3 Structure of C. thermophilumCel6A (PDB: 4A05). (**a**) Divergent stereo cartoon of Cel6A structure in complex with both cellobiose and cellotetraose molecules. (**b**) Observed electron density for the ligands in the active site cleft. The yellow and purple sticks indicate the mutated noncatalytic residues and the catalytic residues, respectively. The active-center Li + ion is shown as a blue sphere and hydrogen bonds are represented as black dashed lines. All of the structural diagrams were drawn using PyMOL software. (Han et al. 2018)

analysis showed that the content of Mn was 2.05 mg/mg of protein and Fe was not detected in the purified enzyme. These results suggested that the SOD in C. thermophilum was the manganese superoxide dismutase type. N-terminal amino acid sequencing (10 residues) was KX (X is uncertain) TLPDLKYD. The Nterminal amino acid sequencing homologies to other MnSod also indicated that it was a manganese-containing superoxide dismutase. The SOD exhibited maximal activity at pH 7.5 and optimum temperature at 60 C. It was thermostable at 50 and 60 C and retained 60% activity after 60 min at 70 C. The half-life of the SOD at 80 C was approximately 25 min and even retained 20% activity after 30 min at 90 C.

In industry, a major requirement for commercial SOD is thermal stability because thermal denaturation is a common cause of enzyme inactivation. Thermostable SOD potentially is useful due to its high stability. In recent years, there has been an increasing interest in SOD of thermophiles, which were expected to produce thermostable SOD. In the 1990s, an antioxidant enzyme—superoxide dismutase (SOD)—was introduced into the market. Although the enzyme initially showed great promise in therapeutic applications (Fig. 16.4), it did not perform up to expectations. Consequently, its use was limited to nondrug applications in humans and drug applications in animals.

Biotechnological Application of Thermophilic Enzymes
Although thermophilic fungi have existed in nature for millennia, they were cultured in the laboratory only in late nineteenth and early twentieth centuries. These fungi comprise a small assemblage among the Eukarya domain that possesses a unique mechanism of growing at elevated temperatures up to 61 °C. During the last five decades, many species of thermophilic fungi sporulating at 45 °C have been reported. The first modern comprehensive account of the biology and classification of thermophilic fungi was published by Cooney and Emerson in 1964; it included the 11 species known at that time, with a few new to science. Since then, several thermophilic fungi have been discovered and documented in scientific literature. Notwithstanding their potential use in industrial processes, studies on thermophilic fungi have been neglected until recently. Further, their uncertain taxonomic affiliation puts them in a state of disarray, often leading to misidentification and confusion.

Thermophilic fungi have been playing a role in the economy of nature ever since they evolved on this earth. Their importance in the human economy has been realized from their ability to efficiently degrade organic matter, acting as biodeteriorants and natural scavengers; to produce extracellular and intracellular enzymes, organic acids, amino acids, antibiotics, phenolic compounds, polysaccharides, and sterols of biotechnological importance; and to produce nutritionally enriched feeds and single-cell protein (SCPs), and as well as from their suitability as agents of bioconversion, for example, their role in the preparation of mushroom compost. The fungal protein, or "mycoprotein," is attracting the attention of food and feed scientists and protein engineers. *Chaetomium cellulolyticum* and *Sprotrichum pulverulentum* are the most widely used organisms for the upgrading of animal feed and producing SCP from lignocellulosic wastes (see Chap. 13). Similarly, investigations on the process of composting municipal solid wastes with thermophilic

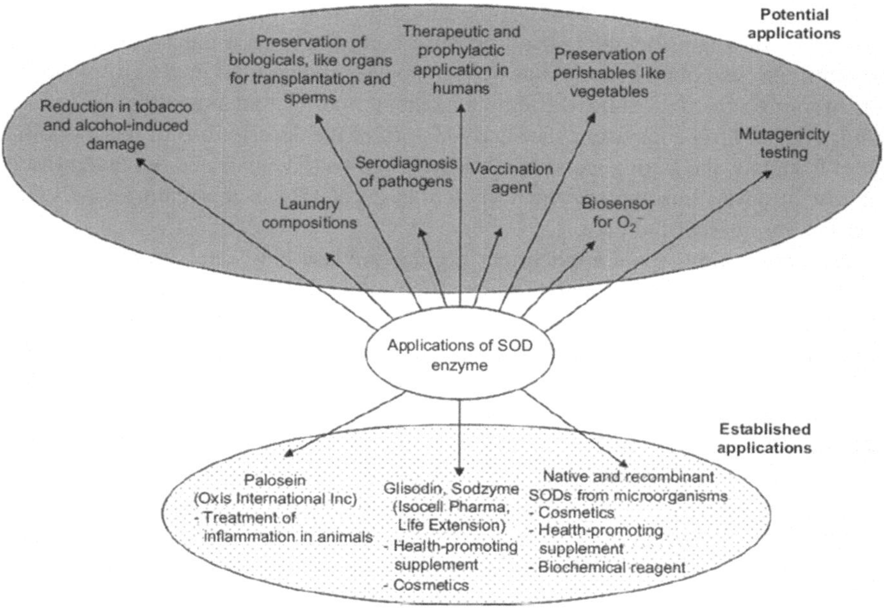

Fig. 16.4 Applications of superoxide dismutase (SOD) enzyme

fungi, that is, *Chaetomium thermophile*, *Humicola lanuginosa*, *Mucor pusillus*, and *Thermoascus aurantiacus*, have revealed that the resulting compost is richer in N, P, and K.

On the industrial front, the use of thermophilic strains can be an effective solution to the maintenance of optimal temperature in industrial fermentation for the entire cultivation period. It is well known that thermophilic activities of microbes are generally associated with protein and enzyme thermostability. The advantages of the use of thermostable proteins and enzymes for conducting biotechnological processes at high temperature include a reduced risk of contamination with mesophilic microbes, a decrease in the viscosity of the culture medium, an increase in the bioavailability and solubility of organic compounds, and an increase in the diffusion coefficient of substrates and products, resulting in a higher rate of reactions. Further, their involvement in genetic manipulations is a much more recent development. Nevertheless, because of these and many more advantages, thermophilic fungi appear to be suitable candidates in biotechnological applications. Additionally, with the paradigm shift in industry as it moves from fossil fuels toward renewable resource utilization, the need for microbial biocatalysts is envisaged to increase, and undoubtedly there will be an unrelenting and increased need for thermostable selective biocatalysts in the near future. Thus, future perspectives relating to their diversity, taxonomy, phylogeny, genome-wide study, and biotechnology, entailing research on thermophilic fungi, are warranted.

The biotechnological potential of thermophilic fungi has been known to microbiologists for a considerable period because composting as a means of providing nutrient-enriched plant material has been in vogue for degradation of agro-residues, mushroom production, solid waste management, and for understanding the role of fungi in plant litter ecosystem. The early history of the descriptions of thermophilic fungi from hay and retting gauyule is too well known to be narrated but this infused interest in unraveling the enzymatic potential especially in the realm of polysaccharases, proteases, and lipases.

The major commercial application of glucoamylase is to catalyze starch and to yield glucose for use in food and fermentation industries. Glucose production from starch, along with glucoamylase requires the synergistic action of a series of amylases. In the first step, ~30–35% dry solid starch slurry is gelatinized (~60–90 °C) and subsequently liquefied at 95–105 °C (pH 6.5) by α-amylase to short-chain dextrins. These dextrins in the next step are saccharified by glucoamylase to release glucose. The fungal glucoamylase being optimally active at around pH 4.0–4.5, the saccharification is essentially carried out under acidic conditions at 60 °C for 3–4 days to achieve a final yield of ~96% glucose (Crabb and Mitchinson 1997; Reilly 1999).

Additionally, debranching enzymes (pullulanase or isoamylase) are used to hasten starch processing by cleaving α-1,6 glycosidic bonds, which allows to attain an early peak in glucose yield with less byproduct formation (Uma Maheswar Rao et al. 2011). Glucose has ~75% of the sweetness of sucrose, while its isomer fructose is twice sweeter than sucrose. Consequently, fructose is preferred especially in so-called low-calorie health/diet foods, where it provides double the sweetness of sucrose at half the weight and can be metabolized without insulin (Uma Maheswar Rao et al. 2011). Commercially fructose is produced by the isomerization of glucose using fungal glucose/xylose isomerase (E.C. 5.3.1.5, D-xylose-ketol isomerase) at 50–60 °C and pH 7–8. Glucose isomerase is the most expensive of all the enzymes involved in starch processing, and thus, is reused until it loses most of its activity. The concentrated glucose syrup is passed through the immobilized glucose-isomerase column or sometime through the glucose isomerase producing cells (Uma Maheswar Rao et al. 2011). The process yields around 40–42% of fructose and the concentration of fructose in the final product is raised to ~55% by chromatographic enrichment of glucose–fructose mixture (Uma Maheswar Rao et al. 2011).

Obtaining new biotechnological products from uncultivable microfungi is quite interesting and a current topic of debate. Both basic research and biotechnological developments require routinely applicable tools for the functional analysis, expression, and manipulation of genes. These techniques include procedures for genetic transformation and selection of the transformants, well-characterized molecular markers, and expression signals. As thermophilic fungi colonize, multiply, and survive in habitats having elevated temperatures, they represent a formidable pool of bioactive compounds and are a strategic source for new and successful commercial products. Recent technological advances made in genomics, proteomics, and combinatorial chemistry show that nature continues to preserve compounds in its metagenome having the essence of bioactivity or function within the host and the environment. Bioprospecting of fungal genomes, such as thermophilic fungi, offers several advantages over their biocatalysts, besides being thermostable. However, studies on fungal distribution and mapping are challenging due to the lack of sufficient knowledge about their taxonomy and the lack of expert mycologists around the world. Nevertheless, the fungal world provides a fascinating and almost continual source of biological diversity, which is a rich source to exploit for human welfare.

References

Abdullah SK, Zora SE (1993) *Chaetomium mesopotamicum*, a new thermophilic species from Iraqi soil. Cryptogam Bot 3:387–389

Adams MWW, Kelly RM (1998) Finding and using hyperthermophilic enzymes. TIBTECH 16:329–332

Alponti JS, Maldonado RF, Ward RJ (2016) Thermostabilization of *Bacillus subtilis* GH11 xylanase by surface charge engineering. Int J Biol Macromol 87:522–528

Ames LM (1963) A Monograph of the Chaetomiaceae. U.S. Army Research and Development Series 2. Bibliotheca Mycologica Cramer, Lehre, Germany

Aneja KR, Kumar R (1994) *Chaetomium senegalense*—a new record from India. Proc Natl Acad Sci India Sect B 64:229–230

Baramee S, Teeravivattanakit T, Phitsuwan P, Waeonukul R, Pason P, Tachaapaikoon C (2017) A novel GH6 cellobiohydrolase from Paenibacillus curdlanolyticus B-6 and its synergistic action on cellulose degradation. Appl Microbiol Biotechnol 101:1175–1188

Beckham GT, Bomble YJ, Matthews JF et al (2010) The O-glycosylated linker from the Trichoderma reesei family 7 cellulase is a flexible, disordered protein. Biophys J 99(11):3773–3781

Bhat MK, Bhat S (1997) Cellulose degrading enzymes and their potential industrial applications. Biotechnol Adv 15:583–620

Biely P (1985) Microbial xylanolytic systems. Trends Biotechnol 3:286–290

Bock T, Chen WH, Ori A, Malik N, Silva-Martin N, Huerta-Cepas J (2014) An integrated approach for genome annotation of the eukaryotic thermophile *Chaetomium thermophilum*. Nucleic Acids Res 42:13525–13533

Cantarel BL, Coutinho PM, Rancurel C, Bernard T, Lombard V, Henrissat B (2009) The Carbohydrate-Active EnZymes database (CAZy): an expert resource for glycogenomics. Nucleic Acids Res 37:D233–D238

Chen J, Li DC, Zhang YQ, Zhou QX (2005) Purification and characterization of a thermostable glucoamylase from *Chaetomium thermophilum*. J Gen Appl Microbiol 51:175–181

Chen X, Li W, Ji P, Zhao Y, Hua C, Han C (2018) Engineering the conserved and noncatalytic residues of a thermostable β-1,4-endoglucanase to improve specific activity and thermostability. Sci Rep 8(1):2954

Coolbear T, Daniel RM, Morgan HW (1992) The enzymes from extreme thermophiles: bacterial sources, thermo stabilities and industrial relevance. Adv Biochem Eng Biotechnol 45:57–67

Cooney DG, Emerson R (1964) Thermophilic fungi: an account of their biology, activities, and classification. W.H. Freeman and Co., San Francisco

Crabb W, Mitchinson C (1997) Enzymes involved in the processing of starch to sugars. TIBTECH 15:349–352

Dagel DJ, Liu YS, Zhong L et al (2011) In situ imaging of single carbohydrate-binding modules on cellulose microfibrils. J Phys Chem B 115(4):635–641

Doveri F (2013) An additional update on the genus *Chaetomium* with descriptions of two coprophilous species, new to Italy. Mycosphere 4:820–846

Enzymes for Industrial Applications (2004) Business Communications Company, Inc; RC-147U.

Ford C (1999) Improving operating performance of glucoamylase by mutagenesis. Curr Opin Biotechnol 10:352–357

Gomes J, Gomes I, Kreiner W, Esterbauer H, Sinner M, Steiner W (1993) Production of high level of cellulase-free and thermostable xylanase by a wild strain of *Thermomyces lanuginosus* using beechwood xylan. J Biotechnol 30:283–297

Grassick A, Murray PG, Thompson R et al (2004) Threedimensional structure of a thermostable native cellobiohydrolase, CBH IB, and molecular characterization of the cel7 gene from the filamentous fungus, Talaromyces emersonii. Eur J Biochem 271(22):4495–4506

Guo F-x, E S-j, Liu S-a, Chen J, Li D-c (2008) Purification and characterization of a thermostable MnSOD from the thermophilic fungus *Chaetomium thermophilum*. Mycologia 100(3):375–380. https://doi.org/10.3852/06-111R

Hakulinen N, Turunen O, Janis J, Leisola M, Rouvinen J (2003) Three-dimensional structures of thermophilic beta-1,4-xylanases from *Chaetomium thermophilum* and *Nonomuraea flexuosa*. Comparison of twelve xylanases in relation to their thermal stability. Eur J Biochem 270:1399_1412

Han C, Li W, Hua C, Sun F, Bi P, Wang Q (2018) Enhancement of catalytic activity and thermostability of a thermostable cellobiohydrolase from *Chaetomium thermophilum* by site-directed mutagenesis. Int J Biol Macromol 116:691–697

Hashimoto H (2006) Recent structural studies of carbohydrate binding modules. Cell Mol Life Sci 63(24):2954–2967

Huy ND, Nguyen CL, Park HS, Loc NH, Choi MS, Kim DH, Seo JW, Park SM (2016) Characterization of a novel manganese dependent endoglucanase belongs in GH family 5 from *Phanerochaete chrysosporium*. J Biosci Bioeng 121:154–159

Kristjansson JK (1989) Thermophilic organisms as sources of thermostable enzymes. TIBTECH 7:349–353

Kumar S, Satyanarayana T (2009) Microbial glucoamylases: characteristics and applications. Crit Rev Biotechnol 29:225–255

Kunze G, Schmidt JK (1817) Mykologische Hefte 1 (1). Leipzig, 16 pp.

Koivula A, Reinikainen T, Ruohonen L, Valkeajärvi A, Claeyssens M, Teleman O, Kleywegt GJ, Szardenings M, Rouvinen J, Jones TA, Teeri TT (1996) The active site of *Trichoderma reesei* cellobiohydrolase II: the role of tyrosine 169. Protein Eng 9:691–699

Knowles J, Teeri TT, Lehtovaara P, Penttila M, Saloheimo M (1988) The use of gene technology to investigate fungal cellulolytic enzymes. In Aubert JE, Béguin E, Millet J (eds) Biochemistry and Genetics of Cellulose Degradation, FEMS Symposium. Academic Press, New York, USA, pp 153–169

Lee CR, Sung BH, Lim KM, Kim MJ, Sohn MJ, Bae JH, Sohn JH (2017) Co-fermentation using recombinant *Saccharomyces cerevisiae* yeast strains hyper-secreting different cellulases for the production of cellulosic bioethanol. Sci Rep 7(1):4428

Li YL, Li H, Li AN, Li DC (2009) Cloning of a gene encoding thermostable cellobiohydrolase from the thermophilic fungus *Chaetomium thermophilum* and its expression in *Pichia pastoris*. J Appl Microbiol 106(6):1867–1875

Li D-C, Li A-N, Papageorgiou AC (2011) Cellulases from thermophilic fungi: recent insights and biotechnological potential. Enzyme Res 2011, Article ID 308730, 9 pages. https://doi.org/10.4061/2011/308730

Maheshwari R, Bharadwaj G, Bhat MK (2000) Thermophilic fungi: their physiology and enzymes. Microbiol Mol Biol Rev 64:461–488

Mamma D, Hatzinikolaou DG, Christakopoulos P (2004) Biochemical and catalytic properties of two intracellular β-glucosidases fromthe fungus *Penicillium decumbens* active on flavonoid glucosides. J Mol Catal B 27(4–6):183–190

Manjunath P, Shenoy BC, Raghavendra Rao MR (1983) Fungal glucoamylase. J Appl Biochem 5:235–260

Mäntylä A, Paloheimo M, Hakola S, Lindberg E, Leskinen S, Kallio J, Vehmaanperä J, Lantto R, Suominen P (2007) Production in *Trichoderma reesei* of three xylanases from *Chaetomium thermophilum*: a recombinant thermoxylanase for biobleaching of kraft pulp. Appl Microbiol Biotechnol 76:377–386

Murray PG, Collins CM, Grassick A, Tuohy MG (2003) Molecular cloning, transcriptional, and expression analysis of the first cellulase gene (cbh2), encoding cellobiohydrolase II, from the moderately thermophilic fungus *Talaromyces emersonii* and structure prediction of the gene product. Biochem Biophys Res Commun 301(2):280–286

Nigam P, Singh D (1995) Enzyme and microbial systems involved in starch processing. Enzym Microb Technol 17:770–778

Papageorgiou AC, Li D (2015) Expression, purification and crystallization of a family 55 β-1,3-glucanase from *Chaetomium thermophilum*. Acta Cryst F 71:680–683

Purkarthofer H, Sinner M, Steiner W (1993) Effect of shear rate and culture pH on the production of xylanase by *Thermomyces lanuginosus*. Biotechnol Lett 15:405–410

Reilly PJ (1999) Protein engineering of glucoamylase to improve industrial properties. A review. Starch-Starke 51:269–274

Rouvinen J, Bergfors T, Teeri T, Knowles JKC, Jones TA (1990) 3-Dimensional structure of cellobiohydrolase-II from *Trichoderma reesei*. Science 249:380–386

Salar RK (2018) Termophilic fungi: basic concepts and biotechnological applications. CRC Press, Boca Raton, FL

Shoseyov O, Shani Z, Levy I (2006) Carbohydrate binding modules: biochemical properties and novel applications. Microbiol Mol Biol Rev 70(2):283–295

Sonnleitner B, Fiechter A (1983) Advantages of using thermophles in biotechnological process: expectations and reality. TIBTCH 1:74–80

Thompson AJ, Heu T, Shaghasi T, Benyamino R, Jones A, Friis EP (2012) Structure of the catalytic core module of the *Chaetomium thermophilum* family GH6 cellobiohydrolase Cel6A. Acta Crystallogr D 68:875–882

Törrönen A, Mach RL, Messner R, Gonzalez R, Kalkkinen N, Harkki A, Kubicek CP (1992) The two major xylanases from *Trichoderma reesei*: characterization of both enzymes and genes. Nat Biotechnol 10:1461–1465

Turunen O, Etuaho K, Fenel F, Vehmaanperä J, Wu X, Rouvinen J et al (2001) A combination of weakly stabilizing mutations with a disulfide bridge in the α-helix region of *Trichoderma reesei* endo-1,4-β-xylanase II increases the thermal stability through synergism. J Biotechnol 88:37–46

Uma Maheswar Rao JL, Boorgula GDY, Leitão AL (2011) Fungal enzymes: present scenario and future perspectives. In: Leitão AL (ed) Mycofactories. Bentham Science Publishers Ltd, pp 1–27. https://doi.org/10.2174/978160805223311110101

Varrot A, Schulein M, Davies GJ (1999) Structural changes of the active site tunnel of *Humicola insolens* cellobiohydrolase, Cel6A, upon oligosaccharide binding. Biochemistry 38:8884–8891

Voutilainen SP, Puranen T, Siika-Aho M, Lappalainen A, Alapuranen M, Kallio J, Hooman S, Viikari L, Vehmaanperä J, Koivula A (2008) Cloning, expression, and characterization of novel thermostable family 7 cellobiohydrolases. Biotechnol Bioeng 101(3):515–528

Wang Q, Qian C, Zhang X-Z, Liu N, Liu N, Yan X, Zhou Z (2012) Characterization of a novel thermostable β-glucosidase from a metagenomic library of termite gut. Enzyme Microb Technol 51:319–324

Wang HC, Chen YC, Huang CT, Hseu RS (2013) Cloning and characterization of a thermostable and pH-stable cellobiohydrolase from *Neocallimastix patriciarum* J11. Protein Exp Purif 90:153–159

Xie T, Song B, Yue Y, Chao Y, Qian S (2014) Site-saturation mutagenesis of central tyrosine 195 leading to diverse product specificities of an α-cyclodextrin glycosyltransferase from *Paenibacillus* sp. 602-1. J Biotechnol 170:10–16

Xu R, Teng F, Zhang C, Li D (2011) Cloning of a gene encoding βglucosidase from *Chaetomium thermophilum* CT2 and its expression in *Pichia pastoris*. J Mol Microbiol Biotechnol 20:16–23

Zhou QZ, Jia JC, Ji P, Han C (2017) A novel application potential of GH6 cellobiohydrolase ctcel6 from thermophilic *Chaetomium thermophilum* for gene cloning, heterologous expression and biological characterization. Int J Agric Biol 19:355–362

Index

A. M. Abdel-Azeem (ed.), *Recent Developments on Genus Chaetomium*,
Fungal Biology, https://doi.org/10.1007/978-3-030-31612-9

The manufacturer's authorised representative in the EU is Springer Nature Customer Service Centre GmbH, Europaplatz 3, 69115 Heidelberg, Germany. If you have any concerns regarding our products, please contact ProductSafety@springernature.com

Printed and bound by CPI Group (UK) Ltd, Croydon, CR0 4YY
05/10/2026
02228145-0014